OCEAN YEARBOOK 8

OCEAN
YEARBOOK 8

Pacem in Maribus

Sponsored by the
International
Ocean Institute

Edited by
Elisabeth Mann Borgese,
Norton Ginsburg, and
Joseph R. Morgan

Assistant Editor: Glenys Owen Miller

The University of Chicago Press

Chicago and London

The University of Chicago Press, Chicago 60637
The University of Chicago Press, Ltd., London

© 1989 by The University of Chicago
All rights reserved. Published 1990
Printed 1990 in the United States of America

International Standard Book Number: 0-226-06611-8
Library of Congress Catalog Card Number: 79:642855

551.46
015
242897

Contents

Coastal Management

Military Activities

Regional Developments

Appendices

Acknowledgment to the East-West Center

The International Ocean Institute is indebted to the East-West Center for its valuable support of and contributions to the *Ocean Yearbook*. The Center houses its editorial offices and provides essential infrastructural support as well as the salary of an Assistant Editor. Even more important are the substantive contributions of the research staff of the Ocean Governance Program of the Center's Environment and Policy Institute (EAPI), with which the *Yearbook* is closely integrated, as well as staff members of the Center's Resource Systems Institute. Two of the coeditors of the *Yearbook* and its Assistant Editor are members of the EAPI research staff.

The East-West Center is a public nonprofit educational institution with an international Board of Governors. Its objectives are research and training concerning important issues in Asia, the Pacific, and the United States, in collaboration with scholars and policymakers in those areas. The Center was established in 1960 by the Congress of the United States, which provides principal funding, although support also comes from more than 20 Asian and Pacific governments as well as private institutions. The Environment and Policy Institute of the Center was established in October 1977, to further understanding of the relationships between resource endowments and developmental polices within Asia and the Pacific and to convert that understanding into practical programs designed to enhance human welfare within the region.

THE EDITORS

Acknowledgments

Many individuals and organizations have contributed to this issue of the *Ocean Yearbook,* but only a few can be cited here. We are grateful to the Intergovernmental Oceanographic Commission of Unesco, particularly for assistance in distributing copies of the *Yearbook* to institutions in developing countries. Warm thanks also go to the Board of Editors for numerous formal and informal contributions and to the University of Chicago Press for its editorial and infrastructural support. Special thanks go to The University of Chicago Press for its competent coordination of *Ocean Yearbook* matters at the Press.

Many of the maps and illustrations in *Ocean Yearbook 8* are the creation of Lynn Indalecio. Nicholas Dunning and David Berner updated the tables, and David Berner compiled the index. Dalhousie University's Department of Political Science and the Lester Pearson Institute for International Development there provided valuable material support.

THE EDITORS

In Memoriam

J. King Gordon
December 6, 1900–February 24, 1989

Ocean Yearbook mourns the loss of its oldest, most active, most helpful, and most beloved patron and friend.

J. King Gordon introduced the nascent *Pacem in Maribus* program to Lester Pearson, then prime minister of Canada, in 1969, and secured Canadian support for the International Ocean Institute. Through its almost two decades, Canada has been the IOI's most generous supporter, and we owe this to King. King Gordon was a member of the Planning Council from the beginning; during the last three years of his life he was a member of our Board of Trustees.

King Gordon was born in 1900, the son of a minister of the United Church of Canada. His own early avocation was that of a minister of that church. He also was a teacher at the United Theological College at McGill University. In the 1930s he joined a social gospel movement to combat the depression. Persecuted for his socialist views, he lost his job in 1933. He became an editor for a New York publisher, then managing editor of the *Nation* (1944–47), and then CBC correspondent at the United Nations, until he joined the UN secretariat, where he stayed until his appointment to teach international relations at the University of Alberta in 1962.

King Gordon was a strong advocate of peacekeeping, and he wanted Canada to play a leading role in its implementation and development. His work took him to many of the trouble spots where peacekeeping was needed: the Congo, Korea, and the Gaza Strip.

From the University of Alberta, he moved on to the University of Ottawa and then served as university liaison adviser with the International Development Research Centre (IDRC) in Ottawa.

In the 1970s he was fully engaged in promoting an international development role for Canada. He was a leader of CUSO (the Canadian equivalent of the Peace Corps) and the United Nations Association. He founded the Group of 78, a group of progressive Canadians dedicated to the task of designing for Canada a policy of peace and ecologically sustainable development. He was considered an elder statesman, even an éminence grise, whose advice would always be listened to with respect by Canada's political leadership.

The aspiration of the IOI and *Pacem in Maribus* to utilize the world's ocean as the Great Laboratory for the building of the kind of order he believed in inspired him as much as he inspired it. "There is a biblical suggestion," he cabled to me on the occasion of my seventieth birthday, "that three score years and ten is a culminating anniversary of a well spent life; after that, the carpet slippers, an easy chair, and a good book. [The oceans] changed all that for me. Just as I was contemplating a leisurely retirement, Ritchie Calder arrived in Ottawa from Malta reporting on a world conference held under the Papal benediction, *Pacem in Maribus*. . . . First thing I knew, I was on my way to an ocean conference in Malta. And how life changed! That easy chair was postponed indefinitely."

The following passages are excerpted from the meditation spoken by the Reverend David MacDonald at King's funeral on Feburary 27, 1989. The Reverend MacDonald was Canada's ambassador to Ethiopia during the famine and is now a member of the Canadian Parliament.

In the sixth verse of chapter 13 in Paul's attempt to describe through various descriptions the power and meaning of love, he says: "But it delights in the truth." These words jumped out at me as catching as much as anything the meaning for me, and I would think for many of us here today.

Delighting in the truth. And as I thought over this past weekend of King's life, the words "delight" and a life committed to exploring and confronting and revealing the truth has certainly been the sum and substance of the man. From earliest days until the final moments, there was interwoven a constancy of search and revelation, in which the truth became revealed, not so much as a serious or weighty or sober fact to be contended with but as that which would liberate the human spirit and the human condition.

Some years ago, in fact quite recently in terms of the total chronology of King's life, he wrote an article in which he attempted to set out his view of the then-hot debate on the New International Economic Order and he concluded it with this sentence. He said: "An ethic is an imperative to put into practice the principles and purposes to which one is committed."

King began the early part of his life following his formal education, as a teacher of ethics. But he was not content merely to echo words or repeat statements. He wanted to make it manifestly real, and his life, within so many and several disciplines, was a reflection of that. Not just what he did, but how he did it. The gospel reading that we've heard this afternoon which is that remarkable conversation on the road to Emmaus is another reminder that we can walk quite close to those who would show to us the truth and not fully see it until there is that moment of discovery.

I remember not too many years ago when we were trying to remind Canadians, as they responded to the African famine, that there was more than just raising of funds and the sending of relief: there was the discovery of what our true responsibilities and relationships were to one another. One of our own attempts as we came to the conclusion of that whole remarkable episode was to sponsor a national conference here in Ottawa in the Parliament buildings called Forum Africa, but it was King and the Group of 78 who came to us with that remarkable publication that Murray Thomson referred to a moment or two ago, *Common Cause*, which in its own way tried to set out what was really behind all of the crisis and agony that was besetting Africa and what was really our long-term need for understanding. That was a vision that went far beyond the issues of the moment.

King's life was a respect for those visions and an honoring of them in this delightful search for the truth. It was a life that was lived with joy and celebration. I tried to think shortly after the call that I received on Friday of King's death, what jumped out first in my own thinking, and I thought, not of all the serious conferences we attended or the times that we have debated or argued several points, but of a rather prosaic—one could almost say boring—meeting that we were attending at Mount Allison University several years ago—in fact it was the annual dinner of the United Nations Association. On that occasion they had invited a leading international figure, whose name I won't mention this afternoon for obvious reasons, because his speech was one of the worst that I have ever heard—and that's saying something. And it happened on this occasion that King and my wife Sandra and I were sitting together near the back of this hall. There then began a series of what I can only regard as counteractivities, engaging in humor not only the members of our table but the chief who was trying to hold the whole thing together and it provided a moment of needed humor and reality in an otherwise totally stuffy event.

King lived his life to the full, not only for himself but for others to the very last moment. When I talked to Charles about how King had passed away, Charles made the remark that the only improvement on the event would have been had it happened in a canoe. That was not to be the case. But there is no doubt among those who are here and any others that the legacy of J. King Gordon is one that we will continue to celebrate to our own life's end. The name Frank Scott has already been mentioned and in your bulletin this afternoon you will see this very simple four-line creed that was meaningful for King as it was for Frank Scott—and is the summing up of the man himself:

> The world is my country.
> The human race is my race
> The spirit of man is my God
> And the future of man is my heaven.

E. M. BORGESE

The UN Convention on the Law of the Sea: The Cost of Ratification*

Elisabeth Mann Borgese, Aldo E. Chircop, and
Mahinda Perera
Dalhousie University

INTRODUCTION

In 1982 the UN Convention on the Law of the Sea was adopted and opened for signature. During the 2 years from December 10, 1982, until December 9, 1984, the Convention gathered 159 signatures (155 States and four Entities). At this date, it has been ratified by 40 States and one Entity. Sixty ratifications are required for its entry into force.[1]

Although some experts have attempted to project dates for the Convention's entry into force, this is a somewhat hazardous undertaking. Linear projections from the past rate of ratification obviously are not very meaningful. Assuming an average of six ratifications per year, one might be led to hope for ratification by the year 1992. This calculation, however, would have to be modified by consideration of the fact that the rate of ratification has slowed down during the last year. At the present rate of ratification, the entry into force of the Convention could be delayed for a very long time, which would encourage divergent State practice, generate confusion, waste 14 years of labor by the international community, and miss an opportunity for establishing a new order that is not likely to recur.

REASONS FOR DELAY

It may be opportune, therefore, to investigate the reasons for the slowdown and to try to eliminate or reduce them as far as possible. The major reasons for the delay are:

*EDITORS' NOTE.—Elisabeth Mann Borgese (Canada) is the chairman of the Planning Council, International Ocean Institute, chairman of the International Centre for Ocean Development, and professor of political science at Dalhousie University. Aldo Chircop (Malta) is academic director of the International Ocean Institute. Mahinda Perera (Sri Lanka) is a doctoral student of political science at Dalhousie University.

1. For a list of states signatory to the United Nations Convention on the Law of the Sea, see App. G in this volume.

a) the staggering novelty of this "Constitution for the Oceans";

b) difficulties arising from Part XI, which establishes the International Sea-Bed Authority;

c) the U.S. refusal to join; and

d) uncertainty as to the cost of ratification.

It is almost a miracle that a convention of this complexity and comprehensiveness could in fact be adopted and signed by as many as 159 States, a feat without precedent in history. Once this was achieved, however, it is as though States were now asking themselves: What have we done? Where are we to go from here and how?

Logically, the most innovative feature of the Convention, Part XI establishing the International Sea-Bed Authority, is causing the most serious hesitations. The refusal by the United States to agree to Part XI certainly has been a powerful slowing agent. Even governments highly favorable to the Convention feel tempted to wait and see what the United States will do, wondering whether the Convention could be meaningful without U.S. participation, whether seabed mining could get off the ground without U.S. technology, or what could be done to change the mind of the U.S. government. The most important element, one capable of inducing this change, is the Preparatory Commission for the International Sea-Bed Authority and for the International Tribunal for the Law of the Sea. By clarifying controversial issues and, possibly, establishing an interim regime useful and satisfactory to all, this commission could dispel many doubts.

THE COST OF RATIFICATION

The cost of ratification can be broken down into two parts: (*a*) costs arising from the new responsibilities that come together with the new rights over extended areas of jurisdiction; and (*b*) payments for the establishment and administration of the new institutions to be set up under the Convention, that is, the International Sea-Bed Authority, the Enterprise, the International Tribunal for the Law of the Sea, and the Commission on the Limits of the Continental Shelf.

Costs under (*a*) may vary, depending on the stage of technical and organizational preparedness of a State at the time the Convention comes into force. Not much additional effort is needed for a state like the United States or Canada, with a more or less well-managed Exclusive Economic Zone (EEZ) already in place. For a developing country without infrastructure and without technology, this cost may be considerably higher. The amount to be spent, however, should not be considered simply as a "cost": it should be accepted as an investment in development. The greatest benefit of the Convention may, in fact, consist in the stimulus it gives to this kind of investment in development: development of human resources, of infrastructure, of technology.

THE MEANING OF "IMPLEMENTATION"

Action by a government that purports to render an international treaty effective within the national scene is generally referred to as "implementation." By ratifying or acceding to a treaty, a government undertakes to fulfill the obligations, duties, and responsibilities stipulated therein, in addition to exercising incumbent rights or powers. Implementation is often narrowly construed as a legalistic exercise by virtue of which a government renders a treaty applicable within its jurisdiction through legislation. As a matter of constitutional law, it is a fact that most constitutions do not provide for the direct applicability of an international instrument as municipal law, so that legislation, being the only recognized source of authority within the State, is necessary. In this sense, the international instrument is "adopted" within municipal law. Several international treaties in marine affairs, such as the International Maritime Committee (Comité Maritime International) and the International Maritime Organization private maritime law conventions require little more than legislation for their implementation.[2] But different treaties entail different implementation measures.

Like any other international legal instrument, the United Nations' Convention on the Law of the Sea requires action by State parties in order to be effective. More than any other international legal instrument, it requires varied and far-reaching action in order to be rendered effective within the national scene. Furthermore, much of the Convention cannot be simply legislated into municipal law. In dealing with interdependence, this modern multilateral treaty is both contract and constitution, but it is not a purely legal instrument. It is much like a social contract. Moreover, there is a close relationship between the Convention and international organizations, so that its implementation is not solely a national affair. Vis-à-vis governments, implementation of the Convention implies the undertaking of a system of internal measures that may include any or all of the following:

a) policy measures (e.g., formulation of a development and management plan);
b) constitutive measures (e.g., creation of new national institutions);
c) administrative measures (e.g., reports and enforcement);
d) legislative measures (e.g., new legislation and amendments to existing laws implementing conventional stipulations);
e) technical measures (e.g., research and monitoring);
f) judicial measures (i.e., prosecution of offenses);

2. For further information about the International Maritime Organization, see "Report of the International Maritime Organization in 1985 and 1986," *Ocean Yearbook* 7, ed. Elisabeth Mann Borgese, Norton Ginsburg, and Joseph R. Morgan (Chicago: University of Chicago Press, 1988), pp. 411–18.

    g) education and training measures;
    h) measures promoting participation (i.e., of all affected interests); and
    i) public information measures.[3]

Some of these measures are one-time events, others are occasional, while yet others are continuous. The actual measures taken by a given government may depend on any combination of variables, such as geography, internal politics, external political affiliations, economic well-being, and ocean management considerations. The extent and degree of implementation cannot be expected to be symmetrical among State parties.

State parties apart, different parts of the Convention vary in extensiveness and intensity in implementation requirements. The Convention's provisions are not always clear in the requirement of national action. The language employed ranges widely, including descriptive, constitutive, distributive, and prescriptive elements.[4]

More than the others, prescriptive language either explicitly demands or implicitly requires national measures. Prescriptive language is sometimes discretionary rather than peremptory. Distributive language is next, although the requirement for measures varies from one provision to another. Thus, in international law, a coastal state need not declare a continental shelf in order to appropriate it, but it must make a specific declaration in order to possess an EEZ. Descriptive or constitutive language would seem to require less national action.

Insofar as implementation involves a range of activities to be undertaken by a government, certain costs are involved, such as the allocation of human resources, funds and other material resources, and time. For instance, in the case of zones of national jurisdiction, insofar as the conservation of living resources and marine environment protection are concerned, the measures which a coastal State is required to take are, at a minimum, of a scientific, legislative, administrative and judicial nature, including surveillance and monitoring. Every one of these measures entails economic costs. In the case of small States, these costs can be reduced through regional cooperation.

IMPLEMENTING PART XI

Part XI is different from the parts with zones of national jurisdiction in that it concerns an international area over which an international authority is to have

---

3. As summarized in Aldo Chircop, *Cooperative Regimes in Ocean Management: A Study in Mediterranean Regionalism* (J.S.D. thesis, Dalhousie University, Halifax, 1988).
    4. For instance, the definition of enclosed or semienclosed seas in Art. 122 is descriptive; the provisions on the International Sea-Bed Authority are constitutive; the allocation of maritime zones and other resource rights are distributive; and prescriptive elements are contained in regulating duties and obligations vis-à-vis other states, international organizations, and the international community.

jurisdiction. The measures that states are required to take are measures supportive of the International Sea-Bed Authority, and they are to desist from taking any other measures that undermine the latter's authority and the common heritage concept. Supportive measures include political and economic and, possibly to a lesser extent, legislative support. Political support consists, at a minimum, of an endorsement of the philosophy that underpins the Authority's policies, insofar as those policies are consistent with the Convention. This "measure" may not entail other than political "costs" in terms of political (to some, ideological) alignment. Economic support involves the monetary contributions that a State party to the Convention has to make according to UN scales of assessment. One may also add that the fulfillment of the obligation to transfer technology may appear as a measure of economic support, although it may equally be seen as bestowing benefits on the transferor as well as on the transferee. It is possible that legislative support may concern some State parties more than others. At this time, the only States that have enacted national legislation concerning deep seabed mining are the Pioneer Investors and the parties to the "Reciprocal Understanding."[5] Even when all deep seabed mining countries are parties to the Convention, legislative resources to regulate the activities of their nationals and vessels or installations flying their flag will be necessary. Insofar as other States, which may participate in seabed mining through the Enterprise, are concerned, there does not seem to be a need for legislative activity of this kind in the foreseeable future.

Payments for the establishment of the new institutions are easier to quantify. In Secretariat document A/Conf.62/L.65 (18 February 1981) (hereafter referred to as L.65), the grand total of nonrecurring costs is estimated as $228,894,000. (The document gives low, medium, and high estimates. The figure quoted here is high. Considering the rate of inflation since 1981, this figure is realistic, if we accept the assumptions underlying the Secretariat study.) Recurring (annual) administrative costs would be $52,759,200.

On the scale of assessments for the contribution of member States to the UN total budget, the poorest States contribute 0.01%, the richest, 25%. Thus, the contribution of the poorest countries for nonrecurrent costs would be $2,288,940, and $527,592 for recurrent costs—up to the time when the Au-

5. Editors' note.—See the statements of the Chairmen of Special Commissions 1 and 2 to the Plenary of the Preparatory Commission, App. B of this volume; S. P. Jagota, "Recent Developments in the Law of the Sea," Elisabeth Mann Borgese, "Implementing the Convention: Developments in the Preparatory Commission," "The International Venture: Study Submitted by the Republic of Colombia to the Preparatory Commission for the International Sea-Bed Authority and for the International Tribunal for the Law of the Sea, Special Commission 2, Fifth Session, Kingston, Jamaica, 30 March–16 April 1987," and "The Sixth Session of Preparatory Commission for International Sea-Bed Authority and International Tribunal for Law of the Sea, Kingston, 14 March–8 April, 1988," all in *Ocean Yearbook 7*, ed. Elisabeth Mann Borgese, Norton Ginsburg, and Joseph R. Morgan (Chicago: University of Chicago Press, 1988), pp. 65–93, 1–7, 469–79, and 513–28, respectively.

thority will be self-sufficient. The richest countries would pay up to $57,223,500 towards nonrecurring costs and $13,189,500 toward recurring costs. This is assuming that all States ratify the Convention. In case of a likely shortfall, during the first years after the entry into force of the Convention, an additional amount of perhaps as much as $100 million might have to be borne by ratifying States and, perhaps, international financing institutions.

In accordance with the Convention, the Secretariat document calculated, another $700 million in interest-free loans would have to be made available over the first 5 years of operations of the Enterprise. According to the same scale of assessment, this would mean $7 million for the poorest countries and $175 million for the richest over a 5-year period.

These calculations were based on the assumption that nodule mining would be a major ongoing concern at the time the Convention comes into force. The Authority is conceived of as almost as big as the United Nations, and the Enterprise as a full-blown multinational concern. And yet, even at this level, the figures are not really all that frightening. Two million and a bit, as a one-time, nonrecurrent cost, for the establishment of all the institutions established by the Convention, the first piece of a new international economic order, is affordable even for the poorest States.[6] Half a million dollars for 5 years is a bearable cost. It is far less than the cost of a patrol boat or what is spent, even by poor States, on arms purchases. Given the political will, even a shortfall of a $100 million could be absorbed by 60 or 80 States and international funding agencies.

The figures presented in the present study, however, are far more modest. They are based on the here and now, with the vision of an integrated mining operation of the Enterprise evolving organically from the present, and projected to begin some time near the end of the century. We assume a simple and smooth transition from the interim regime of the Preparatory Commission to the regime of the Authority. To give any realistic value to present-day figures, we assume that the Convention will come into force in 1990. The Preparatory Commission is, in a way, a preincarnation of the Authority; the activities mandated by paragraph 12 of Resolution 2—exploration of a first mine site for the Enterprise; technology acquisition; and development of human resources for the future Enterprise—necessitate, in fact, an interim or prenatal Enterprise, constituted somehow by agreement between the Preparatory Commission and the Pioneer Investors. The level of operations need not increase significantly upon ratification if economic/technological circumstances do not warrant it.

Building, equipment, and staff are in place in Jamaica. The plenary of the Preparatory Commission becomes the Assembly of the Authority, with no

6. For further information about the new international economic order, see M. C. W. Pinto, "Legal Aspects of North/South Transfer of Marine Technology," in this volume.

additional cost; the general committee is transformed into the council. There is already a group of experts of 15 (on the applications of the Pioneer Investors), whose expenses can help predict the cost of the two commissions of 15 to be established under the Convention. The running cost of the Preparatory Commission is in the order of $1 million per annum, to which another $3 million approximately would have to be added for the cost of the Secretariat. Another interesting figure is the cost of a global, but small, efficiently run specialized agency, like the International Maritime Organization (IMO), with an annual budget of around $18 million. The administration and coordination of EUREKA, which over the past 3 years has generated investments of $5 billion, is reported to cost between $8 million and $12 million a year.[7] This appears to be a reasonable range of figures for the Authority in its initial stage.

As far as the Enterprise is concerned, we base our assumptions on JEFERAD,[8] the Austrian working paper WP10, and the International Venture,[9] Colombian working paper 14, 14 add.1 and 14 add.2,[10] as well as on the feasibility study for the establishment of a Mediterranean Centre for Research and Development in Marine Industrial Technology, published by the International Ocean Institute, which relies on the most advanced concepts and experiences in high technology development and management. We assume a core module that need not cost more than half a million dollars a year, plus variable modules, financed independently, optimally to the rate of about $200 million dollars over 5 years. Half of this, or $100 million over 5 years, would be contributed as research and development (R&D) investment by Pioneer Investors and other investors. The other half should be contributed by the Enterprise. This amount would pay for the participation of developing countries in the exploration, R&D, and training projects of the Enterprise. We have therefore added it to the "recurring costs." This would be the "interest-free loan" States parties are to contribute toward the operational costs of the Enterprise. Research and development would be devoted to the upgrading and upscaling of technology in connection with the exploration and, subse-

---

7. Michel Gauthier, IFREMER, personal communication, 1987.

8. For the text of JEFERAD, see "JEFERAD: A Proposal for a Joint Enterprise for Exploration, Research and Development," *Ocean Yearbook 5*, ed. Elisabeth Mann Borgese and Norton Ginsburg (Chicago: University of Chicago Press, 1985), pp. 362–80; and a summary of the proposal may be found in "Selected Documents," *Ocean Yearbook 6*, ed. Elisabeth Mann Borgese and Norton Ginsburg (Chicago: University of Chicago Press, 1986), pp. 570–71. JEFERAD was the first Austrian working document and contained an analysis of the proposals and studies made during the time of the ad hoc Committee on the Peaceful Uses of the Sea-Bed and the Ocean Floor beyond the Limits of National Jurisdiction and was not included by the Secretariat in document LOS/PCN/SCN.2/L.2. It is, however, available from the Secretariat upon request.

9. Elisabeth Mann Borgese, "The International Venture."

10. Elisabeth Mann Borgese, "Implementing the Convention," The International Venture," and "The Sixth Session of Preparatory Commission."

quently, the exploitation (in an R&D sense only) of the mine site(s) of the Enterprise. Alternatively, this part of the "recurring costs" would be omitted, and $20 million annually could be raised from international financing institutions to match the investments of the Pioneer Investors and other participating industries.

We have not taken into consideration the cost of an integrated mining operation that would amount to about $2 billion at current prices. The date of the start of full-blown commercial seabed mining is too uncertain to make such calculations meaningful today. If the interim phase of exploration, R&D and the development of human resources, is successful, surely ways and means will be found to garner the benefits from the subsequent phase of exploitation. Most likely, the organization and financing of this international high tech venture will follow the pattern of other, already existing, international high tech ventures such as EUREKA. There is indeed a legal space for joint undertakings of this sort in the Convention.

The International Tribunal for the Law of the Sea will inevitably be expensive—costs include the salaries for 21 judges, even though the Tribunal may not have all that much work during the first years. The budget of L.65 is quite plausible. The operating budget of the International Court of Justice is $11,012,100 for the year 1986/87. The International Tribunal for the Law of the Sea will be more expensive, considering that there are 21 Judges rather than 15. Sixteen million dollars per year might be a reasonable figure. The nonrecurring costs for the establishment of the headquarters should be deducted as they would be contributed by the Federal Republic of Germany (FRG).

The cost of the Commission on the Limits of the Continental Shelf might be comparable to the cost of the Group of Experts on the applications of the Pioneer Investors. Here is a practical experience from which to learn and which did not exist when L.65 was drafted.

On these assumptions, the "Summary of Possible Administrative Costs Arising from the Adoption of the Draft Convention on the Law of the Sea, Informal Text" might be redrawn as shown in the Appendix below.[11]

CONCLUSION

If the Convention were ratified by all States members of the United Nations, the poorest States would have to contribute $5,000 to the annual budget, the richest (USA) $12.5 million: a very modest undertaking indeed, which, however, would be sufficient to globalize the most advanced concepts of interna-

---

11. The title quoted is drawn from the "Summary of Possible Administrative Costs Arising from the Adoption of the Draft Convention on the Law of the Sea, Informal Text," p. 113.

tional scientific/industrial cooperation, including the developing countries as equal partners.

APPENDIX[12]

The recurring costs for the first 5 years, as shown in table A1 above, will amount to roughly $50 million per annum. This is assuming that activities will take place in Jamaica, where the infrastructure is already in place, and in Hamburg, where the FRG would take care of establishment costs.

In the following pages, five scenarios are presented. On the basis of the scale of assessments of contributions of States to the UN budget, the cost to States parties will be calculated on the assumption that:

a) all ratifying States are developing countries, similar to the 34 States and one Entity that have ratified thus far (it might be noted here that the level of contributions of "entities" is yet to be established);

b) the 60 ratifying States include the 9 Eastern European socialist states;

c) the ratifying States include the 34 who have already ratified, plus the 9 socialist States of Eastern Europe, the 12 States of the EEC, the 3 Nordic States, plus Canada, Australia, and New Zealand;

d) ratifying States include 37 developing States (the 34 that have already ratified, plus China, India, and Saudi Arabia); 9 Eastern European socialist States and 20 OECD States; and

e) all States members of the UN and of UNCLOS III ratify.

TABLE A1.—INSTITUTIONAL COSTS

Organization	Costs (US$)
The International Sea-Bed Authority:	
Nonrecurrent costs (establishment)	0
Recurring costs (administration, per annum)	12,000,000.00
The Enterprise:	
Nonrecurring costs (establishment)	0
Recurring costs (core module)	500,000.00
Project financing (half of $40 million per annum)	20,000,000.00
The International Tribunal:	
Nonrecurring costs (establishment)	0
Recurring costs	16,000,000.00
The Commission on the Limits of the Continental Shelf:	
Nonrecurring costs	0
Recurring costs	1,752,900.00
Total	50,252,900.00

12. EDITORS' NOTE.—These tables were compiled when the number of ratifications was 34, and all calculations are made on this basis. The number of ratifications today is 41 but this does not at all change the conclusions of this study.

The shortfall will be identified in each scenario, and solutions will be suggested.

SCENARIO 1: 60 MAINLY DEVELOPING COUNTRIES

This scenario begins with contributions from the 34 States that have already ratified the Convention.

Country	Contribution to Total Funds (%)	Contribution (US$)
Bahamas	.01	5,000.00
Bahrain	.02	10,000.00
Belize	.01	5,000.00
Cameroon	.02	10,000.00
Cape Verde	.01	5,000.00
Cuba	.09	45,000.00
Egypt	.08	40,000.00
Fiji	.01	5,000.00
Gambia	.01	5,000.00
Ghana	.02	10,000.00
Guinea	.01	5,000.00
Guinea-Bissau	.01	5,000.00
Iceland	.03	15,000.00
Indonesia	.13	65,000.00
Ivory Coast	.03	15,000.00
Iraq	.15	75,000.00
Kuwait	.28	140,000.00
Malta	.01	5,000.00
Mexico	.97	485,000.00
Nigeria	.22	110,000.00
Paraguay	.01	5,000.00
Philippines	.09	45,000.00
St. Lucia	.01	5,000.00
São Tomé and Principe	.01	5,000.00
Senegal	.01	5,000.00
Sudan	.01	5,000.00
Togo	.01	5,000.00
Trinidad and Tobago	.04	20,000.00
Tunisia	.03	15,000.00
Tanzania	.01	5,000.00
Yemen, P. Dem. Rep.	.01	5,000.00
Yugoslavia	.48	240,000.00
Zambia	.01	5,000.00
Total	...	1,435,000.00

Also add 26 States of the same categories. Their contributions would amount to (1,435,000.00 divided by 34) × 26, or $1,097,353.00. This results in a grand total of contributions at $2,532,353.00, with a shortfall of $47,467,647.00. Suggested solutions:

  a) The shortfall of $47.5 million could be redistributed among the 60 ratifying States; in this case, the poorest would have to pay an additional $93,722.41, the richest (Mexico) would pay $9,091,074.11.

  b) The $20 million for R&D projects might be omitted in this scenario. Industrialized countries or companies and international funding agencies might take care of this. This would reduce the shortfall to $27.5 million. The poorest States would have to contribute an additional $54,297.33, the richest would pay $5,266,840.76.

SCENARIO 2: MAINLY DEVELOPING COUNTRIES PLUS EASTERN EUROPEAN SOCIALIST STATES

This scenario calls for:

  a) the contribution of 34 mainly developing countries who have already ratified—$1,435,000.00;

  b) contribution of 17 additional developing countries of the same category (1,435,000.00 divided by 34) × 17, or $717,500.00;

  c) the contribution of 9 socialist States.

Country	%	Contribution (US$)
Bulgaria	.18	90,000.00
Byelorussian SSR	.36	180,000.00
Czechoslovakia	.74	370,000.00
German Dem. Rep.	1.39	695,000.00
Hungary	.20	100,000.00
Poland	.62	310,000.00
Romania	.19	95,000.00
Ukrainian SSR	1.32	660,000.00
USSR	10.34	5,170,000.00
Total		7,570,000.00

Contributions from (a), (b), and (c) equal $9,722,500.00. This leaves a shortfall of $40,277,500.00, which can be eliminated in the following manner. The Eastern European States, on the same UN assessment rates, would absorb 15.34% of this shortfall, that is, $6,178,568.50. This reduces the shortfall to $34,098,931.50. This remainder could be divided among the 51 developing countries/ratifiers. The poorest would have to pay an additional $17,536.00, the richest would pay $1,701,000.93.

SCENARIO 3: 34 DEVELOPING COUNTRIES, 9 EASTERN SOCIALIST STATES PLUS 12 EEC, 3 NORDIC COUNTRIES, CANADA, AUSTRALIA, AND NEW ZEALAND

Combine the contributions of:

a) 34 developing countries who have already ratified (including Yugoslavia and Iceland) for a total contribution of $1,435,000.00;

b) include 9 Eastern European socialist countries, whose total contribution is $7,570,000.00;

c) add States whose categories and contributions are shown in the following table.

Country	%	Contribution (US$)
12 EEC States:		
Belgium	1.28	640,000.00
Denmark	.75	375,000.00
France	6.51	3,255,000.00
Germany, Fed. Rep.	8.54	4,270,000.00
Greece	.40	200,000.00
Ireland	.18	90,000.00
Italy	3.75	1,875,000.00
Luxembourg	.06	30,000.00
Netherlands	1.78	890,000.00
Portugal	.18	90,000.00
Spain	1.95	975,000.00
United Kingdom	4.67	2,335,000.00
Subtotal		15,025,000.00
3 Nordic countries:		
Finland	.48	240,000.00
Norway	.51	255,000.00
Sweden	1.32	660,000.00
Subtotal		1,155,000.00
Other countries:		
Australia	1.53	765,000.00
Canada	3.01	1,505,000.00
New Zealand	.26	130,000.00
Subtotal		2,400,000.00

The grand total of contributions is $27,585,000.00, with a shortfall of $22,415,000.00. If this shortfall were to be distributed among the 61 ratifying States according to the UN rate of assessments, 58.44% or $13,099,326.00 would have to be paid by the developed countries and $9,315,674.00 would have to be divided among the 34 developing ratifying countries. The poorest would have to pay an additional $1,688.53, while the richest would pay $163,788.36.

SCENARIO 4: 37 MAINLY DEVELOPING STATES (INCLUDING 34 THAT HAVE ALREADY RATIFIED, PLUS CHINA, INDIA, AND SAUDI ARABIA), 9 EASTERN EUROPEAN SOCIALIST STATES, AND 20 OECD STATES

This scenario combines:
    a) contributions of the 34 States that have already ratified ($1,435,000.00);
    b) contributions of 3 other developing states.

Country	%	Contribution (US$)
3 developing States:		
China	.81	405,000.00
India	.32	160,000.00
Saudi Arabia	.91	455,000.00
Subtotal		1,020,000.00
9 Eastern European socialist States	. . .	7,570,000.00
20 OECD States:		
Australia	1.53	765,000.00
Austria	.75	375,000.00
Belgium	1.28	640,000.00
Canada	3.01	1,505,000.00
Denmark	.75	375,000.00
Finland	.48	240,000.00
France	6.51	3,255,000.00
Greece	.40	200,000.00
Ireland	.18	90,000.00
Italy	3.75	1,875,000.00
Japan	10.33	5,165,000.00
Liechtenstein	.01	5,000.00
Luxembourg	.06	30,000.00
Netherlands	1.78	890,000.00
New Zealand	.26	130,000.00
Norway	.51	255,000.00
Portugal	.18	90,000.00
Spain	1.95	975,000.00
Sweden	1.32	660,000.00
Switzerland	1.10	550,000.00
Subtotal		18,070,000.00

The total from the 66 States listed above is $28,095,000.00, leaving a shortfall of $21,905,000.00. This shortfall could be distributed among the 66 ratifying States in accordance with the UN rate of assessment. The poorest would have to pay an additional $3,898.38, the richest would pay $4,027,027.05.

SCENARIO 5: ALL STATES MEMBERS OF THE UNITED NATIONS AND OF UNCLOS III RATIFY

Country	%	Contribution (US$)
Afghanistan	.01	5,000.00
Albania	.01	5,000.00
Algeria	.15	75,000.00
Angola	.01	5,000.00
Antigua and Barbuda	.01	5,000.00
Argentina	.70	350,000.00
Australia	1.53	765,000.00
Austria	.75	375,000.00
Bahamas (R)	.01	5,000.00
Bahrain (R)	.02	10,000.00
Bangladesh	.03	15,000.00
Barbados	.01	5,000.00
Belgium	1.28	640,000.00
Belize (R)	.01	5,000.00
Benin	.01	5,000.00
Bhutan	.01	5,000.00
Bolivia	.01	5,000.00
Botswana	.01	5,000.00
Brazil	1.47	735,000.00
Brunei Darussalam	.03	15,000.00
Bulgaria	.18	90,000.00
Burkina Faso	.01	5,000.00
Burma	.01	5,000.00
Burundi	.01	5,000.00
Byelorussian SSR	.36	180,000.00
Cameroon (R)	.02	10,000.00
Canada	3.01	1,505,000.00
Cape Verde (R)	.01	5,000.00
Cent. African Rep.	.01	5,000.00
Chad	.01	5,000.00
Chile	.08	40,000.00
China	.81	405,000.00
Colombia	.11	55,000.00
Comoros	.01	5,000.00
Congo	.01	5,000.00
Costa Rica	.02	10,000.00
Cuba (R)	.09	45,000.00
Cyprus	.01	5,000.00
Czechoslovakia	.74	370,000.00
Denmark	.75	375,000.00
Djibouti	.01	5,000.00
Dominica	.01	5,000.00
Dominican Republic	.03	15,000.00
Ecuador	.03	15,000.00
Egypt (R)	.08	40,000.00
El Salvador	.01	5,000.00

SCENARIO 5: (Continued)

Country	%	Contribution (US$)
Equatorial Guinea	.01	5,000.00
Ethiopia	.01	5,000.00
Fiji (R)	.01	5,000.00
Finland	.48	240,000.00
France	6.51	3,255,000.00
Gabon	.03	15,000.00
Gambia (R)	.01	5,000.00
German Dem. Rep.	1.39	695,000.00
Germany, Fed. Rep.	8.54	4,270,000.00
Ghana (R)	.02	10,000.00
Greece	.40	200,000.00
Guatemala	.02	10,000.00
Guinea (R)	.01	5,000.00
Guinea-Bissau (R)	.01	5,000.00
Guyana	.01	5,000.00
Haiti	.01	5,000.00
Holy See	.01	5,000.00
Honduras	.01	5,000.00
Hungary	.20	100,000.00
Iceland (R)	.03	15,000.00
India	.32	160,000.00
Indonesia (R)	.13	65,000.00
Iran	.58	290,000.00
Iraq (R)	.15	75,000.00
Ireland	.18	90,000.00
Israel	.23	115,000.00
Italy	3.75	1,875,000.00
Ivory Coast	.03	15,000.00
Jamaica (R)	.02	10,000.00
Japan	10.33	5,165,000.00
Jordan	.01	5,000.00
Kampuchea	.01	5,000.00
Kenya	.01	5,000.00
Kiribati	.01	5,000.00
Korea, Dem. P. Rep.	.05	25,000.00
Korea, Rep.	.21	105,000.00
Kuwait (R)	.28	140,000.00
Laos	.01	5,000.00
Lebanon	.01	5,000.00
Lesotho	.01	5,000.00
Liberia	.01	5,000.00
Libyan Arab Jamahiriya	.28	140,000.00
Liechtenstein	.01	5,000.00
Luxembourg	.06	30,000.00
Madagascar	.01	5,000.00
Malawi	.01	5,000.00
Malaysia	.09	45,000.00

SCENARIO 5: (Continued)

Country	%	Contribution (US$)
Maldives	.01	5,000.00
Mali	.01	5,000.00
Malta (R)	.01	5,000.00
Mauritania	.01	5,000.00
Mauritius	.01	5,000.00
Mexico (R)	.97	485,000.00
Monaco	.01	5,000.00
Mongolia	.01	5,000.00
Morocco	.06	30,000.00
Mozambique	.01	5,000.00
Nauru	.01	5,000.00
Nepal	.01	5,000.00
Netherlands	1.78	890,000.00
New Zealand	.26	130,000.00
Nicaragua	.01	5,000.00
Niger	.01	5,000.00
Nigeria (R)	.22	110,000.00
Norway	.51	255,000.00
Oman	.02	10,000.00
Pakistan	.06	30,000.00
Panama	.02	10,000.00
Papua New Guinea	.01	5,000.00
Paraguay (R)	.01	5,000.00
Peru	.09	45,000.00
Philippines (R)	.09	45,000.00
Poland	.62	310,000.00
Portugal	.18	90,000.00
Qatar	.04	20,000.00
Romania	.19	95,000.00
Rwanda	.01	5,000.00
St. Christopher and Nevis	.01	5,000.00
St. Vincent and the Grenadines	.01	5,000.00
San Marino	.01	5,000.00
São Tomé and Principe (R)	.01	5,000.00
Saudi Arabia	.91	455,000.00
Senegal (R)	.01	5,000.00
Seychelles	.01	5,000.00
Sierra Leone	.01	5,000.00
Singapore	.10	50,000.00
Solomon Is.	.01	5,000.00
Somalia	.01	5,000.00
South Africa	.36	180,000.00
Spain	1.95	97,000.00
Sri Lanka	.01	5,000.00
Sudan (R)	.01	5,000.00
Suriname	.01	5,000.00
Swaziland	.01	5,000.00

SCENARIO 5: (Continued)

Country	%	Contribution (US$)
Sweden	1.32	660,000.00
Switzerland	1.10	550,000.00
Syria	.04	20,000.00
Thailand	.08	40,000.00
Togo	.01	5,000.00
Tonga	.01	5,000.00
Trinidad and Tobago (R)	.04	20,000.00
Tunisia (R)	.03	15,000.00
Turkey	.33	165,000.00
Tuvalu	.01	5,000.00
Uganda	.01	5,000.00
Ukrainian SSR	1.32	660,000.00
USSR	10.34	5,170,000.00
U. Arab Emirates	.19	95,000.00
United Kingdom	4.67	2,335,000.00
United Rep. of Tanzania (R)	.01	5,000.00
U.S.A.	25.00	12,500,000.00
Uruguay	.05	25,000.00
Vanuatu	.01	5,000.00
Venezuela	.58	290,000.00
Vietnam	.02	10,000.00
Western Samoa	.01	5,000.00
Yemen Arab Rep.	.01	5,000.00
Yemen, P. Dem. Rep.	.01	5,000.00
Yugoslavia (R)	.48	240,000.00
Zaire	.01	5,000.00
Zambia (R)	.01	5,000.00
Zimbabwe	.02	10,000.00
Total	100.00	49,847,000.00

No shortfall is found in scenario 5. The marker (R) indicates States that have ratified the Convention.

Piracy and Its Repression under the 1982 Law of the Sea Convention

Bjorn Aune
London School of Economics and Political Science

INTRODUCTION

Piracy is the heinous seizure of vessels so as to plunder them. The act of piracy is as old as antiquity, dating back to the earliest trading vessels. For the seafarer, nautical historian, romanticist, adventure bibliophilist, dreamer, and cinema enthusiast, the word "piracy" brings to mind a myriad of terms, phrases, and images, such as "swashbuckler," "buccaneer," "pieces of eight," Blackbeard, Captain Kidd, and the Spanish Main. In context with these recollections comes the fanciful thought of many of the populace that piracy is part of history, that in our modern age it does not occur. Though such people are woefully wrong, the basis for their ignorance is not entirely unfounded. Rather, it is a misconception derived from the passage of time and the explosive changes in both technology and society. Though of questionable relevance, the *Achille Lauro* seizure in 1985 revitalized society's recognition of the activity.

Nowadays, most people believe that piracy does not occur because the development of the world places the pirate at a severe disadvantage. It is thought that today's "civilized world" precludes the act of piracy. This misconception is based on the following facts: a widespread populace that inhabits virtually all areas of the globe (so there is nowhere to hide); sophisticated telecommunications and technological developments; modern navies and coast guards; and large steel ships. What is not being recognized is that piracy has adapted to the changing world. Piracy endures today, not as expansively as in earlier centuries, but on a smaller, regional scale and in an altered form. The one aspect that has remained unchanged is the ruthlessness and violence of piratic acts. A germane point is that the notion of a civilized populace existing throughout the world is Western in definition. In actuality, areas continue to exist that have not changed morally to any appreciable extent even though technology has intruded. Hence, piratic behavior now exists, utilizing the benefits of technology.

International law repressing piracy has existed for 2 millennia. However, pertinent international doctrine has varied in interpretation and application based on national attitudes and perceptions toward piracy and politics. A fundamental problem has been the definition of piracy and what constitutes piratic acts. For example, one of the former supreme maritime powers, Great Britain, has never defined piracy in an English law statute.[1] It unilaterally suppressed piracy in the heyday of the empire, but not until 1878 was the offence of piracy *jure gentium* incorporated in English law.[2] The definition of piracy has varied from simple robbery on the sea by one vessel against another, to including intent to commit robbery and violent acts, and to further embrace the complex aspects of actions by unrecognized belligerents and recognized insurgents who attack noninvolved parties. Concomitant with the increasing complexity of defining piracy per se is the difficulty of defining it in international law in a comprehensive and concise manner. The international antipiracy doctrine has vacillated from simple and relatively concise laws derived from custom to more complex and purportedly comprehensive doctrines and agreements that suited the signatories at the time but were not universally accepted and led to ambiguity in application. The *travaux prépartoires* on the subject are long and, considering the still-existent debate, inconclusive. As a result, codification is less than desirable. The 1958 Geneva Convention on the High Seas dealt with piracy on a rather simple, straightforward basis,[3] but, regrettably, it became obsolete because the Convention codified "old rules of international law" that the European powers had promulgated.[4] Although it is still in force, the Convention never attained general international acceptance.[5] First, the Convention has only 57 signatories and was formulated in the years prior to the period of proliferation of states with anticolonialist attitudes. Second, the existing ocean regime that the first United Nations Convention on the Law of the Sea (UNCLOS I) codified became void, as rapidly altered perspectives concerning the ocean evolved subsequent to the treaties. States opted for greatly extended sovereignty and jurisdiction over the oceans—a move predicted to an appreciable degree by economic interests and new states' desires. The 1982 United Nations Conven-

---

1. Daniel P. O'Connell, *The International Law of the Sea*, 2 vols. (Oxford: Clarendon Press, 1982 and 1984), 2:979.

2. Territorial Waters Jurisdiction Act of 1878, 41 & 42 Victorian Code, p. 90; for earlier acts, see D. H. N. Johnson, "Piracy in Modern International Law," *Grotius Society Transactions* 43 (1957): 63.

3. Convention on the High Seas, Geneva, April 29, 1958; in force September 20, 1962; *United Nations Treaty Series (U.N.T.S.)* 450:82.

4. Roy S. Lee, "The Law of the Sea and Ocean Development Issues in the Pacific Basin," in *1982 Convention on the Law of the Sea*, ed. Albert W. Koers and Bernard H. Oxman, Proceedings of the Seventeenth Annual Conference of the Law of the Sea Institute, Oslo, Norway (Honolulu: Law of the Sea Institute, 1983), p. 8.

5. R. P. Anand, *Origin and Development of the Law of the Sea* (The Hague: Martinus Nijhoff, 1982), p. 236.

tion on the Law of the Sea[6] echoes the 1958 High Seas Convention in spirit and content concerning the repression of piracy, but further ambiguity is created with the introduction of the Exclusive Economic Zone (EEZ). The status of the international antipiracy doctrine within this zone is contentious, particularly in light of the apparent prevailing notion to interpret the Convention in its narrowest and absolute sense. Because the 1982 Convention virtually copied the text of the antipiracy articles in UNCLOS I, it failed to address the more controversial and complex issues regarding the subject. International law has not progressed in defining and repressing piracy *jure gentium* but has become shrouded in ambiguity and fractionated, even though the 1982 Convention alleges to be the modern, comprehensive legal doctrine governing the ocean regime. Compounding the issue, a new concept is emerging that promises to be contentious, though in practice it may prove more readily suppressed. Recent attacks on ships by terrorist groups and extremist opposition factions whose motives are dubious and who inflict violence indiscriminantly have introduced the concept of maritime terrorism. The *Achille Lauro* incident unduly highlighted the otherwise simmering debate by its worldwide exposure and need for expedient resolution. Arguably, maritime terrorism is a divorced component of piracy evolving tangentially because of problematic aspects of the antipiracy doctrine. The counterpoint is that it is simply a new form of criminal activity hitherto nonexistent—terrorism on the sea.

The paucity of articles in the existing body of literature on the state of the international antipiracy doctrine subsequent to the 1982 Convention necessitates redress. The spurt of articles and debate concerning the *Achille Lauro* seizure in relation to piracy remained limited in scope to that event. Thus, my objective is to examine the 1982 Convention to ascertain how piracy is to be suppressed in the prevailing ocean regime. Specific questions to be addressed include: (1) What does the term piracy actually embrace? (2) How is piracy dealt with in the new multivarious zones? (3) Is the 1982 Convention an effective repressant of piracy? and (4) If not, how is piracy to be contained and suppressed? A brief chronology of piracy today is given imprimis as evidence that piracy is not a dying activity.

PIRACY TODAY

In the last 15 years, piracy in its traditional form has been resurging. The piratic events of today may be categorized into four forms of piracy and then further defined regionally, depending on the type. The four categories include traditional piracy against modern shipping, politically motivated piracy

---

6. *The Law of the Sea: Official Text of the United Nations Convention on the Law of the Sea with Annexes and Index* (New York: United Nations, 1983), sales no. E.83.v.5. (hereinafter cited as the 1982 Convention).

and piratic acts of belligerents, piratic acts of violence against refugees, and yacht piracy. Geographically speaking, piracy tends to be most prevalent off the west coast of Africa, in the Archipelagic Waters of southeast Asia, off the Brazilian coast, and in the Caribbean Sea. Traditional piracy against modern shipping appears to be the dominant form of piracy today, based on the number of ships boarded. There are more than 400 reported incidents of piracy annually, and the figure is climbing.[7] Piracy has resurged to the point where shipowners, mariners, shipping organizations, governments, and international bodies are demanding action. The International Maritime Organization (IMO) has been occupied with the piracy issue for several years. It recently sponsored the Convention on the Suppression of Unlawful Acts against Safety of Maritime Navigation, which includes antipiracy measures.[8] The International Shipping Federation has issued numerous recommendations to its members on how to avoid piracy and minimize the threat.[9] Both the Liberian Shipowner's Council and the Baltic and International Maritime Conference have done likewise. On the national level, Britain, Japan, the Netherlands, Sweden, and West Germany are countries specifically attempting to deal with the problem. The regions where piracy against modern shipping is particularly acute are rather clearly demarcated. The west coast of Africa—specifically, the waters off the coasts of Nigeria, Ghana, the Ivory Coast, and Sierra Leone—and the Straits of Malacca and Singapore and the Phillip Channel in Southeast Asia are the two regions most afflicted by piracy. The Sulu Sea has been identified as particularly troublesome for local commerce. According to Bruneian authorities, several hundred attacks per year occur there.[10] The Natunas, Spratly, and Tawitawi island groups are pirate havens. On a minor scale, vessels off Brazil have been attacked. The pirates tend to be well informed, organized, and equipped for their raids. The piratic attacks usually involve robbery, vandalism, and threat of violence if demands are not met. Murder, torture, kidnapping, rape, and violent assault are generally excluded, though they may occur.

Boat-People Piracy

Piratic acts of violence against the Vietnamese boat people and local fishermen in the South China Sea and Gulf of Thailand have been the other

7. Paul Bartlett, "Counteracting the Piracy Menace," *Seatrade* 14, no. 2 (1984): 69–71.

8. IMO, "Convention on the Suppression of Unlawful Acts against Safety of Maritime Navigation." A conference of plenipotentiaries in Rome adopted the treaty on March 10, 1988, and it was signed by 23 states. It enters into force after 15 ratifications. The Convention is reproduced in Appendix B of this volume.

9. Liberian Shipowner's Council, *Annual Report, 1983* (New York: Liberian Shipowner's Council, 1984), p. 14.

10. Roger Villar, *Piracy Today* (London: Conway Maritime Press, 1985), p. 85.

dc ainant form of piracy today. More than 20,000 pirates are active in the Strait of Malacca and South China Sea.[11] There are 55,000 fishing vessels in the Gulf of Thailand alone, half of them unregistered.[12] The majority of priates attacking the refugees are Thai fishermen. It is believed that about 75% of all vessels carrying Vietnamese refugees have been repeatedly attacked.[13] Between 1975 and 1983, more than 514,000 boat people managed to land somewhere in Southeast Asia or be picked up by passing ships.[14] A simple calculation using the above figures derives a crude estimation of 385,500 refugees having been subjected to piratic attacks. Obviously, the number of people in the boats attacked versus those which eluded attack will affect the calculation, but not significantly. Relative to the other forms of piracy, the attacks against the Vietnamese boat people and local fishermen are of a degree of atrocity unsurpassed. The violent acts committed by the pirates run counter to any sense of humanity. The refugees possess only the few material goods that they escaped with, and if these are of no value to the pirates, the pirates often resort to assault, rape, abduction, and murder. Presumably, the basis for such behavior by the pirates is to vent their frustration at not being able to get anything or to penalize the refugees for their plight or to exercise an inbred barbarism. Since 1981, the UN High Commissioner for Refugees (UNHCR) has maintained statistics on these piracies. Table 1 provides detailed data.

Politically Motivated Piracy

Clear-cut acts of piracy by recognized and unrecognized belligerents are rare today. In contrast, a growing trend has been the attacks and seizures that are being labeled as acts of "maritime terrorism." Between 1975 and 1985 there have been 47 terrorist attacks on ships, of which 8 were hijacked and another 11 sea-going vessels destroyed.[15] The debate is whether these are some form

11. Editors' note.—For information about piracy activities in the South China Sea, see Daniel J. Dzurek, "Boundary and Resource Disputes in the South China Sea," *Ocean Yearbook 5*, ed. Elisabeth Mann Borgese and Norton Ginsburg (Chicago: University of Chicago Press, 1985), pp. 250–284, esp. 279–80; and Daniel J. Dzurek, "Piracy in Southeast Asia," *Oceanus* 32, no. 4 (1989): 65–70.

12. Villar, pp. 33 and 36; and see Magnus Torell, "Thailand's Fishing Industry: Future Prospects," *Ocean Yearbook 7*, ed. Elisabeth Mann Borgese, Norton Ginsburg, and Joseph R. Morgan (Chicago: University of Chicago Press, 1988), pp. 132–44.

13. David Marjoram (personal communiqué from the U.K. Department of Trade, Shipping Policy Division, 1982).

14. UN High Commissioner for Refugees, *Refugees Magazine* (May 1983), p. 13.

15. Eric Ellen, "Piracy and Law Enforcement," *Proceedings of the Nautical Institute* (Seminar on Piracy at Sea held October 31, 1985), pp. 7–13. Editors' note.—Mr. Ellen is Director of the ICC International Maritime Bureau (IMB) and Executive Secretary of the International Association of Airport and Seaport Police (IAASP). He supplied reports on both organizations for *Ocean Yearbook;* see *Ocean Yearbook 7*, pp. 396–404.

TABLE 1.—PIRATIC ATTACKS AGAINST VIETNAMESE BOAT PEOPLE,
1981–1987

Year	Murders	Rapes	Abductions Retrieved	Abductions Unrecovered	Missing	Total Number of Refugees Attacked
1981	529	856 ⎫	⎫	⎫	⎫	
1982	184	341 ⎬	270 ⎬	349 ⎬	700 ⎬	3,470
1983	68	173 ⎭	⎭	⎭	⎭	
1984	61	110	48	82	150	451
1985	73	113	48	63	203	500
1986	18	142	17	47	92	316
1987	2	80	3	12	80	177

SOURCES.—International Maritime Bureau, *A Third Report into the Incidence of Piracy and Armed Robbery from Merchant Ships* (Essex, 1985), pp. 36; UN High Commissioner for Refugees, "Annual Statistics on Attacks" (personal communiqué, Geneva, 1988.

of politically motivated piracy or outright terrorism. Compared to the other three forms, piracies of this category differ in that they are not limited to any one geographic locale, and there is no systemization to the respective incidents. In other words, the incidents and circumstances behind them are solitary, isolated events without connection to other similar events or forms of piracy.

In recent years, only four distinct incidents come to mind. Two occurred in Southeast Asian waters in 1975, one in the Strait of Florida in 1976, and most recently one in the eastern Mediterranean in 1985. In the *Mayagüez* seizure, the Khmer Rouge Navy seized an American ship in the Gulf of Thailand.[16] The Khmer Rouge was the opposition faction to the Cambodian government, but the United States had refused to recognize them. Why the vessel was seized remains unclear, although, allegedly, there was a mutiny. The United States labeled the seizure an act of piracy and used armed force to retrieve the *Mayagüez* from Cambodia. The American position was that Cambodia had committed piracy because the Khmer Rouge rebels were Cambodians who had seized a ship on the High Seas without valid authority.

In the *Sheira Maru* incident, a Filipino rebel group seized a Japanese vessel off Zamboanga as a form of protest against the Marcos regime. The Philippine government labeled the seizure piracy and freed the vessel by force. Because of the questionable status of the Maro National Liberation Front, the incident was considered an act of piracy by unrecognized belligerents. Had the rebels been recognized belligerents, the incident would still have been piracy since the *Sheira Maru* was Japanese and without involvement in the internal affairs of the Philippines.

16. Eleanor C. MacDowell, "Contemporary Practice of the United States Relating to International Law," *American Journal of International Law (AJIL)* 69 (October 1975): 861–86 (with specific emphasis on pp. 875–79 for SS *Mayagüez* incident).

The Cay Salzbank incident involved Cuban exiles who attacked Cuban fishing boats near Cay Salzbank in the Bahamas, presumably on political grounds. Assault and murder were committed by Cubans resident in the United States against Cuban citizens in Bahamian waters. Because of the extraterritorial nature of the incident, this incident was considered a case of politically motivated piracy.[17]

The hijacking of the *Achille Lauro* off Egypt in 1985 by the Palestine Liberation Front (PLF) was deemed a piratic event though it may have fitted more appropriately under the embryonic concept of maritime terrorism.[18] The PLF seized the passenger liner in retaliation for an Israeli raid on the Palestine Liberation Organization (PLO) headquarters in Tunisia. Because the vessel was of Italian registry and the person killed was not a national of either of the conflicting parties, the hijacking was deemed piracy.

Yacht Piracy

Relatively speaking, yacht piracy is considered a minor problem. The small nature of the vessels and the "rich man's toy" notion associated with yachts has led to apathy on the part of authorities in maintaining records. Reliable statistics are unavailable. United States authorities believe more than 200 people have disappeared in the last few years as a result of piratic attacks in the Gulf of Mexico and Caribbean Sea.[19] The predominant and continuous area of occurrence is the Bahamas because of the constant presence of yachts. The narcotics trade between South America and the United States is an important contributing factor to yacht piracy. The modus operandi of yacht pirates who are drug smugglers is to hijack a yacht, dispose of the crew, and use the vessel for a one-time delivery to Florida.[20] The element of piracy is introduced when the yachts are seized on the water and the vessels' owners are murdered outright, left to die in the sea, set adrift in a life-preserving device, or forced to go along with the smugglers. Yacht piracy tends to be more severe in nature because intimidation by threat of violence is always applied, and assault and murder are readily committed. A majority of the yachts seized are American owned. Yacht piracy unrelated to narcotics trafficking does occur, but such incidents are few, sporadic, and less publicized. These piracies are more prev-

17. The extraterritoriality stems from the alien status of the parties involved (Cubans and American alien residents) relative to the juridical status of the site (Bahamian territory).

18. G. Constantinople, "Towards a Definition of Piracy: The *Achille Lauro* Incident," *Virginia Journal of International Law* 26 (1986): 723–25.

19. Jan Fiksdal, "Hvor Sjoroverne Herjer ("Where Pirates Rampage")," *Na* (Now) 28 (1985): 28–30.

20. Emma Cussans, "The Pirate and the Armed Yacht," *Yachting Monthly*, 908 (April 1982), pp. 760–63.

alent in Southeast Asian waters, notably around Indonesia and the Philippines, and in the seas around the Arabian Peninsula.

THE ANTIPIRACY DOCTRINE

Today's composite international antipiracy doctrine is the culmination of an evolutionary process over 2 millennia long. It is partially convention law in that it is written. The definition of piracy contained in both UNCLOS I and the 1982 Convention is narrow and ambiguous. It portrays the international community's abhorrence of piracy, but it reflects the tenuousness that the evolutionary process inherently imputes. The definition embodied in both conventions purportedly codifies the international crime of piracy by embracing the *jure gentium* connotation. Hence, whenever piracy *jure gentium* occurs, it is an international crime, and international law applies.

UNCLOS I

Composed of four conventions, UNCLOS I allegedly represented a culmination in development of the Law of the Sea. The Preamble to the High Seas Convention claimed as much. Vis-à-vis the customary antipiracy doctrine, it meant only partial codification. The adoption of UNCLOS I split the antipiracy doctrine into two separate entities: (1) an antipiracy doctrine promulgated by convention dealing with acts committed for "private ends," and (2) an antipiracy doctrine in customary law applicable to illicit acts associated with belligerencies. Articles 14–22 were taken from the International Law Commission's (ILC) final draft of articles on piracy.[21] The controversy concerning the ILC's rejection of part of the customary antipiracy doctrine simply resurfaced in the UNCLOS I conference. In signing the 1958 High Seas Convention, several states made reservations to the piracy articles because they considered the articles noncomprehensive and in contravention of international law.[22]

The exclusion of politically related actions from the definition, combined with the growing tendency of several states to consider the articles in the High Seas Convention as the sole piracy doctrine, induced other states to challenge the definition. The use of the term "private ends" nourished this rebellion. States countered with their own interpretations.[23]

21. For a review of the International Law Commission's piracy articles, see "73 Draft Articles concerning the Law of the Sea," *Yearbook of the International Law Commission* 2 (1956): 256 ff.
22. The Eastern bloc countries were the notable objectors.
23. Using the term "private ends" was to invite trouble because it means different things to different people. Many believe that "public ends" and "political ends" fall

If the term "private ends" were interpreted in the narrowest sense, then acts of murder, rape, and the like would not fall within the scope of the definition of piracy except if accompanied by theft. Maintaining the historical preoccupation with *causa lucri*, which had been the essence of piracy, would validate this interpretation.[24]

Support for the moderate position came from two sources. The broader historical definition in customary law accepted murder and rape as acts of piracy provided they were committed by persons of the pirate ship against persons of the victimized vessel. The Preamble of the High Seas Convention described its provisions as "generally declaratory of established principles of international law." It was to be construed that acts contrary to humanity were prohibited under the treaty articles. The divisive interpretation of the piracy articles rendered consensus weak, and the UNCLOS I articles were regarded as incomplete doctrine.[25]

Article 15 was the crux of the antipiracy doctrine advanced by UNCLOS I. It was one of the "least successful essays in codification of the Law of the Sea, and the question remains whether it was comprehensive so as to preclude dependence upon customary law where it varied or superseded it."[26] The definition envisaged a category of acts beyond simple theft. Theoretically, violence was the essence of the definition.[27] The big question was what acts of a criminal and violent nature were subsumed therein.

It is generally agreed that an act of piracy must be committed against

---

within the scope of "private ends." Adherents of this philosophy argued that the acts committed or actions coincidental to those acts in politically oriented incidents often contained characteristics of personal gain or, at minimum, were a mix of the two. This meant every incident would need careful examination to ensure the actions effected did not transcend acts of piracy.

24. Sexual gratification, personal satisfaction, and insanity are intrinsic concepts that defy objective analyses. They are not measurable units of private gain. Most writers, however, concede the term "private ends" is not that restrictive.

25. E. D. Brown, "Maritime Commercial Malpractices and Piracy under International Law," *Maritime Policy and Management* 8, No. 2 (1981): 99–107; and the 1982 Convention articles are presented in full later with the variances from UNCLOS I noted, and thus the latter's articles are only referenced here. Analysis of them coupled with some thought and imaginative speculation readily sketches a picture of an international antipiracy doctrine fraught with ambiguity.

26. O'Connell (n. 1 above), 2:970.

27. There is no specificity to the extensiveness of the terms. The terms "violence," "detention," and "depredation" are general nouns that individually and collectively embrace many acts. The syntax of the phrasing suggests that violence does not take precedence over the other two terms. They are equally and mutually recognized. Thus the assumption is nurtured that the purpose for their inclusion in Article 15 was to embrace all the subcategories of acts without creating a long-winded treaty provision. The limit is the maximum scope of the terms, which is subjective. The only restrictions are that the acts must be committed for "private ends" against other ships or persons and property thereon.

another vessel. The list of acts presented in the Appendix were considered acts of piracy provided they were committed against another ship or the persons and property on board. Internal seizure, hijacking, criminal activity wholly within the vessel, and mutiny do not constitute piracy because the dual presence of a pirate ship and victim ship is nonexistent.[28] Internal seizure and hijacking are acts instigated from within a ship. Though they may, and often do, affect innocent parties, these acts are considered a matter for only the affected parties to resolve. Under customary law, internal criminal activity was a matter for the flag state.[29]

Article 15, Subparagraphs 2 and 3, provided for accessories to piratic acts to be deemed guilty of piracy.

A final point about Article 15 is that it required the pirate vessel to be a "private ship." This was contrary to the ILC's recommendation of 1956. The inclusion of the term was to preclude the questionable acts of warships from being labeled "piracy." It did not categorically preclude politically oriented piracies because rebels could conceivably use private vessels to conduct their rebellions.

Article 16 specifically defined when the actions of warships could be construed as piracy. The crew, or part of it, had to mutiny, gain control of the vessel, and then commit the acts of piracy recognized in Article 15. Article 17 provided that a ship be deemed a pirate vessel if the intent of those on board was to commit piracy. Implicit in the article was the concept that persons intending to commit piracy were pirates. By analogy, the article suggested that intent alone constituted an act of piracy.

The geographic limits to where piracy *jure gentium* could occur was the areal extent of the High Seas. Article 15, Subparagraph 1, specifically stated that piratic acts had to be committed on the High Seas or in places outside the jurisdiction of any state. An example of the latter were the waters around Antarctica.[30] Article 1 of the High Seas Convention stated the High Seas were "all parts of the sea that are not included in the territorial sea or in the internal waters of a State." As no other zones were recognized, the matter appeared relatively clear-cut. Unfortunately, reality was not that simple. UNCLOS I

28. Mutiny has a legal concept of its own under international law. Some have argued that the wording of item b in Subparagraph 1 suggests that internal seizure is piracy based on the phrase being construed not to mean location but the same vessel. Under close scrutiny, this contention is not borne out. A careful reading of Subparagraph 1, item b indicates that the acts must be "committed . . . by the crew . . . of a private ship . . . and directed: Against a ship" or something external to the offending vessel. The use of the word "another" in item a was redundant. The wording used in Subparagraph 1 for describing the acts of piracy was tautologous and imprecise. See O'Connell, 2:971.

29. The "French rule" pertaining to jurisdiction over internal criminal activity was incorporated in the 1958 Geneva Convention on the Territorial Sea and Contiguous Zone as Article 19. Refer to *UNTS* 516:205.

30. O'Connell, 2:971.

failed to adopt a definite breadth to the Territorial Sea. Consequently, a multitude of diverse Territorial Sea claims existed.

In the years subsequent to UNCLOS I, a myriad of new and expanded claims emerged. It became difficult to delineate with any degree of certainty where the boundary lay between those waters subject to municipal jurisdiction and those subject to international jurisdiction. Customary law was unable to give any guidance as it was undergoing changes. The only thing certain was that municipal jurisdiction clearly extended to at least 3 miles offshore. In the absence of measurable limits to the High Seas, one was left to presume states would exercise their international right to repress piracy *jure gentium* to the extent they recognized other states' Territorial Sea claims.

Articles 18–22 were technical and procedural provisions and relatively concise. They are elaborated on in the discussion of their 1982 Convention counterparts.

The 1982 Convention Definition of Piracy

When signed, the 1982 Convention was supposed to be a step forward. However, the net effect on the antipiracy doctrine is that it retracted into further ambiguity, and the treaty, as a whole, updated obsolete and contentious issues and expounded on new doctrines. The piracy articles are limited in scope, and the words and terms used continue to be ill defined and imprecise. Many contentious issues pertaining to piracy remain unaddressed and, thus, unresolved. The areal extent of application of the international law repressing piracy remains debatable. This is because the boundary between international juridical areas and areas given to territorial sovereignty, where national law may be exercised, is not clearly established and is often controversial and thus often polemical.

Two distinct events spawned the shortcomings in the antipiracy doctrine of the 1982 Convention. The first was that the piracy articles were virtually the same as those in UNCLOS I. Consequently, the 1982 Convention inherits all the problems of definition and scope plaguing its predecessor. Worthy of note is the fact that in 1980 the 1969 Vienna Convention on the Law of Treaties, which proffers advice on article interpretation, entered into force.[31] Article 31, Paragraph 1, requires that in treaty provisions the "ordinary meaning" of "terms" should be used in context with the doctrine's purposes and objectives. However, vis-à-vis the 1982 Convention, it had little impact on clarifying the ambiguities in definition. Second, the 1982 Convention introduced three realms of unique, but differing, legal statuses. The Exclusive Economic Zone (EEZ) and Continental Shelf Zone (CSZ) are contiguous to the Territorial Sea and, generally speaking, conterminous to each other and with

31. Refer to *United Kingdom Treaty Series (UKTS)*, no. 58 (1980).

the Contiguous Zone. The concept of Archipelagic Waters involves an entirely new legal water body that possesses an identity similar to internal waters. It is from the outer (peripheral) limit of the Archipelagic Waters that all other zones are measured. In these three realms, coastal and archipelagic states exercise varying degrees of sovereign rights and sovereignty. The realms are dealt with individually in the treaty, with each accorded a separate Part.

The problem in delineating with certainty the areal extent of application of the international piracy repression laws is the definition of the EEZ as a zone sui generis. It possesses neither the total juridical nature of High Seas nor the sovereign nature of Territorial Waters but exhibits characteristics of both.[32] The articles establishing the legal applicability of the international antipiracy doctrine (and the High Seas doctrine as a whole) to the EEZ are ambiguous.[33] On the one hand, Article 86 of Part VII explicitly states that the High Seas doctrine does not apply, while Article 58 of Part V allows for lawful uses of the High Seas within the EEZ. Piracy is not a lawful use of the sea, hence, it may be implicitly interpreted as reserving the right to deal with piracy to the coastal state. In other words, piracy and the legal right to combat it is moved from the international domain into the municipal domain in the EEZ. Because of the overlapping relationship of the EEZ with the CSZ and Contiguous Zone, the question arises as to the status of the antipiracy doctrine within the latter two zones. It is this aspect of the 1982 Convention that differentiates it from its predecessor with regard to the piracy articles and creates confusion.[34]

Before these issues are analyzed, however, the piracy doctrine of the 1982 Convention is presented below. Articles 100–107 and 110 of Part VII are reproduced, and the underlined sections denote the variances in wording from the articles in UNCLOS I.

> Article 100 [Art. 14 in the 1958 High Seas Convention (H.S.C.)]
> All states shall co-operate to the fullest possible extent in the repression of piracy on the high seas or in any other place outside the jurisdiction of any state.
> Article 101 [Art. 15 in H.S.C.]
> Piracy consists of any of the following acts:
> (a) any illegal acts of violence *or* detention, or any act of depredation, committed for private ends by the crew or the passengers of a private ship or a private aircraft, and directed:

32. Several publicists and writers cite this point. See, e.g., O'Connell, 1:579; Robin R. Churchill and Alan V. Lowe, *The Law of the Sea* (Manchester: Manchester University Press, 1983), p. 130; Ann L. Hollick, *U.S. Foreign Policy and the Law of the Sea* (Princeton, N.J.: Princeton University Press, 1981), p. 326; and Nasila S. Rembe, *Africa and the International Law of the Sea* (Alphen aan den Rijn: Sijthoff & Noordhoff, 1980), p. 163.
33. O'Connell, 1:577.
34. Brown, p. 102.

(i) on the high seas, against another ship or aircraft, or against persons or property on board such ship or aircraft;

(ii) against a ship, aircraft, persons or property in a place outside the jurisdiction of any State;

(b) any act of voluntary participation in the operation of a ship or of an aircraft with knowledge of facts making it a pirate ship or aircraft;

(c) any act of inciting or of intentionally facilitating an act described in *subparagraph (a) or (b).*

Article 102 [Art. 16 of H.S.C.]

The acts of piracy, as defined in article 101, committed by a warship, government ship or government aircraft whose crew has mutinied and taken control of the ship or aircraft are assimilated to acts committed by a private ship *or aircraft.*

Article 103 [Art. 17 in H.S.C.]

A ship or aircraft is considered a pirate ship or aircraft if it is intended by the persons in dominant control to be used for the purpose of committing one of the acts referred to in article 101. The same applies if the ship or aircraft has been used to commit any such act, so long as it remains under the control of the persons guilty of that act.

Article 104 [Art. 18 in H.S.C.]

A ship or aircraft may retain its nationality although it has become a pirate ship or aircraft. The retention or loss of nationality is determined by the law of the state from which such nationality was derived.

Article 105 [Art. 19 in H.S.C.]

On the high seas, or in any other place outside the jurisdiction of any State, every State may seize a pirate ship or aircraft, or a ship *or aircraft* taken by piracy and under the control of pirates, and arrest the persons and seize property on board. The courts of the State which carried out the seizure may decide upon the penalties to be imposed, and may also determine the action to be taken with regard to the ships, aircraft or property, subject to the rights of third parties acting in good faith.

Article 106 [Art. 20 in H.S.C.]

Where the seizure of a ship or aircraft on suspicion of piracy has been effected without adequate grounds, the State making the seizure shall be liable to the State the nationality of which is possessed by the ship or aircraft for any loss or damage caused by the seizure.

Article 107 [Art. 21 in H.S.C.]

A seizure on account of piracy *may be* carried out *only* by warships or military aircraft, or other ships or aircraft *clearly marked and identifiable as being* on government service *and* authorized to that effect.

Article 110 [Art. 22 in H.S.C.]

1. Except where acts of interference derive from powers conferred by treaty, a warship which encounters *on the high seas a foreign ship, other than a ship entitled to complete immunity in accordance with articles 95 and 96,* is

not justified in boarding *it* unless there is reasonable ground for suspecting *that:*

(a) *the* ship is engaged in piracy; . . .

2. In the *cases* provided for in *paragraph 1,* the warship may proceed to verify the ship's right to fly its flag. To this end, it may send a boat under the command of an officer to the suspected ship. If suspicion remains after the documents have been checked, it may proceed to a further examination on board the ship, which must be carried out with all possible consideration.

3. If the suspicions prove to be unfounded, and provided that the ship boarded has not committed any act justifying them, it shall be compensated for any loss or damage that may have been sustained.

4. *These provisions apply mutatis mutandis to military aircraft.*

5. *These provisions also apply to any other duly authorized ships or aircraft clearly marked and identifiable as being on government service.*[35]

Undoubtedly, the lack (or lack of knowledge) of many incidents of piracy or questionable acts in the years following UNCLOS I maintained the antipiracy doctrine in an isostatic state of ambiguity. Presumably, this was the reason for the old articles being simply transcribed into the 1982 Convention. In 1970 the International Law Association (ILA) reopened the question of what acts constituted piracy, but nothing of substance came of the ILA's inquiry.

Article 100, which is identical to Article 14 of the High Seas Convention, recognizes the equality of states and their competency to deal with pirates. The phrase "any other place outside the jurisdiction of any State" is a reference to Antarctica. Other places within the latter category are islands that are *terra nullius* and their adjacent waters. Article 101, like Article 15 in UNCLOS I, is the crux of the antipiracy doctrine promulgated in the 1982 Convention. Repetition of all that was said earlier in reference to Article 15 is senseless. Instead a brief summary will do.

Acts of piracy must be directed against another vessel; there must be a pirate ship and victim ship (Subpar. a, items i and ii). The pirate vessel must be a private ship. Government vessels and warships are excluded unless their actions follow the guidelines of Article 102. The acts committed must be for "private ends." The acts must be of a criminal or violent nature and directed against persons or property on board the second ship, if not the ship itself. Acts of a criminal and violent nature remain those previously cited in UNCLOS I (see the Appendix). Accessories to acts of piracy are also guilty of piracy. Subparagraph b provides that anyone voluntarily aiding in a piratic operation is a pirate. Subparagraph c provides that anyone causing a piratic act to occur or purposefully facilitating its occurrence is a pirate. The wording of the subparagraph means individuals who induce others to commit piracy

35. Refer to UN document A/CONF.62/122, Part 7.

are guilty of piracy even though they did not physically participate in the act or acts. Article 102 provides the exception to when a pirate vessel need not be a private ship. Under the article, warships and government ships can be equally guilty of piracy if the crew mutinies, assumes command of the vessel, and then embarks on piratic exploits as defined in Article 101.

Article 103 stipulates that intent to commit piracy equally constitutes an act of piracy. The article does not state this outright but in a circumventive manner. It reads that ships under the control of persons intending to use the vessels to commit piratic acts are pirate ships. Hence, by analogy, the persons on board who are intending to commit piratic acts must be considered pirates. Therefore, it follows that intent to commit piracy constitutes an act of piracy. The second part of Article 103 provides that pirate ships continue to maintain their status as such for as long as they remain under the command of pirates. Should the pirates flee from the vessel they used to commit piratic acts and the vessel is under the command of other persons, then it is no longer a pirate ship. The vessel intending to seize it no longer has the right to capture it. This article requires that states intending to capture a pirate vessel first verify if the suspected vessel is still under the pirates' control. Article 110 provides the basis by which a state may undertake the verification process.

Article 104 simply means that pirate ships may continue to belong to a flag while engaged in piracy. The flag state has sole authority in determining whether to disown a pirate vessel or acknowledge it as still under its flag. This article in no way affects the right of all states to board and seize pirate ships.

Article 105 establishes that all states may seize pirate ships, arrest the pirates, and confiscate their spoils. The only limitation to seizure is Article 103's requirement of pirate control. The courts of the state effecting a seizure have full jurisdiction in the matter and may penalize the pirates in any manner that their legal system prescribes. The courts also have jurisdiction over the seized ships and goods and may dispose of them as they deem fit. The only limitation is that they must take into account the rights of the injured parties and rightful owners. It has been argued by some that the actions required of states in effecting their prosecution of pirates and the penalties that may subsequently be imposed should be stipulated in Article 105.[36] Global harmonization of the measures and procedures to be taken by states against pirates would eliminate the debate over the differing penal codes and their varying degrees of severity.

Article 106 is a liability clause for unwarranted seizures of vessels by states. The liability is to the state whose flag is flown by the seized vessels, not the vessel owners, vessel operators, cargo owners, crew, or passengers. Though not explicitly stated, liability is to be in the form of financial remuner-

36. See, e.g., Patricia W. Birnie, "Piracy—Past, Present, and Future," *Marine Policy* 11 (July 1987), pp. 163–83.

ation for any loss or damage that the vessel sustained in being seized. Although this article appears succinct, it is unfortunately subject to interpretation. The use of the adjective "adequate" allows states to use the pretext of "suspicion of piracy" for stopping vessels when in reality the bases for intercepting them were unrelated grounds (i.e., criminal manhunt, drug trafficking, espionage, etc.).

Article 107 mandates that only warships, military aircraft, and other government craft are authorized to conduct seizures of pirate vessels. Article 110 specifies the procedural bases by which the authorized vessels are entitled to board suspect vessels. A cursory analysis of this article suggests it is straightforward. Regrettably, like Article 106, it is open to interpretation. The phrase "reasonable ground for suspecting" gives states latitude such that they may board vessels for other reasons but claim they did so under the purview of this article.

In the 1982 Convention, the definition of piracy provided in Article 101 and the right of all states to exercise jurisdiction over pirates (Art. 105) remains applicable only in the High Seas and places outside the jurisdiction of any state (Art. 100). Article 86 states the extent of the High Seas. Ignoring for a moment the controversy over the EEZ but considering the new breadth of the Territorial Sea and nature of Archipelagic Waters, the only thing that can be said with certainty is that the area of the High Seas has decreased in real terms.

REPRESSING PIRACY IN THE MULTIZONAL REGIME

The 1982 Convention envisages nine different maritime zones, each possessing distinct legal status. Six of these were either developed or introduced in UNCLOS I, whereas the other three were identified for the first time in the 1982 Convention. The nine zones are internal waters, Territorial Seas, Contiguous Zones, EEZs, CSZs, High Seas, Archipelagic Waters, archipelagic sea-lanes, and straits used for international navigation. Under the new ocean regime, states are granted varying degrees of sovereignty or sovereign rights in the respective zones. The extent of sovereignty or sovereign rights the state possesses directly affects the extent of applicability of the international antipiracy doctrine. Consequently, establishing the juridical nature of the respective zones in regard to piracy is paramount.

Internal Waters and Territorial Seas

The international antipiracy doctrine does not apply within the internal waters and Territorial Seas of States. Article 86 specifically states this:

Article 86

The provisions of this part apply to all parts of the sea that are not included in the exclusive economic zone, in the territorial sea or in the internal waters of a State, or in the archipelagic waters of an archipelagic State. This article does not entail any abridgement of the freedoms enjoyed by all States in the exclusive economic zone in accordance with article 56.

Internal waters are water bodies subject to the total sovereignty of the littoral state (Art. 8). Subparagraph 2 of Article 8 provides an exception in the case of newly defined internal waters, but it does not affect the international piracy doctrine. Territorial Seas are adjacent maritime zones where the states have complete sovereignty aside from having to allow innocent passage (Art. 17) and leave internal crime and civil matters to the flag states (Arts. 27 and 28). The littoral state has absolute jurisdiction in its internal waters and Territorial Sea. It has the exclusive right to establish selectively any and all legal and penal codes applicable therein. The form of law a state exercises in internal waters and Territorial Seas is municipal law. This means the state's definition of piracy can be at variance with that of piracy *jure gentium*. States may have more comprehensive definitions of piracy incorporated into their legal codes and include acts not within the scope of piracy *jure gentium*. Alternatively, their definitions may be less comprehensive. Consequently, it is possible for acts of piracy *jure gentium* not to constitute piracy when committed in states' internal waters and Territorial Seas. In any case, all acts of piracy committed within these areas are under the sole jurisdiction of the littoral state.

The Contiguous Zone, CSZ, and EEZ

The conterminous relationship of the Contiguous Zone, CSZ, and EEZ coupled with the contentious nature of the latter necessitates analyses of them together. If the EEZ did not exist, then the applicability of the international antipiracy doctrine to the first two zones would be clear. All three zones are measured from the same baselines used to delimit the Territorial Sea (Art. 33, Par. 2, Arts. 57 and 76). The Contiguous Zone is granted a maximum breadth of 24 miles, the EEZ is granted a maximum breadth of 200 miles, and the CSZ has a maximum variable breadth of 200–350 miles, provided there are no states nearby to force a reduction of the limits. If a state claims a full 12 miles as the breadth of its Territorial Sea, then for each of the three zones the true breadth is the maximum breadths minus 12. Hence the Contiguous Zone is 12 miles, the EEZ is 188 miles, and the CSZ varies from 188 miles to 338 miles as measured from the outer limit of the Territorial Sea. Because of the overlapping relationship of the Contiguous Zone, CSZ, and EEZ, the extent of applicability of the international antipiracy doctrine to all three zones requires that there be identical or at least congruous provisions repressing piracy in each realm. Unfortunately, this is not the case. The applicability of the High Seas doctrine to the EEZ is dubious because of ambiguity in the interrelationship of the two, which is subject to interpretation. It is the large breadth of the EEZ totally engulfing the Contiguous Zone and more than half of the CSZ that precludes the otherwise clear application of the High Seas doctrine to the Contiguous Zone and CSZ.

The Contiguous Zone is a zone where coastal states may exercise limited jurisdiction regarding customs, fiscal, immigration, sanitary, and archaeological matters (Art. 33, Subpar. 1, and Art. 303). The provisions of the Contiguous Zone contained in Part 2, Section 4, are absolute. Since Article 86 does not include the Contiguous Zone as an exclusion, it means that the High Seas regime, including the antipiracy doctrine, applies in the Contiguous Zone. Ignoring for a moment the EEZ concept, this means the international antipiracy doctrine should apply up to the outer boundary of the Territorial Sea. Without discussing all the possible variations in breadth of the CSZ that the 1982 Convention permits, one can see in Article 78 of Part VI that the legal status of the waters over the Continental Shelf is not determined by association with the shelf floor but by some existing surface regime derived from elsewhere:

> Article 78
> 1. The rights of the coastal State over the continental shelf do not affect the legal status of the superadjacent waters or of the air space above those waters.
> 2. The exercise of the rights of the coastal State over the continental shelf must not infringe or result in any unjustifiable interference with navigation and other rights and freedoms of other States as provided for in this Convention.

Article 86 excludes the waters over the Continental Shelf from mention because the intent is that the High Seas doctrine applies to those waters. The purpose of Article 78 is to define implicitly the legal status of the superadjacent waters as High Seas. Obviously, when the CSZ extends beyond the EEZ, the legal status of the waters is High Seas, and the international antipiracy doctrine applies. If a state does not claim an EEZ or if it considers the High Seas regime as existent within its EEZ, then all the waters of the CSZ seaward of the outer limit of the Territorial Sea is High Seas, and the international antipiracy doctrine applies throughout.

The debate over the EEZ centers on the issue of the extent of sovereign rights or sovereignty that coastal states possess therein vis-à-vis the degree of rights that other states have in the EEZ. Article 86 categorically excludes the application of the High Seas regime to the EEZ. Article 58, in contrast, specifically states that the High Seas freedoms listed in Article 87 and other international lawful uses (not cited in the article) that all states enjoy on the High Seas exist within the EEZ:

> Article 58
> 1. In the exclusive economic zone, all States, whether coastal or land-locked, enjoy, subject to the relevant provisions of this Convention, the freedoms referred to in article 87 of navigation and overflight and of the laying of submarine cables and pipelines, and other internationally lawful uses of the sea related to these freedoms, such as those associated with the operations of ships, aircraft and submarine cables and pipelines, and compatible with the other provisions of this Convention.
> 2. Articles 88 to 115 and other pertinent rules of international law apply to the exclusive economic zone in so far as they are not incompatible with this Part.
> 3. In exercising their rights and performing their duties under this Convention in the exclusive economic zone, States shall have due regard to the rights and duties of the coastal State and shall comply with the laws and regulations adopted by the coastal State in accordance with the provisions of this Convention and other rules of international law in so far as they are not incompatible with this Part.

Paragraph 2 of Article 58 specifically includes the antipiracy doctrine as applicable within the EEZ so long as it is not incompatible. The contradiction of Articles 58 and 86 and the issue over terminology used in the former are the core of the controversy. Three lines of thought have dominated debate, but only one seems to be surviving. The first was the concept that the EEZ is a residual High Seas area; the second was the opposing notion that the EEZ is a Territorial Sea area.[37] The third line of theorization is that the EEZ is a zone sui generis—a hybrid of the first two.[38] The phrasing of Article 58 tends to support the notion of the freedoms permitted in the EEZ not being the same as those allowed on the High Seas.[39] The EEZ legitimizes the monopoly of the coastal state and extends the area over which national control may be exercised.[40] Hence, the rights of foreign navies in the EEZ seem likely to be restricted.[41] In practice, there is little activity at sea that cannot be embraced within the term "economic." Piracy, for the most part, is an economic venture, albeit criminal. Unless the 1982 Convention is amended, the conceivable outcome of the controversy and ambiguity is that coastal states will unilaterally decide the extent of their jurisdiction in their EEZs. The extent to which a coastal state exercises jurisdiction over piracy in its EEZ depends largely on its

37. Churchill and Lowe, p. 129.
38. See, e.g., Rembe, p. 163; Hollick, p. 326; and Churchill and Lowe, p. 130.
39. O'Connell (n. 1 above), 1:578.
40. Elizabeth Young, "The EEZ: International Enforcement Problems," *Maritime Policy Management* 4 (1977): 315–24.
41. Ibid.

perspective of the issue and its ability to physically do so. Several states will adhere to the belief that Article 58 means the international antipiracy doctrine applies within the EEZ and grants them the right to deal with pirates in other states' EEZs. Many states will maintain that piracy is not one of the lawful uses of the sea permitted under Articles 58 and 87 and, concomitantly, that the right to capture pirates falls to the coastal state. Other states will argue that piracy is an economic venture, and, under the scope of Articles 55 and 56, only they have the right to deal with pirates in their EEZs. In terms of actually determining which position a state will adopt, one must look to the arguments and opinions they advanced at the 1982 Convention sessions or, if the states are nonparticipants, the proclamations they have made. The United States, Finland, Italy, Japan, Mauritius, Switzerland, the United Kingdom, and other states maintain that they have the right to deal with piracy in other states' EEZs.[42] Argentina, Bénin, Brazil, Congo, El Salvador, Ghana, Guinea, the Khmer Republic, Liberia, Nicaragua, Panama, Sierra Leone, Somalia, and Uruguay, all maintain they have outright 200-mile Territorial Seas, while Ecuador, Guatemala, and Jamaica believe the EEZ is a patrimonial sea.[43] China maintains the EEZ is not High Seas because, if it is, then the establishment of such a zone is ridiculous.[44] Algeria, Nigeria, Peru, Senegal, and Venezuela believe the EEZ is not High Seas but some form of transition zone.[45] The question of whether states will exercise a right to repress piracy *jure gentium* in the EEZs of other states hinges on their perceptions of how much control the coastal states possess in their EEZs. Presumably, there will be mutual concurrence on the status of the international antipiracy doctrine in the EEZ between those states of like attitudes. Where states disagree, each will pursue its own course of action (or lack of action). Undoubtedly, the "might is right" notion will play a role. The powerful maritime states that are advocates of the EEZ being High Seas will simply wield their might to repress piracy in the EEZs of other states as they deem necessary.

Archipelagic Waters and Sea Lanes

Archipelagic Waters have a status somewhere between that of internal waters and Territorial Seas.[46] The international antipiracy doctrine does not apply

42. United Nations, *Third United Nations Conference on the Law of the Sea, Official Records,* summary record of meeting (New York: United Nations, 1975), 2:180–81, 196, 200, and 207; and Rembe (n. 32 above), p. 162.

43. Churchill and Lowe (n. 32 above), pp. 302–10; Hollick (n. 32 above), p. 162; and United Nations, pp. 180 ff.

44. Jeanette Greenfield, *China and the Law of the Sea, Air, and Environment* (Alphen aan den Rijn: Sijthoff & Noordhoff, 1979), p. 111.

45. United Nations, pp. 180 ff.

46. Peter Polomka, *Ocean Politics in Southeast Asia* (Singapore: Institute of Southeast Asian Studies, 1978), p. 11.

within Archipelagic Waters and sea-lanes. Article 86 categorically states the exclusion of Archipelagic Waters from the High Seas regime. Article 49 of Part IV specifically states that the archipelagic state has full sovereignty over the Archipelagic Waters (Pars. 1 and 2) and archipelagic sea-lanes (Par. 4):

> Article 49
> 1. The sovereignty of an archipelagic State extends to the waters enclosed by the archipelagic baselines drawn in accordance with article 47, described as archipelagic waters, regardless of their depth or distance from the coast.
> 2. This sovereignty extends to the air space over the archipelagic waters, as well as to their bed and subsoil, and the resources contained therein.
> 3. This sovereignty is exercised subject to this Part.
> 4. The regime of archipelagic sea lanes passage established in this Part shall not in other respects affect the status of the archipelagic waters, including the sea lanes, or the exercise by the archipelagic State of its sovereignty over such waters and their air space, bed and subsoil, and the resources contained therein.

Consequently, any acts of piracy committed therein are crimes against the state. The archipelagic state may recognize additional acts, other than those of piracy *jure gentium,* as piracy, or it may choose not to recognize the international definition within its own Archipelagic Waters. No other states may claim jurisdiction over piracy in Archipelagic Waters and sea-lanes. States that qualify for archipelagic status and that may claim Archipelagic Waters and establish sea-lanes include the Bahamas, Cape Verde, Fiji, Indonesia, Mauritius, Micronesia, Papua New Guinea, the Philippines, Samoa, São Tomé and Principe, Solomon Islands, and Tuvalu.[47]

International Straits

Though international straits are recognized as separate marine geographical entities that possess a distinct legal character, they in no way alter the legal status of the water forming the straits. Article 34 of Part III specifically states this point:

> Article 34
> 1. The regime of passage through straits used for international navigation established in this Part shall not in other respects affect the legal status of the waters forming such straits or the exercise by the States

47. Churchill and Lowe, p. 97; and O'Connell, 1:246–54.

bordering the straits of their sovereignty or jurisdiction over such waters and their air space, bed and subsoil.

2. The sovereignty or jurisdiction of the States bordering the straits is exercised subject to this Part and to other rules of international law.

Consequently, if the width of a strait is 24 miles or less at its narrowest points, then those areas of the strait are Territorial Seas—half belonging to each state bordering the strait (provided each state claims a 12-mile Territorial Sea). The international antipiracy doctrine does not apply at all. Where the width of the strait is greater than 24 miles, the question of whether the international antipiracy doctrine applies to the water area beyond the Territorial Seas of the bordering states depends on the status of those waters as determined by those states. If the states do not claim EEZs, then the international antipiracy doctrine does apply in the water area between the Territorial Seas of each strait state. But, if the strait states insist on EEZ claims for the water area beyond their respective Territorial Seas forming the middle part of the strait, then the question will depend on the respective states' interpretations of the EEZ doctrine. Both states may recognize the applicability of the international antipiracy doctrine in the EEZ, in which case the question is academic, or they may opt not to. It is also conceivable that one state may recognize the applicability of the antipiracy doctrine in its EEZ, while the other state may not. The dilemma created here is the same as that posed earlier concerning the EEZ in general. Presumably, any dilemma would be resolved more readily since the provisions of the international straits doctrine lend impetus to the formation of any doctrines or treaties that promote safe navigation and counter threats to it (Arts. 42, 43, and 44). If piracy is a problem in a strait, then it is feasible to expect some form of an international antipiracy doctrine to be quickly implemented via the treaty process. The importance of the strait to navigation should mandate such development.

BETWEEN NOW AND THEN AND INTO THE FUTURE

For a variety of reasons, it is conceivable that the 1982 Convention may not come into being—at least not in its present form.[48] An interesting trend in recent literature is the either veiled or open references to an UNCLOS IV treaty being needed. Birnie acknowledges the hopelessness of UNCLOS III and questions whether the 1982 Convention will come into force or, if it does, be effective.[49] She goes further to suggest that a fourth Law of the Sea Conference be held to renegotiate the UNCLOS III package, at which time the

48. As of October 1989, 41 states and one UN-recognized entity have ratified. See Appendix G in this volume.
49. Birnie (n. 36 above), p. 179.

piracy articles could be revamped.[50] Menefee obliquely refers to UNCLOS IV by stating that the time required for a separate treaty on maritime terrorism would take too long, and therefore an antiterrorism doctrine should be included in an upcoming fourth Law of the Sea Convention.[51] The relevant question here is, What is the status of the international antipiracy doctrine during the waiting period?

Categorically speaking, there currently exist four distinct antipiracy doctrines. If the 1982 Convention is ratified, there will then be five antipiracy doctrines even though the new treaty provisions are basically a repetition of the ones in the 1958 High Seas Convention.

The first antipiracy doctrine is the one codified in UNCLOS I. It is international law in the form of convention law. For the 57 states party to it and those nonsignatories that adhere to the conventions, the international law repressing piracy is that which is promulgated in the 1958 High Seas Convention. Because UNCLOS I has not been denounced or superseded, it is still in force, and if the 1982 Convention enters into force, it will not relegate the UNCLOS I doctrine on piracy to obscurity. Instead, a complicated set of treaty relations develop. Signatories of the 1982 Convention are bound by it, but, in their relations with nonsignatories who are UNCLOS I signatories, the latter's treaty provisions take precedence—not the 1982 Convention articles.[52] In other words, the 1982 Convention will only apply between the parties to it. For example, the United States voted against the 1982 Convention but is party to UNCLOS I, while Belize, which was not an independent state in 1958, ratified the 1982 Convention. Any debate between the two states over a piratic incident would be negotiated under the UNCLOS I rules and provisions.

The second antipiracy doctrine consists of the portion of the original customary law that was not incorporated into UNCLOS I or the 1982 Convention. This is the branch of customary law concerned with acts of piracy committed in belligerencies (recognized and unrecognized). Because of the paucity of events and incidents testing this doctrine in recent times, it must be presumed that this customary antipiracy doctrine is still in effect. The *Sheira Maru* seizure in 1976 indicates this to be the case. States that have ratified or adhere to either UNCLOS I or the 1982 Convention but feel that the piracy provisions are incomplete subscribe to this antipiracy doctrine as well. Returning to the above example, both Belize and the United States recognize the concept of belligerencies.

The third doctrine is the customary antipiracy doctrine applicable for

50. Ibid.

51. Samuel P. Menefee, "Terrorism at Sea: The Historical Development of an International Legal Response," *Violence at Sea,* ed. B. A. H. Parritt (Paris: ICC Publishing, 1986), pp. 191–220.

52. D. J. Harris, *Cases and Materials on International Law,* 3d ed. (London: Sweet & Maxwell, 1983), p. 283, n. 20. Also refer to Article 311 of the 1982 Law of the Sea Convention (n. 6 above).

states not party to UNCLOS I and nonsignatory to the 1982 Convention. The proliferation of states in the 1960s and 1970s drastically increased the number of states not party to UNCLOS I. A minority of these new states along with a few older ones either refused to participate in or sign the 1982 Convention. The only Law of the Sea applicable for them is the entire body of customary international law pertaining to the sea. States in this category include Ecuador, Jordan, Kiribati, North Korea, Syria, Taiwan, and Turkey. For these states, the international law repressing piracy is both the branch of customary law concerning piracy committed in belligerencies, cited above, and the body of customary law concerned with the traditional forms of piracy that was predecessor to the provisions embodied in the 1958 High Seas Convention. The entire customary antipiracy doctrine that existed as a whole prior to UNCLOS I retains a spot in the limelight.

The limited conventions and agreements used to deal with particular incidents of a piratical nature constitute the fourth antipiracy doctrine. This is a situational antipiracy doctrine. Acts not within the definition of piracy *jure gentium* but of a character similar to piracy lead to this antipiracy doctrine being utilized. Piratic acts of war can often be defined this way. The antipiracy doctrine established is a response to the incident or event that induced its formation. These treaties adopt measures and laws that are limited in application to the signatories. The 1922 Washington Naval Treaty[53] and the 1937 Nyon Arrangement[54] typify this antipiracy doctrine. The IMO's 1988 Rome Suppression of Unlawful Acts Convention is also of this category.[55]

Though these four types of international antipiracy doctrines constitute the current state of affairs, they are not immune to change. The customary international law of the sea is in flux. While the international antipiracy doctrines have not changed appreciably in recent decades, other aspects of the Law of the Sea have. The antipiracy doctrines need to consider the changes and correspondingly become integrated within the new framework of the Law of the Sea. The 12-mile territorial sea is now accepted. Although the EEZ is not yet defined in internationally accepted language or law,[56] exclusive fishing zones (EFZs) are accepted customary doctrines today. However, they are vulnerable to the debate that plagues the EEZ doctrine concerning the exclusive rights a coastal state has therein. Therefore, all states may exercise their international right to repress piracy *jure gentium* anywhere on the seas up to 12 miles offshore of a littoral state, unless that state has claimed a Territorial Sea breadth of less than 12 miles. Australia, Bahrain, Belgium, Belize,

53. The Washington Declaration of 1922, Washington Naval Treaty, February 6, 1922, *American Journal of International Law* 16, suppl. (April 1922): 57–64.

54. The Nyon Arrangement of 1937, International Agreement for Collective Measures against Piratical Attacks in the Mediterranean by Submarines, September 14, 1937, *UKTS*, no. 38 (1937).

55. "Convention on the Suppression of Unlawful Acts" (n. 8 above).

56. Young (n. 40 above), p. 315.

Chile, Denmark, the Dominican Republic, Finland, Greece, Ireland, Israel, Jordan, Kiribati, the Netherlands, Norway, Qatar, St. Lucia, St. Vincent, Singapore, Tuvalu, the United Arab Emirates (except for Sharjah), and West Germany all claim Territorial Seas of less than 12 miles.[57] Until the 1982 Convention enters into force, it is presumed that Archipelagic Waters do not exist as a legal entity. Hence, a state may exercise its right to repress piracy in waters that an archipelagic state has claimed to be Archipelagic Waters provided such waters are not part of the Territorial Sea under existing rules and custom.

After the Sixtieth Ratification

Should the 1982 Convention enter into force, 12 months after the deposit of the sixtieth ratification or accession, then all the points discussed concerning the definition of piracy and where it may be universally repressed become applicable. As to the question of whether the 1982 Convention provides a good antipiracy doctrine that will curb piracy, the answer is "no." There are three reasons why it will not. The first is that the 1982 Convention does not contain a comprehensive and concise definition of acts of piracy *jure gentium*. It merely repeats the UNCLOS I definition and sustains all the concomitant controversy. The second reason is the EEZ itself and all the controversy surrounding it. The prevailing propensity by states to usurp all the power they can obtain will only serve to enhance debate and promote creeping jurisdiction. The third reason is that the 1982 Convention codifies an increased Territorial Sea and expansive Archipelagic Waters. Most piracy today occurs within 12 miles of the coast or near islands of an archipelago. Because these sites of frequency will be within the Territorial Seas of states, the international antipiracy doctrine no longer applies.[58] A good example is the Strait of Malacca and the Phillip Channel. The majority of pirates attacking shipping therein operate from Indonesia and commit their attacks in what would be Territorial Seas under the 1982 Convention. Indonesia cannot hope to eradicate these pirates, much less control them, because as an archipelagic state, it has countless islands with hidden coves.[59] Singapore would like to do something about the problem but is reluctant to go into Indonesia's waters or offend their neighbor by suggesting it clean up its act.[60] In the case of Nigeria, the problem is more a lack of willingness and effort on the part of the government to prevent piracy off the Nigerian coast. Under the 12-mile Territorial

57. Churchill and Lowe, pp. 302–9.
58. Brown (n. 25 above), p. 103.
59. Susumu Awanohara, "Please Repel our Boarders," *Far Eastern Economic Review* (October 9, 1981), pp. 43 ff.
60. Ibid.

Sea regime, only Nigeria has jurisdiction, and it is incapable of exercising that control. As no other state can intervene, piracy will continue to be a problem, but it will not be an international offence.

There will be a little piracy committed on the High Seas for which the international antipiracy doctrine will apply under the new ocean regime. Only somewhere between 7%–15% of the incidents that occurred in the past 5 years would have been classified as piracy on the High Seas had the 1982 Convention been in force. This specifically holds true for modern ship piracies and yacht piracies. The form of piracy with the highest percentage of incidents occurring more than 12 miles offshore are the boat-people piracies because the pirates prefer to attack at sea, away from the coast, maritime traffic, and patrol craft. However, these piratic attacks are declining. Politically motivated piracies occur, both on the High Seas and in zones subject to municipal jurisdiction. Their infrequency precludes definitive conclusions at this stage, but it may be fair to say that, in the future, maritime terrorism will come to dominate the scene. It will make headlines and preoccupy legislators' thoughts. A maritime version of the 1988 Kuwait Airways hijacking is a plausible scenario.

APPENDIX

PIRATIC ACTS AND OFFENSES UNDER UNCLOS I

larceny	assault	infanticide
plunder	mayhem	fratricide
pickpocketing	rape	matricide
vandalism	sodomy	patricide
blackmail	torture	enslavement
extortion	manslaughter	endangering life
riotry	murder	hijacking (of second vessel)
arson	pogrom	scuttling (of second vessel)
kidnapping	genocide	

The Role of Islands in Delimiting Maritime Zones: The Case of the Aegean Sea[1]

Jon M. Van Dyke
School of Law, University of Hawaii at Manoa

INTRODUCTION

The problem of delimiting the continental shelf and exclusive economic zone between Turkey and Greece is one of the many issues that currently dominate the international relations between the two countries. Although legal principles can be identified that apply to this dispute, the drawing of boundary lines is intrinsically a political process and is usually accomplished by direct negotiations between the states.[2] Increasingly in recent years, however, states have turned to arbitral or judicial tribunals to resolve disputes involving maritime boundaries, and the decisions of these tribunals have identified and developed legal principles that can now be drawn upon to resolve difficult boundary controversies.[3]

The disputes that have been submitted for decision have usually been those in areas with unusual geographical configurations, frequently involving islands. Many of the decisions that have been issued, as will be discussed in detail below,[4] have given islands less stature in generating extended maritime zones than the continental land masses that they are opposite or adjacent to. Article 121(2) of the 1982 Law of the Sea Convention states that islands generate continental shelves and Exclusive Economic Zones in the same manner as "other land territory" except for "rocks which cannot sustain human habitation or economic life of their own," which do not generate these zones at all. The decisions rendered in recent years do not, however, take this all-or-

1. I would like to express deep appreciation to Carolyn Nicol, Class of 1988, University of Hawaii Law School, and Michael Reveal and Dale Bennett, Class of 1989, University of Hawaii Law School, for their assistance in the preparation of this article and to Professor P. John Kozyris for his comments on an earlier draft. This article was originally presented at an international symposium on Aegean issues organized by the Foreign Policy Institute, Ankara, held in Çeşme, Turkey, in October 1987.

2. *The Law of the Sea: Official Text of the United Nations Convention on the Law of the Sea with Annexes and Index* (New York: United Nations, 1983), Sales No. E.83.V.5. (hereafter cited as the Law of the Sea Convention), Arts. 74 and 83.

3. See nn. 58–123 below and the accompanying text.

4. See ibid.

nothing approach and instead have given islands that are within 200 nm of the continental land mass of another nation "half effect" in generating extended maritime zones or—in some cases—no effect at all. The status of islands in generating such zones is thus currently unresolved in international law, and each geographic configuration must be examined individually to determine what effect the islands should have in relation to their continental neighbors. After examining the controversy between Turkey and Greece, this article will analyze the recent arbitral and judicial decisions and explore how the principles used in these decisions might apply to the delimitation of the maritime boundary between Turkey and Greece in the Aegean Sea.

BACKGROUND

The 1923 Lausanne Peace Treaty awarded Turkey the entire Anatolian mainland but awarded Greece sovereignty over almost all islands of the Aegean, which were populated by Greeks.[5] At the time of the negotiation of that treaty, Turkey sought to retain Turkish sovereignty over Imbros (Gokceada), Tenedos (Bozcaada), and Samothrace (Samothraki) and demilitarization of Limnos (Lemnos), Lesvos (Lesbos), Chios, Samos, and Ikaria.[6] Turkey was awarded Imbros (Gokceada) and Tenedos plus the Rabbit Islands because of their proximity to the strategically important Dardanelles.[7] Samothrace and Limnos were demilitarized but awarded to Greece.[8] The Dodecanese group of islands,[9] long under Turkish control, was ceded to Greece in 1947, following decades of Italian occupation.[10]

The islands around which the current marine resource boundary delimitation controversy centers are the Greek islands in the eastern Aegean close to

5. Derek Bowett, *The Legal Regime of Islands in International Law* (Dobbs Ferry, N.Y.: Oceana Publications, 1979), p. 252; the Convention regarding the Regime of the Straits (Lausanne Convention), July 24, 1923, *League of Nations Treaty Series (L.N.T.S.)*, 93:115, reprinted in J. Grenville, *The Major International Treaties, 1914–45* (New York: Methuen, 1987), p. 80.

6. Bowett, p. 250.

7. Ibid., p. 249.

8. Ibid.

9. *Columbia Lippincott Gazetteer of the World*, 1962 ed., s.v. "Dodecanese." The Dodecanese group consists of 14 main islands and about 40 islets and rocks. The main islands are Astypalea, Khalke, Kalymnos, Karpathos, Kasos, Kos, Leros, Lipsi (Leipsos), Nisyros, Patmos, Rhodes, Symi, and Tilos in the southeastern Aegean, and Megisti (Kastellorizo), the easternmost island separated from the rest. Between A.D. 1523 and 1912, the Dodecanese group was controlled by the Turks. The group (except for Kastellorizo) was occupied by Italy after the Italo-Turkish war of 1911–12 and awarded to Italy in 1920. Following the Second World War, the islands were awarded to Greece because of their Greek population.

10. Bowett, p. 255; Gerald Blake, "Marine Policy Issues for Turkey," *Marine Policy Reports* 7, no. 4 (1985): 1.

the Turkish land mass. The islands specified in the 1976 Greek application to the International Court of Justice (ICJ) (see discussion below) were Samothrace, Limnos, Aghios Eustratios, Lesvos, Chios, Psara, Antipsara, Samos, Ikaria, and the Dodecanese group (Patmos, Leros, Kalymnos, Kos, Astypalaea, Nisyros, Tilos, Symi, Khalki, Rhodes, Karpathos, etc.).[11] See table 1 for area and population of the islands and figure 1 for a map of the region.

In 1936, Greece claimed a 6-mile territorial sea; in 1964 Turkey claimed a 12-mile territorial sea in the Black Sea and a 6-mile territorial sea in the Aegean.[12] The 1958 Convention on the Territorial Sea and Contiguous Zone[13] did not define the breadth of the territorial sea, but the 1982 Law of the Sea Convention allows nations to establish the breadth of the territorial sea to a limit of 12 nm.[14] Turkey signed neither the 1958 nor 1982 Conventions; Greece signed both.[15] Greece would like to extend to 12 miles the territorial seas around each Greek island, expanding its territorial sea from 43.7% to 71.5% of the Aegean.[16] If Turkey were to extend its territorial sea claim from 6 to 12 miles, Turkey's gain in share of Aegean territory would be much smaller: from 7.5% to 8.8%.[17] Turkey has declared that a Greek attempt to enforce such an extension would be a causus belli.[18]

In 1973, Turkey granted 27 permits to the Turkish Petroleum Company, an oil exploration company,[19] to explore for petroleum on the continental

11. Aegean Sea Continental Shelf Case, International Court of Justice (ICJ), Interim Measures of Protection, Order of September 11, 1976, *International Court of Justice Reports* (hereafter cited as the *I.C.J. Rep.*) *1976*, p. 3, Par. 15, reprinted in *International Legal Materials* (hereafter cited as *I.L.M.*) 15 (1976): 988–89, citing Greece's request for interim measures of protection dated August 10, 1976 (hereafter cited as the "1976 Interim Protection Order").

12. Blake, pp. 2–3. Turkey and Greece also claim different airspace limits. Greece claims a 10-mile airspace and Turkey claims 6 miles. During the 1974 Cyprus crisis, however, Turkey extended its flight information region to the Aegean median line.

13. Geneva Convention of the Territorial Sea and the Contiguous Zone (done April 29, 1958) (in force September 10, 1964), *United States Treaty Series (U.S.T.)* 15:1606, *United States Treaties and Other International Agreements (T.I.A.S.)*, No. 5639, *United Nations Treaty Series (U.N.T.S.)*, 516:205.

14. Law of the Sea Convention (n. 2 above), Art. 3.

15. Greece signed the 1982 Convention on the first day it was opened for signature. See Blake, p. 3.

16. Alan Cowell, "Greece and Turkey Alert Forces as Tension Builds on Oil Search," *New York Times* (March 28, 1987), p. 1 and p. 4.

17. Ibid.

18. Ibid.; Clive R. Symmons, *The Maritime Zones of Islands in International Law* (The Hague: Martinus Nijhoff, 1979), p. 91; and Bowett (n. 5 above), p. 252.

19. On November 1, 1973, Turkey acknowledged it had issued concessions for a part of the northern Aegean seabed to the Turkish Petroleum Company (TRAO) and in July 1974 made a second concession to TRAO, expanding the western boundary of the November 1983 concession and creating a second one in the southeastern Aegean. See Christos L. Rozakis, *The Greek-Turkish Dispute over the Aegean Continental Shelf*, Occasional Paper No. 27 (Honolulu: Law of the Sea Institute, 1975), p. 1.

TABLE 1.—AREA AND POPULATION OF ISLANDS (Listed in Roughly North-to-South Order)

Island	Area		Population		
	Square Miles	km²	1951 Census	1971 Census	1981 Census
Samothrace (Samothraki)	71	178	3,993[a]	3,012	2,871
Limnos (Lemnos)	186	476	23,842	17,367	15,721
Aghios Eustratios	16.1	43	1,131	N.A.	296
Lesvos (Lesbos)	632	1,630	134,054	114,797	88,601
Chios (Khios)	321	842	72,777	52,487	48,700
Psara	16.4	40	751	N.A.	460
Antipsara[b]	1.5	4	...	...	...
Samos	194	476	56,273	32,664	31,629
Ikaria (Nikaria)	99	255	11,614	7,702	7,559
Patmos	13	34	2,428	2,432	2,534
Leros	21.2	53	6,131	N.A.	8,127
Kalymnos (Kalimnos)	41	111	11,864	N.A.	14,295
Kos	111.4	290	18,545	16,650	20,350
Astypalea (Astypalaia)	37	97	1,791	N.A.	1,030
Nisyros (Nisiros)	16	41	2,605	N.A.	916
Tilos (Telos)	24.3	63	1,085	N.A.	301
Symi (Simi)	22	58	4,083	2,489	2,273
Khalki (Chalki)	11.2	28	702	N.A.	334
Rhodes	542	1,398	55,181	66,606	87,831
Karpathos	111	301	7,396	5,420	4,645
Kasos	25	66	1,322	N.A.	1,184
Lipsi (Leipsos)	6	16	873	N.A.	574
Megisti (Kastellorizo)	3.5	9	800	N.A.	222

SOURCES.—L. Seltzer, ed., *Columbia Lippincott Gazetteer of the World* (1961); F. de Mello Vianna, ed., *International Geographic Encyclopedia and Atlas* (1979); National Statistical Service of Greece, ed., *Statistical Yearbook of Greece* (1984).
NOTE.—N.A. = not available.
[a] In 1940.
[b] Uninhabited.

FIG.—The Aegean Sea

shelf westward of several Greek Islands.[20] Greece claims that these areas are part of Greece's continental shelf[21] and that Turkey's concessions overlapped areas where in 1972 Greece had granted oil exploration concessions.[22] Turkey's action was apparently based on its view that the continental shelf delimitation should be drawn midway between the Greek and Turkish continental

20. Symmons p. 145. The granting of permits or concessions was made known in the November 1, 1973, issue of the *Official Turkish Gazette;* see Rozakis.

21. The Turkish claim conflicted with Greek territorial sea claims around the Greek islands of Samothrace, Lemnos (Limnos), Aghios Eustratios (Ayios Evstratios), Lesbos, Chios, Psara, and Antipsara. See Rozakis, p. 3.

22. Blake (n. 10 above), p. 3.

land masses, with no adjustment whatsoever for the Greek islands in the Aegean.[23]

Table 1 lists, in roughly north to south order, the disputed Greek islands in the Aegean near the Turkish coast.[24] This list, which shows the area in square miles and the population as of 1981,[25] indicates that the islands in question range in size and population from the 1.5-square-mile (4-km^2) uninhabited Antipsara to the 632-square-mile (1,630-km^2) Lesvos (Lesbos) with a (1981) population of 88,601.

The contentious issue is the extent to which the Greek islands very near Turkey's coast entitle Greece to exploit the resources of the continental shelf. Turkey and Greece exchanged *notes verbales* over this question in 1974.[26] Greece claimed that in continental shelf delimitations "islands, as any other part of the coast, are entitled to have *full seabed* area."[27] Turkey rejected this claim and argued that "geographical study of the Aegean Sea . . . does in fact prove the existence of vast submarine spaces of little depth all along and off the Turkish coast, which constitute the *natural prolongation of the Anatolian Peninsula, and thus of its continental shelf, whereas the Greek islands* situated very close to the Turkish coast *do not possess a shelf of their own.*"[28] Turkey's president reiterated the belief that the Anatolian Shelf, which extends midway into the Aegean, belongs to Turkey in his 1976 statement that the Aegean is "an extension of Asia Minor, and we will never allow it to be turned into an internal sea of another country."[29] During the 1974 exchange, Turkey stated that it wanted to settle the dispute through direct negotiation, but the Greek government stated that it preferred to submit the dispute to the International Court of Justice.[30]

In July 1976, Turkey announced plans to begin exploration for oil in

---

23. For a map of the disputed areas showing the November 1973 and July 1974 concessions, see Rozakis.

24. This list contains the islands that are named in the 1976 Greek application to the International Court of Justice and are analyzed in Donald Karl, "Islands and the Delimitation of the Continental Shelf: A Framework for Analysis," *American Journal of International Law* 71 (1977): 642, discussion at 669–72.

25. *Columbia Lippincott Gazetteer of the World* (n. 9 above), various pages.

26. Alona Evans, "Judicial Decisions," *American Journal of International Law* 73 (1979): 493–94. For a discussion of the exchange of notes, see Symmons, pp. 145–47.

27. *Note verbale* from Turkey to Greece, February 7, 1974, cited by Symmons, p. 146 (emphasis added).

28. *Note verbale* from Turkey to Greece, February 27, 1974, quoted by Symmons (n. 18 above), p. 137 (emphasis added). See also the letter from the permanent representative of Turkey to the Secretary General of the United Nations, August 18, 1976 (UN document 5/12182 [1976]), quoted in Leo Gross, "The Dispute between Greece and Turkey concerning the Continental Shelf in the Aegean," *American Journal of International Law* 71 (1977): 31.

29. "The Aegean: Acts of Piracy," *Time* (August 23, 1976), p. 33, quoting Turkish President Fahri Koroturk.

30. Evans, p. 493.

Turkish waters and on the high seas.[31] Turkey's *Sismic I* began conducting seismological exploration on August 6, 1976, in Aegean waters claimed by Greece,[32] concentrating research efforts on the waters adjacent to the islands of Limnos, Lesvos, Chios, and Rhodes, all of which are within 20 miles of the Turkish coastline.[33] On August 10, 1976, Greece simultaneously protested to the UN Security Council[34] and instituted proceedings in the ICJ,[35] requesting interim measures of protection pending judgment on the merits[36] and ultimately seeking a declaration delimiting the continental shelf in the Aegean.[37]

31. Ibid., p. 495.

32. Gross, p. 34.

33. "The Aegean: Acts of Piracy."

34. For a discussion of the appeal to the Security Council, see Gross, pp. 34–39. The Security Council passed Resolution 395 on August 25, 1976. This resolution suggested that Greece and Turkey should resume direct negotiations but should consider submitting to the International Court of Justice any legal differences that remained. See Evans, p. 495.

35. Application Instituting Proceedings, August 10, 1976, cited in Symmons, p. 147.

36. For a discussion of the interim measures of protection requested by Greece, see Gross, p. 40. For a discussion of interim measures in general, see Rainer Lagoni, "Interim Measures Pending Maritime Delimitation Agreements," *American Journal of International Law* 78 (1984): 345. In its application Greece asked the Court to direct that the governments of both Greece and Turkey

(1) unless with consent of each other and pending the final judgment of the Court in this case, refrain from all exploration activity of any scientific research, with respect to the continental shelf areas within which Turkey has granted such licenses or permits or adjacent to the islands, or otherwise in dispute in the present case,

(2) refrain from taking further military measures or actions which may endanger their peaceful relations.

See 1976 Interim Protection Order (n. 11 above), p. 987.

37. Because the court held that it lacked jurisdiction over the matter, this case did not reach the merits; Aegean Continental Shelf Case (Greece v. Turkey) (Jurisdiction), *I.C.J. Rep. 1978*, p. 1. See Evans (n. 26 above), p. 493. The Greek government had requested that court to adjudge and declare

(i) that the Greek islands [specified in the Application] as part of the territory of Greece, are entitled to the portion of the continental shelf which appertains to them according to the applicable principles and rules of international law;

(ii) what is the course of the boundary (or boundaries) between the portions of the continental shelf appertaining to Greece and Turkey in the Aegean Sea in accordance with the principles and rules of international law which the Court shall determine to be applicable to the delimitation of the continental shelf in the aforesaid areas of the Aegean Sea;

(iii) that Greece is entitled to exercise over its continental shelf sovereign and exclusive rights for the purpose of researching and exploring it and exploiting its natural resources;

(iv) that Turkey is not entitled to undertake any activities on the Greek continental shelf, whether by exploration, exploitation, research or otherwise, without the consent of Greece;

The International Court of Justice denied Greece's request for interim protection.[38] Interpreting Article 41 of the ICJ Statute,[39] the court reasoned that its powers to grant interim protection are limited to cases where an injured party would suffer "irreparable prejudice"[40] and that in this case Greece's alleged injury from Turkey's seismic exploration would be "capable of reparation by appropriate means."[41]

In November of 1976, Turkey and Greece signed an agreement in Bern stating that neither country would explore for oil in the continental shelf of the Aegean until the issue of delimitation of the continental shelf was settled.[42] The agreement requires the two countries to hold talks to resolve their differences on the Aegean, but Greece now claims the Bern accord no longer has effect because talks broke down in 1981.[43]

In the meantime, Greece continued to press its claim at the International Court of Justice, but the court decided in 1978 that it lacked jurisdiction over Greece's application for a declaration of rights of the parties in the continental shelf.[44] One basis for rejecting Greek claims to ICJ jurisdiction was that, in becoming a party to the 1928 General Act for Pacific Settlement of Disputes, Greece had made a reservation excluding disputes relating to "territorial status." The court interpreted this phrase to include sea boundary delimitations.[45] The court also rejected Greece's argument that a communiqué issued

---

(v) that the activities of Turkey described [in the Application] constitute infringements of the sovereign and exclusive rights of Greece to explore and exploit its continental shelf or to authorize scientific research respecting the continental shelf;

(vi) that Turkey shall not continue any further activities as described above in subparagraph (iv) within the areas of the continental shelf which the Court shall adjudge appertain to Greece.

1976 Interim Protection Order, p. 986.

38. For a discussion of this decision, see Gross (n. 28 above), pp. 40–48.

39. Article 41(1) of the Statute of the International Court of Justice provides: "The Court shall have the power to indicate, if it considers that circumstances so require, any provisional measures which ought to be taken to preserve the respective rights of either party."

40. *I.C.J. Rep. 1976*, p. 8, Par. 33, as quoted by Gross, p. 41.

41. Gross, p. 41.

42. Bern Agreement on Procedures for Negotiations of the Aegean Continental Shelf Issue, November 11, 1976, *I.L.M.* 16 (1977): 13, cited in Bowett (n. 5 above), p. 260.

43. Cowell (n. 16 above), p. 4.

44. Aegean Sea Continental Shelf Case (Greece v. Turkey) (Jurisdiction), *I.C.J. Rep. 1978*, p. 1. For a discussion of the proceedings, see, generally, K. Jayaramen, *Legal Regime of Islands* (New Delhi: Marwah Publications, 1982), pp. 77–85.

45. Evans, p. 489. See General Act for the Pacific Settlement of International Disputes (done September 26, 1928), *L.N.T.S.* 93:343. The Act provided for conciliation, arbitration, and judicial proceedings for the resolution of international disputes. The Act was subsequently amended in the revised 1928 General Pacific Settlement Act for the Pacific Settlement of International Disputes (done April 28, 1949; in force

to the press jointly by Greece and Turkey was a binding international agreement that required Turkey to submit to the jurisdiction of the ICJ.[46] Years of deliberation on procedural matters at the international court thus left the substantive issues of the delimitation of the Aegean continental shelf undecided.

This dispute heated up again in the spring of 1987 when both nations put their armies on alert after Turkey's oceanographic research vessel, the *Piri Reis,* ventured around the Greek islands of Limnos, Samothrace, and Thasos (northwest of Samothrace), where hydrocarbon deposits are located.[47] The prime minister of Greece warned of "huge dangers" if a second Turkish research vessel, the *Sismic I,* were to enter disputed waters of the Aegean Sea where Greece claims exclusive rights to explore the seabed for oil.[48] Tensions subsided once Turkey's prime minister announced that Turkey would honor the 1976 Bern accord and refrain from oil exploration unless Greece made the first move.[49]

Greece has argued that the issues of delimitation of the continental shelf should be decided by the International Court of Justice as an isolated legal question,[50] an approach some Turks have called "Greek salami tactics."[51] Turkey's position is that, because the rights to exploit the natural resources of the seabed of the Aegean affect economic interests and national security interests of both countries, political and legal issues should not be considered separately.[52] Turkey wants a dialogue with Greece and has agreed to accept

---

September 20, 1950). The amended Act replaced references to the Permanent Court of International Justice and the League of Nations by incorporating references to corresponding organs of the United Nations. For discussion of the current status of the Act, see J. Merrills, "The International Court of Justice and the General Act of 1928," *Cambridge Law Journal* 39 (1980): 137.

46. Evans, pp. 502–3. The joint communiqué stated in pertinent part: "[The Prime Ministers of Turkey and Greece decided that the problems] should be resolved peacefully by means of negotiations and as regards the continental shelf of the Aegean Sea by the International Court at the Hague."

47. *Turkish Daily News* (April 27, 1987), p. 1; "Greek-Turkish Sea Dispute Defused," *Honolulu Star Bulletin and Advertiser* (March 29, 1987), sec. A, p. 9; and Alan Cowell, "Aegean Dispute Worsens Turkish—Greek Ties," *New York Times* (March 24, 1987), p. 5.

48. See ibid. The *Sismic I* had also sparked international tension in 1976 when its explorations led Greece to protest Turkish action to the UN Security Council and appeal to the International Court of Justice (see text accompanying nn. 32 and 33 above).

49. *Honolulu Star Bulletin and Advertiser,* quoting Costil Stefanopoulos, leader of Greece's Democratic Party.

50. *Turkish Daily News.*

51. Personal communication from Joel Marsh, Fulbright Scholar at the Department of International Relations, Faculty of Political Science, University of Ankara, Turkey, April 28, 1987.

52. *Turkish Daily News.*

Greece's demand to take the issue to the ICJ, but only if Greece will also talk about the political aspects of the problem.[53] Turkey has felt that the issues should be resolved through bilateral negotiation[54] to permit trade-offs among the many key issues dividing Greece and Turkey, such as the militarization of Greek and Turkish islands in the Aegean[55] and the division of Cyprus.[56]

The prime ministers of Greece and Turkey agreed in January 1988 at Davos, Switzerland, to improve relations between the two countries by establishing several bilateral committees to define problem areas between the two nations and identify potential solutions. They also agreed that they would meet at least once a year to discuss mutual problems.[57] The prime ministers met again at Brussels in March 1988 and reached an agreement on the Greek property seized by the Turkish government in Istanbul, and Greece agreed to drop its objection to Turkey's becoming a member of the European Economic Community.[58] In June 1988, Turkey's prime minister went to Greece for 3 days, the first time in 36 years that a Turkish prime minister had visited Greece. The joint communiqué issued at the conclusion of the meeting made no mention of the disputes in the Aegean Sea.[59] A further meeting was held in September in Turkey between the two leaders, but again the talks did not deal with the substantive issues regarding territorial claims in the Aegean. Greece again offered to submit the Aegean continental shelf dispute to the International Court of Justice for resolution, but Turkey rejected the offer because of its position that all of the Aegean disputes should be examined together.[60]

53. Ibid.
54. Marsh.
55. A military junta governing Greece in 1974 backed a coup in Cyprus. In response to the coup, Turkey sent an invasion force to Cyprus. An estimated 20,000 Turkish troops remain on the island. In 1983, Turkish Cypriots unilaterally declared independence, but Greece does not recognize their government and refuses to negotiate on any subject until the troops are withdrawn. See Cowell (n. 16 above), p. 4.
56. After Turkey invaded Cyprus in 1974, Greece began to fortify islands in the eastern Aegean, and Turkey fortified Gokceada (Imbros) and Bozcaada (Tenedos), two strategically important islands guarding the approach to the Dardanelles. Turkey's army of 550,000 and U.S. military aid (U.S.$775 million in 1984) indicate Turkey's capacity to invade Greek islands near the Turkish coast, a possibility Greeks fear. See Blake (n. 10 above), p. 4. In 1975 Turkey formed an army with amphibious landing capacity called the "Army of the Aegean." See ibid.
57. Steven Greenhouse, "Chiefs Reach a Greek-Turkish Accord," *New York Times* (February 1, 1988), p. A12, col. 1.
58. "Progress at Greek-Turkish Talks, *New York Times* (March 5, 1988), p. A3, col. 1.
59. Robert Suro, "Few Gains Seen as Greek-Turkish Talks End," *New York Times* (June 16, 1988), p. A15, col. 1.
60. "Negotiators Unable to Agree," Associated Press wire service, September 6, 1988, at 13 hours, 42 minutes, 15 seconds.

TREATY PROVISIONS AND JUDICIAL/ARBITRAL
DECISIONS INVOLVING ISLANDS

The 1958 Convention on the Continental Shelf stated that the boundary
between continental shelves of opposite and adjacent states should be the
median line unless special circumstances dictate another line,[61] but even
under this regime an island in the midst of another nation's geologic continental shelf was considered to be a classic special circumstance.[62] Under the
1982 Convention, the median line/equidistance principle is no longer necessarily even the starting point for boundary delimitations, and nations with
opposite or adjacent coasts are instructed simply to negotiate pursuant to the
principles of "international law, as referred to in Article 38 of the Statute of
the International Court of Justice, in order to achieve an equitable solution."[63]

The language in the 1982 Convention referring to "Article 38 of the
Statute of the International Court of Justice" is widely thought of as a shorthand reference to the 1969 *North Sea Continental Shelf Case*,[64] where the ICJ
applied equitable principles to delimit the continental shelf of the Netherlands, Denmark, and the Federal Republic of Germany. Strict application of
the equidistance principle would have denied Germany all but a small share of
the shelf because Germany's coastline is concave. The Court held that "relevant circumstances" to consider in achieving an equitable solution include the
configuration of the coastline and the proportionality between the length of a
nation's coastline and the area of that nation's continental shelf.[65]

The *North Sea Continental Shelf Case* is probably best known for its reliance
on the principle of the "natural prolongation" of the continental shelf, a view
that sees the undersea shelf as an extension of the continent, which leads to
the conclusion that the islands projecting up from this underlying shelf do not
have the same capacity to generate zones as does the continental landmass
itself. Indeed, the court said in this opinion that "the presence of islets, rocks
and minor coastal projections, the disproportionality distorting effect of
which can be eliminated by other means," should be ignored in continental
shelf delimitation.[66] It is significant that this early boundary decision thus
rejected the notion that all islands should generate equal zones, even though
the only provision defining the role of islands in the 1958 Conventions did not
differentiate among islands.[67]

---

61. Geneva Convention on the Continental Shelf, April 29, 1958 (in force June 6,
1984), Art. 3; *U.S.T.*, 15:471; *T.I.A.S.* 5578; *U.N.T.S.*, 499:311 (1958).
62. Karl (n. 24 above), p. 648.
63. Law of the Sea Convention (n. 2 above), Arts. 74 and 83.
64. North Sea Continental Shelf Case (Federal Republic of Germany v. Denmark; Federal Republic of Germany v. The Netherlands), *I.C.J. Rep. 1969*, p. 3.
65. North Sea Continental Shelf Case, Par. 101(d).
66. Ibid.
67. Geneva Convention of the Territorial Sea and the Contiguous Zone (n. 13
above), Art. 10, defines an "island" as "a naturally-formed area of land, surrounded by

When the negotiations that led to the 1982 Law of the Sea Convention began in earnest in the early 1970s, the question of the role of islands in generating ocean space was a central issue, and a number of countries proposed how this matter should be resolved. The Pacific island states and Greece both introduced draft articles stating that island maritime spaces should be determined by the same rules governing other land territory.[68] The Greek draft and Part A of the Pacific islands draft were substantially identical, and both declared that the provisions should apply to all islands.

Romania, Turkey, and a number of African states submitted draft proposals that would have limited maritime spaces of islands according to various criteria. The Romanian proposal defined "islets" as naturally formed high-tide elevations less than 1 km^2 in area.[69] This proposal also used the words "islands similar to islets," which were defined as "naturally formed elevations

---

water, which is above water at high tide" and then says that the territorial sea of all islands is determined in the same manner as the territorial sea of any other land areas. Article 1(b) of the 1958 Continental Shelf Convention uses the term "island" without defining it further. It could have been argued, therefore, that these two conventions taken together recognized that *all* islands generated continental shelves. The language quoted in the text accompanying n. 66 indicates that the International Court of Justice rejected this possible interpretation.

68. The Greek proposal was as follows:

Article 1

1. An island is a naturally formed area of land, surrounded by water, which is above water at high tide.
2. An island forms an integral part of the territory of the State to which it belongs.
3. The foregoing provisions have application to all islands, including those comprised in an island State.

Article 2

1. The sovereignty and jurisdiction of a State extends to the maritime zones of its islands determined and delimited in accordance with the provisions of this Convention applicable to its land territory.
2. The sovereignty over the islands extends to its territorial sea, to the air space over the island and its territorial sea, to its sea-bed and the subsoil thereof and to the continental shelf for the purpose of exploring it and exploiting its natural resources.
3. The island has a contiguous zone and an economic zone on the same basis as the continental territory, in accordance with the provisions of this Convention.

See United Nations document A/Conf.62/C.2/L.50, 1974. For the text of the Pacific Islands proposal, see Jon Van Dyke and Robert Brooks, "Uninhabited Islands: Their Impact on the Ownership of the Ocean's Resources," *Ocean Development and International Law* 12 (1983): 294–95, n. 3, quoting UN document A/Conf.62/C.2/L.30, 1974.

69. For the text of the Romanian proposal, see Van Dyke and Brooks, p. 296, n. 64, quoting UN document A/Conf.62/C.2/L.53, 1974.

of land" larger than islets that cannot be permanently inhabited or have their own economic life. Both categories would have been allowed in some circumstances to generate security areas and territorial seas as long as they did not prejudice the maritime zones of another nation. "Islets" or "islands similar to islets" in the international zone of the seabed would have been allowed to have such marine spaces as agreed on with an international authority that would be established to monitor maritime boundary delimitation.

The Turkish proposal would not have allowed economic zones for islands under foreign domination or for islands situated on the continental shelf of another state if the island's land area was not at least one-tenth of the total land area of the nation to which it belonged.[70] The Turkish proposal stated

70. Turkey's draft articles on the regime of islands were as follows:

Article 1

[Definitions]

Article 2

Except where otherwise provided in this chapter the marine spaces of islands are determined in accordance with the provisions of this Convention.

Article 3

1. No economic zone shall be established by any State which has dominion over or controls a foreign island in waters contiguous to that island.

The inhabitants of such islands shall be entitled to create their economic zone at any time prior to or after attaining independence or self-rule. The right to the resources of such economic zone and to the resources of the continental shelf are vested in the inhabitants of that island to be exercised by them for their benefit and in accordance with their needs or requirements.

In case the inhabitants of such islands do not create an economic zone, the Authority shall be entitled to explore and exploit such areas, bearing in mind the interests of the inhabitants.

2. An island situated in the economic zone or the continental shelf of other States shall have no economic zone or continental shelf of its own if it does not contain at least one tenth of the land area and population of the State to which it belongs.

3. Islands without economic life and situated outside the territorial sea of a State shall have no marine space of their own.

4. Rocks and low-tide elevations shall have no marine space of their own.

Article 4

A coastal State cannot claim rights based on the concept of the archipelago or archipelagic waters over a group of islands situated off its coast.

Article 5

In areas of semi-enclosed seas, having special geographic characteristics, the maritime spaces of islands shall be determined jointly by the States of that area.

that "islands without economic life and situated outside of the territorial sea of a State shall have no marine space of their own."[71] "Rocks" and "low-tide elevations" would also have been denied marine spaces.

The draft articles introduced by the African states divided the world of land areas surrounded by water into four categories—"islands," "islets," "rocks," and "low-tide elevations"—with the final three being denied jurisdiction over marine space.[72] The definitions offered, however, would have needed additional refinement. An "island" was defined as "a vast naturally formed area of land, surrounded by water, which is above water at high tide," and an "islet" was distinguished simply by substituting the word "smaller" for "vast." A "rock" was defined as "a naturally formed rocky elevation of ground, surrounded by water, which is above water at high tide." The marine spaces of these categories of land protrusions were to be determined by considering equitable criteria such as size, geographical configurations, "the needs and interests of the population living thereon," any conditions that "prevent a permanent settlement of population," and whether it is located near a coast.

These views were controversial and did not command a consensus among the delegates to the Third UN Conference on the Law of the Sea. When the president of the Conference and the chairmen of the three committees prepared the Single Negotiating Text (SNT) in April 1975,[73] they attempted to formulate articles that would represent consensus without prejudicing the position of any delegation. The language on the regime of islands was therefore brief and was designed to be inoffensive to all. Unfortunately, the ambiguity from the 1958 Geneva Conventions was carried forward.[74] Paragraphs (1) and (2) of Article 132 of the SNT[75] were taken directly from Article 10 of

---

Article 6

The provisions of this chapter shall be applied without prejudice to the articles of this Convention relating to delimitation of marine spaces between countries with adjacent and/or opposite coasts.

Article 7

For the purposes of this chapter the term "marine space" implies either the territorial sea and/or continental shelf and/or the economic zone according to the context in which the term has been used.

See UN document A/Conf.62/C.2/L.55, 1974.

71. See ibid., Art. 3.3.

72. For the text of the African proposal, see Van Dyke and Brooks, pp. 297–99, quoting UN document A/Conf.62/C.2/L.62/Rev.1, 1974.

73. UN document A/Conf.62/WP.8/Rev. Pts. 1, II, and III, 1975.

74. See n. 67 above; and Van Dyke and Brooks, pp. 274–76.

75. Article 132 of Single Negotiating Text, n. 73 above:

1. An island is a naturally formed area of land, surrounded by water, which is above water at high tide.

the 1958 Territorial Sea Convention.[76] Paragraph (3) of Article 132 was new, however. This paragraph denied exclusive economic zones and continental shelves to "rocks which cannot sustain human habitation or economic life of their own" and thus added a new ambiguity.

The article on the regime of islands is now numbered Article 121 in the 1982 Convention; its language and ambiguities remained unchanged through the Revised Single Negotiating Text of May 1976,[77] the Informal Composite Negotiating Text (ICNT) of 1977,[78] the Revised ICNT of April 1979, and the Draft Treaty of August 1980.[79] Indeed, no formal substantive discussion of the topic occurred after the 1974 Caracas session. S. H. Amerasinghe, then president of the Conference, noted in his explanatory memorandum to the 1979 Revised ICNT that the regime of islands "had not yet received adequate consideration and should form the subject of further negotiation during the resumed session."[80]

Further consideration of the regime of islands did not take place, however, because of the pressure applied to complete the treaty during the 1980 and later negotiating sessions because of the limited negotiations after the 1981 U.S. announcement regarding its reassessment of the Convention and because many of the major nations saw benefits from Article 121 as it was worded. Consequently, Article 121 remains in its somewhat ambiguous form, and scholars and diplomats have been struggling to give precise meaning to its language.

The 1977 *Anglo-French Arbitration*[81] was the first instance in which a tribunal addressed the effect of islands on delimitation of a continental shelf boundary.[82] This dispute required the tribunal to determine whether the British Channel Islands were entitled to a continental shelf as separate islands and what influence these islands should have on the delimitation of the continental shelf between England and France.[83]

---

2. Except as provided for in paragraph 3, the territorial sea, the contiguous zone, the exclusive economic zone and the continental shelf of an island are determined in accordance with the provisions of this Convention applicable to other land territory.

76. See n. 67 above.

77. UN document A/Conf.62/WP.8/Rev.1/Pts. I, II, III, and IV, 1976.

78. UN document A/Conf.62/WP.10, 1977.

79. The Revised Informal Composite Negotiating Text of April 1979 is UN document A/Conf.62/WP.10/Rev.1, 1979, and the Draft Treaty of August 1980 is A/Conf.62/WP.10/Rev.3/add.1.

80. UN document A/Conf.62/WP.10/Rev.1, 1979, p. 19.

81. Case concerning the Delimitation of the Continental Shelf between the United Kingdom of Great Britain and Northern Ireland, and the French Republic, *United Nations Reports of International Arbitral Awards (R.I.A.A.)* 18 (1977): 74; reprinted in *I.L.M.* 18 (1979): 397 (hereafter referred to as "the Anglo-French Arbitration").

82. See, generally, Bowett (n. 5 above), pp. 193–247.

83. "Whether, and if so, in what manner, the presence of the British Channel

The Channel Islands archipelago consists of four groups of islands, including the main islands of Jersey, Guernsey, Alderney, Sark, Herm, and Jethou, as well as a large number of rocks and islets, some of which are inhabited.[84] The islands are under British sovereignty but are located as close as 6.6 km from the French Normandy coastline,[85] that is, "on the wrong side of the median line."[86] Geological evidence indicates that the Channel Islands are part of the physical land mass of Brittany and Normandy.[87] These islands have a total land area of 195 km² and a population of 130,000.[88] Politically, the Channel Islands are British dependencies, not constitutionally part of the United Kingdom.[89]

The tribunal awarded Britain 12-nm enclaves around the Channel Islands[90] but ruled that otherwise they would not affect the delimitation of boundary and thus that the area around these enclaves would belong to France. As to the median line, the tribunal rejected the British proposal that the median line should "automatically deviate southwards in a long loop around the Channel Islands."[91] The tribunal also explained that the juridical concept of natural prolongation requires consideration of geographical circumstances to be viewed in light of "any relevant consideration of law and equity."[92]

Islands close to the coast of Normandy and Brittany affects the legal framework of a median line delimitation in mid-channel which would otherwise be indicated by the opposite and equal coastlines of the mainlands of the two countries." See the Anglo-French Arbitration, Par. 189; and reprinted in *I.L.M.* 18 (1979): 442.

84. The Anglo-French Arbitration, Par. 6; and reprinted in *I.L.M.* 18 (1979): 408. Editor's note.—For further comments on the Anglo-French Arbitration and for a map of the Anglo-French maritime boundary, see John Briscoe, "Islands on Maritime Boundary Delimitation," *Ocean Yearbook 7*, ed. Elisabeth Mann Borgese, Norton Ginsburg, and Joseph R. Morgan (Chicago: University of Chicago Press, 1988), pp. 14–41, and fig. 2, p. 34.

85. Anglo-French Arbitration.

86. Ibid., Par. 173; and reprinted in *I.L.M.* 18 (1979): 440.

87. Symmons (n. 18 above), p. 138: "Scientific evidence showed clearly the [Channel Islands are] an *integral part* of the amorican area and are included in the French hercynien shelf, [and] truly thus formed a part of the physical mass of Brittany and Normandy" (emphasis added in original).

88. Bowett, p. 195.

89. Anglo-French Arbitration, Par. 184.

90. This solution created the first true total enclave of a continental shelf in state practice. See Bowett, p. 206.

91. Anglo-French Arbitration (n. 81 above), Par. 189; and reprinted in *I.L.M.* 18 (1979): 442. "In the opinion of the Court . . . such an interpretation of the situation in the Channel Islands region would be as extravagant legally as it manifestly is geographically"; Anglo-French Arbitration, Par. 190; and *I.L.M.* 18 (1979): 442.

92. Anglo-French Arbitration, Par. 194; and *I.L.M.* 18 (1979): 443:

The principle of natural prolongation of territory is neither to be set aside nor treated as absolute in a case where islands belonging to one State are situated on continental shelf which would otherwise constitute a natural prolongation of

Another portion of the *Anglo-French Arbitration* concerned the relative weight to be given to the Scilly Isles off the British Coast near Land's End, compared with Ushant off the northwest coast of France. The Scilly Isles, lying some 21 miles (34 km) from the mainland, are "a group of 48 islands of which six are inhabited."[93] France argued that they should be essentially ignored. The tribunal resolved the dispute by splitting the difference. It constructed one set of baselines and equidistance lines using the Scilly Isles and another set that ignored them. The triangle that was hereby created was then divided in half to create the "half-effect" line.[94] The tribunal justified its use of this "half-effect" approach in part because the Scillies are twice as far from Land's End as Ushant is from Finistère,[95] and in part because of the economic and political conditions on the islands.[96]

This "half-effect" idea was apparently taken from other situations where similar results were reached through negotiations. Italy and Yugoslavia, for instance, had a number of very small islands lying between them in the Adriatic Sea that were given partial effect in delimitation.[97] Similarly in the delimitation between Iran and Saudi Arabia, the island of Kharg was given a half effect.[98]

A year after the *Anglo-French Arbitration,* in 1978, Australia and Papua New Guinea negotiated an "imaginative"[99] solution to the problem created by the presence of Australian islands just south of the main island of Papua New

---

the territory of another State. The application of that principle in such a case, as in other cases concerning the delimitation of the continental shelf, has to be appreciated in the light of all the relevant geographical and other circumstances. When the question is whether areas of continental shelf, which geologically may be considered a natural prolongation of the territories of two States, appertain to one State rather than to the other, the legal rules constituting the juridical concept of the continental shelf take over and determine the question. Consequently, in these cases the effect to be given to the principle of natural prolongation of the coastal State's land territory is always dependent not only on the particular geographical and other circumstances but also on any relevant considerations of law and equity.

Significantly, the tribunal ignored altogether the small rocks and islands in the Channel Islands that are not inhabited. See ibid., Par. 184.

93. Anglo-French Arbitration, Par. 227, and *I.L.M.* 18 (1979): 450–51.

94. Anglo-French Arbitration, Par. 249, and *I.L.M.* 18 (1979): 455.

95. Bowett (n. 5 above), p. 215, citing the Anglo-French Arbitration, Par. 251, and *I.L.M.* 18 (1979): 455.

96. Bowett, pp. 223–24.

97. N. Ely, "Seabed Boundaries between Coastal States: The Effect to Be Given Islets as 'Special Circumstances,' " *International Law* 6 (1971): 227–28.

98. Ely, p. 229; and Bowett, p. 215, citing U.S. State Department Office of the Geographer, "Continental Shelf Boundary: Iran-Saudi Arabia," *Limits of the Sea,* Ser. A, No. 24 (July 6, 1970).

99. J. R. V. Prescott, *The Maritime Political Boundaries of the World* (New York: Methuen, 1985), p. 191.

Guinea,[100] which are also on the "wrong" side of the median line. It was agreed by both states that these small Australian islands would produce an "inequitable boundary if given full effect,"[101] and so they decided that these small islands would generate fishing zones, but that they would have no effect on the continental shelf boundary and thus that the Australian islands would sit atop the Papua New Guinea continental shelf.[102] The treaty also creates a protected zone to preserve the traditional way of life for the inhabitants of the islands.[103]

In three ICJ maritime boundary decisions handed down since 1982, the court has held in each case that islands should be given only a partial effect in delimiting the boundaries. The first of these decisions was the 1982 *Tunisia-Libya Continental Shelf Case*, where the ICJ relied on the *Anglo-French Arbitration* decision and gave only half effect to Tunisia's Kerkennah Islands in delimiting the continental shelf between the two nations.[104] The main island of Kerkennah is 180 km^2 (69 square miles) and has a population of 15,000. In drawing a line to represent the general direction of the coast, the court disregarded large areas of low-tide elevation on the islands. The court drew a delimitation line between Tunisia and Libya in two sectors to adjust for a change in the general direction of the Tunisian coastline. The first sector extended seaward from the land boundary between Tunisia and Libya at Ras Ajdir, roughly perpendicular to the coast at an angle approximately 26 degrees east of north.[105] Instead of continuing the line at that angle to the edge of the shelf, the court deflected the line eastward in a second sector to give Tunisia more continental shelf area because of the change in Tunisia's coast-

---

100. Australia–Papua New Guinea: Treaty on Sovereignty and Maritime Boundaries in the Area between the Countries, done at Sydney, December 18, 1978, reprinted in *I.L.M.* 23 (1984): 291. Editors' note.—For further comments and map of the region, see Briscoe (n. 84 above), text and fig. 3, pp. 35–36.

101. Prescott, p. 191.

102. Ibid., fig. 7.5, "Maritime Boundaries in Torres Strait," pp. 194–95. Editors' note.—This figure is reproduced in Briscoe, fig. 3, "Maritime Boundaries in Torres Strait," p. 36.

103. Prescott, p. 191.

104. Case concerning the Continental Shelf (Tunisia/Libyan Arab Jamahiriya), *I.C.J. Rep. 1982*, p. 89, Par. 129. Editors' note.—For the text and a discussion of the ICJ judgment on the continental shelf boundary between Tunisia and the Libyan Arab Jamahiriya, see Nicholas P. Dunning, "International Court of Justice Judgment of February 24, 1982: Case concerning the Continental Shelf (Tunisia/Libyan Arab Jamahiriya)," *Ocean Yearbook 4*, ed. Elisabeth Mann Borgese and Norton Ginsburg (Chicago: University of Chicago Press, 1983), pp. 515–32; and for discussion and map of the maritime boundary between the countries, see Briscoe, pp. 14–41, and fig. 5, p. 38, which is a reproduction of Prescott, fig. 12.1, "The Maritime Boundary between Libya and Tunisia," p. 301.

105. For a map of the area, see D. Christie, "From the Shoals of Ras Kaboudia to the Shores of Tripoli: The Tunisia/Libya Continental Shelf Boundary Delimitation," *Georgia Journal of International and Comparative Law* 13 (1983): 19.

line.[106] The angle of deflection, however, was less than it would have been had the seaward boundary of the Kerkennah Islands been used to represent the direction of the coast. The Kerkennah boundary line angle was averaged together with a hypothetical coastline angle that would properly have represented the coast were no islands present.[107]

Similarly, Canada's Seal Island and Mud Island and other adjacent islets in the vicinity of Cape Sable in Nova Scotia were given only partial effect in a 1984 determination by a chamber of the ICJ of the maritime boundary between Canada and the United States in the *Gulf of Maine Case*.[108] As in the *Libya/Tunisia* delimitation, the chamber used a two-sector line, with the first segment roughly following an equidistance formula. The second sector allocated ocean space between the United States and Canada in a ratio proportional to the relative lengths of their coastlines in the Gulf.[109] Had Seal Island and its neighboring islets been given no effect, the ratio of ocean area belonging to the United States compared to that of Canada would have been 1.38 to 1.[110] The court decided that although Seal Island and its neighbors "cannot be

106. Ibid., p. 20.
107. Ibid., p. 21:

> The Court represented the general direction of the coast as a line of an approximate 42 degree bearing drawn from the most westerly point of the Gulf of Gabes to Ras Kaboudia. . . . This depiction of the coastline, however, did not give effect to the Kerkennah Islands. . . . The Court described a line along the seaward side of the islands as having an approximate 62 degree bearing. . . . In spite of the fact that the 62 degree line disregarded large areas of low tide elevations to the east of the Kerkennah Islands, the Court considered that a delimitation running parallel to the seaward side of the islands would give excessive weight to the Kerkennahs.
>
> Following the example of other delimitations which have given only partial effect to islands, the Court determined that . . . the islands should be given "half effect." By bisecting the angle formed by the 42 degree and 62 degree lines, the Court effectively gave halfweight to the islands in the delimitation by drawing the line in the second sector at an angle of 52 degrees from the meridian.

108. Judgment of the International Court of Justice on the Delimitation of the Maritime Boundary in the Gulf of Maine Area (Canada v. United States of America), October 12, 1984, *I.C.J. Rep. 1984*, pp. 336–37, Par. 222; and reprinted in *I.L.M.* 23 (1984): 1242–43. Editors' note.—For further information on this case, see "Analysis of the Judgment of the International Court of Justice on the Delimitation of the Maritime Boundary in the Gulf of Maine Area (Canada/United States of America), 12 October 1984," *Ocean Yearbook 6*, ed. Elisabeth Mann Borgese and Norton Ginsburg (Chicago: University of Chicago Press, 1986), app. B, pp. 516–26; and Nicholas P. Dunning, "Boundary Delimitation in the Gulf of Maine: Implications for the Future of a Resource Area," *Ocean Yearbook 6*, pp. 390–98, and fig. 1, p. 391.
109. See n. 108 above, *I.C.J. Rep. 1984*, p. 336, Par. 221; *I.L.M.* 23 (1984): 1242.
110. "The ratio between the coastal fronts of the United States and Canada on the Gulf of Maine . . . 1.38 to 1 . . . should be reflected in the location of the second segment of the delimitation line." *I.C.J. Rep. 1984*, p. 336, Par. 222; *I.L.M.* 23 (1984): 1242.

disregarded" because of their dimensions and geographical position,[111] it would be "excessive" to give them full effect.[112] Thus the court decided it was appropriate to give the islands half effect, and, as a result, the U.S./Canada ocean space ratio became 1.32 to 1.[113]

In its most recent decision involving islands, the 1985 *Libya/Malta Continental Shelf Case,*[114] the ICJ ruled that equitable principles required that the tiny uninhabited island of Filfla (belonging to Malta—3 miles (5 km) south of the main island) *should not be taken into account at all* in determining the boundary between the two countries.[115]

Another dispute regarding offshore islands has concerned Argentina and Chile, both of which declared 200-nm territorial seas around all of their mainland and insular coasts.[116] These countries recently settled a century-old dispute concerning islands lying off the coast of Tierra del Fuego in the Beagle Channel based on the proposal of a papal mediator.[117] The larger,

111. The Court stated:

> The Chamber considers that Seal Island (together with its smaller neighbour, Mud Island), by reason both of its dimensions and, more particularly, of its geographical position, cannot be disregarded for the present purpose. According to the information available to the Chamber it is some two-and-a-half miles long, rises to height of some 50 feet above sea level, and is inhabited all the year round. It is still more pertinent to observe that as a result of its situation off Cape Sable, only some nine miles inside the closing line of the Gulf, the island occupies a commanding position in the entry to the Gulf.

See *I.C.J. Rep. 1984,* pp. 336–37, Par. 222; *I.L.M.* 23 (1984): 1242–43.
112. No explanation was given for the determination that "it would be excessive." See n. 111 above.
113. See n. 111 above.
114. Judgment of the International Court of Justice on the Continental Shelf (Libyan Arab Jamahiriya/Malta), June 3, 1985, *I.C.J. Rep. 1985,* p. 13. Editors' note.— For further information on this case, see "Analysis of the 'Judgment of the International Court of Justice on the Continental Shelf (Libyan Arab Jamahiriya/Malta), 3 June 1985,'" *Ocean Yearbook 6,* app. B, pp. 504–15, which was excerpted from UN Office of the Special Representative of the Secretary-General for the Law of the Sea, *Law of the Sea Bulletin,* no. 6 (October 1985); and Briscoe (n. 84 above).
115. Judgment of the International Court of Justice on the Continental Shelf (Libyan Arab Jamahiriya/Malta), p. 48, Par. 64. After referring to the statement in the North Sea Continental Shelf Cases (see n. 64 above), quoted in the text accompanying n. 66 above, the court stated: "The Court thus finds it equitable not to take account of Filfla in the calculation of the provisional median line between Malta and Libya."
116. See Presidential Declaration concerning the Continental Shelf, Article 1, June 23, 1947, reprinted in A. Szekely, *Latin America and the Development of the Law of the Sea, Chile* (Dobbs Ferry, N.Y.: Oceana Publications, 1980), 2:13; and Law No. 17,094–M24 of December 29, 1966, Art. 1, ibid., s.v. "Argentina," 2:20.
117. Treaty of Peace and Friendship (Chile/Argentina) (November 29, 1984), reprinted in *I.L.M.* 24 (1985): 10–28; Papal Proposal in the Beagle Channel Dispute: Proposal of the President, December 12, 1980, reprinted in *I.L.M.* 24 (1985): 7. Editors' note.—The Treaty of Peace and Friendship was published in "Selected Docu-

inhabited islands in the channel are fringed by many smaller uninhabited rocks and islets. The resolution of the dispute limited the Chilean maritime claim by giving less than full effect to the smaller Chilean islets in the Atlantic waters off the Argentine coast of Tierra del Fuego.[118]

In summary, recent arbitrations, judicial decisions, and negotiations have been relatively consistent in refusing to give full effect to islands in delimiting maritime boundaries.[119] The *Anglo-French Arbitration*,[120] this resolution of the long-standing dispute between Argentina and Chile, and the four opinions of the International Court of Justice described above[121] all stand for the proposition that islands do *not* generate extended maritime jurisdiction in the same way that other land masses do. Even inhabited islands (such as Jersey and Guernsey in the English Channel, Kerkennah Island near Tunisia, and Seal Island in the Gulf of Maine)[122] do not generate full extended maritime zones if the impact of such an extension is to interfere with the claim of another nation based on a continental land mass.[123]

---

ments," *Ocean Yearbook 6,* pp. 606–20. For a discussion of the regional implications of the treaty and a map showing the Chilean and Argentine maritime zones, see Michael A. Morris, "EEZ Policy in South America's Southern Cone," *Ocean Yearbook 6,* pp. 417–37. See also Michael A. Morris, "South American Antarctic Policies," *Ocean Yearbook 7,* (n. 84 above), pp. 356–71.

118. Treaty of Peace and Friendship (Chile/Argentina), Art. 7. The uninhabited islands of Evout, Barnevelt, and Horn generate only 12-mile zones. See Papal Proposal in the Beagle Channel Dispute, Art. 4(a)(b)(4).

119. A significant exception would be the recent negotiations carried out by the United States with Venezuela and Mexico in which full effect was given to small islands. See Maritime Boundary Treaty between the United States of America and the Republic of Venezuela, done March 28, 1978, entered in force November 24, 1980, *T.I.A.S.* 9890; Treaty on Maritime Boundaries between the United States and the United Mexican States, S. Exec. Doc. F, 96th Cong., 1st Sess. (1979); Mark Feldman and David Colson, "The Maritime Boundaries of the United States," *American Journal of International Law* 75 (1981): 729, 735, 740. The United States accepted the Venezuelan and Mexican claims, not out of altruism, but because it felt that it had much to gain in other maritime boundary disputes if all small islands were allowed to generate 200-mile zones without limitation. See, generally, Jon M. Van Dyke, Joseph R. Morgan, and Jonathan Gurish, "The Exclusive Economic Zone of the Northwestern Hawaiian Islands: When Do Uninhabited Islands Generate an EEZ?" *San Diego Law Review* 25 (1988): 425–94. For examples of other agreements that have used tiny insular formations as base points for determining equidistance lines in resolving boundary disputes, see Symmons (n. 18 above), pp. 190–91.

120. See text accompanying nn. 81–96 above.

121. See text accompanying nn. 64–67 and 104–15 above.

122. See text accompanying nn. 84–92 and 104–13 above.

123. One recent commentator said that the decision failing to give full effect to the Channel Islands was unjust because it failed to recognize the rights of the sizable population that lives there; see Charles Brand, "The Legal Relevance of South African Insular Formations Off the SWA/Namibian Coast," *Sea Changes* 4 (1986): 101.

HOW THESE PRINCIPLES APPLY
TO THE AEGEAN SEA SITUATION

As the preceding section explains, the practice of tribunals examining maritime boundaries—and of most nations negotiating boundary disputes—has been to give islands less than full effect in generating extended maritime zones. The "power" of the island to generate an Exclusive Economic Zone or continental shelf has been determined by the size of the island, its population, and its location. The closer the island is to the mainland of its country, the greater its power is to generate a full zone; indeed, if it is close enough to the mainland, a baseline can be drawn directly connecting the island to the mainland. If the island is far from the mainland, however, and especially if it is on the "wrong" side of the median line dividing the state's continental land mass from that of its opposite or adjacent state, then the island is likely to be viewed as a "special circumstance" which can generate its own territorial sea but may have little or no effect on the location of the primary maritime boundary between the two nations. In both the Anglo-French dispute[124] and the Papua New Guinea–Australia agreement,[125] for instance, the islands of the United Kingdom and Australia adjacent to the coasts of France and Papua New Guinea, respectively, were viewed as sitting on the continental shelf of the other nation and thus were not allowed to generate any continental shelf of their own. These examples provide support for Turkey's position that it is entitled to the continental shelf extending to the median line between the mainlands of the two countries, save for the territorial seas that surround the Greek islands on the "wrong" side of that median line.[126] The "natural prolongation" theory has not been followed in the geographical sense in which it was first used in the 1969 *North Sea Continental Shelf Cases*,[127] but it has not been abandoned yet as a depiction of the general concept that continental land masses generate continental shelves.

As originally developed in the *North Sea* case, the natural prolongation theory appeared to require a close examination of sea-floor configuration to determine where the continental slope extending from one land mass ends and that of another begins. In its more recent decisions,[128] however, the ICJ has stated that this approach was rejected by the world community at the

124. See nn. 81–96 above and accompanying text.
125. See nn. 99–103 above and accompanying text.
126. See text at nn. 28 and 29 above; see also A. Wilson, *The Aegean Dispute*, Adelphi Papers, No. 155 (London: International Institute for Strategic Studies, 1979).
127. *I.C.J. Rep. 1969*, p. 3; see text accompanying nn. 64–67 above; and see, generally, Keith Highet, "Whatever Happened to Natural Prolongation?" in *Rights to Oceanic Resources: Deciding and Drawing Maritime Boundaries*, ed. Dorinda Dallmeyer and Louis De Vorsey, Jr. (Dordrecht: Kluwer, 1989), pp. 87–100.
128. See esp. Case concerning the Continental Shelf (Libya Arab Jamahiriya v. Malta) (n. 114 above), p. 33, Par. 34; p. 35, Par. 39; and p. 36, Par. 40.

negotiations leading to the 1982 Convention when the negotiators decided that the principles used to resolve boundary disputes involving continental shelves should be the same as those used to resolve disputes involving Exclusive Economic Zones (EEZ).[129] Because these principles appear to exclude the possibility of using a geological or geomorphological approach to resolving EEZ disputes, the court has felt that they should not now be applied to continental shelf disputes.[130]

This shift need not, however, be viewed as a rejection of the more general idea that each continental land mass generates a continental shelf. Solutions to boundary disputes should therefore recognize that each continental land area should be entitled to its fair share of the adjacent continental shelf, and, indeed, Article 83 of the 1982 Convention maintains that approach by stressing that the nations with opposite or adjacent coasts should endeavor to reach an *"equitable* solution."[131]

Among the relevant factors that must be considered when focusing on the islands in the Aegean, as mentioned above,[132] is their size, population, and location. Table 1 lists their size and population and indicates that some of these islands are substantial in size with thriving communities while others are small in size with declining populations. The island that appears to be least deserving of generating an extended maritime zone is Megisti (Kastellorizo), a tiny island (3.5 square miles/9 km^2) with only 222 inhabitants. If allowed to generate an extended maritime zone, Megisti would effectively cut off Turkey's access to the resources of a large part of the Mediterranean because of its location close to Turkey's coast but far from the other Greek islands.[133]

Limnos (Lemnos), Lesvos (Lesbos), Chios, Samos, Kalymnos, Kos, and Rhodes each have more than 10,000 inhabitants and thus—except for their awkward location—meet the usual criteria for being legitimate islands entitled to generate maritime zones.[134] If they are entitled to generate full zones, however, Turkey would be almost completely excluded from access to the resources of the Aegean, a solution that hardly seems "equitable," particularly since the Turkish population along its Aegean coast is many times larger than the population of these adjacent Greek islands.

Another relevant factor might be the historical linkages between the communities involved in this dispute and the disputed ocean area and its resources. Because the islands creating this problem have changed hands so frequently in recent years, however, this factor does not point toward a clear

---

129. Compare the virtually identical Art. 74 and 83 of the Law of the Sea Convention (n. 2 above).

130. Case concerning the Continental Shelf (Libya Arab Jamahiriya v. Malta), p. 13, Par. 40.

131. Law of the Sea Convention, Art. 83(1).

132. See text preceding n. 124 above.

133. Prescott (n. 99 above), pp. 308–9.

134. See text accompanying nn. 68–80 above.

solution. Limnos (Lemnos), Lesvos (Lesbos), Chios, Samos, and their near neighbor islands were governed by the Turkish Empire from the late fifteenth or mid-sixteenth century until the end of the Balkan Wars in the 1913–14 period, when they were transferred to Greece.[135] The Dodecanese Islands in the southeastern Aegean became part of the Turkish Empire in 1522–23, came under Italian control in the Italo-Turkish War of 1911–12, and then were awarded to Greece because of their Greek population in the Allied peace treaty with Italy in 1947.[136] The residents of these islands have been primarily Greeks during all these periods, and they have had a maritime orientation, but the Turks and other occupying powers have participated in the development of the ocean resources during their periods of dominance. It appears difficult, therefore, to sustain any particular claim that the waters surrounding or connecting these islands are akin to "historic waters."[137]

The factor that was quite important in the decision of the ICJ chamber in the *Gulf of Maine Case*[138] was the length of the coastlines of the two countries adjoining the disputed ocean area.[139] The ratio of Greek to Turkish coastlines bordering on the Aegean has been estimated at about two to one in favor of Greece.[140] Decision makers seeking an equitable solution to this dispute might well follow the lead of the ICJ chamber in the *Gulf of Maine Case* and divide jurisdiction over the Aegean waters by giving Greece jurisdiction over two-thirds, with Turkey given jurisdiction over the remaining one-third.

The concept of "equity" relevant to the solution to this dispute has been developed in other papers[141] and thus will not be explored in detail here. This concept clearly should be kept uppermost in the minds of all those addressing this problem because the unique geography of this area requires innovative and creative solutions to this dispute. Among the alternative solutions that have been suggested are the following:

A. *The "enclave" approach* would involve drawing territorial seas around

135. *Columbia Lippincott Gazetteer of the World* (n. 9 above), pp. 399, 1039, 1044, and 1658; see generally C. M. Woodhouse, *A Short History of Modern Greece* (London: Faber, 1968); and J. P. C. Carey and A. G. Carey, *The Web of Modern Greek Politics* (New York: Columbia University Press, 1968).

136. *Columbia Lippincott Gazetteer of the World*, p. 521.

137. The concept of historic waters is acknowledged in Art. 10(b) of the 1982 Law of the Sea Convention, but it is difficult to meet the standards required by international law to achieve this status. See, generally, UN General Assembly, *Official Records*, vol. 14, document A/CN.4/143; Sherry Broder and Jon Van Dyke, "Ocean Boundaries in the South Pacific," *University of Hawaii Law Review* 4 (1982): 12–23.

138. Gulf of Maine Case (n. 108 above).

139. Ibid., pp. 335–37, Pars. 218–22; reprinted in *I.L.M.* 23 (1984): 1242–43.

140. Karl (n. 24 above), p. 672.

141. For example, see Barbara Kwiatkowska, "Maritime Boundary Delimitation between Opposite and Adjacent States in the New Law of the Sea: Some Implications for the Aegean" (paper presented at the International Symposium on Aegean Issues, Çeşme, Turkey, October 15–17, 1987), in press.

each of the populated Greek islands but would otherwise deny them the power to generate extended maritime zones. The division of the area would then be determined by drawing the median line between the opposite and adjacent continental land masses of the two countries, thus giving Turkey significant areas of the continental shelf in the eastern half of the Aegean, reduced only by the territorial seas enclaves generated by the Greek islands. The amount of ocean jurisdiction Turkey would gain under this approach depends on whether the territorial sea around the Greek islands is 6 or 12 nm.[142] As discussed above,[143] this enclave approach should almost certainly be used for Megisti (Kastellorizo)—the easternmost island—no matter what other decisions are reached regarding the other islands.

B. *The "finger" approach* would give Turkey four finger-shaped projections into the eastern Aegean between the islands of Samothrace and Limnos (Lemnos), Limnos (Lemnos) and Lesvos (Lesbos), Lesvos (Lesbos) and Chios, and Chios and Samos.[144] This approach has the advantage of maintaining contiguity among the ocean zones, but it would give Turkey less ocean resource jurisdiction then would the enclave approach (assuming that Greece's territorial sea is 6 nm).

C. *Fishing rights could be separated from continental shelf rights*, as was done in the Torres Strait Treaty between Australia and Papua New Guinea.[145] Territorial sea enclaves would again be drawn around the Greek islands; Turkey would then be given jurisdiction over the resources of the remaining continental shelf in the Aegean Sea east of the median line between the continental land masses, with Greece having rights to the fish in most of the water above. This approach is therefore similar to the "enclave" approach discussed above except that Turkey's access to fishing resources would be greatly reduced. Again, the jurisdiction granted to Turkey would vary greatly depending on whether the territorial sea enclaves had 6- or 12-nm radii.

D. *Joint development* is perhaps the most logical solution to this dispute because it would allow the two countries to postpone the ultimate decision of how to draw the boundary but nonetheless allow them to endeavor to exploit the resources for the benefit of the populations of both countries. Joint-development zones have been created between Saudi Arabia and Kuwait, Saudi Arabia and Sudan, Japan and Korea, Malaysia and Thailand, Norway and Iceland, and most recently Australia and Indonesia,[146] and other nations are actively considering this possibility.

142. For maps illustrating this approach, see Wilson (n. 126 above), pp. 36–37. Note that the caption of Map 1, p. 36, apparently should read, "6 nautical miles" instead of "16 nautical miles."

143. See text accompanying n. 133 above.

144. For a map illustrating this approach, see Wilson, p. 38; and Karl, pp. 671–72.

145. See text accompanying nn. 99–103 above.

146. See, generally, Mark Valencia, ed., *The South China Sea: Hydrocarbon Potential and Possibilities of Joint Development* (New York: Pergamon, 1981).

A joint-development approach usually involves an agency managed by persons nominated by the two countries which supervises development of the area with some degree of autonomy. The resources are then explored and exploited through concessions granted by the joint-development agency with the revenues shared by the two countries according to an agreed-upon formula. Each of the existing schemes has differences in approach, and some difficult questions are always raised regarding the legal regime that should govern both the commercial aspects of the activity and the labor-management and environmental aspects.

Nonetheless, if the political will exists, these problems can be resolved, and a difficult dispute can be set aside for the mutual benefit of all concerned. If successful, a joint-development project will not only expedite the development of the offshore resources but may also promote mutual cooperation and trust between the nations which can enable them to address other problems as well. Because of these advantages, and because none of the other solutions to the maritime boundary in the Aegean seem satisfactory, the joint-development approach deserves additional study by Turkey and Greece.

Baselines along Unstable Coasts: An Interpretation of Article 7 (2)

Sally McDonald and Victor Prescott
Department of Geography, University of Melbourne

INTRODUCTION

The origin of this analysis can be traced to a meeting, held in September 1987 at the headquarters of the United Nations in New York, with the aim of producing a publication designed to assist member states of the United Nations in following a uniform and consistent approach in drawing baselines and marking them on charts or maps of appropriate scales.[1]

The meeting was considering a working paper prepared by Victor Prescott, and the only major disagreement on interpretation concerned the second paragraph of Article 7. Prescott, following a pattern set over the previous 9 years, had asserted that this paragraph contained a justification for drawing straight baselines entirely separate from indented coasts and coasts fringed with islands. He was surprised to learn from people who had attended the meetings of the Law of the Sea Conference and who had been involved in drafting this article that they considered the second paragraph to be subordinate to the first paragraph. This means that the special provisions for highly unstable coasts can only be invoked on coasts that are deeply indented or fringed with islands. That might have been the end of the matter, but experts from three major countries possessing large deltas expressed the conviction that they regarded the two paragraphs as being quite separate.

It therefore seemed worthwhile to explore this question further, and it was suggested as a topic for an honors thesis to one of us (McDonald) at the end of 1987. This analysis is based largely on that thesis.[2] The following sections deal with the background and purpose of Article 7 (2), its textual and teleological interpretation, examination of some baselines drawn around deltas, and the possible future applications of this provision.

1. The meeting was attended by 20 experts on baselines from 18 different countries and eight members of the Secretariat of the Office for Ocean Affairs and the Law of the Sea headed by Mr. Satya Nandan.
2. Sally McDonald, "Difficulties in Interpreting the Rules for the Use of Straight Baselines on 'Unstable' Coasts" (B.A. Honours Thesis, Department of Geography, University of Melbourne, 1988).

BACKGROUND AND PURPOSE OF ARTICLE 7 (2)

The vague wording of Article 7 (2) raises a number of questions about the purpose of the provision, the coasts that qualify to use it, and the rules to which the straight baselines must adhere. In the following section the documentation of the drafting of the provision is examined in order to pinpoint when the ambiguities came into being and to discover the basic purpose of the provision.

The first suggestion aimed at dealing with "highly unstable" coasts was made by Bangladesh on July 3, 1974, to the Second Committee of UNCLOS III at Caracas. The suggestion went as follows: "In localities where no stable low-water line exists along the coast due to continual process of alluvion and sedimentation and where the seas adjacent to the coast are so shallow as to be non-navigable by other than small boats and pertain to the character of inland waters, baselines shall be drawn linking appropriate points on the sea adjacent to the coast not exceeding the 10 fathom line."[3]

Though deltas were not specifically mentioned, it is obvious that coasts dominated by river processes were being referred to. The mention of enclosed waters pertaining to the character of inland waters appears to have been a deliberate effort to make this suggestion appear to conform with the spirit behind the original straight baseline provision, which stipulated that only waters closely integrated with the land territory deserve to become internal waters. In this case, the waters were shallow, rather than being enclosed by indented coasts or fringing islands.

There were two innovative features in this proposal. First, it was proposed that straight baselines should be applied to "unstable" coastlines and, second, that these baselines have basepoints in the sea rather than on land. Bangladesh had shown its prior confidence in the outcome of the negotiations by proclaiming straight baselines with basepoints corresponding closely to the ten-fathom line in the Bay of Bengal on April 13, 1974[4] (fig. 1).

The main arguments or justification for this proposal by Bangladesh were summarized as follows.

1. The estuary of Bangladesh is such that no stable water line or demarcation of landward and seaward area exists.
2. The continual process of alluvion and sedimentation forms mud-banks and the area is so shallow as to be non-navigable by other than small boats.
3. The navigable channels of land [*sic*] through the aforesaid banks are continuously changing their courses and require soundings and demarcation

3. Renate Platzöder, *Third United Nations Conference on the Law of the Sea: Documents* (New York: Oceania, 1982), 3:213.

4. S. Lay, R. Churchill, and M. Nordquist, eds., *New Directions in the Law of the Sea: Documents* (London: Dobbs Ferry, 1974), 2:290.

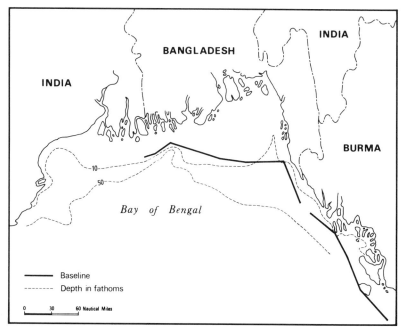

FIG. 1.—Baseline proclaimed by Bangladesh on April 13, 1974

so that they pertain to the character of the river mouths and inland waters.[5]

These points appear rather repetitive, but the crucial features are that the low-water line is rapidly changing and that, due to the navigational obstacles, only local navigation occurs; in other words, the waters have the character of internal waters. It is uncertain whether problems of surveillance were supposed to be inferred. Another point that was made was that at some stage Bangladesh might wish to reclaim some of this shallow water.

Bangladesh's revised proposal incorporated part of the original baseline provision in order to emphasize its conformity with the basic principles underlying the straight baseline concept. "*The [sic] localities where the coast line is deeply indented and cut into or if there is a fringe of island [sic] in its immediate vicinity* or if the water adjacent to the coast is marked by continual process of alluvion and sedimentation creating a highly unstable low water line the method of the straight baseline joining appropriate points on the coast or on the coastal waters may be employed in drawing the baseline from which the breadth of territorial sea is measured."[6]

5. Platzöder, 4:181.
6. Ibid., emphasis added.

The draft text the committee produced did not recognize basepoints related to a specific isobath. This is confirmed by the report of the Australian delegation.[7] The committee stipulated that the original basepoints must be on the low-water line but may be maintained in the sea if the low-water line has retreated.

> In localities where the coastline is deeply indented and cut into, or if there is a fringe of islands along the coast in its immediate vicinity, the method of straight baselines joining appropriate points may be employed in drawing the baseline from which the breadth of the territorial sea is measured. Where because of the presence of a delta or other natural conditions the coastline is highly unstable, the appropriate points may be selected along the furthest seaward extent of the low-water line and notwithstanding subsequent regression of the low water line such baselines shall remain effective until changed by the coastal State in accordance with the present Convention.[8]

The committee categorized Bangladesh's numerous problems as being those characteristic of a delta coast. So at least it is clear that delta coasts qualify for this provision. However, this draft raises the question of whether highly unstable coasts are a third justification for using straight baselines or a special application of baselines along indented coasts fringed with islands. The ambiguous phrase "other natural conditions" made its first appearance in this draft, and there was mention of changing the baselines at some unspecified stage.

In the next draft of May 1976 two more significant changes had been made. The one paragraph had been divided into two apparently for aesthetic reasons, and "or other natural conditions" had changed to "and other natural conditions," which gives the impression that only delta coasts may qualify. This wording found its way into the final text unchanged. The information about the aesthetic change was provided by S. Nandan in September 1987. He explained that it was considered that the first paragraph was too long.

The wording of the remaining paragraphs of Article 7 remained unchanged from the original 1958 straight baseline provision, apart from allowing low-tide elevations to be used when they had received international recognition.

Despite a clear rejection of the depth method, Bangladesh persevered, suggesting in 1977 an even more radical solution: "Where because of the presence of a delta and other natural conditions, the waters adjacent to the

7. Department of Foreign Affairs, *Third United Nations Conference on the Law of the Sea: Third Session, Geneva* (Canberra: Government Printer, 1975), p. 65.

8. United Nations, *Informal Single Negotiating Text*, A/CONF. 62/WP (New York: United Nations, 1975).

coast are marked by continual fluvial erosion and sedimentation creating a highly unstable baseline, the baseline may be delimited by a straight line or series of straight lines connecting appropriate points of such adjacent coastal waters."[9]

The suggestion of a single baseline was unlikely to gain support, as it is well known that a single line cannot adequately follow the general direction of a coast and has the effect of enclosing waters that do not resemble internal waters. Bangladesh made it clear that only coasts influenced by fluvial processes should qualify.

Bangladesh's last proposal cleverly attempted to detach the baseline from the committee's preferred low-water line along the edge of the subaerial delta, by emphasizing that the delta does not cease to exist below low tide but continues under water as the subaqueous part of the delta. This proposal would overcome the shallow waters mentioned earlier by Bangladesh, but it was not accepted. "Where most part of a coastline of a State is constituted by a continuous process of sedimentation and fluvial deposits rendering the low-water line highly unstable, the method of straight baseline [*sic*] joining appropriate points may be employed along the furthest seaward extent of submerged sedimentary delta in drawing the baseline from which the breadth of the territorial sea is measured."[10] A search of the meetings concerned with the territorial sea has found evidence of only one other state's opinion regarding Bangladesh's proposals. Vietnam said it supported Bangladesh, "a country in a situation rather similar to that of Viet-Nam, on the methods to be used for drawing the baselines."[11] On April 30, 1982, Burma and India protested against the baselines declared by Bangladesh in 1974.[12]

Summary

The official records give a fairly clear picture of Bangladesh's explicit reasons for proposing straight baselines for unstable coasts. Along a major part of Bangladesh's coast sedimentation creates shallow waters, while at the same time marine processes redistribute these sediments, causing changes in position of the low-water line and submarine banks. It appears, therefore, that there were two benefits Bangladesh was hoping to gain from the provision: Bangladesh wanted an artificial construction that would be easier to survey and that would not change frequently, and, more important, perhaps, Bangladesh wanted to be able to claim as internal waters shallows where naviga-

9. Platzöder, 4:389.
10. Ibid., 5:11.
11. Ibid., 5:255.
12. J. R. V. Prescott, *The Maritime Political Boundaries of the World* (London: Methuen, 1985), p. 166.

tion and surveillance was difficult. The first benefit was granted for a retreating deltaic coast, but despite Bangladesh's persistence the second benefit clearly was not granted.

METHODS OF INTERPRETATION

As there is scope for debate about the meaning of this provision, there are two main methods of interpretation that the International Court of Justice (ICJ) or a State might choose to take. These are the textual and the teleological. The first places great emphasis on the actual wording, while the second attempts to give effect to the purpose of the provision. In the discussion that follows, both interpretations will be outlined after geographical definitions of the less debatable aspects of the provision are given.

The interpretation of the word "delta" is not as straightforward as might be thought by those unfamiliar with geomorphological terms. Some writers are inclined to restrict the term "delta" to depositional landforms that protrude from the coast, such as the Nile. A broader definition is used by Wright: "Subaerial and subaqueous accumulations of river-derived sediment deposited at the coast when a stream decelerates by entering and interacting with a larger receiving body of water."[13] This includes what are defined more specifically as "estuarine deltas." These occur when sedimentation is in process but has not yet filled formerly submerged valleys, and as a result the deltas do not protrude from the coast. Whether the committee realized that the term "delta" could refer to estuarine deltas is uncertain.

The word "coastline" is clear within the context of the article. Because the purpose of Article 7 is to overcome the problems of intricate or changeable low-water lines, it is the low-water line, rather than the high-water line, that is being referred to.

The word "unstable," when referring to coasts, most commonly refers to tectonic movement, but Shepard uses it in the sense of rapid change in the outline of the coast.[14] When placed in the context of this article, it is obvious that the second interpretation is intended. In the remainder of this discussion the word will be used in this second sense.

Textual Interpretation

Reliance on the wording of the provision means that the phrase "and other natural conditions" requires that a delta alone is insufficient to qualify under

13. L. D. Wright, "Deltas," in *The Encyclopaedia of Beaches and Coastal Environments,* ed. M. L. Schwartz (Pennsylvania: Hutchinson Ross, 1982), pp. 358–68.
14. F. P. Shepard, "Coastal Classification and Changing Coastlines," *Geoscience and Man* 14 (1976): 53–64.

this provision. Presumably a delta alone would not be "highly" unstable. Therefore, it is necessary to find out what other "conditions" exist with deltas to make the coastline "highly unstable."

By taking a very strict interpretation of "other," the conditions would have to refer to landforms because a delta is a landform. Landforms that are often associated with deltas along the coast are spits, barriers, alluvial islands, river mouth bars, shoals, and distributary levees.

Bangladesh placed much emphasis on the effect of shoals on the coastal waters with frequently shifting channels, so it might be suggested that these conditions were considered by the committee to be "other natural conditions." By restricting initial basepoints to the low-water line, however, the provision clearly does not cater for shallow waters and shoals.

An alternative textual interpretation might be that "natural conditions" refers to the processes affecting the delta. This idea is supported by Bangladesh's statement that "the cumulative effects of river flood, monsoon rainfall, cyclonic storms and tidal surges have contributed to a continuous process of erosion and shoaling."[15]

If this view is accepted, the opinions of some analysts suggest that only processes of erosion are involved.[16] They presumably base this on the idea that the sole purpose of the provision is reflected by the following phrase: "Notwithstanding subsequent regression of the low-water line, the straight baselines shall remain effective."[17] In other words, the provision deals only with maintaining the claims made from an eroding or retreating delta. Advancing deltas are thought not to need such a provision because their claims are not at stake. Claims from them will advance seaward with the normal baseline.

The mention of changing the straight baselines in Article 7 (2) has caused some interpreters to suggest that change is mandatory, and thus the question is one of when to do so.[18] The concern is that, if a coast is retreating, the baseline will become further and further detached from the low-water line (fig. 2). This leads to questions of whether the provision was to cater for long-term or short-term instability.[19] An indication may be gained from the requirements of paragraph 3, which will be discussed shortly.

15. Platzöder (n. 3 above), 4:180.

16. R. D. Hodgson and R. W. Smith, "The Informal Single Negotiating Text (Committee II): A Geographical Perspective," *Ocean Development and International Law (ODIL)* 3 (1976): 225–59; and J. R. V. Prescott, "An Analysis of the Geographical Terms in the United Nations Convention on the Law of the Sea" (Department of Foreign Affairs, Canberra, 1983), p. 47.

17. *The Law of the Sea: Official Text of the United Nations Convention on the Law of the Sea with Annexes and Index* (New York: United Nations, 1983), Sales No. E. 83. v. 5, Art. 7 (2).

18. K. Kittichaisaree, *The Law of the Sea and Maritime Boundary Delimitation in South-East Asia* (Singapore: Oxford University Press, 1987), p. 31.

19. Prescott, "An Analysis of the Geographical Terms," p. 50.

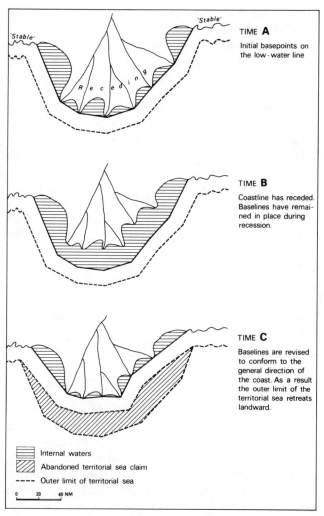

FIG. 2.—One interpretation of the application of Article 7 (2)

Retreating deltas must also be indented and have a fringe of islands according to the textual interpretation. This is apparently backed up by the words "the appropriate points," which appear to relate back to "appropriate points" in paragraph 7 (1). The implication of this is that the second paragraph is merely an application of paragraph 1, where the basepoints on fringed and indented coasts are unstable.

Bangladesh mentioned in its early baseline proposal and justifications that its coastline was among the most heavily indented in the world.[20] But in

20. Platzöder, 4:179.

the same document the proposal stated that straight baselines should be used where a coast is fringed or indented *or* highly unstable.

A textual interpretation that suggests that the unstable provision is subordinate to paragraph 1 would have to continue in this vein and say that baselines drawn under paragraph 2 must comply with the requirements of the remaining paragraphs. Hodgson and Smith imply this is so.[21]

Paragraph 3 requires that the baselines not depart from the general direction of the coast and enclose only waters able to be subject to the regime of internal waters—that is, that they are *inter fauces terrarum* (between the jaws of the land). This paragraph would require baselines around unstable coasts to be altered as soon as they depart from the general direction of the coast, difficult as this may be to determine (fig. 2). There could be a difference of opinion as to whether paragraph 5 applies to paragraph 2, but, using a textual interpretation, it probably would apply. This means that long-recognized economic interests in the region of an unstable coast may be taken into account in determining particular baselines.

Presumably, paragraph 6 is applicable to paragraph 2 because it ensures that baselines do not cut another state off from high seas or an Exclusive Economic Zone (EEZ). However, it is difficult to think of an actual case where this might occur. Possibly it could only happen in the vicinity of a delta shared by two countries.

Teleological Interpretation

A teleological interpretation would first require determination of the purpose of the provision and then argue that a textual interpretation does not give effect to these purposes.

The purpose of Article 7 on straight baselines is to simplify the boundaries of the zones measured from them in order that areas of different levels of jurisdiction are clearly presented to other states and the vessels of other states. Straight baselines also make surveillance by the coastal state easier.

The straight baseline concept embodied in Article 7 (1) originated with the *Anglo-Norwegian Fisheries Case* (1951), in which the ICJ found that Norway was justified in using straight lines to represent the low-water mark. "Where a coast is deeply indented and cut into, as is that of Eastern Finnmark, or where it is bordered by an archipelago such as the 'skjaergaard' along the western sector of the coast here in question, the base-line becomes independent of the low-water mark, and can only be determined by means of a geometric construction. In such circumstances the line of the low-water mark can no longer be put forward as a rule requiring the coast line to be followed in all its

21. Hodgson and Smith, p. 238.

sinuosities."[22] In its proposal that straight baselines be applicable to unstable coasts, Bangladesh claimed that the ICJ in 1951 had recognized the need to adapt international law to local geographical conditions: "A State must be allowed the latitude necessary in order to be able to adapt its delimitation to practical needs and local requirements."[23] The documentary evidence shows Article 7 (2) was created primarily to solve Bangladesh's problems. Instead of great indentedness, the main feature of Bangladesh's coast was that there was "no stable water line or demarcation of landward and seaward area," leading to a "highly shifting and unstable baseline."[24]

Geomorphological evidence has shown that Bangladesh's coast undergoes both erosion and accretion. Therefore, it could be argued that, rather than being intended to preserve eroding claims, the unstable provision is intended to reduce the confusion and surveying difficulties of unstable coastlines. The Australian delegation's report suggests just this: "Bangladesh, which is concerned about its changing delta *information,* sought a provision."[25] A teleological interpretation might argue that "or other natural conditions" would make better sense than "and other natural conditions" for a number of practical reasons. (1) On large deltas, geomorphologically speaking, the stability of the component landforms cannot be separated from that of the delta itself because, as Wright notes, large deltas are complexes of many coastal depositional features.[26]

If processes are being referred to, then the processes would not merely be erosion but sedimentation as well because Bangladesh's coast experiences both. However, because these processes are found on many deltas, such a requirement is ineffective in determining which deltas could qualify. Therefore, interpreting the phrase as "or" would apply it more effectively to other unstable landforms and the processes associated with those than to conditions associated with deltas. (2) If the purpose is to overcome confusion over the position of maritime zone limits and surveying difficulties, then these problems are likely to occur on coasts other than delta coasts. (3) A provision with wider applicability would be more likely to receive more support during the negotiating and drafting process than one with only very limited application. (4) The initial draft text, the Informal Single Negotiating Text (*ISNT*), used the wording "or other natural conditions." The change to "and" may have been overlooked or considered to mean the same thing. (5) The Russian final

22. International Court of Justice, "Anglo-Norwegian Fisheries Case," *1951 International Court of Justice, Reports* (The Hague: International Court of Justice, 1951), p. 16.

23. Platzöder, 4:179.

24. Ibid., pp. 180–81.

25. Department of Foreign Affairs (n. 7 above), p. 65; emphasis added.

26. Wright (n. 13 above), p. 358.

text utilizes the word "or," and this text has equal standing with those of the other languages.

The teleological interpretation, like the textual, cannot make the term "highly unstable" any clearer. A teleological interpretation would regard "notwithstanding subsequent regression" as merely a safeguard that does not mean that only retreating deltas may use the provision. As mentioned previously, Bangladesh said it had a "constantly fluctuating" coast, and Polcyn's results, discussed later, suggest the coast is advancing. Bangladesh made no mention or inference that its important claims were at stake. A coast may be unstable whether it is advancing, retreating, or doing both.

The provision's mention of change could be considered as enabling revision rather than prescribing it. It may be purposefully open as to when it should occur because rates of coastline change vary from place to place and some states have the resources to resurvey their coastlines more frequently than others. Unstable coasts would not be required to be deeply indented or fringed with islands because this requirement would be seen as unnecessarily restricting the application of the provision which aims to cater for unstable coastlines. The phrase "the appropriate points" in this case would be regarded as a poor choice of words.

Bangladesh's coast is in a continuous state of instability; in other words, the problems are not short-term, so it seems reasonable to expect that baselines could be left in place over long periods of time—for example, 50 years. If a state had to change the baseline as soon as it deviated from the direction of the coast to comply with paragraph 3, then little benefit would be gained from being able to leave baselines in place when the coast retreated.

Highly Unstable

Neither the textual nor the teleological interpretation can clarify the meaning of "highly unstable." The only way to gain an idea of what rates of coastline change could be considered as highly unstable is to compare coastline changes worldwide. In this respect, the findings of the International Geographical Union (IGU) 1972 Working Group on the Dynamics of Shoreline Erosion and its 1976 Commission on the Coastal Environment are useful.

As summarized by Bird, it seems that only a very small part of the world's coastline has changed by more than 1 m over the past century and that gain or loss of more than 10 m per year has been exceptional.[27] In the absence of a large body of research on shoreline changes as opposed to coastline changes, this is the best indication of change rates available worldwide. It has been suggested that practical considerations such as the width of a line on a map or

27. E. C. F. Bird, *Coastline Changes: A Global Review* (Chichester: Wiley & Sons, 1985), p. 158.

the distance error in locating foreign vessels can be an indication of when it is necessary to resurvey a baseline. However, the distance error in locating a ship from a surveillance vessel may vary from 15 to over 1,000 m depending on the weather conditions and equipment used, so this is not a very useful indicator. If the width of a line is 100 m on a chart at a scale of 1:100,000, which is suitable for navigation, it would only be worth resurveying the baseline if the coast has changed by at least 200 m.[28]

A change of 200 m can occur either over a long or a short period of time, so if it is accepted that only rates of coastline change greater than 10 m a year produce a "highly unstable" coast, then the maximum length of time over which a total change of 200 m would occur would be 20 years. This would appear reasonable, as states such as Australia and Britain with fairly stable coastlines have resurveyed their low-water line at 40- or 60-year intervals.[29] Hence, a state with a coast retreating at 5 m a year would not be justified in adopting this provision. The coast may reach the 200 m total change mark in 40 years, but this rate is too slow to be highly unstable. Of course, if the coast is already deeply indented or fringed with islands, rapid rates of change would have to be occurring on the most seaward points of the low-water line, or else they would not affect the straight baselines to be drawn.

Of the delta coastlines surveyed for which rates of coastline change were available, the majority had rates of change of 10 m a year or greater. This might reflect the concentration of studies on highly changeable deltas rather than others and the consequent paucity of rates of change on the latter. Other types of coast that exhibit change greater than 10 m a year and that might be considered as highly unstable include volcanic, glaciated, and periglaciated coasts.

Summary

A strictly textual interpretation would mean that only deeply indented and fringed coasts with receding deltas caused by natural conditions would qualify for use of Article 7 (2). Baselines would continually have to conform to the general direction of the coast. This interpretation has an advantage in that it does not challenge the existing rules and concepts relating to straight baselines around indented and fringed coasts.

It could be argued, however, that by conforming so closely with the original straight baseline requirements, the provision's effectiveness is greatly reduced. Therefore, the alternative is that Article 7 (2) should not be considered as subordinate to paragraph 1 or be required to conform with some of

28. Prescott, "An Analysis of Geographical Terms" (n. 16 above), p. 49.
29. E. C. F. Bird, personal communication, 1988.

the requirements of the remaining paragraphs. A teleological interpretation could apply the provision to a range of coast types experiencing rapid change.

USE OF ARTICLE 7 (2)

Due to the fact that countries rarely state the legal justifications for their use of straight baselines, it is often necessary to guess the reasons for their use from indicators such as coastal configurations, dates of declaration, and positioning of baselines. The following section discusses the use of straight baselines around deltas, the probability of these baselines having being drawn with Article 7 (2) in mind, and whether a body of state practice is emerging. The possible adoption of Article 7 (2) for other types of unstable coasts has not been explored due to a conceivably large range of unstable coast types and the difficulty in attributing the use of straight baselines to instability.

A considerable number of deltas have been enclosed by straight baselines, but only those baselines declared after Article 7 (2) was officially accepted along with the rest of UNCLOS III in 1982 are likely to have been drawn to conform with that article. Bangladesh's baselines are an exception, having been declared in 1974. They are included here because Bangladesh would probably seek to justify them with Article 7 (2). States declaring straight baselines in the vicinity of deltas in 1982 or later are: Bangladesh, USSR, Vietnam, and Guinea-Bissau.

Bangladesh

The dominant feature of Bangladesh's coast is the delta that extends for 400 km if indentations are ignored. The Ganges, Brahmaputra, and Meghna rivers converge in eastern Bangladesh to empty the largest sediment load of the world's rivers into the Bay of Bengal.[30] Polcyn found that over a 7-year period, accretion only slightly outweighed erosion; 932 square km were gained, and 717 square km lost.[31] This can be roughly translated into an average annual net rate of advance over the 400-km-long coastline of 77 m. This is close to the figure of 60 m a year recorded by Ahmad.[32]

As Polcyn's findings indicate, deposition does not cause constant advance. Shoals, mudflats, and estuarine islands in the Meghna delta region change outline or position frequently due to strong tidal action and storm surges.

---

30. T. D. Milliman and R. Mead, "Worldwide Delivery of River Sediment to the Oceans," *Journal of Geology* 91 (1983): 1–21.

31. Bird, *Coastline Changes*, p. 112.

32. E. Ahmad, "India, Coastal Morphology," in *The Encyclopaedia of Beaches and Coastal Environments*, ed. M. L. Schwartz (Pennsylvania: Hutchinson Ross, 1982), pp. 484–86.

There is no mention of the effects on the delta coastline of river and coastal embankments designed to protect against floods and storm surges.[33] However, the diversion of the Meghna south from its former course east to Sandwip Island has caused erosion to Hatia and other islands in the estuary, according to Johnson.[34]

The most significant evidence for Bangladesh having adopted Article 7 (2) for this coast is that it was Bangladesh that initiated the development of the provision for unstable coasts. Because Bangladesh could not possibly justify its baselines under Article 7 (1), it is highly likely that Bangladesh would cite Article 7 (2).

Bangladesh had drawn its radical baselines in 1974, close to the 10-fathom isobath, in expectation of its proposals being fully accepted. The baselines, with eight basepoints all placed on the water, clearly violate the rules set out in the 1958 and 1982 Conventions on the Law of the Sea. Comments made by M. Habibur Rahman, associate professor of law at the University of Rajshahi in Bangladesh, indicate Bangladesh might try to dispute this fact, but there is clearly no room for argument.[35]

Bangladesh's interpretation of the basepoint requirement does not aid the interpretation of Article 7 (2) because it is obviously illegal. The fact that Bangladesh's coast is fringed with islands (fig. 3) does not indicate any particular stance on the question of whether Article 7 (2) is subordinate to Article 7 (1). Nevertheless, it could be reasonable to assume that Bangladesh believes its coastal conditions qualify for straight baselines. This means that constantly fluctuating deltas with overall advance may qualify.

Soviet Union

The Lena delta is a major feature of Russia's coast on the Laptev Sea, with a coastline approximately 480 km long when indentations are ignored. This delta is experiencing erosion. Zenkovich believes it is a relict formation, which implies that it is no longer receiving sediment.[36] The north and eastern sides are being "destroyed" by the effects of the periglacial activity and ice pressure.[37] The relief of the shore, according to Zenkovich, is highly variable, and the coastline is indented and fringed with islands.

Much of the Arctic coast of the USSR has been enclosed by straight

33. H. Rasid and B. Paul, "Flood Problems in Bangladesh: Is There an Indigenous Solution?" *Environmental Management* 2 (1987): 155–73.

34. B. L. C. Johnson, *Bangladesh* (London: Heinemann, 1975), p. 14.

35. H. Rahman, "Delimitation of Maritime Boundaries: A Survey of Problems in the Bangladesh Case," *Asian Survey* 25 (1984): 1302–17.

36. V. P. Zenkovich, "Arctic USSR," in *The World's Coastline*, ed. E. C. F. Bird and M. L. Schwartz (New York: Rheinhold, 1985), pp. 863–69.

37. Ibid.

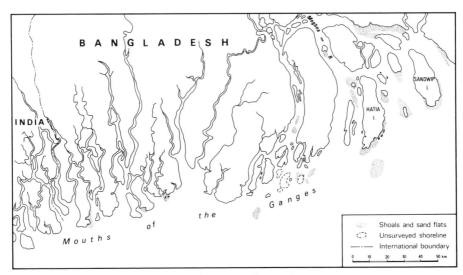

FIG. 3.—The coast of Bangladesh

baselines declared in 1985. These lines enclose deep indentations and join fringing islands and have a number of basepoints on the small islands around the Lena delta (fig. 4).

The USSR has also drawn baselines in the Black Sea, which enclose the Chilia lobe of the Danube. There are no basepoints on the Danube delta itself; the baseline is at least 5 nm from the delta, so, unless the USSR has imitated Bangladesh's illegal stance, it is unlikely that Article 7 (2) has been adopted here. This baseline system satisfies the requirements of both Articles 7 (1) and 7 (2), so it is difficult to say which provision the USSR would cite.

Vietnam

The Mekong has a coastline length of approximately 250 km. The only available rate of advance is based on figures for the 1960s, which record progradation of 61 m a year.[38] More recent reports say that it continues to prograde, with shoals at the distributary mouths growing into swampy islands.[39]

There is no information on whether the extensive water control programs of the 1970s have slowed the rate of advance at all. Though Bardach predicted the lower Mekong would undergo many changes, he did not

38. A. S. Zakaria, "The Geomorphology of Kelantan Delta," *Catena* 2 (1975): 337–50.
39. Bird, *Coastline Changes* (n. 27 above), p. 117.

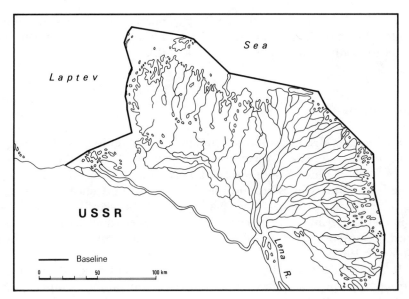

Fig. 4.—Soviet straight baseline in the vicinity of the delta of the Lena River, published on February 7, 1984.

specifically refer to the delta.[40] The *Sailing Directions* for the approach to the Mekong describe the land in the vicinity of the mouths as low and subject to frequent changes as a result of deposition.[41]

Vietnam is officially recorded as having shown support for Bangladesh's proposal, but whether it would attempt to justify its baselines on this basis is uncertain.[42] The Vietnamese delegation mentioned that the Mekong River "left heavy alluvial deposits on the coast," but the baselines drawn show little relation to the delta.

The baselines have four basepoints off the delta, but the northernmost is on an island approximately 120 nm from the coast, while those in the south are on an island 60 nm from the delta. There are shallow areas within these lines, but they are never uncovered.[43] The high rate of advance could be considered as being highly unstable, but the lines drawn by Vietnam do not satisfy either Article 7 (1) or 7 (2). The delta does not have a fringe of islands,

40. J. E. Bardach, "Some Ecological Implications of Mekong River Development Plans," in *The Careless Technology: Ecology and International Development*, ed. M. Farar and J. Milton (New York: Natural History Press, 1972), pp. 236–44.

41. The Hydrographer of the Navy, *China Sea Pilot*, 4th ed. (Taunton, Somerset: Her Majesty's Stationery Office, 1978), 1:122.

42. Platzöder (n. 3 above), 5:255.

43. J. R. V. Prescott, "Delimitation of Maritime Boundaries by Baselines," *Marine Policy Reports*, 8, no. 3 (1985): 1–5.

and the islands Vietnam has chosen for basepoints are too far seaward to be considered part of the unstable delta. Thus, it is unlikely that Vietnam would attempt to justify its baselines with Article 7 (2).

Guinea-Bissau

Much of Guinea-Bissau's coast is occupied by the Arquipelago dos Bijagos, a deltaic formation created by a multitude of alluvial islands in the mouth of an estuary. When Guinea-Bissau was still a Portuguese colony, that country proclaimed straight baselines around this deltaic coast in August 1966, long before Bangladesh's proposals were made.[44] Since the islands form a fringe in the immediate vicinity of the coast, these baselines could have been drawn according to Article 4 of the 1958 Convention, which Portugal had ratified in September 1964.

When Guinea-Bissau became independent in September 1974, it quickly redrew part of these baselines in the southern section of the archipelago. Most surprisingly, the reconstruction resulted in a significant move landward of one section of the 1966 baseline.

In May 1978, the third version of the baselines was published and showed a marked seaward movement. This new baseline is anchored at Cabo Roxo, on the boundary with Senegal, and the mouth of the Rio Cajet, on the boundary with Guinea, and the two intervening points are located in the sea. This baseline corresponds exactly to the description of part of the boundary that defined the islands that belonged to Portugal in the Franco-Portuguese boundary agreement of May 12, 1886.[45] An examination of this treaty shows that this line was simply drawn to avoid naming all the islands that belonged to Portugal. This was a common technique in the latter years of the nineteenth century, when colonial powers were drawing intercolonial boundaries in the Pacific Ocean. There has never been any suggestion in any of the treaties using this geographical shorthand that the lines were meant to serve as maritime boundaries.

It is true that seaward of the Arquipelago dos Bijagos there are areas of a submarine delta and submarine shoals. However, this delta is not retreating at present, and it will be a long time before any aggradation will move the delta front to the vicinity of the baselines which are now established in the sea.

44. Bird, *Coastline Changes,* p. 101; and for a map of Guinea-Bissau's baselines, see Prescott (n. 12 above), pp. 316–17.

45. I. Brownlie, *African Boundaries: A Legal and Diplomatic Encyclopaedia* (London: Hurst, 1979), pp. 351–53.

Summary

It appears that only Bangladesh has drawn straight baselines around the delta of the Ganges in accordance with what the Bangladesh authorities regard as the spirit of the proposals that led to Article 7 (2). Since those baselines were drawn before Article 7 (2) was finally drafted, it is not surprising that they breach the terms of that paragraph. In one sense Bangladesh was unfortunate. A number of specific provisions were incorporated in the 1982 Convention for the implicit benefit of countries such as Norway, the Bahamas, and Malaysia. Only in the case of Bangladesh did such insertions apparently fail to deliver the relief requested. It is now necessary to look at the way in which this rule about unstable coasts might be applied in the future.

THE FUTURE APPLICATION OF ARTICLE 7 (2)

The discussion of Article 7 (2) to this point has been academic. The records have been searched, and words and phrases have been carefully examined to discover shades of meaning. But this and other baseline rules are not applied by academics. They are applied by public servants, who usually occupy a fairly senior position and will normally do their best to expedite policies that have been fashioned on the basis of national self-interest. Hence, this examination of the future application of this baseline provision adopts a pragmatic political stance.

We can begin by considering those states that possess deltas on their coasts and that have not drawn any straight baselines. Such states might decide simply to enclose deltas in systems of straight baselines drawn in accordance with Article 7 (1). They would be following the example set by 14 other countries including Burma, Colombia, Iran, Spain, the Soviet Union, Venezuela, and Vietnam. One of the advantages of this arrangement is that there is no need to redraw those baselines at some stage in the future if the coastline retreats.

If a country decided to enclose its delta with straight baselines under the provision of Article 7 (2), we can be sure that the issue of it being subordinate to Article 7 (1) will not pose a problem. This is because the rules in Article 7 (1) have been interpreted in such a liberal fashion it is difficult to think of any delta that would not be able to satisfy that liberal interpretation. The delta of the River Ebro might be considered by some to be neither deeply indented nor fringed with islands, but Spanish authorities have already enclosed it within straight baselines. Even if there are deltas that could not be considered to be deeply indented or fringed with islands according to the very liberal standards set by Senegal and Vietnam, respectively, it was noted in the introduction that experts from three countries that possess major deltas believed that the second paragraph of Article 7 provides a third and separate justification for drawing straight baselines.

It also seems likely that the concept of a highly unstable coast will also be interpreted in a liberal fashion in two distinct senses. First, it is likely that rates of erosion far less than 10 m per annum will be considered to justify the description of being highly unstable. Second, it is possible that coastal retreat in one small area, which can be styled as being highly unstable, will be used to justify the construction of straight baselines along a much longer section of coast.

It is also probably safe to predict that, if baselines are established around highly unstable coasts that then retreat, leaving the straight baselines in the sea, there will be a tendency to leave the baselines in their original position for a long time.

It is now necessary to look at states that possess coasts that become highly unstable but where there are no deltas. This might become a more common situation than at present if predicted rises in sea levels occur over the next century, due to the warming of the earth's atmosphere because of the effects of so-called greenhouse gases. These predictions vary, but the average is for a rise of 1 m in the sea level by 2090.[46]

Although most attention has been given to the new level of high-water lines, the low-water marks will also generally move landward and not always by the same amount as the high-water mark. For the purposes of making maritime claims, it is the low-water line that is important. The horizontal movement of low-water marks will usually be largest on low-lying coasts on unresistant geological formations. The tundra coasts around the Arctic Ocean appear to be particularly vulnerable. In some of these areas there will be deltas. For example, on the Arctic Ocean's coast there are the Colville, Mackenzie, Lena, and Indigirka deltas, and only the Colville delta has not been surrounded by straight baselines. However, there will be some highly unstable coasts, perhaps of volcanic or coral formation, where there will not be any deltas. In such cases, it is quite possible that an attempt will be made to apply Article 7 (2) to any highly unstable coast whether a delta is present or not. This could be dubbed the "Russian version" because the Russian translation of the 1982 Convention has preserved the original phrasing that made "other natural conditions" an alternative to a delta.

CONCLUSIONS

This study leads to the following conclusions. First, Article 7 (2) was designed to satisfy the concerns raised by Bangladesh about its fluctuating coastline

46. J. S. Hoffman, "Estimates of Future Sea-Level Rise," in *Greenhouse Effect and Sea-Level Rise,* ed. M. C. Barth and J. G. Titus (New York: Van Nostrand Reinhold, 1984), pp. 79–103.

along the Ganges delta. Bangladesh wished to be able to claim as internal waters coastal seas where patterns of sedimentation and erosion created hazardous conditions for navigation in very shallow seas. In fact, Article 7 (2) in its final form does not give Bangladesh what it wanted. Instead it allowed states with deltaic coasts to maintain their baselines in the original position of the low-water line after the delta had retreated.

The second conclusion is that, in common with so many other baseline provisions, Article 7 (2) was worded in an ambiguous manner, which means there are uncertainties about when it can be used and how it should be applied.

The third conclusion is that thus far there is no evidence that any country has drawn straight baselines according to the provisions of Article 7 (2). Bangladesh has drawn offshore baselines in accordance with its original proposal, but that was done before the final form of the provision was drafted.

Several states have drawn straight baselines that enclose deltas on coasts that are believed to be either deeply indented or fringed with islands.

The fourth conclusion is that there is every reason to expect that, if states start to use Article 7 (2), it will be interpreted as freely as the other provisions of Article 7. There is also a growing chance that Article 7 (2) will attract more attention if predicted rises in sea levels occur and normal baselines begin to retreat landward.

Fishing Technology*

John Fitzpatrick
Fishing Industries Division, Food and Agriculture Organization, Rome

INTRODUCTION

In 1970, the Third Congress on Fishing Gear, organized by FAO and held in Reykjavik, Iceland, debated three main subjects: fish finding, purse seining, and aimed trawling. The results were published as *Modern Fishing Gear of the World No. 3*[1] and are technically valid today. Many of the fishing systems discussed are still used in development plans for fisheries.

An important part of the congress was devoted to the future of fisheries, and it foresaw not only technological development but a need to find answers to legal, economic, and social questions on an ever-increasing scale. It also thought that there might be a transition of fisheries from a hunting procedure to one of range management using the community of animals available.

Since the last congress, we have seen rapid development of the ideas expressed at that time in the field of electronics and a new awareness of the legal, economic, and social questions arising from the UN Convention on the Law of the Sea.[2] Extended economic zones have brought with them attendant responsibilities for coastal states and the need for assistance in the management of resources. In particular, the focus on shared stocks and regional fisheries bodies has been sharpened.

Within this ambient, the technologist faces the future knowing that the growth rate in fish production has decreased in recent years to less than the growth rate of the world population.

In fact, the explosive growth in world fish production during the 2 de-

---

*EDITORS' NOTE.—Mr. Fitzpatrick, Chief, Fishing Technology Service, Fishery Industries Division, Food and Agriculture Organization, Rome, presented this paper at Pacem in Maribus XVI, Halifax, August 22–26, 1988. The views expressed in this article are those of the author and do not necessarily reflect that of the Food and Agriculture Organization.

1. FAO, *Modern Fishing Gear of the World No. 3* (London: Fishing News Books, 1979).

2. *The Law of the Sea: Official Text of the United Nations Convention on the Law of the Sea with Annexes and Index* (New York: United Nations, 1983).

TABLE 1.—WORLD FISH PRODUCTION, BY ORIGIN, 1950–85
(million tons)

Year	Marine	Freshwater	Total
1950	17.6	3.2	20.8
1960	32.8	6.6	39.4
1970	59.5	6.1	65.6
1975	59.2	7.2	66.4
1980	64.5	7.6	72.1
1985	74.8	10.1	84.9

SOURCE.—Food and Agriculture Organization.

cades after 1950 ceased in the early seventies (table 1). Underlying the early expansion in fish production was the growth in the world economy, restricted largely to the developed countries in the fifties but extending to developing countries in the sixties. It was based mainly on increased income levels and, consequently, on increased demand for food products of the fisheries. Concomitantly, an increase in the demand for fish meal (and thus for shoaling pelagics) stemmed from a shift of intensive raising of livestock in North American and Western Europe, which entailed feeding with nutritionally balanced rations.

From table 1 it can be seen that world fish production in 1985 totaled some 85 million tons, an increase of 2.2% above that of the preceding year. This, following an increase of almost 7.5% in 1984, represented the eighth consecutive year of growth in physical output.

These increases resulted almost entirely from increases in catches of shoaling pelagics that are notoriously subject to fluctuations in abundance. The recent (1984–85) surge in production is attributable to increased catches in the southeastern Pacific (the west coast of South America). Elsewhere, for the most part, either production was relatively stable or marginal increases in some areas were offset by marginal declines in others.

FUTURE DEMAND

The total annual demand for fish beyond the year 2000 might well exceed 100 million tons. The increase in supply, based on potential catches of more or less conventional species, is theoretically feasible, but only part of it could be realized through more extensive or intensive fishing effort (table 2).

Demersal Stocks

According to evidence presently available, almost all important stocks of demersal species are either fully exploited or overfished. Many of the stocks of

TABLE 2.—FISH FOR FOOD: PROJECTED DEMAND AND SUPPLY
(million tons)
A. INCREASE IN DEMAND 1980–2000

Location	Quantity
Developing countries	+ 22.5
Developed countries	+ 5.9
Total	+ 28.4

B. ESTIMATED POTENTIAL FOR INCREASED PRODUCTION (1985)

Category	Quantity
Demersal	1.0– 8.0
Shoaling pelagic	3.0–10.0
Other marine	4.0– 6.0
Freshwater and aquaculture	5.0–10.0
Total	13.0–34.0

SOURCE.—Food and Agriculture Organization.

more highly valued species are depleted. Reef stocks and those of estuarine/littoral zones are under special threat from illegal fishing and environmental pollution. Year-class (recruitment) variation may have a marked effect on catches, and there is evidence that some species may be subject to long-run changes in abundance due to climatological or other factors.

There is little prospect, therefore, of increasing the catch of demersal species. It is necessary to distinguish between landings and catch, as quantities of demersal species are taken and discarded, for example, in shrimping operations, and this is likely to continue until incentives are provided for landing and marketing these fish. Scattered stocks exist elsewhere but, in general, the cost of harvesting them would not be justified at present by the revenue obtainable from the derived products.

Pelagic Stocks

There seems to be a greater possibility of increasing the harvest of small, shoaling pelagic species. Stocks of such species are subject to periods of high and low abundance, extending over decades. Prediction of potential availability is complicated further by evidence that when the abundance of one species is low that of other species may increase. Additional harvesting may be feasible, however, by more intensive exploitation in some areas and improved fishery management through regulation of fishing effort in others.

Crustacean Stocks

Crustacean species are generally heavily exploited and many, if not most, stocks are depleted. Some local increases in catches may be expected, for example, from stocks of the smaller crabs. Most stocks of shrimp in potentially productive areas are already being exploited and no major increase in the harvest of capture fisheries can be foreseen. These fisheries generally have reached a stage of economic overfishing.

Other Marine Stocks

Very few untapped resources of conventional species remain anywhere. Resources of the continental slope in tropical seas tend to be sparse and the cost of their exploitation relatively high. Squid and other cephalopods are lightly exploited in some parts of the world but are regarded as conventional foods in some countries. The harvest of common squid has been increasing, as a result of better targeting and (possibly) of depletion of predator-species stocks.

Among unconventional marine resources, those that might support fisheries in the future include mesopelagic species and krill. The latter already supports a fishery, but there are economic constraints on the development of mass production and marketing.[3] Mesopelagic species are widely distributed throughout the oceans and are found locally in considerable abundance. For the foreseeable future, because of the nature and size of these fish, they would have to be used for fishmeal production and, at present, incentives for commercial investment for this purpose are lacking.

Aquaculture

The production of marine and freshwater aquaculture (finfish, crustaceans, and mollusks) amounted to 7.7 million tons (1985). Finfish production represented 65% of this total with about 50% produced in China. In addition, 2.8 million tons of seaweeds and aquatic plants were cultivated.

Therefore, to attain the expected increase, the industry needs further technological development, as well as better management, to reduce postharvest losses, to harvest currently underutilized and nonutilized stocks, to make fishing gear effective and selective, and to increase aquaculture production.

3. Editors' note.—See John E. Bardach, "Fish Far Away: Comments on Antarctic Fisheries," *Ocean Yearbook 6*, eds. Elisabeth Mann Borgese and Norton Ginsburg (Chicago: University of Chicago Press, 1986), pp. 38–54.

TECHNOLOGY DEVELOPMENT

Conservation Fishing

To fish a selected species and not to take incidental catches of mammals, unwanted species, or juveniles is the aim of most fishermen (unless the incidental catch is a high-value species); to the resource manager, selectivity is an essential component of a management program.

Investigations into the efficiency and selectivity of passive and active fishing gear have benefited in recent years from the use of flume tanks, from new developments in measuring devices, and from direct and indirect underwater observations.

Development has been promoted by world opinion, which has brought about the use of turtle escape devices in shrimp trawls and has caused scrutiny of purse-seine fisheries and the entanglement of mammals in gill nets. Development has also been prompted by the large volume of unwanted by-catch taken in shrimp trawl fisheries.

Some countries have already imposed legislation on the fishery sector, banning certain fishing methods in some cases and insisting on the use of exclusion devices. These actions are not always favored by the fishermen, who consider themselves to be conservative by nature and think that more research into fishing-gear design and fish behavior should be carried out by administrations.

Minimum-mesh-size regulations have been in use for a long time and, for target species, the mesh sizes are set by legislation. In many cases, however, the measurements were determined at a time when the rate of growth in landings was encouraging enough not to give cause for concern for the future. Today, it is apparent that not only is the minimum mesh size important, but so is the very shape of the mesh, such as the square mesh recently introduced in Europe. This maintains its square shape during fishing, permitting small fish to escape.

Allowing fry and undersized fish of commerically important species to escape *and survive* is vital to conservation. Recent research and commercial testing of special nets in Norway suggest that acceptable levels of success can be achieved in its coastal shrimp fishery. Scientists have proved that up to 90% of all fish can escape from the net alive at a loss of less than 10% of the shrimp.

This living after escape is the key to success. Some exclusion devices are extremely efficient, but, after escaping, the species targeted for protection might be damaged in the release process or by the very action of the trawl. This seems to be the case where types of crab are excluded but even slight damage leaves them more vulnerable to their natural predators.

Research on selectivity extends to other forms of fishing, such as traps and longlines. Recent experiments have shown that fewer young cod are taken by longlines when a certain type of artificial bait is used.

Given the progress to date, administrations should be encouraged to support further research programs and to make use of new technologies so developed for the better management of fisheries.

Fishing Gear

Industrial fisheries rely more than ever before on high-tech systems, and this applies equally to vessel and fishing gear. Modern trawls approach the expectations of the fishing-gear congress in 1970 and have required complete cooperation between the fisherman, the net maker, and the engineer.

Advanced aimed mid-water trawling is worthy of note in this connection. The concept is to collect and process data in the wheelhouse from the sonar, echo sounder, net-mounted transducers, and navigation systems, in order to adjust the position of the net in the water relative to the target. The winch control system also reacts to changes in tension in the warps to maintain the net in the best fishing mode even when the vessel is altering course.

Once the gear has been shot under manual control to a preselected depth, the system is placed under electronic control and the skipper has a visual presentation of the position of the net relative to the sea bed and the school of fish and of information on the catch in the cod end. In this way, judging the trawling time is less of a hit-or-miss exercise, and damage to the net through overfilling of the cod end when big schools are encountered can be avoided.

These systems also use sensors on the net to relay information on sea temperature to the data processor. This information is important when the reaction of the targeted species to temperature changes is known.

However, it should be noted that such systems were developed for single-species fisheries and may not prove to be appropriate technology in tropical, mixed-species fisheries, at least not without further refinement.

Large purse seines can also be controlled in a manner similar to that of the mid-water trawl. Modern sonars process data on the target school—its position, speed, direction, mass, and so on—presenting the skipper with a visual display of the total pursing operation. Here again, temperature sensors are used to trigger the pursing moment at the correct depth, depending upon the species being fished.

In the area, once again, of trawl fisheries, the fishing-gear technologist continues to improve the efficiency and selectivity through careful studies of fish behavior in the harvesting process, netting materials, and net designs. Technical developments in flume tanks and underwater observation have greatly assisted the technologist with, in some cases, much lower towing speeds than previously thought possible, lowering the energy requirement and increasing catching efficiency.

The development of fishing gear, whether passive or active, could not have been accomplished without the introduction of new netting materials, which are a product of technology advances in the petrochemical and textile industries.[4]

While the chemical processes for the manufacture of the synthetic fibers are not so new, the presentation of the twine for nets and ropes has changed. Monofilaments and multimonofilaments have greatly improved the performance of gill nets, and their strength has facilitated the deployment of the gear through the use of hydraulic hauling systems. Since the filaments are stronger than natural fibers, thinner twines can be used which, if the color choice is correct, render the nets almost invisible in water.

Purse-seine development has been made possible only through the use of synthetic materials. High-density fiber, smooth surface, and the small diameter of the netting yarn accelerate the sinking speed of the net, and the latter two reduce resistance to water flow. Knotless netting is also commonly used in purse seines (for the main body), which was made possible by synthetics. Purse seines are bulky, are heavy when fully rigged, and are handled by powered net haulers. They also catch massive amounts of fish and, therefore, must be strong and flexible. Perhaps these seines will benefit from further improvements in design and the makeup of the twine to increase strength and flexibility.

Trawl nets demand yarns of high extensibility combined with high elasticity and high wet knot-breaking loads. Toughness is required to withstand rough treatment during fishing and to safely haul in large catches. Of all fishing gears, trawls suffer the most wear and tear on the bottom during towing and when being dragged on board. Probably more research in fishing gear has been applied to trawling than any other gear, no doubt due to the need to meet the exacting fishing specifications, to increase efficiency and selectivity, and to reduce resistance.

Further developments in the design of towed-gear innovations, such as multiple trawls and split or double cod ends, are likely to emerge, and the methods of encouraging fish to enter the net will have to be improved.

Other fishing gear, such as longlines, continues to improve through refinement of the construction of the line. A medium-sized vessel can set as much as 30 km of line and up to 20,000 hooks in automated systems. The length and mechanized handling (particularly in deep water) under adverse weather conditions, and the species to be fished, are major considerations in the selection of the line material, which can fall into the categories of twines and ropes.

In general, strong and relatively indestructible materials are being sought for fishing gear made out of textile materials. However, nonbiode-

---

4. The following chemical groups or classes of synthetic fibers are used for fishing nets: polyamide (PA), polyester (PES), polyethylene (PE), polypropylene (PP), polyvinyl chloride (PVC), polyvinylidene chloride (PVD), and polyvinyl alcohol (PVA).

gradable fibers have a negative side, and this is shown in some areas of the world by lost or abandoned fishing gear. Lost nets often continue to "ghost fish," and concern is felt in some countries due to the mortality level of marine mammals, turtles and seabirds as a consequence of becoming entangled in the netting. Further investigations need to determine the impact of this problem before solutions can be put forward.

Natural fibers are, of course, biodegradable and may still be found in use, but these are less strong, more difficult to handle with mechanized equipment, and require more maintenance; consequently, twines such as cotton are not likely to regain favor.

Synthetic materials have also displaced natural products for other parts of fishing gear. Ropes and floats are but two examples, and further development in this area, and in fishing gear design in general, will be necessary as fishing is explored at greater depths and for currently nonutilized stocks.

Fish Location

The sighting of fish schools by the naked eye is not yet a thing of the past in some types of fisheries, but it is a redundant method in the highly sophisticated fishing machines found in many developed and some developing countries. Competition has seen to it that technology must remove as much of the chance factor in the hunting process as possible.

More knowledge is available on fish resources and their distribution than ever before, and space technology is helping to enhance that knowledge. Satellites are being mobilized by the fisheries community to provide data, for example, on sea surface temperatures, upwelling, and areas of phytoplankton concentrations in real time. Given the historical data on the oceans, predictions can be made of probably productive fishing areas at that moment. Such information can be made available to fleets of vessels, cutting searching time and decreasing costs. The tuna industry is one such beneficiary.

Satellite-generated data is also being used to identify appropriate sites for aquaculture and harbors. Potential and actual reef fisheries are also readily identified, and in the not-too-distant future penetration will be increased to assist in the preparation of fishing charts. This method of charting in fisheries is being researched in Australia with some success, and FAO's own work in the Maldives gave interesting and encouraging results.

Weather prediction has gone through an extensive refinement period and weather charts are now in greater use by fishing fleets in the major fishing areas of the world. Navigation charts have also moved into the electronic age, and these are now displayed electronically in a variety of forms. Data previously found in little black books is now stored electronically and retrieved at will, adding information to the display from previous fishing trips.

With such information systems providing an overview which can be transmitted directly to the vessel by facsimile receivers, local identification of the schools becomes a hydroacoustic exercise backed up by improved navigation aids and experience.

Echo sounders and sonars have improved considerably in recent years due to technological advances in the electronics industry and research into underwater sound. Long-range sector-scanning sonars with ranges up to 4,000 m and automatic search programs are in ever-increasing demand. Such sonars have the ability to read fish concentrations, depth, range, bearing, and movement of up to nine separate schools. They even provide species identification through color patterns (for a limited number of species) and give real-time presentation of vessel, net, and fish. The information can be displayed on a radar-type screen, a plan position indicator (PPI), a television (CRT), or an echogram, or in audible form. Such sonars, as data processors, also form part of the integrated computerized control systems.

Like sonars, echo sounders have developed, and production technology has led to relatively lower costs for accurate high-resolution instruments. Knob controls have been replaced by microcomputer-based control systems with easy-to-read menus. They can even instruct the user if an operational error has been made. Recent sounders for trawling have evolved from advances in the design of scientific echo sounders and have time-varied gain (TVG) functions. Dual channels, one for fish finding (40 Log R) and one for bottom detection (20 Log R) are being calibrated according to standards used for scientific purposes. These can be used individually or in combination, giving good fish-finding echograms and good bottom discrimination.

Although the designer can make ready use of data-processing technology, the application of acoustic technology demands an in-depth knowledge of fish anatomy, and the fisheries biologist is intimately associated with ongoing research. This is of particular importance in determining target strengths, which are related to the size of the swim bladder of the fish. Some large fish, such as tuna, have a low target strength, and in developing long-range sonars, especially for the tuna industry, this affected the performance of earlier models.

Other factors affecting the performance of hydroacoustic equipment include noise emission from the vessel, transducer technology, power to the transducer, and, under certain conditions, the thermocline. Ongoing research programs by fisheries institutions and industries are indications of the progress yet to be made in fish finding. Nevertheless, progress to date is significant, as illustrated below in a block diagram representative of systems currently in use (fig. 1).

From the foregoing it may appear that the use of hydroacoustic technology is confined to large-scale industrial fishing. This is far from the truth, and small-scale fisherman place an equal reliance on sounders and sonars.

In developing countries there is a great need for a low-cost and relatively sensitive acoustic device for fish finding and bottom identification. At first

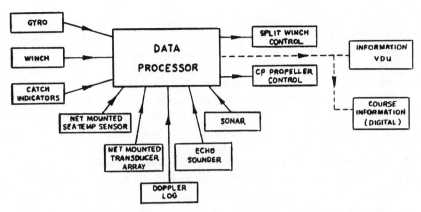

F‌IG. 1.—Computerized trawling system (source: John Fitzpatrick, 1989)

glance this may seem to be a relatively simple exercise, but these fisheries are conducted under a variety of conditions in shallow waters, over reefs, and also in very deep waters. In some cases, the deployment of fish-aggregating devices (FADs) in depths of 3,000–4,000 m is not uncommon and requires an assured measurement method.

Gradually, the cost is coming down, and echo sounders have even been fitted to the Ghanaian-type dugout canoe in West Africa. The FAO experience in this area is encouraging, but much work has to be done to ensure reliability and/or low enough cost to make it economical to replace rather than repair. One constraint to be overcome is in relation to the power to the transducer, especially for slope fisheries and FAD deployment from small boats.

Postharvest Conservation

242897

In order to reduce postharvest losses and obtain the best price for the fisherman, the catch must be properly cared for as it is being brought on board and when it is taken from the net, hook, or pots. The attention to be given to the catch varies considerably from one type of fishery to another and more often between developed and developing countries.

However, this is one area in which the transfer of technology can play a vital part in the future needs of the world for fish and fisheries products through reducing postharvest losses; this applies to artisanal and industrial fisheries alike. It may not be high technology to handle fish without bruising, to keep it out of the direct sunlight, to store it in clean conditions, or to use içe, but in some countries it is a new way of thinking.

Much progress has already been made in developing countries at the small-scale fisheries level by FAO, and in many countries in Africa, for example, even the dugout canoe has been equipped with insulated boxes for storage of the catch. In many cases, however, basic precautions are not taken, and the catch is often not landed in the best conditions.

Fortunately, the use of ice is on the increase, but it often brings with it costs which may be out of proportion to the amount received for the catch. There are a number of reasons for this, and at times low demand price is a major economic factor. This is particularly relevant when energy costs are high. Although research and field tests have been carried out, there is still much to be done in developing solar- and wind-powered ice makers.

Developing countries are also slowly adopting boxing and shelving (on ice) for fresh food fish, and some have accepted the need for hygienic conditions, through regular cleaning of fish holds and the use of antifungal paints. They have also been attracted to impervious plastic fish boxes, and there are manufacturing facilities for these in a number of developing countries.

In general, to determine the storage method to be adopted, the intrinsic characteristics and quality changes taking place after death must be fully understood. Naturally, the variables are numerous with different species, and the applications are also variable, since species with similar characteristics may be taken from the sea at greatly different body temperatures.

Much of the technology currently in use is the result of extensive research and field tests worldwide. FAO, with the support of the Danish government, has made significant progress in transferring the new technology; the work done on the use of small pelagics for human consumption is a case in point.

Methods of storage and the end use to which the material is to be put influence the method of handling on board. With bulk storage, for example, in chilled water tanks, the catch is usually brailed or pumped from the net into the tanks. These methods cause bruising and loss of scales despite the developments being made in pumping systems, and there is a further complication when discharging takes place, although recent improvements with new types of pumps look more promising. An example of this is a wet vacuum unloader of Canadian design and manufacture, which can handle up to 100 tons of dead or live fish per hour.

However, the need to adopt higher-level technologies by the industry as a whole is more apparent where high levels of quality control are a requirement in law. Shrimp is an example, and much of the world market is supplied from the waters of developing countries. More care is now being taken in the handling and sorting of the catch, with freezing and packaging on board becoming an accepted practice.

The tuna fisheries, on the other hand, present industry with more complex handling, preservation, and conservation needs due to the variety of consumer demands with regard to the end product. Repeated handling of fish must be kept to a minimum to prevent bruising and to keep the fish in "as fresh a condition as when caught." Technology transfer in such cases is from the biochemist to the design engineer, to the manufacturers of refrigeration

TABLE 3.—PRACTICAL STORAGE LIFE FOR FISH

	Storage Life (months)		
	−18°C	−25°C	−30°C
Fatty fish, sardines, salmon, ocean perch	4	8	12
Lean fish, cod, haddock	8	18	24
Flatfish, flounder, plaice, sole	9	18	24
Lobster, crab	6	12	15
Shrimp	6	12	12

SOURCE.—Food and Agriculture Organization.

equipment, and to shipbuilders. At the end of the chain are the crews to ensure quality, often through the maintenance of temperatures down to −57°C.

For conservation, the importance of low-temperature storage is clearly illustrated in table 3, which refers not only to the extended length of storage life but also to the quality at any given moment during storage.

Other factors that determine whether or not to freeze on board include the length of time the vessel is to be at sea and the target market. Many fishing industries, including small-scale fisheries working the EEZs of developing states, are having to extend their range of operations, and there is a greater need to freeze on board. The refrigeration systems in use (generally freon) vary from the basic to rather more complicated installations where high condensing temperatures are encountered; the systems become more complex the lower the evaporating temperature is. This tends to complicate transfer of technology, since it must extend to operators and servicemen with probably a lesser technical background than their counterparts in developed countries.

Together with research institutes and other agencies, the FAO has a program of workshops in quality control and preservation, and its fellowship program is geared towards the training of trainers.[5]

Catching Methods

While catching methods have become more efficient, the basic categories have changed little over hundreds of years; the development within some of these categories has been concurrent with advances made in seamanship and navigation. However, while advances have been made, we are still using nets, hooks, spears, and traps, as was done thousands of years ago.

5. Werner Kley, *Training Skills: A Manual for Fisheries Trainers* (Rome: Food and Agriculture Organization, 1987), chap. 8, sec. 830, "Fellowships and Study Tours: A Guideline for FAO Representatives and Project Personnel."

TABLE 4.—LABOR PRODUCTIVITY WITH VARIOUS FISHING METHODS

Gear Types	Annual Catch per Fisherman (tons)
Traps, hooked lines, and nets from small unpowered boats	1
Inshore longlines, entangling nets, and trawls from small vessels	10
High-seas trawls from large vessels	100
Purse seines from superseiners	400

Source.—Food and Agriculture Organization.

As mentioned earlier, many types of fishing gear are more efficient through the introduction of new materials, improved designs, and better gear handling. These improvements have also contributed to better working conditions for crews.

Some of the improvements in fishing methods have been covered in the earlier discussions. What has not been mentioned elsewhere in this article is the relationship of labor productivity to the various fishing methods. Table 4 gives an approximation.

Notwithstanding the high productivity from the large superseiners, small-scale fishermen still catch the bulk of the food fish landed throughout the world. This community uses virtually every fishing method in one form or another, and it also uses more hardware than the large-scale sector. Indeed, there are 30,000 small fishing vessels in the United States alone. This figure exceeds the total number of fishing vessels in the world registered as being over 100 tons. Indonesia alone boasts some 250,000 sail-powered fishing boats.

In small-scale fisheries, gillnetting has benefited greatly from the use of synthetic materials, particularly the monofilament and multimonfilament. Catches have increased because the gear fishes better, and the overall efficiency is increased with the use of net haulers which allow the deployment of a great number of nets. For bottom-set nets, deeper waters can be fished. New designs and a better understanding, through training courses, of how to make and set nets have also helped to improve techniques.

There has been an increase in the number of FADs employed, and they are being deployed in some areas for the first time. New materials have allowed FADs to be anchored in deep water, and, although this is not an easy task, they are being used in the South Pacific, the Maldives, and in the Caribbean with some success.

Pots and traps have improved due to the use of new materials and designs. Small boats can now carry a greater number of the new collapsible pots, and hydraulic pot haulers are in common use in developed countries. This simple aid is gradually filtering through to developing countries.

Some types of trawling which do not require the high installed horsepower associated with the industrial sector have been introduced to small-scale fisheries. Pair trawling, both bottom and mid-water, has been successfully used by low-powered artisanal craft. The technique requires initial training and practice to perfect, and it is one of the subjects of the FAO Training Series publications.[6]

With regard to the future, new fishing methods may be slow to evolve. Some progress might be made in the continuous-trawl method, which carries a pump in the cod end transferring fish directly to a surface receiving unit, but this would appear to have limited applications. The harvesting of krill and mesopelagics, of which there are large unexploited stocks, might be such an application, if there are further improvements of the slow-speed mid-water trawls with extra-large mesh in the wings that have a herding effect. Any advance, however, in the use of these species must be concurrent with processing developments to make full use of the end product.

More development is likely in deep-water fishing of underutilized species, and the great array of jigging machines may be replaced by other methods in the exploitation of cephalopods. But most certainly there will be advances in fish-attraction systems.

Fishing Vessels

In general, naval architecture, marine engineering, and shipbuilding technologies have developed independently of the fishing industry, with the state of the art being applied by those specializing in fishing-vessel design. In this way, the fishing industry has benefited from research and development activities sponsored by other sectors, without which some of the development in fishing methods would not have been achieved.

An example is a series of 33-m steel-hulled stern trawlers built in 1966 by a French shipyard specializing in warship construction. They had a full shelter deck, bulbous bow, propeller nozzle, controllable-pitch (CP) propeller of unique design, warp-tension meters, and winch control from the wheelhouse and at the split winches. The wheelhouse operator could observe the winch operation through a periscope system. A hydraulically operated watertight shutter sealed off the stern ramp when steaming, and the stern gantry was also hydraulically operated. It took 15 years for others to follow some of the design characteristics of these superb fishing machines, and only now are all of these innovations being incorporated and improved upon.

On a larger scale, the fourth generation of Dutch trawlers, with 10,000 HP and capable of freezing 250 tons/day, have enormous mid-water trawls with mesh sizes in the wings of up to 20 m. Consider also the massive 179-m

6. The list of FAO Training Series publications has been deleted from this paper but may be obtained by writing to the Fishing Industries Division, FAO.

Russian Crab factory ship and its flotilla of catchers, which has 12,750 kVA of generator power and can accommodate 520 persons. Such mass-production units are self-contained communities making full use of technological changes in shipping, fishing methods, conservation, and processing, and they require substantial resources at their disposal.

With regard to basic design criteria, all fishing vessels must meet certain safety standards and minimum stability requirements, they must be sea-kindly and provide adequate accommodation for the crew. In this respect, there are the Torremolinos International Convention for the Safety of Fishing Vessels (1977)[7] and the FAO/International Labour Office (ILO)/International Maritime Organization (IMO) *Voluntary Guidelines for the Design, Construction, and Equipment of Small Fishing Vessels.*[8] There are also the *Code of Safety for Fishermen and Fishing Vessels*[9] and the rules of classification to aid the designer.

It is, however, the end use and the environment in which the vessel is to be used that the designers must consider—such things as the different fishing methods, local customs, weather conditions, and sea state.[10]

Most fishing vessels have pure displacement hulls, and unlike merchantmen with assigned freeboards, the draft and trim deviate dramatically from the designed water line in some fisheries, leaving only the superstructure above water. This has been seen in the herring and capelin fisheries of Iceland and Norway and the sprat fishery of Scotland. To retain an adequate range of stability under such conditions is the task of the designer.

Smaller vessels, however, adopt a variety of hull forms through semidisplacement to planing, with hard chines, vee bottoms, and flat bottoms depending upon use, materials of construction, and ship- or boatbuilding techniques.

Fortunately, the naval architect and marine engineer have the software specially developed for computer-aided ship design, and these programs cover virtually all sizes of fishing vessels.[11] Adoption of computer technology

7. The International Convention for the Safety of Fishing Vessels, 1977, was adopted in Torremolinos, Spain, in 1977 and by March 1, 1984, was adhered to by 11 contracting states. It was to enter into force after ratification by 15 states whose fishing fleets constituted not less than 50% of the number of the world's fishing vessels of a length of 24 m or more. The convention applies only to new fishing vessels exceeding that length. It remains unratified.

8. FAO/ILO/IMO, *Voluntary Guidelines for the Design, Construction, and Equipment of Small Fishing Vessels* (London: International Maritime Organization, 1980). (This is an amplification of Part B of the Code of Safety for Fishermen and Fishing Vessels).

9. FAO/ILO/IMO, *Code of Safety for Fishermen and Fishing Vessels* (London: International Maritime Organization, 1975), Parts A and B.

10. As a guide to the range of vessel types, a selection from the FAO Fisheries Technical Paper No. 267 (1985) "Definition and Classification of Fishery Vessel Types" is attached as an Appendix for further information.

11. FAO uses a Circe program with a Hewlett-Packard 9000 for its small-fishing-boat-design activities as well as a Fairline 2 on an IBM PC 18.

has not only saved time but has presented the designer with a new outlook on hull performance under wave action and on stability in general.

In addition to computers, recent developments in tank testing techniques have greatly reduced design errors related to resistance, turbulence, and cavitation. While resistance is related to the energy demand to power the hull, an air-free flow of water over transducers is important to obtain optimum performance from the high-tech hydroacoustic equipment on which so many fishermen depend. Reducing cavitation also cuts noise, which affects hydroacoustic equipment and greatly improves propeller efficiency.

Such attention to detail is extremely important, since the efficient operation of the vessel will affect its economic rentability irrespective of the cost of the unit. In some cases, however, the investment is substantial, as can be seen from table 5.

All of the vessel types mentioned in table 5 would be well equipped by present-day standards, with the medium-sized and larger vessels fitted with an extensive array of electronic equipment. The Appendix lists typical specifications, but it should be noted that such aids as satellite navigation systems, color echo sounders, and hydraulic systems are common to all.

The electronic boom has influenced the design of deck machinery, in particular the related control systems, and while some countries prefer electric drives, the majority rely on hydrostatic transmissions. Hydrostatics, commonly referred to as hydraulics, can be applied to all types of vessels, provided there is a power source to drive the pumps.

The main thrust in the use of hydraulics was probably through the introduction of earth-moving techniques and the flexibility and lightness of the medium- to high-pressure equipment. As a result, power blocks, net haulers, gantries, pot haulers, and cranes are in common use, with the lower-pressure

TABLE 5.—COSTS OF VESSELS

Vessel Type	Length (m)	Material	Cost ($US)
Supertrawler	100	Steel	Up to 80,000,000
Tuna seiner	65	Steel	15,000,000
Freezer trawler	50	Steel	11,000,000
Purse seiner	45	Steel	5,800,000
Stern trawler	35	Steel	4,200,000
Scottish seiner	25	Steel	2,250,000
Scottish seiner	23	Wood	1,900,000
Shrimp trawler	25	Steel	900,000
Shrimp trawler	23	FRP	700,000
Gill-netter	15	FRP	600,000
Trawler	13	Ferrocement	350,000
Fast potter	10	FRP	120,000

SOURCE.—John Fitzpatrick, 1989.

systems being strong contenders for main winches. The ready control over infinitely variable speed units is a prerequisite for use in hauling purse seines and the computerized trawl systems already mentioned.

The benefits accruing to the small-scale fisheries sector are substantial, allowing maximum use of the developments in fishing-gear design. FAO has introduced a number of new designs of boats but has found that this alone does not automatically mean increased catches. In most cases, improved fishing methods had to be adopted, and the vessels had to have the capability of deploying the new gear, which often called for improved gear-handling equipment.

This overall development is spreading to the developing countries, which are demanding a transfer of technology to the shipbuilder, serviceman, and fisherman on an ever-increasing scale. FAO, in its contribution to development, has held special regional workshops on fishing-vessel design.[12] It has also taken the lead in the technique of ferrocement construction,[13] and this technology has already been transferred to a number of developing countries, as have various construction methods in fiber-reinforced plastic (FRP).

Most development, of course, implies higher costs and a higher energy demand, but in many developing countries energy sources such as fuel oil are not only costly but scarce. FAO has reduced operation costs through its energy-saving program. This makes full use of recent developments in easy-to-manufacture propellers and propeller nozzles to be applied to existing vessels as well as new ones.

In a number of cases, fuel is almost nonexistent, and the use of the sail has had to be reintroduced in some countries. Improved sail designs and prototypes of boats developed to maximize the use of the sail in powered boats have been introduced. This has the added advantage of improving the safe operation of small boats in the event of an engine breakdown. Such prototypes developed by FAO include the aluminum multihulled *Alia* in the South Pacific, the FRP *Oru* in India, and, more recently, a 9-m fishing launch for Mexico, also in FRP.

The importance of sail was highlighted in May 1983 at the International Conference on Sail-assisted Commercial Fishing Vessels, which was held in Florida.[14]

12. The results were published by Fishing News Books, London. The list has been deleted from this article. For a list of publications contact the publisher: Fishing News Books Ltd., 1 Long Garden Walk, Farnham, Surrey GU9 7H England.

13. *FAO Investigates Ferro-cement Fishing Craft*, (London: Fishing News Books). This publication was derived from the First FAO Seminar on the Design and Construction of Ferro-cement Fishing Vessels held in Wellington, New Zealand, October 9–13, 1972.

14. The conference was sponsored by the Florida Sea Grant Program, University of South Florida, the Virginia Sea Grant Program, and The Society of Naval Architects and Marine Engineers, assisted by Sail Assist International Liaison Associates.

Development Necessitated by Depletion of the Natural Resources

Other developments are brought about not necessarily through the lack of technology but through the depletion of a natural resource. Such is the case in many developing countries that previously enjoyed an abundance of hardwoods.

West Africa is a classical example only recently brought to the forefront due to the shortage of standing timber suitable for making dugout canoes. Recent estimates put the number of suitable trees under growth at about 6,000 and a demand over the next 10–15 years to replace some 20,000 canoes. Since it takes about 80 years for a tree to grow to maturity, no reforestation program can cope with the immediate needs.

Different materials can be used, and many have been tried, but the requirement for beach landing through the surf defies even the best computer-aided design program available today. Much more development work is essential to deal with the problem of which the West African case is but an example. This will require additional funding, and FAO is currently pursuing possible sources.

FUTURE ACTION

In November 1988, the Institute of Fisheries and Marine Technology, St. John's, Newfoundland, hosted the World Symposium on Fishing Gear and Fishing Vessel Design.[15] The main themes were (1) conservation-oriented fishing, (2) energy optimization, (3) monitoring of gear and vessel performance, (4) on-board fish handling and preservation, (5) small-scale fisheries, and (6) industrial fisheries.

The Symposium dealt with the change in emphasis in fish harvesting from the minimization of harvest to the maximization of benefits from the resource.[16] This expresses the concern of the technologists and their responsibility to provide food from the seas for years to come.

15. The Symposium was cosponsored by the Food and Agriculture Organization (FAO), the International Centre for Ocean Development (ICOD), Massachusetts Institute of Technology (MIT), and the International Council for the Exploration of the Sea (ICES).

16. During the week-long symposium, a workshop was held as a follow-up to the FAO Expert Consultation on Selective Shrimp Trawls, Mazatlan, Mexico, 1986, to determine the future work programs of FAO and interested institutions. In this respect, the Institute's flume tank was placed at the disposal of the participants to demonstrate recent developments. With regard to transfer of technology, FAO convened a meeting on fisheries education and training towards the end of the symposium. This was prompted by the concern of some developing countries in Asia at the imbalance in higher-level fisheries subjects in the region.

Fisheries Advisory Services

Bearing in mind that transfer of technology may be accomplished in a number of ways, the UNDP and FAO have embarked on a program of Fishery Advisory Services in fish-capture and aquaculture technology. The long-term objectives are to have greater self-efficiency within the regions.

Fishery Advisory Services is a global program incorporating the following UNDP/FAO networks:
Marketing information services
   INFOPECHE (Côte d'Ivoire): West Africa
   INFOPESCA (Panama): South America
   INFOSAMAK (Bahrain): Middle East
   GLOBEFISH (FAO Headquarters, Rome): Global
Aquaculture
   MEDRAP: Mediterranean Regional Aquaculture Project
   ADCP: Aquaculture Development and Coordination Programme
   CERLA: Latin American Regional Aquaculture Center, Brazil
   ARAC: African Regional Aquaculture Center, Nigeria
   NACA: Network of Aquaculture Centers in Asia
      RLCC: Regional Lead Center in China
      RLCI: Regional Lead Center in India
      RLCP: Regional Lead Center in the Philippines
      RLCT: Regional Lead Center in Thailand
   IRAC: Inter-Regional Aquaculture Center, Hungary
Cooperative networks:
   INFOFISH (Malaysia): INFOFISH is an Intergovernmental organization that was set up with UNDP assistance and whose projects were run by FAO. FAO still assists and close contact is maintained between the two. INFOFISH cooperated in the design of the data bases and will hold copies of the main FAO program so that it can respond directly to inquiries. INFOFISH activities are not limited to aquaculture.
   ICLARM (The Philippines): The International Center for Living Aquatic Resources (ICLARM) has a strong working relationship with FAO and FAO supports, financially and otherwise, the maintenance of a data base on aquaculture training centers. ICLARM will eventually hold copies of data on aquaculture subjects for direct reference.

Mapmaking for Aquaculture

FAO is engaged in a number of projects for the use of geographic information systems (GISs) and remote sensing to obtain a more comprehensive and integrated treatment of aquaculture-development criteria than is usually possible with manual analytical and mapmaking techniques. Developments in this field will be made available through publications and the Fisheries Advisory Services system.

Resource Mapping

Developments are required in fisheries sciences and in the methodologies currently in use in marine research. Resource mapping will certainly be improved and new technology will be adopted by designers of research vessels. In this respect, hydroacoustic technology will play an important part in stock-assessment work with an emphasis on inshore resources in shallow water.

Research Vessels

Developing countries will continue to need assistance in the immediate future to meet their requirements for research-vessel services. One way in which this might be done is through the Cooperative Use of Vessels program in which IOC and FAO will continue to cooperate. Another is through arrangements similar to the UNDP/NORAD (Norwegian Agency for Development Cooperation)/FAO Global Resources Programme using the Norwegian research vessel *Dr. Fridtjof Nansen*.

Technology Transfer and Training

All further development in fisheries will call for improvements in the education and training systems if technology is to be transferred in an adequate manner. In this respect, the newly published *Document for Guidance on Fishermen's Training and Certification,* prepared jointly by IMO, ILO, and FAO, will greatly assist the fisheries community to absorb new technology.[17] Indeed, these guidelines meet very closely the recommendation made at the Joint UNIDO/FAO First Expert Consultation on the Fisheries Industry, Gdansk, Poland, in June 1987, that in programs for the transfer of technology, training should be tailored to meet the actual needs of recipient countries.

The consultation also called for further studies of the use of alternative materials to hardwoods where these are scarce, and noted that an integrated approach to development should be followed.

17. FAO/ILO/IMO, *Document for Guidance on Fishermen's Training and Certification* (London: International Maritime Organization, 1988).

Safety of Life at Sea

On matters of safety of life at sea, FAO will continue to work closely with IMO. It will also give full support to the promotion of search-and-rescue (SAR) units and ensure that the fisheries sector is kept informed of developments. In particular, fisheries administrations will be encouraged to participate in local arrangements.

With regard to conditions of work at sea, the recommendations of the ILO Committee on Conditions of Work and Service in the Fishing Industry, at its meeting in Geneva, May 4–11, 1988, will be pursued by FAO with special attention given to the transfer of technology.

CONCLUSION

FAO will continue to cooperate with other agencies and organizations in the quest for further technological development to the benefit of users of the seas. In particular, it will follow the strategy adopted by the World Conference on Fisheries Development and Management[18] to facilitate cooperation and institutional coordination in the monitoring and studying of natural fluctuations and of those caused by man (through fishing activities, pollution, and urban development) that affect the production from both marine and inland waters.

APPENDIX

TYPICAL VESSEL SPECIFICATIONS[19]

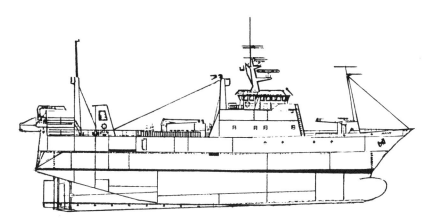

18. The conference was held in Rome, June 27–July 6, 1984.
19. Source.—FAO, "Definition and Classification of Fishery Vessel Types," FAO Fisheries Technical Paper no. 267, 1985.

FIG. A1.—Factory stern trawler (bottom trawling)

Main particulars:

Length overall	59 m
Length between perpendiculars	51.6 m
Breadth	13 m
Depth	8.85 m
Main engine	4,080 HP
Speed	14.5 knots

Capacities:

Fuel oil	500 m^3
Diesel oil	80 m^3
Fresh water	65 m^3
Fish room	950 m^3
Chilled water tanks	40 tons
Freezing	54 tons/24 hours
Accommodation	35 men

Auxiliary power	1,290 kVA
Steering	Rotary vane
Propeller	CP with nozzle
Side thruster	200 HP
Freshwater generation	15 tons/24 hours

Deck machinery (all low-pressure hydraulic):

2 trawl winches	32 tons pull
2 Gilson winches	10 tons pull
4 sweepline winches	10 tons pull
1 outhaul winch	6 tons pull
1 general purpose winch	10 tons pull
1 anchor windlass	8 tons pull
2 deck cranes	4 tons pull at 12-m radius
1 deck crane	2 tons pull at 9-m radius

Refrigeration:

2 tunnel freezers	24-hour capacity
3 horizontal-plate freezers	30 tons
1 vertical-plate freezer	18 tons
Fish hold to −30°C	6 tons
Chilled water tanks	40 tons at +1°C

Electronics:
2 color video sounders
1 net recorder
1 hydroacoustic catch-monitor system
1 seawater-temperature recorder
1 video plotter
1 Loran C navigator
1 satellite navigator
1 Doppler log
1 gyrocompass
2 radars, long range
2 single-sideband radiotelephones and telex facility
1 VHF radiotelephone
1 radio direction finder
1 autopilot
1 weather facsimile receiver

Remarks:
Vessel outfitted for prawn and finfish trawling and extended voyages.

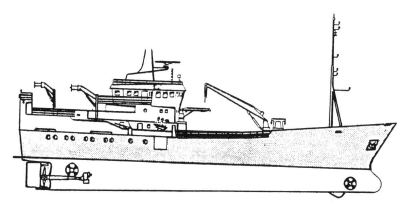

FIG. A2.—Steel purse seiner (small pelagics)

Main particulars:

Length overall	43.5 m	Triplex net	
Length between		hauler and	
perpen-		transport	
diculars	35.5 m	roller	
Breadth	8.9 m	Hydraulic crane	
Depth	5.9 m	and brailing	
Main engine	2,100 HP	winch	
Speed	12.7 knots	Fish pump	14-inch diameter
		Net drum	25 tons pull
Capacities:		Anchor windlass	4 tons pull
Fuel oil	60 m³	Safety line	
Fresh water	31 m³	winch	12 tons pull
Refrigerated		End wire winch	5 tons pull
seawater tanks	300 m³ to 0°C		
Accommodation	14 men		

Electronics:

Low-frequency color scanning sonar
High-frequency short-range scanning
   sonar
Color echo sounder
2 color video sounders
Net-depth monitor
Shipmate navigator and plotter
Racal-Decca navigator
2 marine radars
3 single-sideband radiotelephones
1 VHF radiotelephone and direction
   finder
SR 120 gyrocompass
Autotrawl monitor system
Doppler speed log

Generator

sets	2 × 200 HP,
	169 kVA
Bow thruster	285 HP
Stern thruster	395 HP
Harbor generator	25 kVA
Generator on	
main engine	1,250 kVA

Deck machinery (hydraulics):

Split purse/trawl	
winches	2 × 25 tons pull

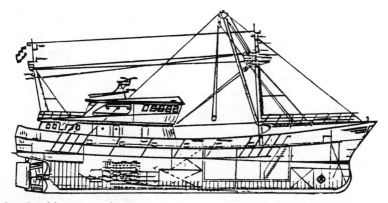

Fig. A3.—Steel-beam trawler

Main particulars:

Length overall	41 m
Length between perpendiculars	36.3 m
Breadth	8.5 m
Depth	4.7 m
Main engine	2,140 HP
Propeller	3 m in kort nozzle
Speed	13.5 knots

Capacities:

Fuel oil	95 m³
Fresh water	45 m³
Fish hold	210 m³
Refrigeration hold to	−5°C

Auxiliaries:

Generators	2 × 300 HP, 221 kW
Bow thruster	90 kW

Deck machinery:

8-drum trawl winch (capacity 500 fathoms 34-mm wire)	6.5 tons pull
Anchor windlass	4 tons pull
Cargo winch	2 tons pull

Electronics:
- Low-frequency echo sounder
- Color video sounder
- Color video track plotter
- Decca navigator
- Satellite navigator
- Single-sideband and VHF radiotelephone
- Marine radar
- Electronic log
- Electronic chart

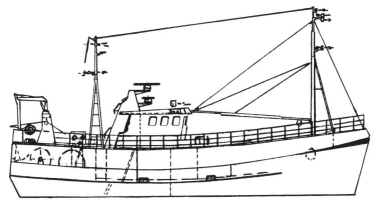

Fig. A4.—Steel/wood Scottish seiner/trawler

Main particulars:

Length overall	26.5 m	Split trawl	
Beam	7.35 m	winch, 12 tons	
Depth	3.9 m	pull	2 × 700 fathoms
Main engine	850 HP	Net drums	2 × 5.5 tons pull
Speed	11.5 knots	Power block	28-inch × 3 tons pull

Capacities:

Fuel oil	32 m³	Hydraulic crane	
Other oil	2.1 m³	Capstan	4 tons pull
Fresh water	7.3 m³	Landing winch	0.5 tons pull
Fish room	180 m³	2 fish washers	
Chilling	+1°C	Conveyor	
Accommodation	10 men	fish-transfer	
		system	

130 kVA auxiliary power
Rotary-vane steering
2 air compressors
Hydraulic steering
Fuel meter

Deck machinery (medium-pressure hydraulic):

Seine winch	4 tons pull
Rope reels	2 × 2,400 fathoms of 4-inch rope

Electronics:
Color echo memory sounder
Chromascope video sounder
Computerized data trawl unit
2 Decca navigators
2 Decca plotters
1 color radar (short-range)
1 long-range radar
2 single-sideband radiotelephones
1 VHF radiotelephone and scrambler
1 autopilot
1 electromagnetic log
Closed-circuit television monitors

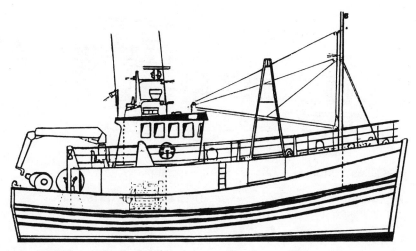

Fig. A5.—Wooden pair trawler

Main particulars:

Length overall	18.8 m
Breadth	6.7 m
Depth	2.7 m
Main engine	400 HP
Propeller	Fixed 4-blade

Capacities:

Fuel	12 m^3
Fresh water	2 m^3
Fish hold	60 m^3
Refrigeration (chilled hold)	+1°C

Deck machinery:

Split trawl winches	12 tons pull
Net drum	8 tons pull
Hydraulic crane and power block	3 tons pull
Cod-end capstan	3 tons pull

Electronics:
Dual-frequency color sounder
Color memory video sounder
Video track plotter
Decca navigator and plotter
2 marine radars, 64 × 24 miles
Single-sideband radiotelephone
2 VHF radiotelephones
Doppler speed log

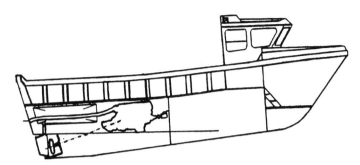

FIG. A6.—Glass-reinforced plastic (GRP) high-speed potter

Main particulars:

Length overall	10.3 m	Deck machinery (hydraulic):	
Length between		Pot hauler	
perpen-		Line hauler	
diculars	9.2 m		
Breadth	3.9 m	Electronic:	
Main engine	320 HP	2 color echo sounders	
Speed	24 knots	Decca navigator	
		Electromagnetic log	
Capacities:		Satellite navigator	
Fuel	0.5 m^3	VHF radiotelephone	
Water	0.2 m^3		
Fish room	2.2 m^3	Remarks:	
Accommodation	2 crew	High quality product on ice. Rapid	
Steering	Hydraulic ram	transit to and from fishing grounds.	

Biomass Yields of Large Marine Ecosystems

Kenneth Sherman
Northeast Fisheries Center

LME CONCEPT

A new era in ocean use was initiated when, in 1982, the United Nations Law of the Sea Convention established Exclusive Economic Zones (EEZ) up to 200 nm from the baselines of Territorial Seas, granting coastal states sovereign rights to explore, manage, and conserve the natural resources of the zones.[1] Within the boundaries of the new economic zones are large marine ecosystems (LMEs) that are being subjected to increased stress from growing exploitation of renewable resources, the dumping of urban wastes, and the fallout from aerosol contaminants. Large marine ecosystems are defined as relatively large regions of the world ocean, generally on the order of $\geq$ 200,000 km^2, characterized by unique bathymetry, hydrography, and productivity within which marine populations have adapted reproductive, growth, and feeding strategies.[2]

Although the designation of LMEs is, at present, an evolving scientific and geopolitical process,[3] sufficient progress has been made to allow for useful comparisons to be made of the different processes influencing large-scale changes in the biomass yields of living marine resources in LMEs. To facilitate the comparisons, a series of symposia have been convened at the annual meeting of the American Association for the Advancement of Sciences

---

1. *The Law of the Sea: Official Text of the United Nations Convention on the Law of the Sea with Annexes and Index* (New York: United Nations, 1983), Sales No. E.83.v.5. (cited as the Law of the Sea Convention).

2. Kenneth Sherman and Lewis M. Alexander, eds., *Variability and Management of Large Marine Ecosystems*, American Association for the Advancement of Science (AAAS) Selected Symposium 99 (Boulder, Colo.: Westview Press, 1986), 319 pp., hereafter called *Variability and Management*, and *Biomass Yields and Geography in Large Marine Ecosystems*, AAAS Selected Symposium Series 111 (Boulder: Westview Press, 1989), hereafter called *Biomass Yields and Geography*.

3. Lewis M. Alexander, "Introduction to Part Three: Large Marine Ecosystems as Regional Phenomena," in *Variability and Management*, pp. 239–40; and Joseph R. Morgan, "Large Marine Ecosystems an Emerging Concept of Regional Management," *Environment* 29, no. 10 (1988): 4–9, 29–34.

(AAAS). Reports presented to the first AAAS symposium on LMEs argued that they were tractable global units for the conservation and management of living marine resources.[4] The second symposium provided additional information on the nature of variability in biomass yields within LMEs and extended the designation of LMEs globally with the application of legal, political, and geographic criteria.[5] The third AAAS symposium on LMEs demonstrated the utility of the comparative ecosystem approach for determining the principal sources of variability in biomass yields.[6]

The movement toward total ecosystem management has been growing slowly within the international community for several decades. The trend began with the deliberations of the International Council for the Exploration of the Sea (ICES) in its first meetings conducted at the turn of the century. The meetings were prompted by the realization that the capacity of the oceans to produce an inexhaustible yield of fish biomass was finite and that overfishing could result in the serious depletion of the stocks. The first attempts to deal with the management of regional ecosystems took place in Kristiania, Norway, in 1901, where representatives from Denmark, Germany, Norway, Russia, and the United Kingdom set the course for the establishment of ICES with a series of resolutions directed to the establishment of joint international biological and hydrographic studies. The ICES, over the past 75 years, has become a vital force in the development and implementation of joint international studies of marine ecosystems. In the process, scientists participating in ICES programs have helped focus on the advantages of coordinated multidisciplinary studies of LMEs (e.g., North Sea, Baltic, Norwegian Sea, U.S. Northeast Continental Shelf, and Iberian Coastal ecosystems). As the trend for management of living resources moves from single-species to multispecies assemblages, it becomes increasingly important to encompass entire ecosystems as management units. This approach will ensure that management measures designed to optimize the natural productivity of target species assemblages will also include consideration for related competitor/predator populations and their environments.

By matching sampling effort to the time and space scales of the processes that are of most direct influence to growth and survival of living marine resource populations, forecasts of biomass yield trends among the species can

4. Sherman and Alexander.

5. Lewis M. Alexander, "Introduction to Part Two: Geographic Perspectives of Large Marine Ecosystems"; Joseph Morgan, "Large Marine Ecosystems in the Pacific," J. R. Victor Prescott, "The Political Division of Large Marine Ecosystems in the Atlantic Ocean and Some Associated Seas," and Martin H. Belsky, "Developing an Ecosystem Management Regime for Large Marine Ecosystems," all in *Biomass Yields and Geography.*

6. Kenneth Sherman and Lewis M. Alexander, eds., *Patterns, Processes, and Yields of Large Marine Ecosystems*, AAAS Selected Symposium Series (Washington, D.C.: AAAS, in press).

be improved for LMEs.[7] Studies of changes in abundance and population renewal of resource species in general and fish stocks in particular on a large marine ecosystem scale is in agreement with the recent proposition by Ricklefs that ecologists should begin to address critical community processes on a regional basis.[8] Ricklefs argues that "the regional historical viewpoint provides a fundamental challenge to ecologists. Broadened concepts of the regulation of local community structure, incorporation of historical, systematic, and biogeographic information into the phenomenology of community ecology, and expanded investigations that address global variation in local species richness will help unite local and regional perspectives."

LMEs IN THE U.S. ECONOMIC ZONE

Greater emphasis has been focused over the past decade within the National Marine Fisheries Service of the National Oceanic and Atmospheric Administration (NOAA) on approaching fisheries research from a regional ecosystem perspective in seven LMEs within the Exclusive Economic Zone of the United States—the Northeast Shelf, the Southeast Shelf, the Gulf of Mexico, the California Current, the Gulf of Alaska, the East Bering Sea, and the Insular Pacific, including the Hawaiian Islands. These ecosystems, in 1986, yielded 3 million metric tons (Mmt) of fisheries biomass values at $2.8 billion and approximately $10 billion to the national economy.

A description of the sampling programs providing the biomass assessments within the U.S. Exclusive Economic Zone has been previously described in folio map 7, produced by the Office of Oceanography and Marine Assessment of the NOAA's National Ocean Service. The map (fig. 1) depicts the seven ecosystems under investigation. The following account of program activity within the seven LMEs has been adapted from NOAA's folio map 7.

Sampling programs supporting biomass estimates in LMEs are designed to (1) provide detailed statistical analyses of fish and invertebrate populations constituting the principal yield species of biomass, (2) estimate future trends in biomass yields, and (3) monitor changes in the principal populations. The information obtained by these programs provides managers with a more complete understanding of the dynamics of marine ecosystems and how these dynamics affect harvestable stocks. Additionally, by tracking components of the ecosystems, these programs can detect changes, natural or human induced, and warn of events with possible economic repercussions. Although

7. Simon A. Levin, "Physical and Biological Scales, and the Modeling of Predator-Prey Interactions in Large Marine Ecosystems," in Sherman and Alexander, eds., *Patterns, Processes, and Yields of Large Marine Ecosystems.*

8. Robert E. Ricklefs, "Community Diversity: Relative Roles of Local and Regional Processes," *Science* 235 (1987): 167–71.

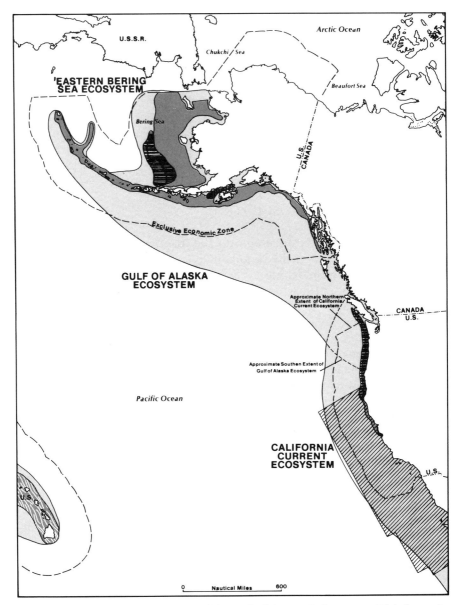

Fig. 1.—Large marine ecosystems of the United States. Reference.—This figure is a modified version of folio map no. 7 in *National Atlas: Health and Use of Coastal Waters, United States of America* (Washington, D.C.: Department of Commerce, National Oceanic and Atmospheric Administration, National Ocean Service, Office of Oceanography and Marine Assessment, 1988). Modified and reproduced with permission. Figure drawn by Lyn Indalecio.

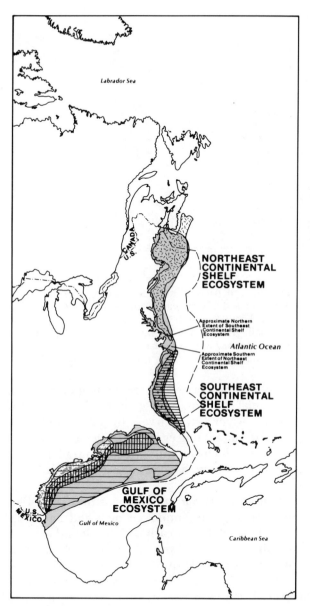

NORTHEAST
CONTINENTAL
SHELF
ECOSYSTEM

Approximate Northern
Extent of Southeast
Continental Shelf
Ecosystem

Atlantic Ocean

Approximate Southern
Extent of Northeast
Continental Shelf
Ecosystem

SOUTHEAST
CONTINENTAL
SHELF
ECOSYSTEM

GULF OF
MEXICO
ECOSYSTEM

Labrador Sea

CANADA
U.S.

U.S.
MEXICO

Gulf of Mexico

Caribbean Sea

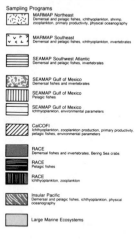

Fishery Resource Assessment Programs

Sampling Programs

MARMAP Northeast
Demersal and pelagic fishes, ichthyoplankton, shrimp, zooplankton, primary productivity, physical oceanography

MARMAP Southeast
Demersal and pelagic fishes, ichthyoplankton, invertebrates

SEAMAP Southwest Atlantic
Demersal and pelagic fishes, invertebrates

SEAMAP Gulf of Mexico
Demersal fishes and invertebrates

SEAMAP Gulf of Mexico
Pelagic fishes

SEAMAP Gulf of Mexico
Ichthyoplankton, environmental parameters

CalCOFI
Ichthyoplankton, zooplankton production, primary productivity, pelagic fishes, environmental parameters

RACE
Demersal fishes and invertebrates, Bering Sea crabs

RACE
Pelagic fishes

RACE
Ichthyoplankton, zooplankton

Insular Pacific
Demersal and pelagic fishes, ichthyoplankton, physical oceanography

Large Marine Ecosystems

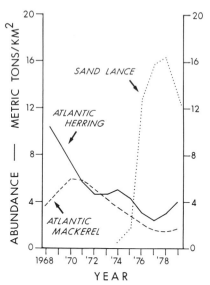

FIG. 2.—Decline of Atlantic herring and Atlantic mackerel and apparent replacement by the small, fast-growing sand eel in the Northeast Continental Shelf Ecosystem (measured in mt per km², 1968–79). From Kenneth Sherman, Reuben Lasker, William Richards, and Arthur W. Kendall, Jr., "Ichthyoplankton and Fish Recruitment Studies in Large Marine Ecosystems," *Marine Fisheries Review* 45, nos. 10–12 (1983): 1–25. Note that the names "sand eel" and "sand lance" are used interchangeably.

sampling schemes and efforts vary among programs (depending on habitats, species present, and specific regional concern), they generally involve use of NOAA vessels for bottom and midwater trawl surveys for adults and juveniles; ichthyoplankton surveys for larvae and eggs; measurements of zooplankton standing stock, primary productivity, nutrient concentrations, and important physical parameters; and, in some habitats, measurements of contaminants and their effects.[9]

A critical feature of each program is the development of a consistent long-term data base for understanding interannual changes and multiyear trends. For example, during the late 1960s and early 1970s, when there was intense foreign fishing within the Northeast Continental Shelf Ecosystem, marked alterations in fish abundances were recorded. A sharp decline in the abundance of various commercial species and an increase in sand lance (fig. 2) led to the conclusion that overall fish biomass, primary productivity, and

9. Kenneth Sherman, Reuben Lasker, William Richards, and Arthur W. Kendall, Jr., "Ichthyoplankton and Fish Recruitment Studies in Large Marine Ecosystems," *Marine Fisheries Review*, 45, nos. 10–12 (1983): 1–25; J. R. Dunn, *A Catalog of Northwest and Alaska Fisheries Center Ichthyoplankton Cruises, 1965–1985*, NWAFC Report 86-08 (Seattle, Wash.: Government Printing Office, 1986).

zooplankton biomass and composition did not undergo any large-scale changes; data analysis was critical in implicating severe overfishing as the cause of the shift in relative abundances.[10]

Northeast Continental Shelf Ecosystem[11]

National Marine Fisheries Service (NMFS) Northeast Fisheries Center has been conducting groundfish and pelagic fish surveys from Nova Scotia to Cape Hatteras since 1967 as part of the Marine Resources Monitoring, Assessment, and Prediction (MARMAP) Program.[12] The objective of this semiannual cruise series is to study the variation of fish distributions and abundances on the northwest Atlantic Continental Shelf and to relate these to changing oceanographic parameters and fishing pressure. The sampling scheme involves 450 stations randomly selected within 76 different areas, based on bottom type and depth strata, extending from the coast to 400 m depth. In 1977, the program was expanded to a full-scale fishery ecosystem study. An average of six MARMAP survey cruises per year were added to sample up to 200 fixed stations for ichthyoplankton, zooplankton, phytoplankton, primary productivity, nutrients, and critical physical parameters.[13] This program is producing an atlas series.

10. Steve H. Clark and Bradford E. Brown, "Changes of Biomass of Finfishes and Squids from the Gulf of Maine to Cape Hatteras, 1963–74, as Determined from Research Vessel Survey Data," *Fishery Bulletin, U.S.* 75 (1977): 1–21; Kenneth Sherman, "Measurement Strategies for Monitoring and Forecasting Variability in Large Marine Ecosystems," and Michael P. Sissenwine, "Perturbation of a Predator-controlled Continental Shelf Ecosystem," both in *Variability and Management*, pp. 203–36 and pp. 55–85.

11. Marine Resources Monitoring, Assessment, and Prediction (MARMAP) Northeast participating institutions are the State of Maine, Department of Marine Resources, Boothbay Harbor, Maine; New Hampshire Fish and Game, Concord, N.H.; Northeast Fisheries Center, National Marine Fisheries Service, Gloucester, Mass.; the State of Massachusetts, Division of Marine Fisheries, Salem, Mass.; the State of Massachusetts, Division of Marine Fisheries, Sandwich, Mass.; Northeast Fisheries Center, NMFS, Woods Hole, Mass.; Northeast Fisheries Center, NMFS, Narragansett, R.I.; Northeast Fisheries Center, NMFS, Milford, Conn.; Northeast Fisheries Center, NMFS, Sandy Hook, N.J.; National Marine Fisheries Service, Washington, D.C.; and Northeast Fisheries Center, NMFS, Oxford, Md.

12. Marvin D. Grosslein, "Groundfish Survey Program of the Bureau of Commercial Fisheries, Woods Hole," *Commercial Fisheries Review* 31 (1969): 22–35.

13. John D. Sibunka and Myron J. Silverman, *MARMAP Surveys of the Continental Shelf from Cape Hatteras, North Carolina, to Cape Sable, Nova Scotia (1977–1983)*, Atlas No. 1: *Summary of Operations*, NOAA Technical Memorandum NMFS-F/NEC-33 (Woods Hole, Mass.: Government Printing Office, 1984).

Southeast Continental Shelf Ecosystem[14]

The sampling program for the Southeast Continental Shelf Ecosystem has two components: MARMAP Southeast and SEAMAP (Southeastern Area Monitoring and Assessment Program) South Atlantic. The South Carolina Department of Wildlife and Marine Resources is conducting a series of MARMAP studies of the Continental Shelf from Cape Hatteras to Cape Canaveral, including two to five cruises and various special studies annually. Based on analysis of bottom types, the study area was divided in 1980 into three regions, each of which is sampled and analyzed separately. The three types delineating the boundaries of the regions supporting commercially important species are nearshore sand, shelf-edge ridges and reefs, and deep rocky and muddy sediments.

The second component, SEAMAP South Atlantic, complements MARMAP Southeast by expanding the study region south through the Florida Keys and including the full width of the EEZ, rather than stopping at the continental slope. The full program is conducted by a consortium including NMFS's Southeast Fisheries Center, the Atlantic States Marine Fisheries Commission, and relevant state agencies in North Carolina, South Carolina, Georgia, and Florida. Present studies are concentrating on shrimp and shrimp-associated groundfish and will expand to include surveys of commercial pelagic and demersal species as well as ichthyoplankton and environmental parameter studies.[15]

Gulf of Mexico Ecosystem[16]

The sampling program for the Gulf of Mexico Ecosystem, SEAMAP Gulf, covers the area from the Texas-Mexico border to the Florida Keys. The

14. Marine Resources Monitoring, Assessment, and Prediction (MARMAP) Southeast participating institutions are the Atlantic States Marine Fisheries Commission, Washington, D.C.; and the South Carolina Department of Wildlife and Marine Resources, Charleston, S.C. Southeastern Area Monitoring and Assessment Program (SEAMAP) Southwest Atlantic participating institutions are Southeast Fisheries Center, NMFS, Beaufort, N.C.; North Carolina Division of Marine Fisheries, Morehead City, N.C.; South Carolina Department of Wildlife and Marine Resources, Charleston, S.C.; Southeast Fisheries Center, NMFS, Charleston, S.C.; Georgia Department of Natural Resources, Brunswick, Ga.; Southeast Fisheries Center, NMFS, Miami, Fla.; and Florida Department of Natural Resources, St. Petersburg, Fla.

15. SEAMAP, *Southeast Area Monitoring and Assessment Program (SEAMAP) Strategic Plan* (Pascagoula, Miss.: NMFS, Southeast Fisheries Center, 1981).

16. Southeastern Area Monitoring and Assessment Program (SEAMAP) Gulf of Mexico participating institutions are Southeast Fisheries Center, NMFS, Miami, Fla.; Southeast Fisheries Center, NMFS, Panama City, Fla.; Alabama Department of Conservation and Natural Resources, Gulf Shores, Ala.; Southeast Fisheries Center, Pas-

program provides information for assessing the status of commercially important species, especially shrimp and Gulf menhaden. Shrimp and groundfish trawl surveys are conducted within predetermined depth strata from the shoreline to the 100-m isobath.[17] Plankton and environmental parameters are sampled annually during 10 sets of cruises at about 500 stations located on a one-half degree grid. The Gulf SEAMAP Program is a cooperative effort involving NMFS's Southeast Fisheries Center, the Gulf States Marine Fisheries Consortium, and relevant state agencies and interested universities in Florida, Alabama, Mississippi, Louisiana, and Texas.[18]

California Current Ecosystem[19]

The California Cooperative Oceanic Fisheries Investigations (CalCOFI), begun as a joint federal-state effort to investigate the biology of the Pacific sardine to explain the collapse of the fishery, have been in progress since 1949. Emphasizing pelagic species, the program samples the eastern Pacific from the California-Oregon border to the tip of Baja California using a basic fixed-station plan of 36 transects perpendicular to the coast, each with 14–16 sampling stations that follow a strict station plan.[20] A complementary groundfish sampling program is in the planning stage. Initially, cruises sampled part or all of the station grid monthly; however, at present, cruises are conducted quarterly and sample a reduced area.

Three institutions cooperate in the program. The NMFS's Southwest Fisheries Center studies distribution of fish eggs and larvae, concentrating on northern anchovy, sardine, and other pelagic species. The California Department of Fish and Game studies juvenile and adult fish distributions and abun-

cagoula, Miss.; Gulf States Marine Fisheries Commission, Ocean Springs, Miss.; Gulf Coast Research Laboratory and Mississippi Department of Wildlife Conservation, Long Beach, Miss.; Southeast Fisheries Center, NSTL Station, Miss.; Louisiana Department of Wildlife and Fisheries, Baton Rouge, La.; and Texas Department of Parks and Wildlife, Austin, Tex.

17. Warren E. Stuntz, C. W. Bryan, K. Savastano, R. S. Waller, and P. A. Thompson, *SEAMAP Environmental and Biological Atlas of the Gulf of Mexico, 1982* (Ocean Springs, Miss.: Gulf States Marine Fisheries Commission, 1984).

18. See n. 14 above; and SEAMAP, *Annual Report to the Technical Coordinating Committee, 1984* (Pascagoula, Miss.: NMFS, Southeast Fisheries Center, 1985).

19. California Cooperative Oceanic Fisheries Investigations (CalCOFI) participating institutions are Southwest Fisheries Center, NMFS, La Jolla, Calif.; Scripps Institution of Oceanography, La Jolla, Calif.; California Department of Fish and Game, Long Beach, Calif.; and Southwest Fisheries Center, NMFS, Tiburon, Calif.

20. David Kramer, M. J. Kalin, E. G. Stevens, J. R. Thrailkill, and J. R. Zweifel, *Collecting and Processing Data on Fish Eggs and Larvae in the California Current Region*, NOAA Technical Report NMFS CIRC-370 (La Jolla, Calif.: Government Printing Office, 1972).

dances. The University of California's Scripps Institution of Oceanography carries out the zooplankton, physical, and chemical data programs. Scripps also publishes the CalCOFI Atlas series, now numbering over 30 volumes, and the *CalCOFI Reports,* a research annual in its twenty-eighth year.

Gulf of Alaska and Eastern Bering Sea Ecosystems[21]

Large-scale sampling on the Pacific coast north of Santa Barbara, California, began in the 1940s. In the 1970s, the program evolved into a series of regional surveys to acquire information on the condition of various fish and shellfish resources. The Northwest and Alaska Fisheries Center provides this information to the North Pacific and Pacific Fishery Management Councils in support of resource management. Data from these programs also form the basis for discussions with foreign nations having interests in transboundary or international resources. Sampling is done in a systematic pattern in extensive regions of trawlable bottom (eastern Bering Sea) or in a random pattern based on depth strata in regions of irregular bottom topography.[22]

Three or four survey cruises are conducted annually in each of three geographic regions—eastern Bering Sea, Gulf of Alaska, and Pacific West Coast—with an emphasis on one region every third year. For example, investigations during one year focus on fish and shellfish resources in the eastern Bering Sea. Extra emphases include expanding survey areas to cover peripheral regions of resource distribution, intensifying sampling effort in critical habitat regions for various life stages or species, and implementing other sampling techniques, such as midwater trawling and hydroacoustic surveying, to improve stock definition. During the following 2 years, emphasis shifts to the Gulf of Alaska and then the West Coast before returning to the Bering Sea.

Insular Pacific Ecosystems[23]

Recruitment and population maintenance in the Pacific islands ecosystems are poorly understood and in need of study. As the problem seems to center on

21. Resource Assessment and Conservation Engineering (RACE) Programs participating institutions are Northwest and Alaska Fisheries Center, NMFS, Seattle, Wash.; Northwest and Alaska Fisheries Center, NMFS, Kodiak, Alaska.

22. Robert S. Otto, R. A. MacIntosh, T. M. Armetta, W. S. Meyers, J. McBride, and D. A. Somerton, "United States Crab Research in the Eastern Bering Sea during 1982," *International North Pacific Fisheries Commission Annual Report, 1982* (Vancouver: International North Pacific Fisheries Commission, 1983), pp. 141–60; and T. M. Sample, "Groundfish Surveys Conducted by the United States in 1983," *International North Pacific Fisheries Commission Annual Report, 1983* (Vancouver: International North Pacific Fisheries Commission, 1984), pp. 114–127.

23. A participating institution of the Insular Pacific Programs is the Southwest Fisheries Center, NMFS, Honolulu, Hawaii.

successful larval settlement from the plankton, the Southeast Fisheries Center is planning a series of intensive physical and biological oceanographic studies around oceanic islands (the seamount/atoll program) to elucidate the mechanisms controlling recruitment from the plankton. The center has initiated general surveys in the northwest Hawaiian Islands and northern Marianas/Guam areas. It has also sampled the slipper and spiny lobster stocks in the Hawaiian Islands for the last 5 years to measure stock size and interactions among species.

Other Programs

A number of other programs provide useful ancillary information but are more limited in scope because of different program objectives. They range from federal programs lasting for several years, to various state monitoring programs, and finally to small seasonal studies conducted by academic researchers. An example of a federal program is the Outer Continental Shelf Environmental Assessment Program (OCSEAP), which studies the marine mammals, seabirds, and benthic organisms around the Alaskan coast. This is sponsored by the Minerals Management Service and administered by NOAA. An example of a state program is Maryland's study of striped bass abundance and recruitment in the Chesapeake Bay. Academic studies include resource-specific programs at university affiliated institutions, often conducted with funding from the NOAA National Sea Grant Program and the National Science Foundation.

The Future

The NMFS-NOAA regional programs concerned with forecasting trends in biomass yields are being conducted in relation to the dynamics of large marine ecosystems. The studies are now yielding hitherto unavailable information on large-scale between-year changes in ecosystem productivity and fish abundances. they will continue to provide increasingly improved information for understanding and managing the nation's living marine resources. Recent programs initiated by the National Science Foundation will expand research in the scientific community on problems of recruitment and provide critical knowledge of the principal factors controlling productivity within LMEs. There is also a need to understand better the relationship of LMEs and adjacent estuaries, many of which provide critical habitat for the early life stages of fishes and invertebrates found in the LMEs as adults. Research on this topic, coupled with the integration of large-scale and small-scale research programs, could greatly improve the management of the marine fisheries of the United States.

A GLOBAL PERSPECTIVE ON LMEs

Nearly 95% of usable marine biomass is produced in LMEs located within the newly expanded Exclusive Economic Zones of coastal states around the globe. Populations of LMEs can be altered significantly by natural and anthropogenic perturbations, leading to severe economic consequences in some coastal states and increasing yields in others.

Increasing attention has been focused over the past few years on synthesizing available biological and environmental information influencing the natural productivity of the fishery biomass within LMEs from a global perspective. Of the 18 LMEs for which recent syntheses have been reported, initial determinations indicate that in four—the Yellow Sea, the Gulf of Thailand, the Great Barrier Reef, and the Northeast Continental shelf—the controlling variable appears to be predation, including both the influence of natural predation and predation expressed as excessive fishing mortality (fig. 3). Major changes in the Great Barrier Reef Ecosystem have been attributed to the predation effect of the crown-of-thorns starfish.[24] The principal variable in the other three is recruitment overfishing, considered for purposes of this discussion as human predation.[25] For six other LMEs, the predominant variable driving the systems appears to be environmental change—Oyashio, Kuroshio,[26] California Current,[27] Humboldt Current,[28] Iberian Coastal,[29] and the Benguela Current.[30] The dominant influence on the Baltic has been reported as coastal pollution,[31] for the remaining LMEs, the information for

24. Roger H. Bradbury and C. N. Mundy, "Large-Scale Shifts in Biomass of the Great Barrier Reef Ecosystems," *Biomass Yields and Geography* (n. 2 above).

25. Qisheng Tang, "Changes in the Biomass of the Huanghai Sea Ecosystem," and Twesukdi Piyakarnchana, "Yield Dynamics as an Index of Biomass Shifts in the Gulf of Thailand Ecosystem," both in *Biomass Yields and Geography;* and Sissenwine (n. 10 above).

26. Takashi Minoda, "Oceanographic and Biomass Changes in the Oyashio Current Ecosystem," and Makoto Terazaki, "Recent Large-Scale Changes in the Biomass of the Kuroshio Current Ecosystem," both in *Biomass Yields and Geography.*

27. Alec D. MacCall, "Changes in the Biomass of the California Current Ecosystem," *Variability and Management* (n. 2 above), pp. 33–54.

28. R. Canon, "Variabilidad ambiental en relacion con la pesqueria neritica pelagica de la zona Norte de Chile," in *La pesca en Chile,* ed. P. Arana (Valparaiso, Chile: Universidad Catolica de Valparaiso, Facultad de Recursos Naturales, Escuela de Ciencias del Mar, 1986).

29. Tim Wyatt and G. Perez-Gandaras, "Biomass Changes in the Iberian Ecosystem," in *Biomass Yields and Geography.*

30. Robert J. M. Crawford, L. V. Shannon, and P. A. Shelton, "Characteristics and Management of the Benguela as a Large Marine Ecosystem," in *Biomass Yields and Geography.*

31. Gunnar Kullenberg, "Long-Term Changes in the Baltic Ecosystem," in *Variability and Management,* pp. 19–31.

LARGE MARINE ECOSYSTEMS

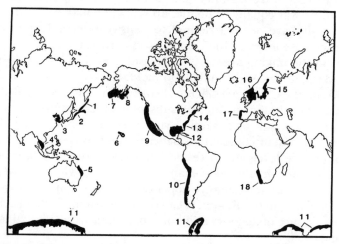

FIG. 3.—Predominant variables influencing changes in fish species biomass in large marine ecosystems. Predominant variable—predation (X); environment (0); pollution (P); inconclusive information (+). From Kenneth Sherman, "Large Marine Ecosystems as Global Units for Recruitment Experiments," In *Toward a Theory on Biological-Physical Interactions in the World Ocean,* ed. B. J. Rothschild (Dordrecht: Kluwer Academic Publishers, 1988), pp. 460–76. Legend: 1 = Oyashio Ecosystem (0); 2 = Kuroshio Ecosystem (0); 3 = Yellow Sea Ecosystem (X); 4 = Gulf of Thailand Ecosystem (X); 5 = Great Barrier Reef Ecosystem (X); 6 = Insular Pacific Ecosystem (+); 7 = East Bering Sea Ecosystem (+); 8 = Gulf of Alaska Ecosystems (+); 9 = California Current Ecosystem (0); 10 = Humboldt Current Ecosystem (0); 11 = Antarctic Ecosystem (+); 12 = Gulf of Mexico Ecosystem (+); 13 = Southeast Continental Shelf Ecosystem (+); 14 = Northeast Continental Shelf Ecosystem (X); 15 = Baltic Sea Ecosystem (P); 16 = North Sea Ecosystem (+); 17 = Iberian Coastal Ecosystem (0); 18 = Benguela Current Ecosystem (0).

making an·initial determination of principal driving force is considered inconclusive.

Management of stocks responding to strong environmental signals will be enhanced by improving the understanding of the physical factors forcing biological changes. The most dramatic of changes in species biomass attributed to large-scale changes in ocean productivity prompted by physical perturbation is the increase in the abundance of sardines of the Kuroshio/Oyashio ecosystems off Japan and in the Humboldt Current Ecosystem off Chile and Peru. The global impact on fishery yields from the responses of these ecosystems to physical change is significant. In 1974, the annual total yield of the global fisheries was approximately 65 Mmt; by 1986, it had risen to 91.5 Mmt. The additional biomass resulted from increases in yields (landings) of sardines and anchovies. The yields of herrings, excluding the Peruvian anchoveta, increased from an annual level of 10 Mmt in 1974 to 23.9

Mmt in 1986. Of this increase, Japanese sardines from the Oyashio/Kuroshio ecosystems increased from 400,000 mt to 5.2 Mmt annual yield. The yield of Chilean sardine was equally dramatic, up from 500,000 mt to 4.3 Mmt. The increased yields have been attributed to density-independent processes involving an increase in lower food chain productivity, made possible by coastward shifts in the mixing areas of the Oyashio and Kuroshio ecosystems off Japan[32] and water mass shifts in the Humboldt Current Ecosystem off Chile.[33] Changes within the California Current Ecosystem in the decline in abundance level and yields of Pacific sardine and subsequent increase in anchovy biomass is considered the result of natural environmental change rather than from any density-dependent competition between the two species.[34] The alternation in abundance levels of horse mackerel and sardine within the Iberian Coastal Ecosystem is attributed to changes in natural environmental perturbation rather than to any density-dependent interaction between the two species.[35] Similarly, in the Benguela Current Ecosystem, the long-term fluctuations in abundance of anchovy and horse mackerel are attributed to environmental changes.[36]

In contrast, there are well-documented examples of excessive fishing mortality and natural predation as the principal sources of decline in biomass yields of several species around the globe. The example of predator interaction among mackerel, herring, and sand lance has been reported for stocks in the Northeast Continental Shelf of the United States.[37] The biomass of commercially important fish stocks of the Northeast Continental Shelf Ecosystem declined by approximately 50% between 1968 and 1975. The principal cause of the loss of biomass is thought to be excessive fishing mortality in juvenile and adult stocks.[38] The predominance of the Atlantic herring, *Clupea harengus*, of Georges Bank "flipped" with sand eel, *Ammodytes* species. A biomass flip occurs when the population of a dominant species rapidly drops to a very low level and is replaced by a second species. Sand eel, herring, and mackerel inhabit, at least for part of the year, the same areas on Georges Bank and the Southern New England Continental Shelf. Evidence of herring predation on sand eel, and mackerel predation on the early developmental stages of both species has been observed on the Northeast Continental Shelf where the

32. Minoda, Terazaki.

33. Canon.

34. MacCall.

35. Wyatt and Perez-Gandaras.

36. Crawford et al.

37. Kenneth Sherman, Cynthia Jones, Loretta Sullivan, Wallace Smith, Peter Berrien, and Leonard Ejsymont, "Congruent Shifts in Sand Eel Abundance in Western and Eastern North Atlantic Ecosystems," *Nature* 291 (1981): 486–89; Kenneth Sherman, "Large Marine Ecosystems as Global Units for Recruitment Experiments," in *Toward a Theory on Biological-Physical Interactions in the World Ocean*, ed. B. J. Rothschild (Dordrecht: Kluwer Academic Publishers, 1988), pp. 459–76.

38. Clark and Brown (n. 10 above).

distribution of the three species overlap. Also it is known that sand eel prey on herring larvae. Christensen, in 1983, demonstrated in laboratory studies that sand eel fed preferentially on herring larvae. Fuiman and Gamble, in 1988, showed that both herring and sand eels preyed on herring larvae in large enclosures. More recently, in 1989, Rankine and Morrison reported sand eel feeding on herring eggs and larvae on an important coastal spawning ground for herring in the North Sea.[39] It is possible, in the absence of any prolonged environmental signal, that the decline in both herring and mackerel stocks during the mid-1970s released predation pressure on sand eel and allowed the population to explode.[40] Fishing mortality has been reduced in herring and mackerel stocks since the mid-1970s. No fishery exists for sand eel. It appears that the reduction of fishing mortality in mackerel and herring has allowed the stocks to begin a recovery trend. The present biomass of mackerel is estimated at 1.4 Mmt and is increasing. Also, evidence of herring returning to Georges Bank has been reported from the recent discovery of juveniles in stomachs of spiny dogfish, *Squalus acanthius,* and other predators. Unlike the North Sea, for which recent climatic changes have been reported,[41] no long-term climatic change has been observed for the Northeast Continental Shelf Ecosystem, and no declining trend in primary production, or zooplankton biomass, has been detected.[42] The energetics of the shelf ecosystem appear tightly bound in relation to fish production. Recent estimates on the Georges Bank subarea place fish predation on fish at 70% of annual production, fol-

39. R. Maurer, "A Preliminary Analysis of Inter-specific Trophic Relationships between the Sea Herring *Clupea harengus* Linnaeus and the Atlantic Mackerel *Scomber scombrus* Linnaeus," *International Commission for the North Atlantic Fisheries,* Research Document 76/VI/121, ser. 3967 (1976), pp. 22; M. D. Grosslein, R. W. Langston, and M. P. Sissenwine, "Recent Fluctuations in Pelagic Fish Stock of the Northwest Atlantic Georges Bank Region, in Relationship to Species Interactions," *Rapports et procès-verbaux des Réunions Conseil International pour l'Exploration de la Mer* (ICES) 177 (1980): 374–404; V. Christensen, "Predation by Sand Eel on Herring Larvae," *International Counsel for the Exploration of the Sea* (ICES), Pelagic Fish Committee C.M. 1983/L:27 (1983); L. A. Fuiman and J. C. Gamble, "Predation by Herring, Sprat, and Sand Eels on Herring Larvae in Large Enclosures," *Marine Ecology Progress Series* 44 (1988): 1–6; P. W. Rankine and J. A. Morrison, "Predation on Herring Larvae and Eggs by Sand Eels (*Ammodytes marinus* [Rait] and *Hyperoplus lanceolatus* [Lesauvage]), *Journal of the Marine Biological Association of the United Kingdom* 69, no. 2 (1989): 493–98. EDITORS' NOTE—For information about North Sea herring see James R. Coull, "The North Sea Herring Fishery in the Twentieth Century," *Ocean Yearbook 7,* eds. Elisabeth Mann Borgese, Norton Ginsburg, and Joseph R. Morgan (Chicago: University of Chicago Press, 1987), pp. 115–131.

40. Sherman, Jones, et al.

41. D. J. Garrod and J. M. Colebrook, "Biological Effects of Variability in the North Atlantic Ocean," *Rapports et procès-verbaux des Réunions Conseil International pour L'exploration de la Mer* 173 (1978): 128–44.

42. Kenneth Sherman, John R. Green, Julien R. Goulet, and Leonard Ejsymont, "Coherence in Zooplankton of a Large Northwest Atlantic Ecosystem," *Fisheries Bulletin, U.S.* 81 (1983): 855–62.

lowed by approximately 10% as fisheries yield, 10% consumed by marine mammals, 5% by marine birds, and 5% by apex predators, including sharks, tunas, and billfish.[43]

The 40% decline in the demersal finfish yields of the Yellow Sea have been attributed to excessive fishing mortality, considered as fishing predation.[44] Within the Yellow Sea Ecosystem, Chinese scientists are presently successfully experimenting with the grow-out of juvenile fleshy prawn into a system where the natural predator field has been reduced through excessive fishing predation.[45]

On a global scale, the loss of sustained biomass yields from LMEs from mismanagement and overexploitation has not been fully investigated. However, it is clear that "experts" have been off the mark in earlier estimates of global yield of fisheries biomass. Projections given in *The Global 2000 Report* indicated that the world annual yield was expected to rise little, if at all, by the year 2000 from the 60 Mmt reached in the 1970s.[46] In contrast, estimates given in *The Resourceful Earth* argue for an annual yield of 100–120 Mmt per year by the year 2000.[47] The trend is upward; the 1986 level of global fishery yields are at 91.5 Mmt.[48] The lack of a clear definition of actual and/or potential global yield is not unexpected, given the meager efforts presently underway to improve the information base on living marine resource yields from an ecosystem perspective. The topic is of particular relevance in relation to a growing awareness of global change.[49]

ECOSYSTEM MANAGEMENT AND THE ANTARCTIC

On a global basis, the LMEs represent geographically distinct units for managing living marine resources.[50] The Convention of the Conservation of Antarctic Marine Living Resources (CCAMLR) is an international agreement

43. Sissenwine (n. 10 above).

44. Tang (n. 25 above).

45. Ibid.

46. U.S. Council on Environmental Quality and the Department of State, Gerald O. Barney, director, *The Global 2000 Report to the President: Entering the Twenty-first Century* (Washington, D.C.: Government Printing Office, 1980), vols. 1–3.

47. John P. Wise, "The Future of Food from the Sea," in *The Resourceful Earth,* J. L. Simon and H. Kahn eds. (New York: Basil Blackwell, 1984), pp. 113–27.

48. Food and Agriculture Organization of the United Nations (FAO), *Yearbooks of Fishery Statistics* (Rome: FAO, 1988), vol. 62.

49. P. H. Abelson, "The International Geosphere-Biosphere Program," *Science* 234 (1986): 657 ff.

50. Alexander, "Introduction to Part Three" (n. 3 above); Martin Belsky, "Legal Constraints and Options for Total Ecosystems Management of Large Marine Ecosystems," in *Variability and Management* (n. 2 above), pp. 145–74; Morgan, "Large Marine Ecosystems" (n. 3 above); and Prescott (n. 5 above).

that supports an ecosystem approach to the conservation and management of living resources found in ocean areas surrounding Antarctica.[51] The convention mandates a management regime committed to applying measures to ensure that harvesting of Antarctic species, such as finfish and krill, is conducted in a manner that considers ecological relationships among dependent and related species. The implementation of CCAMLR is carried out against a background of enlightened international activities in Antarctica that, in recent decades, have been concerned with scientific research and cooperation, demilitarization, denuclearization, resource utilization, and environmental protection. The parties to the convention have conducted their activities under the system of legal, political and scientific relationships established by the Antarctic Treaty of 1959.[52]

The CCAMLR was negotiated from 1977 to 1980, entering into force in 1982. The CCAMLR convention area includes the marine area south of the Antarctic Convergence, the boundary between 48 and 60 degrees south, separating the cold Antarctic waters and the warmer subantarctic waters. South of this boundary is defined as the Antarctic Marine Ecosystem. The convention applies to "the populations of finfish, mollusks, crustaceans, and all other species of living organisms, including birds, found south of the Antarctic Convergence."

Member countries of the CCAMLR have established an organizational structure to assist them in the conservation and management of the Antarctic marine ecosystem. The major operational units of the CCAMLR system are the Commission for the Conservation of Antarctic Marine Living Resources (the "commission"), and the Scientific Committee for the Conservation of Antarctic Marine Living Resources (the "scientific committee"). A secretariat resides at CCAMLR headquarters in Hobart, Tasmania, Australia. Its function is to serve the commission and the scientific committee of the CCAMLR, including organizing the annual meetings and acting as a clearinghouse for communication with member countries.

Members of the Commission for the Conservation of Antarctic Marine Living Resources are Argentina, Australia, Belgium, Brazil, Britain, Chile, the European Community, East Germany, France, India, Japan, New Zealand, Norway, Poland, South Africa, South Korea, the Soviet Union, Spain, the United States, and West Germany.

The functions of the commission are to (*a*) facilitate study of Antarctic marine living resources and the ecosystem of which they are a part; (*b*) compile data on the status of, and changes in, the distribution, abundance, and

51. *International Legal Materials*, 19:841. For CCAMLR information, contact Dr. Darry L. Powell, Executive Secretary, CCAMLR, 25 Old Wharf, Hobart, Tasmania 7000, Australia.

52. The Antarctic Treaty, signed December 1, 1959, *United Nations Treaty Series*, 402:71.

productivity of harvested and dependent or related species and populations of Antarctic marine living resources; (*c*) ensure the acquisition of catch and effort statistics; and (*d*) formulate, adopt, and revise conservation measures on the basis of the best scientific information available.

The commission has met seven times (at the time of writing). The first and second meetings were largely organizational. The third meeting, in September 1984, produced the first conservation measures for depleted stocks of finfish, and a program of data gathering and consideration of conservation options was initiated. The fourth meeting, convened in September 1985, followed initial mesh-regulation measures for aiding the recovery of fish stocks with the adoption of more stringent regulations prohibiting all directed fisheries for the bottom-living species of Antarctic cod, *Notothenia rossii*, in the waters of South Georgia, the South Orkneys, and the Antarctic Peninsula. The fifth meeting in 1986 adopted conservation measures prohibiting fishing for the severely depleted Antarctic cod and permitting the commission to fix catch limitations as a management technique.

The sixth meeting in 1987 established new conservation measures to address the serious depletion of fish stocks. Three measures of significance were taken for the first time—an overall total allowable catch, a reporting system, and a closed season. In addition, a new working group was established to implement, coordinate, and evaluate research on the distribution and abundance of krill.[53]

The seventh meeting of CCAMLR in 1988 reached a milestone based on information received during the meeting, indicating that the total allowable catch levels for *Champsocephalus gunnari* were exceeded by 121 mt. The commission took immediate steps to close the fishery within 24 hours, and this was achieved in large part because of the excellent fish stock assessments produced by the scientific committee.

CONCLUDING REMARKS

As nations move from single-species management to multispecies fisheries management, it will become necessary to provide greater consideration of the resources and the impacts of natural and human perturbations on the resources within marine ecosystems. The management regime presently in place in CCAMLR reflects this trend.

Within several of the LMEs, important hypotheses concerned with the growing impacts of pollution, overexploitation, and environmental changes on sustained biomass yields are under investigation (table 1). By comparing

53. Editors' note.—For more information about krill, see John E. Bardach, "Fish Far Away: Comments on Antarctic Fisheries," in *Ocean Yearbook 6*, ed. Elisabeth Mann Borgese and Norton Ginsburg (Chicago: University of Chicago Press, 1986), pp. 38–54.

TABLE 1.—SELECTED HYPOTHESES CONCERNING VARIABILITY IN
BIOMASS YIELDS OF LARGE MARINE ECOSYSTEMS PRESENTLY UNDER
INVESTIGATION

Ecosystem	Predominant Variables	Hypothesis
Oyashio[a] Kuroshio[b] California Current[c] Humboldt Current[d] Benguela Current[e] Iberian Coastal[f]	Density independent: natural environmental perturbations	*Clupeoid population increases:* Predominant variables influencing changes in biomass of clupeoids are major increases in water-column productivity resulting from shifts in the direction and flow velocities of the currents and changes in upwelling within the ecosystem.
Yellow Sea[g] U.S. Northeast Continental Shelf[h] Gulf of Thailand[i]	Density dependent: predation	*Declines in fish stocks:* Precipitous decline in biomass of fish stocks is the result of excessive fishing mortality, reducing the probability of reproductive success. Losses in biomass are attributed to excesses of human predation expressed as overfishing.
Great Barrier Reef[j]	Density dependent: predation	*Change in ecosystem structure:* The extreme predation pressure of crown-of-thorns starfish has disrupted normal food-chain linkage between primary production and the fish component of the reef ecosystem.
East Greenland Sea[k] Barents Sea[l] Norwegian Sea[m]	Density independent: natural environmental perturbations	*Shifts in abundance of fish stock biomass:* Major shifts in the abundance levels of fish stock biomass within the ecosystems are attributed to large-scale environmental changes in water movements and temperature structure.
Baltic Sea[n]	Density independent: pollution	*Changes in ecosystem productivity levels:* The apparent increases in productivity levels are attributed to the effects of nitrate enrichment resulting from elevated levels of agricultural containment inputs from the bordering land masses.

TABLE 1.—(*Continued*)

Ecosystem	Predominant Variables	Hypothesis
Antarctic Marine[o]	Density dependent:	*Status of krill stocks:* Annual natural production cycle of krill is in balance with food requirements of dependent predator populations. Surplus production available to support economically significant yields is (is not?) within acceptable fisheries effort levels.
	Density independent: natural environmental perturbations	*Shifts in abundance in krill biomass:* Major shifts in abundance levels of krill biomass within the ecosystem are attributed to large-scale changes in water movements and other physical and chemical perturbations including ozone depletion and global warming.

[a] Takashi Minoda, "Oceanographic and Biomass Changes in the Oyashio Current Ecosystem," in *Biomass Yields and Geography of Large Marine Ecosystems*, ed. Kenneth Sherman and Lewis M. Alexander, AAAS Selected Symposium Series 111 (Boulder: Westview Press, 1989), hereafter called *Biomass Yields and Geography*.

[b] Makoto Terazaki, "Recent Large-Scale Changes in the Biomass of the Kuroshio Current Ecosystem," in *Biomass Yields and Geography*.

[c] Alec D. MacCall, "Changes in the Biomass of the California Current Ecosystem," in *Variability and Management of Large Marine Ecosystems*, ed. Kenneth Sherman and Lewis M. Alexander, AAAS Selected Symposium 99 (Boulder: Westview Press, 1986), pp. 33–54.

[d] J. R. Canon, "Variabilidad ambiental en relacion con la pesqueria neritica pelagica de la zona Norte de Chile," *La pesca en Chile*, ed. P. Arana (Valparaiso, Chile: Universidad Catolica de Valparaiso, Facultad de Recursos Naturales, Escuela de Ciencias del Mar, 1986), 359 pp.

[e] Robert J. M. Crawford, L. V. Shannon, and P. A. Shelton, "Characteristics and Management of the Benguela as a Large Marine Ecosystem," in *Biomass Yields and Geography*.

[f] Tim Wyatt and G. Perez-Gandaras, "Biomass Changes in the Iberian Ecosystem," in *Biomass Yields and Geography*.

[g] Qisheng Tang, "Changes in the Biomass of the Huanghai Sea Ecosystem," in *Biomass Yields and Geography*.

[h] Michael P. Sissenwine, "Perturbation of a Predator-controlled Continental Shelf Ecosystem," in *Variability and Management*, pp. 55–85.

[i] Twesukdi Piyakarnchana, "Yield Dynamics as an Index of Biomass Shifts in the Gulf of Thailand Ecosystem," in *Biomass Yields and Geography*.

[j] Roger H. Bradbury and C. N. Mundy, "Large-Scale Shifts in Biomass of the Great Barrier Reef Ecosystem," in *Biomass Yields and Geography*.

[k] Erik Buch and Holger Hovgaard, "Decadal Biomass Yields and Environmental Perturbations in the West Greenland Sea Ecosystem," in *Patterns, Processes, and Yields of Large Marine Ecosystems*, AAAS Symposium Series (Washington, D.C.: AAAS, in press).

[l] Hein Rune Skjoldal and Francisco Rey, "Pelagic Production and Variability of the Barents Sea Ecosystem," in *Biomass Yields and Geography*.

[m] Bjornar Ellertsen, Petter Fossum, Per Solemdal, Svein Sundby, and Snorre Tilseth, "Environmental Influence on Recruitment and Biomass Yields in the Norwegian Sea Ecosystem," in *Patterns, Processes, and Yields of Large Marine Ecosystems*, AAAS Symposium Series (Washington, D.C.: AAAS, in press).

[n] Gunnar Kullenberg, "Long-Term Changes in the Baltic Ecosystem," in *Variability and Management*, pp. 19–31.

[o] Scientific Committee for the Conservation of Antarctic Marine Living Resources, Report of the Seventh Meeting of the Scientific Committee, Hobart, Australia, 24–31 October, 1988, SC-CAMLR-VII.

the results of research among the different systems, it should be possible to accelerate an understanding of how the systems respond and recover from stress. A hierarchical approach to comparative ecosystem studies is given in the Appendix. The comparisons among LMEs should allow for narrowing the context of unresolved problems and capitalizing on research efforts underway in the different ecosystems.

APPENDIX

ECOSYSTEM RESEARCH AND DEVELOPMENT STRATEGY FOR U.S. FISHERIES

	Spatial	*Temporal*	*Unit*
1.	Global (world ocean)	Millennia-decadal	Pelagic biogeographic
2.	Regional (Exclusive Economic Zones)	Decadal-seasonal	divisions
3.	Local	Seasonal-daily	Large marine ecosystems
			Subsystems

4. *Biological elements:*
    4.1 Spawning strategies
    4.2 Feeding strategies
    4.3 Productivity, trophodynamics
    4.4 Stock fluctuations/recruitment/mortality
5. *Environmental elements:*
    5.1 Natural variability:
        Hydrography
        Currents
        Water masses
        Weather
    5.2 Human Perturbations:
        Fishing
        Waste disposal
        Petrogenic hydrocarbon impacts
        Aerosol contaminants
        Eutrophication effects
6. *Options and advice, international, national, and local:*
    6.1 Bioenvironmental and socioeconomic models
    6.2 Predictions to optimize fisheries yields
7. *Feedback loop:*
    7.1 Evaluation of ecosystem status
    7.2 Evaluation of fisheries status
    7.3 Evaluation of management practices

Planning for Offshore Petroleum Development in Newfoundland[1]

Irene M. Baird
Consultant, Newfoundland

INTRODUCTION

Newfoundland is Canada's most easterly province and consists of two areas, the island of Newfoundland, and the larger, less densely populated mainland portion, Labrador, which shares a common border with the Province of Quebec. It is located approximately 2,800 km west of Europe and 1,600 km northeast of New York.

Newfoundland was the first part of North America to be colonized by Europeans and Britain's oldest colony up to April 1, 1949, when the Dominion of Newfoundland became the tenth province of the Dominion of Canada.

Union with Canada brought about many changes in Newfoundland. Gone was independence as a dominion and the protective customs tariff. Many small manufacturing industries that had provided local employment did not survive mainland competition. The new province provided a captive market for central Canadian manufacturers, and the steady inflow of federal funds to the provincial government made possible considerable improvements in living standards and public services. Trade patterns changed, and, over time, Newfoundlanders were introduced to family allowances, old-age pensions, unemployment insurance, and other Canadian social welfare programs.

Newfoundland had enjoyed prosperity during and after the Second World War, largely as a result of defense spending by the United States and Canada and a revival of its export industries. These activities created a strong economy and a favorable government financial position at the time of confederation in 1949. After confederation prosperity continued, but the new prosperity was related not to economic development but rather to large infusions of transfer payments from the national government. Unemployment was per-

1. Irene Baird, Assistant Deputy Minister (Petroleum and Energy Programs), Newfoundland Department of Energy, presented a paper on which this article is based at the International Ocean Institute, Dalhousie University, Halifax, Nova Scotia, on August 3, 1988.

sistently higher and incomes lower than in other provinces of Canada. Economic development in the first 2 decades of confederation tended to ignore natural resources and concentrated instead on efforts to develop an industrial/manufacturing sector. Federal economic policies in the meantime did little to enhance development. The fishery, which had been the backbone of the Newfoundland economy for generations, was allowed to decline, and trade policies favored the manufacturing industries concentrated in central Canada.

It was against this background that in the seventies there was a growing awareness that all was not well within the province and that a steady deterioration in the social fabric of the province had commenced. New directions were needed if Newfoundland was to achieve financial and economic viability within the Canadian confederation. Resource ownership and control and sound management of natural resources became the cornerstone of a planning effort to address the twin evils of high unemployment and lower per capita earned incomes. That planning effort began in the oil-and-gas sector in the mid-1970s, and a white paper was published in 1977.[2] The planning effort in relation to other sectors of the economy became more focused in the 1978 *Blueprint for Development* and was further expanded in a 5-year development plan covering the period 1980–85.[3] Negotiations with the federal government concerning ownership and management of oil-and-gas resources began earnestly in 1981, and that planning effort is reviewed in detail in this article. First, the white paper and other planning documents are reviewed.

WHITE PAPER

Despite its great natural resource endowment, Newfoundland has had a bleak economic history. Energy in the form of oil and gas was a new resource, and the government and public servants involved in the planning were determined that the benefits of the resource would accrue to the people of the province. It is interesting to note that the premier had been the minister responsible for mines and energy resources in the mid-1970s, and it was during his term of office as minister that the Department of Mines and Energy set the tone and themes that have prevailed throughout all the negotia-

2. "A White Paper Respecting the Administration and Disposition of Petroleum Belonging to her Majesty in Right of the Province of Newfoundland," issued under the authority of A. Brian Peckford, Minister of Mines and Energy, May 1977 (no government reference number issued). See table 1.

3. *Blueprint for Development: Supplement to the 1978 Budget,* Province of Newfoundland, March 1978 (no government reference number issued). *Five-Year Development Plan: Managing All Our Resources,* a development plan for Newfoundland and Labrador, 1980–85, October 1980 (Saint John's: Newfoundland Information Services, 1980).

TABLE 1.—GOALS AND ACHIEVEMENTS: PETROLEUM PLANNING PROCESS

	Goals		Achievements	
	1981 Framework for Agreement	1982 Proposal for Settlement	1985 Atlantic Accord	1988 Hibernia Agreement
Provincial ownership[a] Provincial management:	Province offered to set aside Joint management and revenue sharing	Prepared to set aside Joint management and revenue sharing	X[b] Joint management	X
(a) Province to decide when, where, and under what conditions exploration and production rights to be granted.			Joint management	
(b) In view of need for employment and increased public revenues to promote rapid rate of offshore exploration and start-up of a moderate rate of production at the earliest possible, environmentally safe date.				
(c):				
(i) Maximize revenue from highly profitable fields yet allow marginal fields to remain attractive.				X
(ii) Utilize cash bonus, rental, or royalties sparingly in favor of flexible revenue generating mechanisms as government participation.				
(iii) Back-end loading				X

(d):

(i) Commitments in percentage terms for use of local labor, goods and services and further processing within the province where they are competitive in price, quality, and delivery

(ii) Minimum expenditures on oil and gas related R&D — Center of excellence for R&D related to offshore activity

(iii) Establishment of headquarters-type activities within the province

(iv) Ensure that local labor and contractors are given a fair opportunity to participate.

Preference for Newfoundland labor, goods, and services

Training and R&D programs in the province

Landing in province of oil and gas produced offshore

Minimum targets for expenditures within the province

Preference for local refining, processing, and consumption

Provincial control of the rate of development

Preference for a political rather than a legal settlement — Negotiated rather than legal statement

Prevention of inflationary and social disruptive rate of development

TABLE 1.—(*Continued*)

	Goals		Achievements	
	1981 Framework for Agreement	1982 Proposal for Settlement	1985 Atlantic Accord	1988 Hibernia Agreement
Environmental protection:				
(a) Full examination and public hearings for drilling			X	
(b) Full examination and public hearings for development			X	
Local involvement in planning of offshore activity			X	
Application of province's social and fiscal legislation		Understood	X	
Proper administration favored instead of proposed Federal Regulations (1977):				
(a) High degree of financial security			X	
(b) Little ministerial discretion			X	
(c) Arbitration procedure			X	
(d) Smaller bureaucracy			X	
Provision of all pertinent information to government			X	
State participation (i.e., profit sharing)			X	
Split of total government take 75%/25% in favor of the province	Province revenues as if resource on land	Equitable revenue sharing with 75% to province until fiscal and economic maturity is achieved, then the percentage of revenue sharing to the province will decrease.		X

Item	Treatment
Ensure that offshore development fits into the province's general industrial priorities	Emphasized X
Newfoundland government favored:	
(a) Lifting the restrictions on domestic oil and gas prices	(Done outside the accord)
(b) Permitting large-scale exports to United States when there is a surplus to Canadian needs	(Done outside the accord)
Self-reliance for Canada in energy production	
Application of provincial corporate income tax	X
Stable regulatory environment	X
Compensation to fishermen	X
Equalization reduced more slowly	X
Agreement be entrenched in Constitution	X
Single set of regulations	X
Federal government to provide funds to remove constraints in social capital	X
1. Fiscal self-sufficiency	
2. Income and employment consistent with mature economy	X
3. Social capital and services equal to that in other provinces	X
4. Development of human resources	X
5. Reasonable rate of development	X
6. Distribution of economic growth	X
Identical legislation by each order of government	X

The future will judge our success relative to these goals

TABLE 1.—*(Continued)*

	Goals		Achievements	
	1981 Framework for Agreement	1982 Proposal for Settlement	1985 Atlantic Accord	1988 Hibernia Agreement
Management system		(a) Joint mgmt. agency	X	
		(b) Joint legislation and made permanent	X	
		(c) 7-member board	X	
		(d) Regulations agreed to by both governments	X	
		(e) Separate authority for key decisions	X	
		(f) Agency impartial while government would retain necessary degree of control	X	
		(g) Regulations changed only by mutual consent	X	
		(h) Budget jointly approved	X	
		(i) Geological and other information to be shared with governments	X	
		(j) Agency to be located in the province	X	
		Newfoundland and Labrador Petroleum Corporation to share up to 40% interest		
		Equal assignment of "ownership type" privileges between federal and provincial crown corporations	X	

[a] Federal ownership confirmed by the Supreme Court of Canada.
[b] X indicates the presence of qualities described.

tions with the federal government and, more recently, with the oil companies as well.

The white paper outlined Newfoundland's policy objectives for the development of the offshore petroleum resource. Among the more important of these objectives were the control of the rate of development, preference for local goods and services, maximization of revenues to the province, and provision of training, research, and development programs, as well as the encouragement of further processing in the province.

In order to ensure that these objectives were attained, the province promulgated appropriate oil and natural gas regulations in October 1977 to set the terms and conditions under which the oil industry would be granted rights to explore and exploit these offshore resources. Up to that time no discovery had been made. That did not occur until August 1979, but Newfoundland was quietly forging ahead, getting ready for the day when the discovery would be made.

BLUEPRINT FOR DEVELOPMENT, 1978

The *Blueprint for Development* was also prepared prior to the Hibernia discovery in 1979. It noted the high probability that commercial quantities of oil and gas would be found and developed on the province's Continental Shelf. It also pointed out that it was essential that the province have sufficient control to insure that development and production proceed in a manner and at a rate that would optimize the economic benefits arising out of exploration of the resource, while minimizing the social and environmental disruptions that had occurred elsewhere in the world concurrent with developments of this kind.

The *Blueprint for Development* also discussed the linkage between oil and gas and an ocean-related industry base in the province. It noted that exploitation of both the fishery and oil-and-gas sectors in Northern areas, in ice-infested deep waters, and under conditions unique to Canada, would require sophisticated and innovative technology developments. The government recognized the role that science and technology would inevitably play in the development of a strong ocean industry sector and was determined that the province would participate in the technological challenges that development of these resources would present, not only to achieve economic benefits but also to provide rewarding job opportunities for an increasingly better educated and more demanding labor force. To this end, the government has established or supported such institutions as the Newfoundland Oceans Research and Development Corporation, the Centre for Cold Oceans Resources Engineering, the Marine Institute, a naval architectural program at the Memorial University Engineering Department, and model facilities for testing ice-breaking vessels. Today the province is already recognized internationally for its cold oceans research, and efforts continue to make Newfoundland a center of excellence in cold oceans research and engineering.

THE 5-YEAR PLAN (1980–85)

A document entitled *Managing All Our Resources* described the government's concerns and plans for the period 1980–85. With respect to the oil-and-gas sector, the policies of the white paper were reaffirmed, emphasis was placed on control and ownership of offshore petroleum resources, and attempts were made to influence Canadian petroleum pricing policies. The latter was an important objective because if Canada chose to maintain Canadian crude oil prices below the world price, potential provincial revenues would be slashed drastically. Furthermore, the lower the price the less the amount of petroleum that would be commercially recoverable offshore, resulting in lower economic rent. As well, the price of oil has a comparatively greater impact on high-cost offshore oil producers than it has on lower cost onshore operations. The higher cost offshore oil has proportionately less economic rent to be divided among the company, the owner of the resource, and corporate taxes. The objective of influencing Canadian petroleum pricing policies was part of an overall objective to influence national policies that affected natural resources.

The 5-year plan established the Offshore Petroleum Impact Committee (OPIC), which was charged with coordinating the activities of five special committees, each examining a separate aspect of development. These were:

1. the Development Planning Committee;
2. the Education and Training Committee;
3. the Fisheries and Environment Committee;
4. the Social and Cultural Committee;
5. the Finance Committee.

These committees drew their membership from the most senior levels of the relevant departments and agencies of the provincial public service. To support the work of these committees and to ensure an input from the private sector, a separate advisory council was appointed for each committee. The members of these advisory councils represented all regions of the province and were drawn from all walks of life.

The government's position in the 1980 5-year plan relative to oil and gas was essentially unchanged from 1977, but increased emphasis was being placed on certain key objectives, and structures were beginning to crystallize to deal with the effects of what was perceived to be imminent development.

Planning for Offshore Negotiations

When the negotiations began in 1981, the senior public service was mobilized to carry out an intensive research effort and to flesh out the objectives set out in the white paper 4 years before. It was recognized that the negotiations regarding ownership and joint management would have far-reaching implications irrespective of the outcome. On the one hand, a settlement that embodied the province's goals and objectives would have a positive impact on the fabric of Newfoundland society in general. Conversely, should the province's aspirations be put aside, it was likely that Newfoundland would not be able to overcome its economic problems, and economic self-sufficiency would become an unattainable goal.

Petroleum is a nonrenewable resource and it cannot, in the long term, sustain economic well-being. This point was fundamental to the province's stance in the negotiations.

The federal government of the day had clearly espoused an overriding energy objective that sought to achieve energy self-sufficiency on a national basis by 1990. It was also evident that the government of Canada was seeking to rectify its fiscal imbalance through the development of Canadian sources of petroleum. On the other hand, Newfoundland's objectives were much broader and encompassed social, economic, financial, and environmental considerations.

In Canada, the economic policy tools available to the federal government are numerous and allow that body to exercise a great deal of leverage over both the national economy and regional components. The federal government can establish targets and achieve these targets through the use of fiscal, monetary, and other policy instruments. Provinces, however, have few policy instruments. In Newfoundland's case, social and economic objectives could be achieved only through the wise management of provincial natural resources. Unless the province could exercise management prerogatives with respect to offshore petroleum, it was unlikely that it would be able to achieve its social and economic objectives because other policy instruments were, and still are, lacking. In an effort to reach a settlement on the offshore resources, the province agreed to put its claims to ownership aside in favor of a political settlement. It sought to achieve its objectives through a joint management agreement with the federal government and by increased efforts to influence national policies.

Development Phase
The direct impact of offshore expenditures during the development phase will be examined. Opportunities will be open to the Newfoundland private sector to seek economic benefits. These benefits will be in the form of higher incomes and greater job opportunities for both private firms and individuals.

A large number of business opportunities will exist and new jobs will be created. If Newfoundlanders are to take full advantage of these opportunities, they must compete with other business enterprises on a worldwide basis. Similarly, Newfoundland workers must be prepared to deliver a high level of efficiency and productivity in order to compete with experienced personnel available throughout the world.

In a small, underdeveloped economy like Newfoundland, expansion in a high-value, capital-intensive sector such as the petroleum industry will substantially increase the gross domestic product (GDP), but unless special measures are taken, it will not, in the short-to-medium term, lead to significant expansion in other sectors of the economy. Indeed, certain competitive pressures, particularly for labor, will be brought to bear upon other sectors of the economy, especially the fishing industry. A great deal of wealth in the oil sector will leave the province, and it was felt that if economic growth was narrowly based in that sector, underdevelopment and unemployment would continue to exist in other sectors of the economy unless specific measures were taken to address these long-standing problems.

The Production Phase

The impact during the production phase will come primarily through government revenues, both federal and provincial. In magnitude, the revenue impact associated with the production phase is much greater than the direct impact associated with the development phase. Ongoing activity in the private sector associated with production will be minimal. However, the government will have access to much larger revenues and will have the freedom to reduce the level of public debt, to reduce taxes and perhaps even build up a heritage fund. The reduction in levels of taxation will, in itself, create a more favorable economic climate and will encourage the expansion of private investment.

COMPREHENSIVE DEVELOPMENT

The Newfoundland government concluded that development and production of offshore oil required a commitment at the outset, by both levels of government, to an economic development program designed to maximize the absorptive capacity of the Newfoundland economy. The industrial development and industrial benefit package to be negotiated with the federal government would, therefore, have to be one that enhanced the long-term viability of the province. It was felt that an industrial benefit package should be sought that would lend support to all sectors of the provincial economy and not simply to those sectors involved directly in oil field development. Newfoundland has the least developed economy in Canada, and, given the economic disparities that existed then (and now) between the province and the rest of the country, it was clear that petroleum developments should be used as the impetus for a program of economic growth that would pervade all sectors of the economy and continue beyond the oil era.

Fundamental to this approach was the assumption that both governments would commit themselves to a comprehensive program that would cover the development and production phases of offshore oil fields. It was felt that this comprehensive program should extend for a period of 20 years or longer, and that the revenues and direct economic opportunities should be used to create full employment. Development would be predicated upon the need to achieve economic viability without creating an inflated and swollen public sector and without subsidizing the Newfoundland economy through the use of provincial oil revenues.

ECONOMIC RENEWAL PROGRAM

To this end, the framework of an Economic Renewal Program was formulated to assist in determining the kinds of socioeconomic benefits that the province should pursue in the negotiations.[4] The program consisted of the following:

1. human resource development;
2. industrial benefits;
3. social capital; and
4. environmental concerns.

Economic Renewal Program[5]

Objectives
The objectives sought through economic renewal encompassed financial, economic, and social progress; they were stated as follows:

1. to achieve fiscal self-sufficiency in the long term through a viable diversified economy;
2. to attain a high level of income and employment consistent with a mature economy;
3. to provide a level of social capital and services equal to that already attained in other provinces, and hence already enjoyed by other Canadians;

4. For economic indicators, see table 2, "Earned Income per Capita—Canada and Newfoundland," and table 3, "Comparative Economic Indicators—Canada and Newfoundland."

5. "Socioeconomic Dimensions of Offshore Oil Development," an internal document prepared for Newfoundland's Offshore Negotiating Team, December 1981.

TABLE 2.—EARNED INCOME PER CAPITA, CANADA AND
NEWFOUNDLAND (SELECTED YEARS) 1981 CONSTANT CANADIAN
DOLLARS

Year	Canada ($)	Newfoundland ($)	Newfoundland as % of Canada
1949	3,795	1,891	49.8
1950	3,833	1,905	49.6
1955	4,384	2,186	49.8
1960	4,738	2,402	50.6
1965	5,794	3,098	53.4
1970	7,015	4,063	57.9
1975	9,152	5,284	57.7
1980	10,219	5,759	56.3
1981	10,610	5,850	55.1
1982	10,324	5,893	57.0
1983	10,089	5,722	56.7
1984	10,445	5,829	55.8
1985	10,816	5,985	55.3
1986	11,031	7,220	65.4

4. to develop our human resources so that the skills of our labor force match the employment available in all sectors of the economy;
5. to achieve a reasonable rate of development, economic growth, and expansion in order to permit local business and indigenous industry to grow and respond to the opportunities that a revitalized economy will present; and
6. to distribute economic growth throughout the province so that existing towns and communities remain viable entities in an economically healthy Province, even after the depletion of nonrenewable offshore mineral resources.

Human Resource Development

Manpower Supply and Demand
By Canadian standards, the Newfoundland labor market has historically been characterized by extremely high, and sometimes rapidly changing, unemployment rates and extremely low labor-force participation rates. In addition, there is much seasonality of employment and a dependence on unemployment insurance and other forms of income maintenance. In 1980, the proportion of Newfoundland's unemployment insurance contributors who drew benefits at some point during the year was 55.3%, compared to 23.9% for Canada as a whole. By 1986 these figures have increased to 59.6% for Newfoundland and 25.7% for Canada.[6]

6. See Table 4, "Per Capita Expenditure and Indicators for Health, Education, and Social Welfare."

TABLE 3.—COMPARATIVE ECONOMIC INDICATORS: NEWFOUNDLAND AND CANADA, 1949, 1980, AND 1986

	1949			1980			1986		
	Newfoundland	Canada	Newfoundland as % of Canada	Newfoundland	Canada	Newfoundland as % of Canada	Newfoundland	Canada	Newfoundland as % of Canada
Gross Domestic Product at market prices per capita ($)	600	1,279	46.9	6,317	12,445	50.8	11,937	20,030	59.5
Average family income (Census Families) ($)	3,592[a]	4,906[a]	73.2	19,052[c]	26,104[c]	73.0	27,678	39,589	69.9
Earned income per capita ($)	459	926	49.6	4,586	8,655	53.0	8,202	14,605	56.1
Transfer payments per capita ($)	55	70	78.6	1,757	1,267	138.7	3,412	2,444	139.6
Total personal income per capita ($)	507	996	50.9	6,343	9,913	64.0	11,620	17,060	68.1
Gross investment per capita ($)	136	374	36.4	2,512	3,506	71.6	4,538	4,931	92.0
Retail sales per capita ($)	421[b]	763[b]	55.2	2,583	3,514	73.5	4,235	5,522	76.6
Employment ratio (employed as a % of population 15 years or older)	45[c]	55[c]	81.8	45.9	59.2	77.5	53.0	65.7	80.6
Unemployment rate (%)	12.4	2.0	620.0	13.5	7.5	180.0	20.0	9.6	208.3

NOTE.—($) indicates Canadian dollars.
[a] 1961 data—represents family earnings.
[b] 1951 data.
[c] These are for estimated nos.

TABLE 4.—PER CAPITA EXPENDITURE AND INDICATORS FOR HEALTH,
EDUCATION, AND SOCIAL WELFARE

Indicators	Newfoundland	National Average
Education:		
Education expenditures/capital ($):		
1985/86	1,362	1,325
1980/81	790	975
1949/50	13	14
Full-time university enrollment as a % of population aged 18–24:		
1985/86	12.8	14.5
1980/81	9.2	11.6
1949/50	.7	4.7
Health:		
health expenditure per capita ($):		
1985/86	944	1,151
1980/81	515	607
1949/50	21	14
Hospital beds per 1,000 in population:		
1985/86[a]	6.00	6.70
1980/81	5.73	6.61
1953/54	4.77	4.96
No. of doctors per 1,000 in population:		
1985/86	1.71	2.06
1980/81	1.49	1.85
1949/50	.42	1.03
Social welfare:		
Per capita expenditure ($):		
1983	327	364
1980/81	224	224
1949/50	31	14
Per capital direct financial assistance ($):		
1980/81	152	222
% of population dependent on direct financial assistance		
1980	8.4%	5.6%
1986	8.2%	. . .[b]
Unemployment insurance compensation beneficiaries as a % of contributors:		
1978	55.3%	23.9%
1986	59.6%	25.7%

SOURCES.—*Statistics Canada* Catalogue nos. 81–220, 68–202, 63–217 and *Historical Statistics of Newfoundland and Labrador*.
NOTE:—Prepared July 29, 1988. ($) indicates Canadian dollars.
[a] Preliminary.
[b] . . . indicates data is not available.

The level of "disguised" unemployment, especially in the primary industries, is uncertain but is probably substantial compared to the size of the labor force. In particular, a high level of disguised unemployment is often attributed to the fish-harvesting sector, which acts as a sponge, absorbing unemployed workers, who, because of low job vacancies, are unable to secure employment elsewhere, even though they possess the necessary skills. In this case, the addition of one more worker to fishing adds little or nothing to the total output of the fishery. Where extreme disguised unemployment exists, a transfer of labor out of that industry can take place without a decline in the total output of the industry. In this sense, the disguised unemployed may be viewed as a source of surplus labor.

The levels of unemployment and underemployment in Newfoundland and Labrador, combined with the sizeable potential for increasing the labor force through increasing participation rates, is a source of flexibility in the province's labor force. These various sources of surplus labor will be a labor supply should a considerable number of job vacancies open up.

Our forecasts in 1981 indicated that a sizeable surplus labor force would persist in the province for at least the next decade and perhaps much beyond. This conclusion was based on various demographic projections to the year 2000, including the size of the labor force. An astounding finding was that even if development of Hibernia had commenced in 1984, the province would not approach the 1980 national unemployment rate until the early 1990s. Newfoundland, therefore, sought to have the federal government declare Newfoundland's unemployment problem a special situation requiring extraordinary concessions, and the province was successful during the constitutional reforms in negotiating a special provision whereby provincial labor preference could continue until Newfoundland achieves the national employment rate.

Education and Training
The ability of Newfoundland's people to participate in expanded economic activity will depend on the education, training, and employment opportunities open to them. In essence it means that their skills must match those required by the evolving needs of the petroleum-related industries and other new and revitalized traditional industries. Training and skill will be the key to employment opportunity. Without stepped-up efforts to provide Newfoundlanders with opportunities to improve their qualifications, it was felt that the economic activity generated in the province would have little impact on our unemployment rate.

To address these problems the province proposed a human resource development effort that would include the creation of a new institute of marine technology; accelerated funding for postsecondary education; funding for an expanded school of business administration and an expanded department of geology and geophysics at Memorial University, the province's only university; and an industrial development program, which would encourage entrepreneurship and innovation through the use of the incubator mall concept and the development of a capability for new services related to offshore activities such as air-related industrial activities, topsides module fabrication and assembly, supply vessel construction, and supply and services activities. This additional infrastructure would go a long way in enabling the province to achieve its twofold training goals: (1) that sufficient numbers of Newfoundlanders and Labradorians be prepared to assume the skilled and professional roles that the economy would need; (2) that members of the labor force have an opportunity to acquire new skills to meet changing needs and take full advantage of emerging employment opportunities.

It was considered important that all the workers in the province, including those working offshore, should be subject to the same labor laws. The province, therefore, sought to have the social legislation of the province apply to offshore activities.

Industrial Benefits

Newfoundland Petroleum Industry

The development of offshore petroleum resources will give the private sector in the province an opportunity to launch into a wide range of new business opportunities to support the operating requirements of the oil industry. Some of these services are already available on a small scale, particularly those related to marine services; others need to be developed. The aim of the government of Newfoundland and Labrador was to develop an integrated oil industry with a substantial amount of goods and services being supplied from within the province and at least some refining and processing of oil products carried out within the province so that the maximum benefits would be derived from this new resource.

The development phase presents opportunities in engineering design and management, platform construction, manufacturing of topsides equipment, and subsea completion systems. It was felt that, unless Canada, and Newfoundland, participated in the design of the engineering system, a great part of the onshore and offshore fabrication would be engineered around existing U.S. capability. Therefore, it was resolved to get as much of the engineering design work as possible done in the province in order to create a strong national and Newfoundland presence in this area. The object was to increase the chances that the resulting manufacturing would spin off to Newfoundland-owned or Newfoundland-based companies. The whole issue of industrial and employment benefits received a great deal of attention.

High priority was given to the engineering design work, construction of the main support frame and topsides modules, the location of administrative offices in the province, preferences for local goods and services, and the establishment of an industrial benefits office within the structure of the joint management agency.

Research and Development

Industry in Newfoundland is generally labor-intensive and suffers from the absence of an established diversified manufacturing industry requiring a highly skilled industrial labor force. The resource orientation of the New-foundland economy creates a large number of unskilled jobs where New-foundlanders participate at the lower end of the income scale. There are few employment opportunities at the upper end of the income spectrum associated with more sophisticated resource processing or with research and development (R&D). The province has no substantial history of performing and utilizing R&D or the creation or promulgation of particular kinds of industry. This is reflected in the low levels of technology and productivity in several industrial sectors.

One of the greatest opportunities open to Canadians is the development and exploitation of advanced technology and high productivity goods and services. It was felt that if Newfoundland was to get its share of the economic benefits of technological change, it must have facilities for research and development to address the needs of local industry. Ideas developed at home would also allow the province to capitalize on the growing markets for improved technology in other parts of the world. It was considered essential that petroleum-related R&D be located in the province in order to build up a high level of expertise in the oil industry, which could compete worldwide in areas such as equipment design and manufacture. Moreover, it was agreed that research and development should not be restricted to petroleum-related activities, but these should be the springboard for the province to engage in other research and development activities.

Therefore, provisions encompassing oil- and non-oil-related research and development were included in the joint management agreement, specifically that both the federal and provincial governments would take such measures as were necessary to encourage and develop the province as a "center of excellence" for research and development related to offshore activity.

Social Capital

Large construction projects, particularly the size of Hibernia, result in population increases and shifts in populations, creating additional demands for social infrastructure and services. The social service system in the province is limited to essential services and is constrained by inadequate funding. It is expected that oil development will place additional strains on some social services, creating a need for substantially greater expenditures. At the time it was felt that new schools, extra hospital beds, increased law enforcement, and the like, might be needed in the areas where construction activity would take place, thus making demands for substantially increased expenditures before any revenues were received. It was decided that the province would look for an allocation of funds for development purposes as part of the negotiated settlement. To demonstrate its need for infrastructure funding, the province prepared a list of proposed projects totaling $1 billion. In response, Newfoundland received a $300 million Offshore Development Fund.

Environmental Concerns

Prevention of Damage

The governments of Canada and the province share mutual concerns for environmental quality and the protection of offshore marine resources. Experience elsewhere in the world has shown that the development of petroleum resources under the sea can adversely affect such resources. Consequently, measures must be taken to control and monitor the activities and facilities of offshore operators as a means of preventing environmental damage. Regulation within resource management legislation, which sets conditions for license, permit, or lease, is the only means whereby this control can be achieved.

The province proposed that this regulatory control be carried out through the joint management agreement, which would enumerate the specific regulatory areas where the province and the federal government would cooperatively share responsibilities and thus provide for a comprehensive and coordinated effort to prevent environmental damage. This has been done through a memorandum of understanding between the federal and provincial environment departments, the two energy departments and the Canada-Newfoundland Offshore Petroleum Board.

Compensation Funds

It is almost inevitable that the fishing industry will be exposed to damage or loss of fishing grounds as a result of petroleum development, although Hibernia is located 200 miles offshore. The experience of British and Norwegian fishermen was such that, following exploration and development of North Sea oil fields, many British and Norwegian fishermen abandoned the fishery. The main difficulty was the debris discarded by the oil industry, which littered the sea floor and fouled and destroyed expensive fishing gear. Some traditional fishing grounds were encroached upon by oil rigs, and shipping lanes of supply vessels either denied access to fishing grounds or presented hazards to navigation. In addition, manpower losses to the oil industry were substantial, and port development and use presented considerable difficulties to the fishing industry. Appropriate regulatory controls have since been put in place in the North Sea,[7] and such will be the case in the Newfoundland offshore

---

7. See Steinar Andresen, "The Environmental North Sea Regime: A Successful Regional Approach?" in this volume.

area. However, experiences in resolving problems caused by offshore development of petroleum in Britain, Norway, and the United States indicate that compensation schemes are essential and early planning for them is imperative.

The need for two compensation funds was identified—one for compensation to fishermen and the fishing industry and the second to other individuals and governments for environmental damages resulting from offshore oil and gas development. Enabling legislation was passed by the provincial government in its petroleum and natural gas regulations, but this has since been replaced by appropriate clauses in the Atlantic Accord Implementation Acts providing for oil pollution and fisheries compensation regimes.

CONCLUSION

Table 1 sets out the province's goals as they evolved through the planning and negotiating stages. It also shows the goals that were achieved through the Atlantic Accord with the government of Canada, and in the tripartite 1988 Hibernia Agreement with Mobil Oil Canada, Ltd. and the government of Canada. Virtually all of the aspirations described in the white paper in 1977 have been fulfilled, but the excellent results that have been achieved have not come to pass without a struggle and a great deal of tenacity.

Developing countries may be able to benefit from our experience. The first step is to decide to become part of the process rather than just letting things happen. Set your sights on where you want your country to be in the long term. Do not be discouraged by where it is today. The social and cultural aspirations of the people and the preservation of the cultural environment must be part of economic planning. Once the goals are crystallized, careful planning is required to map out the most effective route to reach those goals.

Place requirements on developers so that as much economic activity as desirable can take place in your country. Do not allow them to tell you that your people and industries cannot meet the requirements. Insist that they help you develop the capability that is needed. Human resource development is an essential element in helping a country achieve social, cultural, and economic goals. When the goals and the game plan have been formulated, success can be achieved if commitment and tenacity of purpose are present. Persistence and perseverance are often the deliverers of success.

International Shipping: Developments, Prospects, and Policy Issues[1]

Bernhard J. Abrahamsson[2]
University of Wisconsin–Superior

INTRODUCTION

World trade is overwhelmingly moved by sea. Hence, shipping is an inevitable and inescapable adjunct to international trade; indeed, it is an integral part of such trade and must adjust to, and change with, trade developments.

Throughout the postwar period and, particularly, in the last few decades, international trade has been a major factor in world economic growth. The quest for economic development in some parts of the world, and continued growth in others, has relied heavily on trade; and policies to cope with the current world economic situation are likely to focus again on trade expansion, notwithstanding existing protectionist tendencies in many countries.

As trade policies become increasingly important and elaborate, shipping issues are highlighted and increasingly subjected to governmental scrutiny and involvement. Consequently, political forces have gradually become dominant factors in the international shipping industry. It appears often that shipping has become more of an instrument than an object of policies for the achievement of broad economic and strategic objectives. This is mainly seen in the decline of the fleets of the traditional shipping nations, the emergence of new national merchant marines among the less developed countries, and the participation in international carriage by the fleets of the socialist countries. The shift in ownership of tonnage has been accompanied by attempts to secure cargoes for the national fleets. This is manifested in political arrangements for bilateral and multilateral cargo-sharing agreements. The latter has been achieved for the liner trades in an international convention, the UN Conference on Trade and Development (UNCTAD) Code of Conduct for

---

1. AUTHOR'S NOTE.—At the time of writing, I was working at the Federal Maritime Commission on the 5-year review of the U.S. Shipping Act of 1984.

2. Editors' note.—This is the second article Dr. Abrahamsson has contributed to *Ocean Yearbook*. Readers are referred to his first article, "Merchant Shipping in Transition: An Overview," *Ocean Yearbook 4*, ed. Elisabeth Mann Borgese and Norton Ginsburg (Chicago: University of Chicago Press, 1983), pp. 121–39.

Liner Conferences, which entered into force on October 6, 1983.[3] It calls for the division of conference cargoes between trading partners in such a way that each one carries 40% of the trade, and the remaining 20% is left for third-party carriers. Attempts are also being made by the UNCTAD to reach agreement on the sharing of bulk cargo carriage, the international registration of ships, and the international regulation of multimodal transport.

In addition to these political forces that shape the industry's environment, there have been structural changes resulting from successful attempts to improve the economic efficiency of shipping. Not only have the shapes and sizes of ships changed almost beyond recognition since the 1950s, but the pattern of world trade, as well as its carriage by sea, has changed dramatically. More specifically, we have seen technological progress in shipbuilding, cargo handling, and management. The manifestations of these developments are the very large ships used today in the transport of bulk cargoes, both dry and liquid; the introduction of specialty ships, including various concepts of containerization; and changing market structures in all segments of the industry. These developments have necessitated far-reaching changes in port facilities, transportation patterns, and labor requirements—all of which have had repercussions on economic and social patterns, thus necessitating policy responses.

Finally, since demand for shipping is derived from international trade, structural changes in that trade are of utmost importance for the industry's long-term prospects.

The purpose of this article is to provide an overview of the current situation, what major forces brought this situation about, what future tendencies appear significant given the present perspective, and what consequent policy issues may be identified. Because of their prominent role in the current shipping debate, the discussion is partly focused on the less developed countries (LDCs).

THE CURRENT SITUATION

There has been a general depression in the shipping industry since 1974. The prospects for a turnaround are difficult to assess in the best of circumstances. In this case, they are compounded because the cyclical developments are exacerbated by long-term structural changes that make adjustment a more lengthy, difficult, and complex process than in the past. That is, the industry is adjusting more to permanent, structural change than to cyclical events.

The rapid postwar expansion of world trade, and its seaborne portion, came to an end with the oil crisis in 1974. After an immediate decline in 1975,

3. UNCTAD Code of Conduct for Liner Conferences, entered into force in 1983 (TD/CODE/11/Rev., April 6, 1974).

seaborne trade recovered slowly, but, after reaching a peak in 1979, a sustained decline set in so that, in 1983, it was somewhat below 1974 levels. Preliminary data from UNCTAD for 1986 indicate that the downward phase of a long trade cycle may have ended. This perception is strengthened by the International Monetary Fund's estimate that world trade will grow substantially, at about 7%, in both 1988 and 1989.[4]

The world fleet has expanded over the years to meet the needs of the trade. However, the basic equilibrium between demand for, and supply of, tonnage was disrupted after 1974. Orders placed before the new trading trends were perceived caused the fleet to continue its rapid growth through 1977. Thereafter, the growth rate slowed substantially, but not enough to prevent the development of a very large surplus of tonnage. After reaching a peak in 1983, the world deadweight tonnage began to decline. By the end of 1988, there were indications of renewed investment in shipping, particularly bulk carriers, in response to growing trade and corresponding increases in freight rates.[5]

The current situation is, therefore, characterized by a growing, but still relatively low, level of seaborne trade and a continued, but declining, surplus of tonnage in all types and sizes of ships. While freight rates are still generally low, lay-ups are decreasing, prices for secondhand ships are up, and scrappings are tapering off—all of which reflect an improvement in, but no end to, serious financial difficulties for owners and consequent pressures on governments to take remedial actions.

While the depression is general, the impact varies from market to market as prospects for shipping vary with the industry's response to long-term factors of structural change. Thus, while a major U.S. container operator, the U.S. Lines, went bankrupt in December 1986, other carriers in the U.S. trades have expanded in intermodal services encouraged by the U.S. Shipping Act of 1984.[6]

In the bulk markets, a freight futures contract, Biffex, was introduced in 1985.[7] It allows charterers and operators to hedge against fluctuations in

---

4. Statistical data to support the narrative are not included in this article. Their reproduction from readily available sources is deemed to be a marginal value. Common sources include the World Bank, UNCTAD, OECD, United Nations, and various trade journals.

5. See reports by INTERTANKO, IMIF (International Maritime Industries Forum), and Drewry Shipping Consultants, Inc. as reported in *Fairplay* and *Seatrade* throughout the year; *Lloyd's Shipping Economist* provides monthly tonnage surplus data by ship type.

6. U.S. Shipping Act of 1984, *United States Code Annotated* (*U.S.C.A.*), vol. 46, Secs. 1701–20 (West Suppl. 1985), signed by President Reagan March 10, 1984.

7. Futures Trading (Biffex) opened on the Baltic Exchange on May 1, 1985. See "Futures Trading on Ocean Freight Fees Starts This Week at a London Exchange," (*Wall Street Journal* [April 30, 1985]).

freight rates. Wide and successful use of the contract will impart some stability to bulk shipping markets and ease adjustment. Clearly, economic imperatives of the market are bringing about structural changes in shipping patterns and operations.

STRUCTURAL CHANGE[8]

Change is occurring, first, in the volume, composition, and direction of trade, and this will affect shipping directly as there may be fewer and different cargoes available for sea transport; certainly seaborne trade will not experience the high growth rates of the pre-1974 period. Second, there has been, and will continue to be, structural change in the shipping industry itself because of technological change that will affect, and be affected by, the changes in world trade. Third, there has been far-reaching structural change in the institutional setting of the industry as political concerns have become more pronounced and have effected shifts in the pattern of ownership of shipping and have promoted various measures for international regulation of operations and cargo allocations.

Each aspect is discussed below, albeit in different order than listed above. No attempt is made to provide a detailed description of the developments. Instead, the focus is on some major effects of importance for understanding their implications for policy.

Technological Change

The first aspect is technological change, which has resulted in larger and faster ships and new cargo-handling techniques. In addition to effects on manpower, management, ports, and trade flows, four major consequences have followed. First, flexible combination carriers, such as oil-bulk-ore ships, have made the carriage of dry and liquid cargoes interchangeable. These two segments of bulk carriage are now closely tied together so that surplus tonnage in the oil market can easily move into dry bulk transport and vice versa. Such a move occurred when the oil trade declined substantially after 1974.

Use of Containers
Structural change has affected the second consequence, namely the use of containers. In their early stages of use, containers promoted a very clear

8. Part of this segment was delivered by myself in a paper, "Commercial Sea Lanes of Communication," at a workshop on International Navigation, Honolulu, January 13–15, 1986, and published in J. Van Dyke, L. M. Alexander, and J. R. Morgan, eds., *International Navigation: Rocks and Shoals Ahead,* Workshop of the Law of the Sea Institute (Honolulu: Law of the Sea Institute, 1988), pp. 39–53.

differentiation between the carriage of general cargo and bulk. At a later stage, however, the two sectors were brought closer together as general cargo put in containers became homogeneous and some bulk commodities were suitable for movement in containers. As a result, liner conferences are today meeting increased competition from bulk operators using bulk-container ships; this is in addition to the growing competition from independent liner operators. The final result is oversupply in container tonnage and what appears to be a weakening of conferences in terms of share carried (volume and value). Reports indicate that, on a worldwide basis, conferences carry on the average 60% of the available liner cargo;[9] this compares to some 85% in earlier years.

It is interesting to note that tramps were always considered to be potential competitors to conferences and were so identified in all major studies of the conference system. Yet, it was also assumed that the sophisticated technological change embodied in the container system would promote specialized carriage and thus remove tramp service and bulk ships from the liner scene. Although this was the tendency at the very beginning, it has clearly not been the outcome. While there is much specialized tonnage, most of it is proprietary and serves neobulk trades.[10] Traditional competitive relationships between various shipping markets are basically unchanged, or even sharpened, by the technological changes we have seen. The whole industry is closely tied together so that all major types of carriage—dry and liquid bulk, conference liners and independents—affect each other competitively, but they do so on higher levels of technological sophistication and efficiency than was previously the case.

All this means that heavy demands are made on land-based support facilities, both physical and human. It also means that we have to consider carefully the place of these sophisticated transport systems in economies where resources for their support are scarce. In a concrete sense, technologically advanced ships require similarly advanced ports; but ports cannot be divorced from their hinterlands and the necessary domestic transport infrastructure. To maximize the benefits of advances in cargo handling and transport, the whole system—ships, ports, internal distribution—must be technologically consistent. Such consistency is not the usual condition, particularly in the LDCs, and its achievement must be a basic concern when assessing shipping investments and policy.

9. Statement by Ian Ross-Bell, Secretary-General of the Council of European and Japanese National Shipowners' Associations (CENSA), Haifa University, Haifa, August 1984.

10. These are trades where, in recent years, the size of the shipment lot of a commodity has increased so that a whole ship, rather than a part, is used for its transport.

Operational Capacity

The third consequence has been that the efficiency, or operational capacity of ships, has increased dramatically, thus allowing for more cargo to be carried in fewer ships. The result, even with growing trade, is general surplus tonnage, a situation that is exacerbated by the continuous improvement of the efficiency of these ships. In a period of stable or declining trade, operators see the capturing of cargo share as a zero-sum game, and there is a distinct movement toward mergers and increasing scale of operations—a trend particularly noticeable in, but not limited to, liner trade. To keep at the forefront of the containerization-cum-intermodal movement and to organize services on a large international scale in a highly competitive environment requires large capital resources as well as extraordinary expertise. Since few individual companies have sufficient amounts of such resources, they have resorted to cooperation. Consortia, joint services, joint ventures, space charters, and mergers are common; and leasing arrangements for major pieces of equipment, including ships, are increasingly part of the scene. One consequence of this, combined with the large surplus of ships, is an emerging tendency for vessel operators to charter, or lease, ships, rather than own them outright. While this arrangement has always been possible, it appears to be easier and more common than in the past. The significance of this arrangement, should it become widespread, lies in the potential for increasing separation of ownership and operations. The environments for decision making for the two parties would be different, with the owners focusing on the acquisition of ships—that is, on shipyard policies—and the operators focusing on trade policies that determine the prospects for the operation of ships. Because of the large employment and linkage effects of shipyard activities, governments are more inclined to support their shipyards than their fleets, and owners, separated from the actual operation, may make their decisions more on the basis of the financial incentives to build, provided by governments, than on actual carrying prospects. The outcome may be a perpetuation of the tonnage surplus. On the whole, demands on managerial ability are heavy as we are rapidly moving to a situation in which high-technology shipping will be concentrated in a few very large and financially strong operators. While most of these, such as American President Lines and Sealand, are from the advanced industrial countries, at least one major operator, Evergreen, comes out of the newly industrialized countries, namely Taiwan.

By necessity, these large liner operations will be focused on a few, high-volume trade routes with feeder services to the main loading ports; the round-the-world services provided by Evergreen, K-Line, and, until recently, the U.S. Lines, are examples of such routes. The feeder services supplement the "natural" rate of cargo accumulation in the main port (the load-center port) so that the capacity of these large ships can be reasonably well utilized without calling at a large number of ports. In this "load-center" system, cargo moves from the feeder ports to the load-center ports in one trading area, or range, then to a load-center port in another range, and on to that port's feeder ports. The analogy is the "hub-and-spoke" system in the U.S. air transportation. This system's logical extension is to spawn new services to and from intermediate volume ports in competition with the feeder services. That is, cargo from a feeder port in one range destined to a feeder port in another range would move directly from the former to the latter, bypassing their respective load centers—again, the analogy from the U.S. air transportation is the "point-to-point" system that has developed as a supplement to the "hub and spoke." More will be said about this when the discussion turns to prospects for liner trade.

Intermodalism
The fourth consequence of technological change is that conditions for intermodalism were created. This has had far-reaching and long-term effects on shipping—a notable example is the U.S. Lines case—and is an aspect to be more fully discussed in connection with changes in world trade.

Institutional Change

The second aspect of structural change pertains to institutional change, seen mainly in four ways: new de facto maritime boundaries, a shifting pattern of tonnage ownership and consequent measures for international regulation and cargo allocations, safety issues, and environmental issues. The first of these is of minor importance to shipping at this stage but may have future impacts that are difficult to perceive at this time. The other three are connected, but the shift in ownership and allied measures poses more political problems and issues than do the other two.

New Maritime Boundaries
New maritime boundaries have been established in accordance with the UN Law of the Sea Convention, which has not been globally accepted and is unlikely to enter into force soon.[11] The boundaries have been redrawn with reference to new definitions for Territorial Waters (12 nm), a Contiguous Zone (another 12 nm), an Exclusive Economic Zone (add 176 nm for a total of 200 nm from shore baselines), international straits, and rights of passage. The extended jurisdictional powers of coastal states carry some implications for navigation, particularly in terms of vessel traffic separation schemes and

11. *The Law of the Sea: Official Text of the United Nations Convention on the Law of the Sea with Annexes and Index* (New York: United Nations, 1983), Sales No. E.83.v.5; and see "Marine Jurisdictional Claims, by Country," App. G, in this volume.

safety, including terrorism and piracy,[12] and pollution aspects. To implement these powers, the demands on a country's technical and administrative resources are great and not unlike those required for the administration of a national fleet.

International Maritime Conventions
Safety and environmental issues are manifested in various international conventions formulated in the International Maritime Organization (IMO),[13] mainly the conventions for Safety of Life at Sea (SOLAS), LOADLINE, Standards for Training Certification and Watchkeeping (STCW), and Maritime Pollution (MARPOL),[14] all of which are in force. These conventions affect both shipboard procedures and operating cost and have implications for manpower training on all levels. The implementation, monitoring, and administration of them on the national levels require, as in the case of the Law of the Sea, substantial technical and administrative resources that are often not available in sufficient quantities yet are essential for the maintenance of national fleets that have emerged in the last 2 decades. How to provide for these resources is a matter of urgent policy concern in the international arena. In this context, the World Maritime University,[15] sponsored by the IMO, is an important response to a manifest need.

Tonnage Ownership Changes and Measures to Secure Cargoes

The shifting pattern of tonnage ownership and the consequent measures to secure cargoes are the most significant and sensitive of the structural changes that have taken place in the last 20 years because these are developments that

12. Editors' note.—See Bjorn Aune, "Piracy and Its Repression under the 1982 Law of the Sea Convention," in this volume.
13. Editors' note.—See "IMO: Report of the International Maritime Organization in 1987 and 1988," App. A, in this volume.
14. For SOLAS, see the 1974 International Convention for the Safety of Life at Sea, done November 1, 1974, *United States Treaties and Other International Agreements* (*T.I.A.S.*) 9700, *International Materials* (*I.L.M.*) 14 (1975): 959. For LOADLINE, see the 1966 International Convention on Load Lines, done April 5, 1966, *U.S.T.I.A.S.* 6331. For STCW, see IMO, International Convention on Standards of Training, Certification and Watching for Seafarers, entered into force in 1984 (STW/CONF/13, July 5, 1974). For MARPOL, see the 1973 Convention for the Prevention of Pollution from Ships, done November 2, 1973, *U.S.T.I.A.S.* 10561, *I.L.M.* 12 (1973): 1319; and for more information about MARPOL, see Edgar Gold, "New Directions in Ship-generated Marine Pollution Control: The New Law of the Sea and Developing Countries," *Ocean Yearbook 7*, eds. Elisabeth Mann Borgese, Norton Ginsburg, and Joseph R. Morgan (Chicago: University of Chicago Press, 1988), pp. 191–204.
15. Editors' note.—See "Report of the World Maritime University: 1989," App. A, in this volume.

are more political than market-induced; they constitute a focal point in the "North/South" dialogue related to the call for a "new international economic order" (NIEO) and have brought, anew, explicit national security concerns into the shipping debates. A fuller discussion follows.

The general pattern of ownership shows a significant decline in the share of world tonnage registered in the Organization for Economic Cooperation and Development (OECD) countries (the traditional shipping nations)— almost 65% in 1970 compared to 39% in 1986. Open registries have gained tonnage, and so have the socialist countries and the LDCs. The latter group, which accounted for some 7% in 1970, had raised its share to almost 20% in 1986. The distribution of tonnage between LDCs is uneven; excluding the "flags of convenience" (FOCs), the major part is accounted for by a few major shipping nations, namely, India, South Korea, Brazil, China, Singapore, Saudi Arabia, and Kuwait.

Whereas ownership is likely to continue to shift toward the LDCs, it is not likely to do so without impediments. Indeed, there is reason to believe that the shift will slow down substantially. The Falkland War showed clearly the utility of a merchant fleet in military operations; and in the last few years the policy debates in most maritime countries, be they developed countries (DCs) or LDCs, have focused more on the defense needs of a fleet than its commercial aspects. The defense needs are cast in terms of "economic and military security." In essence, continued flagging out and loss of nationally controlled tonnage is increasingly seen as a threat to national security, thus bringing noneconomic arguments to the forefront.[16]

Security Concerns

These security concerns present at least two possible scenarios. First, developed maritime states may hold back on shipping investments abroad, particularly in LDCs, and encourage FOC registry on the assumption that effective control is more likely under these flags. Second, policy incentives may be used to prevent any flagging out or to bring ships back to home flag. Indeed, recent years have seen the development of "captive" or "off-shore" registries. These are found in the traditional maritime nations. Such registry allows a shipowner to derive benefits as if under an FOC while flying a non-FOC flag. Under the Norwegian International Shipping Registry (NIS), vessels are registered under the Norwegian flag, but they need not be incorporated in Norway. Hence, the shipowner is not liable for Norwegian corporate taxes. Tech-

---

16. See, e.g., Harlan K. Ullman and Paula J. Pettavino, *Forecasts for US Maritime Industries: Balancing National Security and Economic Considerations*, Significant Issues Series, vol. 6, no. 17 (Lanham, Md.: University Press of America, 1984); R. B. Garware, "Outlook for Our National Shipping," *Indian Shipping*, no. 11 (1984); Robert W. Kesteloot, *America's Vanishing Merchant Mariners: Diagnosis, Prognosis, and Prescriptions for a Strong National Defense* (Washington, D.C.: Transportation Institute, September 1986); and various issues of *Seaways*.

nically, the ships must be managed by Norwegian nationals, but they need not be so crewed. The United Kingdom has a similar registry on the Isle of Man, and several others are considering their use.[17] If FOCs are phased out, this development is certain to gain further momentum; there is little reason to expect reflagging to LDCs in general.

Regardless of the actions of the DCs, the LDCs' own security concerns argue for further expansion and development of their fleets. Hence, we are likely to see continued surplus tonnage as well as shipbuilding capacity throughout the world. When and how, not to say if, government policies will balance security and commercial needs for merchant ships cannot be but speculation. The two policy objectives of "economic efficiency" and "national security" are mutually exclusive in terms of policy analysis. Yet, while there is a professed adherence to commercial criteria, it is the security need that is the prime mover in most countries. Attempts to achieve it ensure that surplus of tonnage will be virtually the permanent condition in the industry. Concomitantly, it is certain that governments will attempt to secure cargoes for their fleets and that the industry will be increasingly subject to government scrutiny and actions.

The growing tonnage under LDC flags has been accompanied by measures to secure liner cargoes, resulting in the UNCTAD Code. The Code is an important element in the changing liner market and is a manifestation of the issue that brought it about, namely, the emergence of national merchant marines in the LDCs. One basic reason these fleets have been created is the widespread feeling that nonshipping countries are treated unfairly and arbitrarily by the liner conferences. The problem is not new; it arose shortly after conferences were established about a century ago. At that time, countervailing organizations in the form of shippers' councils were considered the proper solution, and in the postwar period they again became prominent. However, they appeared to be effective mainly when backed by governments, and the concept of a "conference code of conduct" evolved, basically, to defuse growing government involvement. Such a code, with implementation on a voluntary basis, was issued by the Council of European and Japanese National Shipowners' Associations (CENSA) in 1971.[18]

The following year UNCTAD issued its own code, which was accepted in

17. Le T. Thoung, "From Flags of Convenience to Captive Ship Registries," *Transportation Journal* 27 (Winter, 1987), pp. 22–35.

18. The Code of Practice for Conferences was prepared by the Council of European and Japanese National Shipowners' Association (CENSA) and the European Shippers' Councils (ESC) and was endorsed in November 1971 by the governments that were members of the Consultative Shipping Group (CSG). It was intended to set out general principles to be adopted by all liner conferences and similar associations, and it was prepared with due regard to the relevant UNCTAD resolutions unanimously agreed on.

March 1979 and entered into force in October 1983.[19] The Code calls essentially for a system of international regulation of conference trade. It accepts the conference system as a proper mechanism to render efficient liner service but establishes the principles that (1) governments will have a role in relations between shippers and carriers; (2) conference membership shall be granted on the basis of some noncommercial criteria, one of which is the development of national shipping lines; and (3) cargo allocations to aid national lines are acceptable—that is, the 40-40-20 guideline.

Several countries have not signed the Code, notably the United States; and others, notably the European Economic Community (EEC) countries that in December 1986 agreed to pursue a joint shipping policy, will do so only with reservations expressed in the so-called Brussels Package. Under these reservations, the LDC partner's appropriation of its 40 percent share is accepted, but the EEC partner's share is left free for competition among other EEC members. This free access is extended to any OECD member who offers reciprocity. There are some constraints on this "free access."

Many studies have been made to assess the impact of the Code or bilateral shipping arrangements in general.[20] Under some special circumstances, bilaterals may decrease the degree of noncompetition. In general, however, there is likely to be upward pressure on the relevant freight rates because a given volume of trade is restricted to the trading partners' conference tonnage, which in all cases must be smaller than the world supply of shipping.[21] To prevent this, rates are likely to become part of the trade agreements and to come increasingly under government control, as is seen in various countertrade arrangements, bilaterals, and the establishment of agencies to negotiate rates. It should be noted that there have always existed numerous methods for allocating cargoes within the existing conference system. Therefore, the Code is not achieving anything new. What it does, however, is codify into international and national laws these methods of restricting competition. The government, instead of monitoring and controlling anticompetitive behavior, now becomes a partner in it. The logic of the situation argues for an extension of this behavior into other areas of shipping, and such attempts have been made.

With the liner Code in place, UNCTAD concerns and efforts have turned to achieve a similar convention for bulk cargoes. This "bulk-sharing scheme" has surfaced frequently at UNCTAD meetings, but no agreement

19. Editors' note.—For a discussion of UNCTAD's Liner Code of Conduct, see Abrahamsson, "Merchant Shipping in Transition" (n. 1 above), pp. 133–35.

20. Manalytics, Inc., *The Implementation of the UN Code of Conduct for Liner Conferences* (Washington, D.C.: Federal Maritime Administration [MARAD], 1981); and E. Frankel, *Impact of Cargo Sharing* (Washington, D.C.: MARAD, 1981).

21. Manalytics found that if a bilateral agreement is a defensive measure against a unilateral measure reserving a very large proportion, say 100%, then it does increase competition.

has been reached. Such a scheme, coupled with the closing of the open registries, would, at least in some early views expressed by UNCTAD, provide LDCs with both ships and a cargo base. Several preparatory meetings have been held on the potential phasing out of the FOCs. In 1986, UNCTAD accepted an international convention for the registering of ships based upon the "genuine-link" concept—that is, there should be a "genuine link" between the ownership of a vessel and the flag under which it is registered.[22] While the "genuine link" appears weak in the convention's text, its inclusion will, presumably, make open registries unattractive and, perhaps, diminish their role in shipping or, ultimately, phase them out. The discussions are gathering force from the very real abuses that can be attributed to ships under FOC flags. These abuses include several cases of fraud, scuttlings, and cargo theft, as well as violations of safety, environmental, and training standards.

The strongest current argument against the FOCs is that they must be controlled, or phased out, to ensure safety and reliable services. A salient feature of the debate is that the FOCs are also LDCs and that some LDC owners have ships under FOC flags. Nevertheless, when this convention may enter into force can only be speculation. At the present time, there is, as mentioned above, a growing move toward "flagging out" and, instead of closing, new open, as well as captive, registries are emerging.[23]

The facts are that the UNCTAD Code applies only to conference cargoes, that the EEC will apply the Brussels Package, that the United States has not signed the Code, and that independents are taking a larger share of the liner trade. These facts would indicate that the LDCs may not obtain the expected benefits from the Code. Their 40% share of a shrinking pie may not support even existing tonnage of the major LDC shipping nations; certainly, questions may be raised as to the need for additional ships. Hence, there is a dilemma for these nations: the times are auspicious for the acquisition of tonnage, but its eventual use raises serious concerns. These issues are likely to be the focus of the Code's mandated 5-year review to take place at the end of 1988.[24] In particular, India is expected to put forth proposals to bring all shipping, nonconference and tramping, under the Code's provisions and to increase the share of cargo reserved for the trade partners.

In this context, the U.S. Shipping Act of 1984 is relevant because it affects the conference system and the employment opportunities of liner tonnage in general. Under the Act, several conferences serving the U.S.

22. International Convention on the Condition for Registration of Ships, adopted by UNCTAD, 1986.

23. As mentioned in the text, the Isle of Man has a rapidly expanding captive registry, and such unlikely places as Spitsbergen, Kerguelen Island, and Luxembourg have been mentioned in the press.

24. The meeting at the end of 1988 reached no conclusion on this issue. The committee was preoccupied with the issue of whether they should permit voting of nonsignatories, especially the United States.

trades have been able to consolidate into superconferences. Although the conferences must still remain open, the Act gives U.S. flag operators wide scope for cooperation and streamlining of services (rationalization) and easier access than before to participate in the conference system. These particular features, by themselves, ought to strengthen the conference system and provide for a certain stability hitherto absent from U.S. liner trades.

Two other provisions in the Act, however, have the opposite effect and have played havoc with the system. One gives a conference member the mandated right to take independent action without leaving the conference. In essence, if a carrier wants to offer rates and services that are different from those set by the conference, he can do so provided certain procedures are met. The conference must then include, in its tariff, this new rate and service for use by the carrier that initiated the action. Other carriers may follow suit. This breaks the common tariff and, thus, weakens the rate-setting powers of the conference. In a situation of surplus tonnage, the common tariff becomes a ceiling rather than a floor, and the more independent action there is, the lower the de facto rates will be, thus further eroding the conference's power and usefulness.

These tendencies are amplified by another of the Act's provisions that permits carriers and shippers to enter into service contracts. Under these contracts, a conference carrier and a shipper negotiate rates, services, cargo volumes, and the time frame for the contract at terms that are substantially different from those offered by the conference.

The effects of these contracts are the same as under independent action: cargo is removed from application of the common tariff and the actual rates are lowered. There are, however, differences between the two. Under independent action, prior notice, albeit short (10 days), is given, and the applicable terms are published by the conferences. Consequently, both the contemplated action and its timing, and the resulting rates, are known to carriers and shippers alike—this weakens the competitive advantage expected by the carrier taking the action. Service contracts, in contrast, require no prior notice, and the specifics of the contracts need not be publicized. Although the main features of these contracts are known and readily available to other shippers, there is a measure of confidentiality that strengthens the competitive position of both parties to the contract.

The consequences are predictable. While independent action has been taken, it is insignificant compared to the service contracts—some 5,000 were reported in the U.S.-Pacific trades at the end of 1986. In subsequent years, as conferences in these trades restricted the use of service contracts, independent action became very common, amounting to several thousand per year. Under these circumstances, there have been questions as to the viability of the conference system in the U.S. trades. The issue is complex and centers on the common carrier concept in international shipping. In domestic transport, the common carrier functions, and is treated, like a public utility. It is compensated for its legal responsibilities and obligations by regulatory assurances of "fair" return on investment. Not so in international shipping. Although the liner carrier is considered a common carrier, with all it entails, it is denied the protection given to the domestic common carrier. The conference system can be seen as the shipping industry's own mechanism to provide that protection and assure "fair returns." The closed conference certainly is able to do this; the open conference faces problems because it cannot control who, or how many common carriers, it has to protect. Today, these problems of the open conference are accentuated, and some question whether the ocean common carrier can continue to operate in a conference system that provides progressively weaker protection and assurance of "fair return." This problem offers weighty policy issues to be considered when this Act's mandatory review takes place in 1989.

All this, in turn, may have some effects on the UNCTAD Code's impact. Barring changes that may come from the Code's review in 1988,[25] its present guideline is that 20% of the trade is to be made available to third-party carriers—so-called crosstraders. Since, at the present, third parties carry more than their suggested share on most trade routes, much tonnage would become surplus if the rule were strictly implemented. It has been assumed that the displaced ships would seek employment in the open U.S. trades. However, the new U.S. Shipping Act opens opportunities for U.S. operators to rationalize services and, thus, better meet this potential competitive force. As a result, these trades will not easily absorb the displaced tonnage. In addition, the 1984 Act retains provisions for direct retaliation against countries that interfere with U.S. cross-traders.

Another initiative—yet to affect the shipping environment—is UNCTAD's Convention on Multimodal Transport, adopted in 1980 but a long way from ratification.[26] In essence, this convention calls for multimodal transport to be arranged by licensed and regulated multimodal transport operators who would act as agents for shippers, not carriers. On the face of it, this is not consonant with the observed structural changes in the industry, which indicate that carriers are taking the initiative and large multimodal transport companies are being created—at least in the United States.

25. The main issue of the 1988 review of the UNCTAD Code was whether to extend the Code's coverage to all liner cargoes and not merely the conference cargoes. This issue was not resolved but, rather, met with severe resistance from the traditional maritime nations.

26. Editors' note.—For more information on multimodal transport, see R. Vogel, "Multimodal Transport: Impact on Developing Countries," pp. 139–48, and G. Levikov, "Multimodal Container Transport Tariff Rules: Impact on Developing Countries," pp. 149–59, both in *Ocean Yearbook 6*, ed. Elisabeth Mann Borgese and Norton Ginsburg (Chicago: University of Chicago Press, 1986).

Changes in International Trade

Changes in international trade constitute the third and final aspect of structural change. While the proximate cause of the development of trade is the state of the world economy, it is possible to discern some long-term trends that will affect shipping. One is that the composition of trade has shifted in favor of more manufactured goods and partially processed raw materials—that is, higher value-added goods.

Traded Products
The oil trade, in particular, has seen dramatic changes in volume, product mix, and trade routes. High prices and the economic slump have cut demand in a general sense, and conservation efforts and improved technology have changed demand in a fundamental sense so that, on balance, renewed growth will start from a lower level and is not likely to proceed at the high rates of the past.

New sources of supply have shortened the sea routes. The overall result is a surplus of tankers, particularly very large crude carriers (VLCCs). Demand for smaller tankers is being bolstered by producers' supplying products rather than crude—an example of the general shift to more value-added commodities in trade, but also an example of structural market change.

In the wake of the 1974 oil crisis, the oil-shipping market changed rapidly. As producer governments took control of the supplies, they encouraged entry of numerous buyers in order to free themselves from dependence on the few major oil companies. The large number of new buyers resulted in many relatively small shipments to many destinations, and the need for voyage charters became more important than the long-term charters previously demanded by the large oil companies. Hence, the major oil companies let go of the larger part of the fleet under their control. That is, the tanker fleet, which previously had been, largely, centrally controlled by the major oil companies through ownership and long-term charters, leaving only a rather small fraction free for spot chartering, now reversed its composition: a major part was free for the spot market.

In essence, the former, concentrated control of the tanker market had become fragmented. It remains so, although there is a tendency toward some concentration in producing countries. Their buildup of tanker capacity fits into their assuming control of "downstream operations." However, given the need to draw on several sources in order to maximize the security of oil supplies and the sensitivity of the transport link, plus the control the buyer has in designating the flag of transport, there is a general argument for countries to carry their oil imports in their own tankers—that is, an argument for economic and military security. Whether that argument will further affect the world's tanker fleet and its market structure is an open question.

The tendency toward more finished goods trade favors liner and neobulk shipping; most bulk trades tend, however, to be hurt by this development. Also, the growing use of recycled scrap, particularly in steel, copper, and aluminum, will affect bulk trades in these ores and metallurgical coal. Steam coal seems destined to grow in importance as oil prices seem bound to rise again in the future.

The Process of Trading
In addition to these changes in traded products, there is a major development in the process of trading. Barter, or countertrading, is rapidly becoming an important trade feature. While it has always been the preferred arrangement in East-West trade, it has in recent years gained popularity also in other trades, specifically those of the LDCs. Detailed data are not available, but volumes appear to be small, although rapidly growing. The OECD has esti- mated that, in 1980, about 1% of world trade was countertraded; other esti- mates range to 20%.[27] The effects of these arrangements are, in essence, the same as those of any bilateral agreement, but, if shipping services are specified as a counterpurchase, we have a new type of cargo-reservation scheme. These schemes are different in both nature and scope from arrangements entailed in the UNCTAD Code and the usual bilateral agreements because they are totally removed from market forces in terms of rates, and they apply, poten- tially, to any cargo, route and ship.

Intermodalism
Although these trends have had, and will continue to have, heavy impacts on shipping, intermodalism gives broader and bolder insights into the current situation and future prospects. The postwar period has seen a vast develop- ment and expansion of roads and railroads throughout the world. Together with the technological developments in shipping, transport has become inter- modal. In the past, there was little competition between ships and land-based transport systems; even the shortest sea trade flourished. With intermodal transport, these trades are increasingly replaced by ferries that are basically parts of highway or railroad systems. In other words, intermodalism has brought shipping and land-based transportation into both competition and complementarity in ways never seen before. The phenomenon is seen in many, if not most, parts of the world in the concept of the land bridge. The effects are far-reaching. Intercoastal sea trade in North America is virtually gone. The same holds for Europe, where the British short-sea trades are

27. Using this proportion and quoting the OECD and the U.S. International Trade Commission, the value was estimated at $23 billion in 1980—see *Finance and Development* (December 1983), p. 15, and the International Monetary Fund's 1983 *Annual Report on Exchange Arrangements and Restrictions* (Washington, D.C.: IMF p. 46. "Growth in Global Swaps," *New York Times* (May 12, 1985), sec. 3, p. 1, quotes the director of the International Commodities Export Group estimating 1984 barter trade at U.S.$400 billion, or some 20% of world trade.

severely affected; its trade with the European continent is largely by land-based transport, the trucks using the ferry service across the Channel. The realization of plans for a Channel tunnel (the Chunnel) will also affect the ferry services. In addition, the Scandinavian trade with the Continent is by land, as is its trade with the Mediterranean and the Near East. Much European trade with the Far East is by rail through the USSR. The situations in Africa, South America, and the Pacific Rim area are not clear, as both economic and geographic factors seem to hold back these developments.

In short, what were once major traditional liner routes have largely disappeared as the cargoes have increasingly moved by land, facilitated by the intermodal nature of transport. The areas where shipping encounters competition from land transport are growing as technology makes the system increasingly intermodal. While in the past the growth in world trade has been a good indicator of seaborne trade, intermodalism has weakened this relationship. Also, with the tendency toward more value added in traded goods, the higher unit value and the smaller lot sizes may argue for overland transport rather than shipping. In the short term, therefore, world economic recovery will result in a resurgence in international trade, but it may not have as strong an impact on seaborne trade as expected.

There are some very broad policy implications in this view. First, as mentioned, growing international trade need not, as in the past, result in a commensurate growth in seaborne trade. Much depends on where the growth occurs. Second, intermodalism brings international shipping onto the domestic scene in a very explicit manner: domestic economic and development policies must consider the land interface with international shipping. Third, we need to give attention to the identification of areas, be they geographic or economic, where the scope is greatest for continued or growing sea transport. Clearly, such areas exist where there is no contiguous land mass allowing for land transport—for example, the U.S. trades, Pacific and Atlantic, will continue to rely on shipping. So will Australia, Japan, New Zealand, and other island nations (the latter may be physical or political islands). Sea transport will also prevail where land transport is not, or cannot be, sufficiently developed to provide competition. The same holds where the trade is not suitable for land vehicles—this would seem to apply to major bulk cargoes. To be able to say something specific beyond these broad, commonsensical implications, in-depth analyses are needed that are beyond the scope of this article. One may reflect, though, on the prospects for shipping if the USSR and China enter more actively the international trading community.

CONCLUSION

World shipping continues the adjustment process set in motion by economic and technological developments in the post–World War II period. Adjustment is to basic, rather than cyclical, change and entails operational and structural responses that, with a substantial time lag, prompt institutional response and change. The general effects include division of the world fleet into conventional and high-tech ships with substantial overall surplus tonnage; a growing tendency toward intermodalism and intermodal competition that erodes the base for many traditional sea routes—these developments result in fierce competition for cargoes that are growing at lower rates than in earlier years and have a more elastic demand for sea transport—the emerging new service configurations seen in circular and "hub-and-spoke" service systems; and the introduction of political considerations and government involvement to a greater extent than before. It appears that the industry is often used as a policy instrument for the achievement of nonshipping objectives.

Within this general picture are the more specific issues that are symptomatic of the adjustment process: (*a*) distribution of tonnage ownership between nations—this includes the issues of financing and the many arguments for a national fleet; (*b*) access to cargoes—that is, the U.S. Shipping Act of 1984, the UNCTAD Code, the upcoming review of both, bulk sharing, bilateralism in shipping, and counter trade; (*c*) competitiveness—convention on ship registration, flagging out, FOCs and captive registries, and manning; (*d*) safety and environment—IMO conventions, the Law of the Sea, and jurisdictions; and (*e*) operational guidelines and rules—the multimodal convention, documentary practices, insurance, and liability.

Considering these aspects, it is certain that the adjustment process will be lengthy, costly, and continuing. The process would be facilitated if policy objectives were formulated clearly and unambiguously. Political, defense, and commercial objectives ought to be separated as much as possible so that appropriate policy measures can be identified, implemented, and evaluated. The desired policy impact will occur only if a clear distinction is made between the use of the fleet as a foreign-policy instrument and its being the policy objective.

Indian Ports: Problems of Management[1]

Satyesh C. Chakraborty
Indian Institute of Management, Calcutta

INTRODUCTION

With the world's second largest population, a lengthy coastline, and an economy requiring both imports and exports for future growth, India has an undeniable need for efficient ports. This article examines the port structure of the country, the capabilities of the principal ports, and the management problems involved in ensuring that Indian ports best serve the objectives of the government.

Although the terms are often used synonymously, it is useful to distinguish between ports and harbors. Harbors are coastal locations that provide shelter for ships from wind, currents, tides, and coastal hazards such as rocks, reefs, and shoals. Ports, in contrast, are places specifically designed for the handling of cargoes brought to the area by ships. The port function requires specialized facilities, such as cranes, warehouses, and ground transportation. Natural harbors can be developed into efficient ports. When natural harbors do not exist, it is possible to produce them by means of breakwaters and similar engineering structures designed to protect the land from the sea. The capability of a port is dependent on both the facilities available to perform the traditional port functions and the natural attributes of the harbor, such as the depth of water and the degree of protection from troublesome currents, excessive tidal ranges, and strong winds. Ports also serve purposes not directly related to cargo handling. They provide repair facilities, fresh water, ships' food supplies, communications facilities, and fuel.

1. AUTHOR'S NOTE.—The preparation of this article profited immensely from a period of uninterrupted study while I was a research fellow at the East-West Environment and Policy Institute, East-West Center, Honolulu, during the summer of 1987. Grateful thanks are due to Norton Ginsburg, Joseph Morgan, Toufiq Siddiqi, and Glenys Owen Miller for their assistance in reviewing and editing this article and to Hope Sullivan for the typing.

THE NATURE OF INDIAN PORTS

There are all kinds of notions prevailing within the international maritime community about the ports of India. For example, around 1974 in some academic quarters in the United States, it was believed that India had 195 ports.[2] Lloyd's Shipping Register listed only 30 ports in India. The government of India claims that there were 172 ports in India as of 1983, 11 of which were major ports (see fig. 1).[3] One other major port is being created now. While the government of India uses the term "major port," there are obviously many variations in the facilities considered necessary to handle cargoes from different kinds of vessels at the major ports. Some of the vessels may be traditional fishing boats or those used in coastal traffic, and such vessels do not require the infrastructure needed for handling present day trade across the oceans. Evaluations of a port by the international maritime community must be weighed against what the society that created the port envisaged its role to be. An example will clarify.

Bhavnagore is a port located in the Gulf of Cambay on the coast of the Kathiawad peninsula in the State of Gujarat (21°47'N, 78°08'E). It is an old port visited by ships of many nationalities. The Lloyd's Shipping Register stated that citizens of all nations, excepting the People's Republic of China and Pakistan, were allowed ashore against passes issued by the Customs Department. There are several berths and anchorages fitted with cargo-handling gear of diverse sorts. Fresh water is said to be available at alongside berths and anchorage points. But shipmasters evaluated the facilities differently. A report issued in April 1981, stated: "For many years anchorage was used for discharging, but is now closed as about 300 anchor chains remain on bottom. In 1981, another anchorage opened, near to the port, and up until April, 4 ships lost anchors due to strong currents and irregular bottom. Contact with shore is very difficult and at night time, impossible."[4] In another report of a shipmaster, the complaint was as follows: "Water is often in short supply, and ships may be unable to obtain any at all at the anchorage. Ships should not rely upon an evaporator to make water at the anchorage, due to the high mud content of sea water."[5]

There are numerous examples of (1) navigational difficulties of diverse

2. R. F. Nyrop et al. *Area Handbook for India,* Foreign Area Studies (Washington, D.C.: American University, 1975), p. 56.

3. The provincial governments of India have the power to establish ports. These are called "minor" ports since the equipment used is simple and the facilities are moderate. The Government of India reserves the right to upgrade a "minor" port into a "major" port by accepting the obligation of providing better equipment and enlarged facilities. The Government can also develop a major port anew.

4. *Lloyd's Register of Shipping* (London: Lloyd's of London Press, Ltd., 1987), entry Bhavnagore.

5. *Lloyd's Register of Shipping,* entry Bhavnagore.

Fig. 1.—India: location of major ports

kinds that confront the Indian ports, (2) available facilities in Indian ports that fall short of expectations, and (3) Indian port managers who are aware of such limitations and are eager to make improvements, funds permitting. Not all of the 172 ports of India can receive the investments required to meet international expectations, nor is it necessary. In many instances, natural environmental conditions are too limiting, and investments would not achieve the desired ends. For many ports, the nature of the hinterland may not warrant such investments. Therefore, India has decided to develop some ports for international maritime trade and others to handle coastal traffic carried by less modern vessels. The port of Bhavnagore was seen as unworthy of handling international traffic except under emergency conditions, when, for example, congestion in the original port of call makes waiting unprofitable.

The international, or major, ports are required to maintain a variable mix of facilities for servicing modern as well as older vessels. They act as points of collection as well as of distribution of goods handled by coastal traffic. The problem of the policymakers has been to evaluate the potential of individual ports for upgrading. The development of ports has not been a function of any process of economic development in the adjacent hinterland. On the contrary, policy decisions on the development of major ports have influenced the process of economic development in nearby regions in some ways. Investments made on infrastructure at the ports have not always been the products of judicious decisions. However, the available facilities, in many ports, are well utilized.

In 1947, India had five major ports handling international trade: Calcutta (22°32'N, 88°20'E), Vishakhapatnam (17°41'N, 83°18'E), and Madras (13°06'N, 80°18'E) on the east coast; Cochin (08°45'N, 76°15'E) and Bombay (18°56'N, 72°49'E) on the west coast. The patterns of international trade envisaged by the national policymakers required creation of additional ports. Thus, Kandla (23°00'N, 70°12'E) on the west coast was made fully operational in 1959. Two other ports on the west coast, Mormugao (15°25'N, 73°47'E) and New Mangalore (12°57'N, 74°48'E) were commissioned in 1963 and 1974, respectively. On the east coast, Paradeep (20°14'N, 86°42'E), Haldia (22°00'N, 88°05'E), and Tuticorin (08°45'N, 78°13'E) were established in 1965, 1970, and 1975, respectively. Another major port on the west coast, Nava Sheva on the mainland across from Bombay Bay, is being developed; it should be operational by 1990. The bias has been in favor of the west coast, if we note that Tuticorin is approachable only through the Gulf of Mannar and not from Palk Bay. Including Tuticorin, a port on the east coast but not approachable from the east, there are seven major ports on the west coast of India, and five on the east coast. If we further consider that, by virtue of being on the same sea-lane through the Hooghly River, Calcutta and Haldia are managed as a single port, the total number of port complexes on the east coast is four. This relationship between the west and the east coasts reflects the locational orientations of the countries with which India conducts international trade. These countries are in Western Europe or along the Atlantic coasts of North and South America.

In policy formulation and implementation, development of major ports touches on sensitive issues connected with federal-state relationships. All major ports are administered under the port trust act of the federal government. Developmental investments in the major ports are also met by the federal government. Other ports are controlled and financed by the state governments. For this reason, all state governments are eager to see their ports elevated to the status of major ports. An additional reason is that international trade creates opportunities for developing industries in the neighborhood. This occurred when ports at Kandla, Mangalore, Tuticorin, and Paradeep were converted into major ports from their earlier status. However, the national policymakers now believe that, except for Porbandar (21°38'N, 69°36'E) on the west coast, none of the minor or intermediate ports have the potential to become major ports before the end of the present century.

The British Raj in India ended in 1947. At that time, the major ports were riddled with problems caused by the overuse of available facilities during the Second World War. This was a period when minimal investments were made on hardware, and the demands for increased traffic to support war efforts were largely met by employing additional labor. When the war ended and traffic slackened, all the major ports were burdened with surplus labor. Although the Dock Labor Board undertook to remove this problem from the individual ports, the existence of surplus labor conditioned judgments on choices of technology for transport management for many decades afterward. The national policymakers carried out their task of strengthening port facilities in stages according to set priorities. During the first 10 years or so after gaining independence, the priorities were to rehabilitate the overused facilities in the major ports and to augment berthing capacities. After 1960, new needs were attended to, such as the provision of additional capacities, modernization of equipment and related facilities, and so forth. More recently, emphasis has been on the optimal utilization of installed capacity in the ports. Although all decisions on port development are geared to managing international trade and the provisioning of modern facilities for the technologically advanced vessels used for this purpose, the major ports, in combination with the nearby minor and intermediate ports, have the potential of generating and supporting many other kinds of maritime activities. Marine fishing is getting considerable support. Offshore operations for hydrocarbon exploitation are another type of activity supported by Indian ports.

FACILITIES AT MAJOR PORTS OF INDIA

The facilities available at the 11 major ports of India are varied and not strictly comparable among themselves. Age after commissioning, navigational difficulties in approaches, distance from the industrially developed regions, linkages with regional transport systems, and so on, are some factors governing the observed differences between ports. A review of the facilities available in the major ports can provide a basis for reviewing their comparative strength or weakness.

Kandla is located on a tidal creek that joins the Gulf of Cutch in the State of Gujarat. This creek has bars that impose restrictions on the passage of vessels during ebb tides. Vessels longer than 740 feet and with drafts more than 32 feet cannot pass. Nighttime navigation is generally avoided. Dredging is being carried out to deepen the port. Pilotage is compulsory. Anchorages at the entrance of the creek and near the port are available.

There are five alongside berths to handle dry cargo and containers. The port has marine loaders to discharge fertilizer in bulk. The facilities to handle tankers consist of exclusive berths fitted with separate pipelines for white oil, black oil, soybean oil, acetone, formaldehyde, and phosphoric acid. In addition, there are three wharves, one of which is used exclusively by country vessels. A fishing harbor is another available facility.

Medical facilities are limited at the port site but are adequate in the nearby port township. Supplies of fresh water and fuel are available. A floating dry dock handles ships up to 500 mt. For all vessels, facilities for chipping, painting, and minor repairs are available. The port does not permit discharge of tank washings and residues within its area.

The Kandla port is connected to the national network of broad-gauge railroad tracks and to the national highway system. The nearest airport is about 25 km distant. Being a new port, it has not yet stimulated significant industrial development in the neighborhood. But there are favorable signs that a considerable degree of industrial development, using the imported raw materials, will occur shortly. However, being located in a semiarid region, prospects of developing industries that discharge large amounts of effluent are limited. This has led many to believe that assembly-type industry for export stands a better chance of development.

Bombay is located in a sheltered bay on the east side of Bombay Island. There are 60 safe anchorage points, but pilotage is necessary. The port has three docks—Indira, Victoria, and Princess, with a total of 44 berths with capacity to handle 5,000–20,000-deadweight-ton (d.w.t.) vessels. In addition, the Ballard Pier can handle 22,000-d.w.t. ships carrying dry cargo. There are two berths capable of handling container traffic. Tanker services are available at Butcher Island Marine Oil Terminal and at Pir Pau Oil Berth, which have drafts ranging from 28 to 38 feet and capacity to handle tankers of 36,000–54,000 d.w.t. carrying liquid bulk cargo. Dirty ballast tankers are available.

Good medical services and supplies of fresh water and fuel are available at all berths. All categories of repairs are possible. Two dry docks can handle 525 × 65-foot and 1,000 × 100-foot vessels, respectively. The international airport is located 24 km away. The port is serviced by broad-gauge railroad and national highways.

Bombay is the largest port in India measured by the volume of traffic handled. Its immediate neighborhood is very well developed industrially and is one of the fastest growing regions of India. It grew up as a major center of the cotton textile industry starting in the mid-nineteenth century. With the passage of time, it has emerged as a major center of heavy chemicals, electrical engineering, machine tools, and pharmaceuticals.

Mormugao is located on a submerged rift valley close to the mountains on the west coast. It is fairly well sheltered, except during the stormy monsoon season. Several rivers that are navigable by barges debouch into the bay. Pilotage for oceangoing vessels is essential. There are many anchorage points, as well as several alongside berths, some of which are fitted with bulk-handling facilities for iron ore. Tanker service is available for discharging liquid cargo in bulk.

The port was a part of the Portuguese possessions in India and had remained rather isolated from the mainstream of economic development in the rest of India. While Mormugao is not yet connected by a broad-gauge railroad system, its meter-gauge tracks are gradually being replaced by broad-gauge lines. The port is connected with the national highway system. A domestic airport is 7 km away. No significant degree of industrialization has as yet taken place in the neighborhood. The port has no dry dock or facilities for major repairs.

New Mangalore is an artificial harbor that was developed 9 km north of a minor port by the name of Mangalore on the west coast. Within the harbor, several alongside berths handle imported raw materials for a fertilizer factory located in the neighborhood. There is a mechanized iron ore berth, an oil terminal for shore discharge, and facilities for general cargo of all sorts. Available draft is between 30 and 41 feet. Entry into the harbor requires pilotage. Anchorage in the open sea beyond the breakwaters is available.

Freshwater and fuel supplies are available. The port runs its own workshop to carry out all types of repairs. It is connected to both broad-gauge and meter-gauge railroad systems. The domestic airport is located about 30 km away.

The neighborhood is receiving public-sector investments for industrial development. Emphasis is put on heavy chemicals, including fertilizer factories. Very little industrial growth has taken place through private investment.

Cochin is located on a promontory at the head of a very large backwater (lagoon) formed through submergence of the coast. Maintenance dredging is necessary. There are several points for anchorage in the open sea. Pilotage is necessary. Berths fitted with modern cargo-handling equipment deal with coal, oil, general cargo, containers, and so forth. There is a large shipbuilding yard near the port, which undertakes all types of repairs. The port has its own dry docks. Medical service, fuel, and freshwater supplies are available.

The port is connected to a broad-gauge railway line and a national highway. The airport is about 5 km away. Noticeable growth of industries has taken place in the neighborhood, especially petroleum refining and petrochemicals.

Tuticorin has become safe in all seasons since the construction of a sturdy breakwater and a harbor. However, pilotage is compulsory within the port. The old port is located about 7 km to the east. Within the harbor, berths with deep water alongside and efficient cargo-handling gear deal with both break-bulk and dry-bulk cargo. One oil jetty is used for discharging furnace oil and naptha. A considerable amount of industrial development has taken place in the neighborhood, especially in heavy chemicals. Repairs for small craft are possible. The dry dock is used primarily for servicing vessels belonging to the port, but use by private parties is allowed when available. Tuticorin is connected to the meter-gauge railroad system, and national highway plans are afoot to connect it to the broad-gauge railway system. The nearest airport is 160 km away.

Madras is an artificial port with a breakwater and harbor where many anchorage sites are available. Pilotage for inbound, outbound, and within-harbor movements is compulsory. Vessels berth at all times. It was chosen for development as a major container-handling port, and ore handling is becoming less important. All berths are well appointed. Tankers are handled in the outer harbor from an oil berth with five arms. Both freshwater and fuel supplies are available. Although it has no dry dock, all types of deck and engine repairs are possible.

The port is connected to both meter-gauge and broad-gauge railroads as well as national highways. The international airport is about 24 km away. The neighborhood, well established as an industrial region of India, has developed textile, heavy chemical, mechanical engineering, and petrochemical industries.

Vishakhapatnam has a natural harbor within a submerged rift valley. Maintenance dredging is necessary. Safe anchorage, except during the stormy monsoon months, is available just outside the harbor entrance in the open sea. Pilotage is necessary. The inner harbor is open to ships of 193 m length overall (LOA), and 10.21 m draft. The outer harbor admits vessels of 270 m LOA with 15.3 m draft. The inner harbor generally handles containers and general cargo, but it also has two berths for discharging and loading oil. The outer harbor handles bulk cargoes of iron ore and liquids. All berths and jetties are well appointed with modern cargo-handling facilities. Freshwater and fuel supplies are available, as is medical service.

The port is connected to the broad-gauge railroad network and national highways. The domestic airport is about 14 km away. There is a shipbuilding yard next to the port. All types of engine and deck repairs are possible. The neighborhood has experienced rapid industrialization in recent years. A large iron and steel industry is being set up, and petroleum-based industries have developed.

Paradeep is located on the edge of a delta on the east coast. Pilotage is necessary but is available during daylight hours only. Maximum available draft is 39 feet. There are facilities for bulk handling of iron ore and general cargo berths but no tanker- or container-handling facility. It is connected to a broad-gauge railroad and national highway. The nearest domestic airport is 120 km away. The neighborhood has yet to experience any mentionable order of industrialization. Minor repair services to vessels are available. The dry dock has a 500-ton slipway.

Haldia is considered, administratively, a subsidiary port of Calcutta. But the nature of installed technology is so different from Calcutta that it deserves to be described as a separate port. It is a deep-water port located on the Hooghly River, which can admit vessels of 885-foot length. Almost all cargoes are handled by mechanical plants. There are facilities for handling tankers, ore and coal carriers, containers, and general cargo. The port is connected by broad-gauge railroad and national highways with Calcutta and the rest of India. The neighborhood is experiencing considerable industrialization as an extension of the Calcutta metropolitan area and petroleum-based industries have been established.

Calcutta is a river port, located 120 river-miles upstream on the Hooghly, a river that has been experiencing rapid siltation. Maintenance dredging is essential, and pilotage is required for all vessels. The sharp bends in the river restrict the lengths of vessels to 535 feet. Vessels are brought in for berthing only at favorable tide stages. Anchorage at many points in the river is available. Alongside berthing is at Budge Budge and Garden Reach for oil tankers carrying liquids of all kinds. Containers and general cargo are handled at Netaji Subhas Dock and Kidderpur Dock. Medical facilities, fresh water, fuel, and repair service are fully provided. There is a shipbuilding yard next to the port, which includes five dry docks. All types of repairs can be carried out.

The port is served by a broad-gauge railroad and national highways. The international airport is 20 km away. The largest network of inland navigation connects this port, as well as Haldia, with Assam, Bangladesh, Bihar, and Uttar Pradesh. The neighborhood is highly industrialized and has been a major center of the jute textile industry since the middle of the last century. More recently, it became the leading center of the engineering industry, with emphasis on transport equipment and rolling stock. However, industrial growth in this part of the country has been rather sluggish since 1965.

Nava Sheva is being developed on the west coast near Bombay, primarily to handle container traffic. It has not yet been commissioned. It is likely to boost further industrialization within the Bombay metropolitan region.

FUNCTIONS OF MAJOR PORTS AND ASSOCIATED PROBLEMS

In the mid-sixties, almost all major ports of India suffered from congestion and consequent delays in turnaround time of vessels. This was an irksome situation for shippers engaged in international trade. It was generally felt that the prevailing cargo-handling practices were contributing to delayed turnaround time. Since then, substantial efforts have been made to increase the number of major ports and to replace labor-intensive practices in the older ports by installing mechanized facilities. This has not, however, made the management of ports easier.

Table 1 is a convenient starting point for a discussion of problems confronting Indian port managers set against the roles that their ports play. It presents data on types of cargo handled by the major ports of India in 1983–84 and 1984–85. It shows that more than three-fourths of the total cargo handled by the major ports of India is explained by petroleum products (46.18%), iron ore (22.57%), and miscellaneous cargo (7.95%). Both petroleum products and iron ore are being handled by mechanized facilities. But, miscellaneous cargo arrives generally in break bulk and requires labor-intensive methods of handling. Some other types of cargo are unloaded from

TABLE 1.—TYPES OF CARGO HANDLED BY THE MAJOR PORTS
(in thousand tonnes)

Cargo Type	1983–84	1984–85	Average of 1983–84 and 1984–85 Traffic	Average as % of Traffic
Petroleum products	39,189	39,523	39,356	46.18
Iron ore	18,133	20,345	19,239	22.57
Other ores	566	945	756	.88
Fertilizers	1,122	2,878	2,000	2.35
Raw materials	2,077	2,386	2,232	2.61
Food grains	2,792	1,161	1,977	2.33
Coal	3,436	3,289	3,363	3.95
Salt	561	148	355	.42
Sugar	733	572	653	.78
Cement	2,156	531	1,344	1.58
Iron and steel	1,723	1,841	1,782	2.09
Newsprint	97	98	98	.11
In containers	1,903	2,566	2,235	2.62
Vegetable oil	1,182	1,663	1,423	1.67
Miscellaneous goods	6,385	7,163	6,774	7.95
Transhipment goods	1,808	1,456	1,632	1.92
Total	83,863	86,565	85,219	100.00

NOTE.—Data on the cargo handled have been collected from "Basic Port Statistics" (Government of India, Ministry of Shipping and Transport, Director of Transportation Research, 1987, mimeographed). Only a small part of the data collected have been used in the article to highlight the simpler issues.

ships with machines such as vacuum evacuators but are bagged, stitched, and stored through manual operations; food grains and fertilizers are examples. In short, port managers in India are required to pursue two different styles of labor management, one for skilled workers with mechanized systems and the other for unskilled or semiskilled labor engaged in the manual systems.

In table 2, the work force at the major ports in 1984 is shown. Most striking is the large size of the work force, which is partly due to the fact that introduction of mechanized facilities was not necessarily followed by a large-scale retrenchment of employees. This is particularly true of the older ports like Bombay, Calcutta, Madras, Vishakhapatnam, and Cochin. We should also note that a distinction exists between cargo-handling workers on shore and shipboard workers. This means that, for transshipment of a consignment, different labor-gang supervisors oversee shore and shipboard operations. Coordination is frequently difficult.

In addition to the work force administered by the port authorities, there are workers supervised by the dock labor boards, as shown in table 3. Where such boards do not exist, stevedoring companies are permitted to engage

TABLE 2.—WORK FORCE AT MAJOR PORTS, 1984

Major Ports	Officers	Noncargo-handling Workers	Shipboard-Cargo-handling Workers	Shore Workers Handling Cargo	Total
Bombay	510	21,498	2,989	7,343	32,340
Calcutta*	1,090	24,356	1,457	3,195	30,098
Madras	588	9,698	735	1,282	12,303
Vishakhapatnam	259	9,351	992	1,112	11,714
Cochin	211	5,280	659	677	6,827
Kandla	130	3,914	123	665	4,832
Mormugao	173	2,713	1,060	116	4,062
Paradeep	156	2,705	. . .	. . .	2,861
Tuticorin	104	2,262	. . .	. . .	2,366
New Mangalore	92	1,283	. . .	. . .	1,375
Total	3,313	83,060	8,015	14,390	108,778

*Includes Haldia.

their own workers. This further complicates management procedures. Since labor is in oversupply, fresh recruitment of workers is not encouraged. Premature retirements with various incentives have been encouraged. Gradual introduction of productivity-increasing machines, such as bag stitchers, forklifts, and so on, are measures being adopted in all ports. Training of workers is also encouraged.

The total tonnage of all types of cargo handled by all major ports of India has been increasing over the years. This is true for cargo destined for export or received through import. For example, 1984–85 recorded a 3.22% increase in tonnage over the previous year. It is believed that, by the turn of the

TABLE 3.—WORK FORCE IN DOCK LABOR BOARDS, 1984

Major Ports	Officers	Noncargo-handling Workers	Cargo-handling Workers	Total
Bombay	37	734	10,719	11,490
Calcutta	70	1,156	7,754	8,980
Madras	32	511	2,381	2,924
Vishakhapatnam	21	413	2,455	2,889
Cochin	25	182	1,131	1,338
Kandla	3	56	2,452	2,511
Mormugao	10	100	1,120	1,230
Total	198	3,152	28,012	31,362

NOTE.—Haldia, Paradeep, Tuticorin and New Mangalore do not have dock labor boards, but these permit engagement of workers by the stevedores.

TABLE 4.—SHARE OF MAJOR PORTS IN TOTAL CARGO
(as % of sum of all major ports)

Major Port	1983–84	1984–85	Average of 1983–84 and 1984–85
Bombay	25.59	24.21	24.89
Kandla	13.90	15.30	14.61
Madras	12.75	14.30	13.53
Mormugao	12.37	12.19	12.27
Vishakhapatnam	11.76	11.91	11.84
Haldia	6.30	6.12	6.21
Cochin	5.14	3.61	4.37
Calcutta	4.00	3.60	3.80
Tuticorin	3.56	3.60	3.58
New Mangalore	2.92	3.39	3.16
Paradeep	1.71	1.77	1.74
Total	100.00	100.00	100.00

century, the major ports of India will handle about 155 million tons of cargo. This projection is used as a basis for determining needed capacity of Indian ports. Some of the newly installed capacity may remain underutilized in the initial years, and there are reasons to believe that many ports presently have underutilized facilities. The problems faced by port managers in this regard are not simple or straightforward. This is revealed by an examination of table 4.

The increase in gross tonnage between 1983–84 and 1984–85 was not shared by all the major ports proportionately. While the tonnage increased in ports like Haldia, Paradeep, Vishakhapatnam, Madras, New Mangalore, Mormugao, and Kandla, it decreased in Bombay, Cochin, Tuticorin, and Calcutta. International trade is a function of many factors, and choices of the ports used may vary between years. Port managers should, therefore, accept the fact of underutilization of facilities available in the ports as something normal. The problem of capacity utilization should then be the responsibility of an authority other than the port managers themselves. The job would be to see that the aggregate capacity of all the major ports together is optimally utilized. The Indian Ports Association (IPA), a voluntary association, can play such a role. The Ministry of Shipping and Transport of the government of India might assume such a function. However, neither of these two organizations has been able to deal with this problem in the past, nor have they settled which of them should play the primary role in this regard.

If tonnages of individual classes of goods, such as petroleum products, iron ore, containers, and so forth, are considered, rather than simply gross tonnage, additional problems become evident. For example, there has been an increase in aggregate tonnage for petroleum products between 1983–84 and 1984–85. This increase is reflected in all ports except Cochin and Bombay, where the tonnage decreased. The same situation exists at Haldia with regard to iron ore, at Vishakhapatnam and Cochin for ores other than iron, and at Paradeep and Madras for raw materials. The pattern of aggregate growth or decline in the tonnage of one or the other commodity may not be reflected in all ports in the same way over the years. Under these circumstances, individual port managers have to be reconciled to year-to-year variations in utilized capacities of some or all of the facilities available to them. Generally speaking, a port manager feels happy if 67% of the installed capacity gets utilized each year. Such a criterion is, however, easier to apply to mechanized systems than to manual systems as an indicator of performance.

The indication of the randomness of cargo handled by the individual ports should be qualified. Table 4 shows that, by tonnages of cargo handled, Bombay ranks first and Paradeep last. More important is the fact that the ranks of the individual ports have remained unchanged between the 2 years considered. This suggests that the pattern of commodity movement is not random in absolute terms, over short periods at least. Long-term changes are indeed taking place. For example, the position of Haldia as a port for iron ore has been steadily declining. Its high-speed mechanized plant has remained largely unutilized, as the expected increase in the volume of export has not materialized over the years. Stagnation in the regional economy of the hinterland, in contrast, has produced consistent declines in the tonnage in the port of Calcutta. The movement of a cargo such as salt is increasingly promoted through railways. These are some of the varied factors influencing port functions. The managers of ports must take note of these, and ports experience variations in the order of capacity utilization between years. A port, being a facilitator, has to continually readjust its strategy in order to acquire as well as to discard facilities in tune with emerging or declining demands for service. Within the ambit of port administration, there is no specific department that is known to be conducting the required analyses in this regard.

The relative importance of the ports, measured by tonnage of cargo handled, varies between commodities. In table 5, their ranks are shown against each type of commodity. This shows that no port enjoys the premiere rank in all commodities, which suggests that certain ports specialize in handling certain sets of commodities. By computing a coefficient of association between commodities, it has been possible to identify pairs whose distribution between ports are nearly identical or concordant. The closest pairs are noted in table 6 along with their indices of association, which varies between values of 1.00 and zero, representing the range between the most to least concordant distributions. These pairs establish five sets of commodity groups, on which the specialized roles of each port are based and are noted in table 7.

The distinctive roles of the individual ports can be better observed from table 8, which was produced from table 7 by converting the relative incidence

TABLE 5.—RANK OF PORTS BY COMMODITY TYPE,
(Average of 1983–84 and 1984–85)

Rank	Petroleum Products	Iron Ore	Other Ores	Fertilizer
1	Bombay	Mormugao	Mormugao	Madras
2	Kandla	Vishakhapatnam	Madras	Tuticorin
3	Madras	Madras	Vishakhapatnam	Kandla
4	Haldia	New Mangalore	Paradeep	Bombay
5	Cochin	Paradeep	New Mangalore	New Mangalore
6	Vishakhapatnam	Haldia	Tuticorin	Vishakhapatnam
7	Mormugao	Kandla	Kandla	Calcutta
8	Calcutta	. . .	Cochin	Mormugao
9	Tuticorin	. . .	. . .	Haldia
10	New Mangalore	. . .	. . .	Cochin
11	. . .	. . .	. . .	Paradeep

of each set of cargo into units of corresponding incidences for the total of major ports. For example, by dividing 17.35 (which is the relative incidence of set 1 for Calcutta) by 47.85 (which is the relative incidence of the same set for all the major ports), we get 0.36. This has been shown for Calcutta in table 8. Here, sets having values of more than one demonstrate the distinctiveness of the concerned ports in the sense that such ports handle the given sets of commodities in a conspicuous manner compared to the datum given by total trade through major ports. Some of these sets of commodities are amenable to mechanized handling, while others are not. We can appreciate why in Calcutta emphasis on manual and mechanical operations emerges as a special managerial responsibility in contrast, for example, to Paradeep, where set 2 commodities permit mechanical handling.

There are two other types of problems, the effects of which concern port managers. The first is related to the worldwide trend toward containerized transshipment of maritime cargo. The other arises from the relationship between traders and vessel owners. Both are complex, related to problems of sociotechnical changes that India is now undergoing.

Containerized vessels can be gainfully employed if both the consigner and the consignee can make use of a full container. Since the production base of India is composed of small exporters and importers, a number of consigners and consignees together can make effective use of a container. This requires that port managers develop or encourage development of facilities to load and unload containers within ports or at suitable inland locations. Such facilities are gradually being developed in India, but port managers still have much responsibility. Loss of articles within ports is a common problem faced by the users of such services, and insurance and handling charges are therefore high.

The relationship between consigners and vessel owners is also com-

TABLE 6.—PAIRS OF COMMODITIES HAVING CONCORDANT
DISTRIBUTION BETWEEN PORTS
(based on average of 1983–84 and 1984–85)
A. COEFFICIENTS

Pairs of Commodities	Coefficient of Concordant Distribution	Pairs of Commodities	Coefficient of Concordant Distribution
Petroleum products and vegetable oil	.7844	Sugar and fertilizer	.7690
Iron ore and other ore	.7758	Iron and steel and miscellaneous goods	.8491
Fertilizer and cement	.8772	Newsprint and containers	.6100
Raw material and cement	.7082	Containers and miscellaneous goods	.6410
Foodgrains and cement	.7635	Transhipment and iron ore	.4568
Coal and salt	.4586		

B. RANK OF EACH PORT BY COMMODITY

Rank	Containers (1)	Miscellaneous Goods (2)	Vegetable Oil (3)	Transhipment Goods (4)
1	Bombay	Bombay	Bombay	Vishakhapatnam
2	Madras	Calcutta	Kandla	Mormugao
3	Calcutta	Madras	Madras	Paradeep
4	Cochin	Kandla	Calcutta	Tuticorin
5	Haldia	Tuticorin	Vishakhapatnam	. . .
6	Kandla	New Mangalore	New Mangalore	. . .
7	Tuticorin	Cochin	Haldia	. . .
8	New Mangalore	Vishakhapatnam	Cochin	. . .
9	. . .	Mormugao	. . .	. . .
10	. . .	Paradeep	. . .	. . .
11	. . .	Haldia	. . .	. . .

Rank	Raw Materials (5)	Food Grains (6)	Coal (7)	Salt (8)
1	Bombay	Calcutta	Tuticorin	Tuticorin
2	Vishakhapatnam	Madras	Haldia	Kandla
3	Cochin	Kandla	Vishakhapatnam	Calcutta
4	Madras	Vishakhapatnam	Calcutta	Madras
5	Kandla	Bombay	Madras	. . .
6	Calcutta	New Mangalore	Paradeep	. . .
7	Tuticorin	Cochin	. . .	. . .
8	Haldia	Tuticorin	. . .	. . .
9	Mormugao	Paradeep	. . .	. . .
10	. . .	Mormugao	. . .	. . .
11	. . .	Haldia	. . .	. . .

TABLE 6.—(*Continued*)

	Sugar (9)	Cement (10)	Iron and Steel (11)	Newsprint (12)
1	Vishakhapatnam	Madras	Bombay	Bombay
2	Kandla	Kandla	Calcutta	Madras
3	Madras	New Mangalore	Kandla	Cochin
4	Tuticorin	Bombay	Vishakhapatnam	. . .
5	Bombay	Tuticorin	New Mangalore	. . .
6	Mormugao	Calcutta	Madras	. . .
7	New Mangalore	Vishakhapatnam	Cochin	. . .
8	Calcutta	Cochin	Paradeep	. . .
9	Haldia	Mormugao	. . .	. . .
10	Cochin	Haldia	. . .	. . .
11	. . .	Paradeep	. . .	. . .

NOTE.—From part A, the following five sets of commodities are identifiable as having high coefficients of concordant distribution: set 1: Petroleum and other liquids and vegetable oil; set 2: iron ore, other ores, and transhipment goods; set 3: fertilizer, cement, raw materials, foodgrains, and sugar; set 4: coal and salt; and set 5: iron and steel, miscellaneous goods, containers, and newsprint.

plicated. A vessel often has to wait at a port for the cargo to arrive for loading. The exporter is reluctant to take the risk of pilferage at the port warehouses, even if the consignment is fully insured. The reverse is the case with imports.

Here the consignee finds it profitable to leave his goods in the custody of the ports for an unusually long time. The difference in attitude toward import and export cargo is related to current procedures to obtain the insured value of lost consignments. Since the insurance houses are located in the countries of origin of the respective consignments, it is easier for the consign-

TABLE 7.—RELATIVE INCIDENCE OF SETS OF COMMODITIES BY PORTS
(in %, based on average of 1983–84 and 1984–85)

Ports	Set 1	Set 2	Set 3	Set 4	Set 5
Bombay	70.62	. . .	5.99	. . .	23.39
Kandla	81.22	.25	9.14	.76	8.63
Madras	47.06	29.58	11.59	1.02	10.75
Mormugao	5.45	90.32	3.15	. . .	1.08
Vishakhapatnam	21.58	60.28	10.92	3.40	3.81
Haldia	67.71	.25	4.38	25.33	2.33
Cochin	71.16	.12	16.34	. . .	12.38
Calcutta	17.35	. . .	27.65	6.77	48.23
Tuticorin	12.52	.81	22.33	51.69	12.69
New Mangalore	12.65	50.54	18.75	. . .	18.06
Paradeep	. . .	84.25	7.85	. . .	6.17
Total of major ports	47.85	25.37	9.65	. . .	12.77

NOTE.—Figures represent percent share of total tonnage handled by the given port. The bottom line is a summation for all ports.

TABLE 8.—DISTINCTIVENESS OF PORT FUNCTION
(based on average of 1983–84 and 1984–85)

Port	Set 1	Set 2	Set 3	Set 4	Set 5
Bombay	1.48	. . .	.62	. . .	1.85
Kandla	1.70	.01	.95	.17	.68
Madras	.98	1.17	1.20	.23	.84
Mormugao	.11	3.56	.33	.00	.08
Vishakhapatnam	.45	2.38	1.13	.78	.30
Haldia	1.42	.01	.45	5.80	.18
Cochin	1.49	. . .	1.69	. . .	.97
Calcutta	.36	. . .	2.87	1.55	3.78
Tuticorin	.26	.03	2.31	11.83	.99
New Mangalore	.26	1.99	1.94	. . .	1.41
Paradeep	. . .	3.32	.81	. . .	.48

NOTE.—Table 8 is a derivative of table 7. The entries of each column in table 7 have been divided by the entry at the bottom line of respective column. The unit is simply an algebraic number.

ers to recover damages. An added reason is absurdly low warehouse charges at the ports. Whatever the reason, such practices adversely affect the turn-around time of vessels and make the ports less efficient.

INSTITUTIONAL EFFORTS AT PROBLEM SOLVING

As mentioned earlier, the major ports of India are individually autonomous institutions. There is a trust for every port, which is composed of representatives drawn from the different interest groups using the port. The trust makes policy decisions, and its chairman acts as the chief executive of the port. The specific problems confronting the various ports differ, and the autonomous trusts can generally deal with them. However, purely operational issues are increasingly burdening the trusts, leaving little time for performance appraisal and remedial planning. This is to be expected as the economies in the hinterlands are rapidly changing, calling for changes in methods to deal with demands from port users.

Introduction of change is not free from risk. A change appearing logical in the context of a given port has an impact on the labor force. Since the port workers are strongly unionized, numerous norms adopted by one trust may be contested by the affected union and by the labor unions of the other ports. In contrast, if the changes are considered advantageous to workers at one port, unions at other ports may demand similar advantages.

In the decade of the sixties, the chairmen of the major ports felt the need for regular consultations and set up the IPA, which is recognized by the Ministry of Shipping and Transport of the government of India as a useful body to resolve through consensus the common problems confronting the major ports. The IPA meets regularly either at its headquarters in New Delhi or in one of the major ports. Over the years, the IPA has expanded its activities by stimulating the member ports to explore new opportunities for development and to appraise performance. The IPA also offers consulting services to the major ports.

The major ports are presently concerned with adopting policies acceptable to the common membership of the IPA in three areas: (*a*) developing computer-based management information systems to facilitate quick decision making, (*b*) applying priorities between areas requiring modernization of equipment or facilities and their optimal usage, and (*c*) providing on-the-job training for staffs to improve their skills.

Management Information Systems

All the major ports and the IPA now have computer facilities tied to mainframe systems. But the individual systems do not necessarily match, which has hampered information exchanges between ports and with the IPA. This has not contributed toward the emergence of the IPA as a centralized repository of information or to establishment of the IPA as a centralized decision-making body. Hence, the individual ports still enjoy considerable autonomy.

Close examination of the uses made of the computer facilities by the individual ports reveals some emerging common areas of interest as well as differences. Wage and salary administration through in-house or contracted service is common in almost all the major ports. Another area of common usage is the compilation of data on cargo handled. But, these data are not used to analyze emerging trends in the economy of port hinterlands or for measuring changes in service areas. Hardly any data on the usage made of the equipment or facilities are stored. Some ports use computers for inventory control, but this control system is rarely integrated with the functions of maintenance management. Some ports are using computers for financial management and occasionally for monitoring construction projects. Attempts are underway to computerize income and expenditure accounts.

The above observations should not be seen as a critique by an unhappy onlooker with an unbounded enthusiasm for computer technology. They are mentioned only to highlight the very issues that many middle-level managers of the ports have themselves raised. Nevertheless, they are cautious about the process of introducing a new technology for management decision making, feeling that it would entail adoption of a new "culture" throughout the organization. Many of the functioning managers have no idea about what decisions they can make based on the information processed by a computer. Hence, formatting data for transmission from the workplaces to the appropriate decision maker is generally seen as a technological problem, but in reality the problem may very well be social or psychological. Generally, the prospect of participation of the work force in the computer culture is fraught with pessimism, and anxieties overshadow real-life problems. India is at the nascent stage of adoption of computer culture in a society that is burdened with a labor-surplus economy. Nevertheless, there persists a considerable degree of willingness among many managers to explore the options and opportunities of computers.

Allocation of Priority in Investments

Autonomy generally promotes insular perceptions of needs, which may be strongly advocated by Indian ports. Decisions to acquire new facilities came from all ports simultaneously, based on interpretation of announcements of new policy from the national government. For example, all ports wanted to set up mechanized ore-handling plants when the national government expressed a desire to export iron ore. The trade did not meet expectations, and the installed capacity in the ports is now far in excess of need. There were similar desires to install container-handling facilities in ports. The Ministry of Shipping and Transport wishes to avoid excess capacity in this regard. It has been proposed that only two major ports, Bombay on the west coast and Madras on the east coast, be equipped to handle international container traffic. This would minimize heavy investments in port facilities by the Ministry of Shipping and Transport, which is the single source of funding of developmental works in all major ports. There are limitations to the above plan. If investments in container-handling facilities are denied to all ports other than Bombay and Madras, these two would be expected to function as entrepôts. Unless loading and unloading of containers is handled exclusively by these two ports, which may not be the most cost-effective solution for all exporters and importers in a large country like India, all other ports have to function as supporting systems for container traffic and must have some facilities incorporated. Moreover, shippers prefer to deliver or pick up loaded containers from ports nearest to their clients. For these reasons, container handling is an unavoidable task for almost all the major ports. Providing funds only to Bombay and Madras has stimulated the other ports to make ingenious use of traditional cargo-handling gear at their disposal, sometimes leading to accidents.

The IPA is examining a number of options for better utilization of the installed cargo-handling capacities, with or without marginal supportive investments. These include improvement of access, better arrangements of trucking, quicker allocation of railway wagons, and increased use of riverside cargo handling, where opportunities are available, such as in Calcutta, Haldia, Cochin, and Mormugao. Most ports are streamlining procedures to assist the users by providing less cumbersome modes of dispatch of cargo, resulting in quicker turnaround of vessels. The emphasis is on timely maintenance of cargo-handling gear, navigational channels, berthing facilities, and remodeling of warehouses. Labor deployment and supervision are also considered important. Each port decides for itself on how to achieve its objectives, and the IPA assists by acting as a consultant.

Skill Formation through Training

The importance of training was recognized by some ports earlier than others. In the mid-sixties, the port of Calcutta took the lead by setting up an institute to provide managerial training. Shortly afterward, it began catering to similar needs of other ports in India. The IPA took over in the mid-seventies and formed the Indian Institute of Port Management. For reasons not altogether clear, in the mid-eighties, the IPA decided to set up another such training center, now called the National Institute of Port Management, located at Madras. This is designed to cater to managerial training exclusively. The aggregate capacities of these two institutions provide considerable redundancy, but closing of one of the institutes would pose decision-making problems within the IPA.

In addition to training managers, many ports feel the need to train their blue-collar workers. Since it is inappropriate to take such trainees out to distant locations, training centers are being located within the individual ports. Access to properly prepared training materials is a problem. This may be addressed by the institutes of port management at Calcutta and Madras, which will also provide instructor training. The International Labor Organization could be a helpful resource if the port authorities can "Indianize" imported training materials.

CONCLUDING REMARKS

Indian ports are in a transitory state; they are adopting new technologies, hinterlands are changing in character, and management procedures and training of personnel are in a state of flux. The Indian Ports Association and the individual port trusts must adapt to these processes of change if India is to realize its potential as a commercial sea power.

Transportation and Communication

Coal Ports and the Environment*

Toufiq A. Siddiqi
East-West Center

INTRODUCTION

A common basis for classifying ports is the relative importance of the services they provide.[1] One thus refers to fishing ports, naval ports, and free ports, for instance, while realizing that these ports serve a number of other purposes as well.[2]

In this paper, the term "coal port" will be used to refer to ports that handle substantial amounts of coal (at least several million tons per year), whether or not this is their main activity. In addition to the environmental impacts associated with the construction and operation of all ports, there are some specific environmental concerns that need to be addressed when large amounts of coal are to be handled at a port. This paper summarizes some of these.

COAL USE AND SHIPMENTS

During the 1950s, it appeared that coal as an energy source might be on the way out, to be replaced by oil and natural gas—sources that were cleaner and easier to transport. During the 1960s, nuclear power also became a commercial source of electricity, gradually displacing coal in countries where air quality had become a serious concern. The jump in oil and gas prices during the 1970s revived interest in coal, and the fall from grace of nuclear power during the 1980s accelerated this trend.

* EDITOR'S NOTE.—The author would like to express his thanks to Dr. John Wiebe, Dr. Herbert Webber, and other members of the international team that prepared *Environmental Guidelines for Coal Port Development.* Much of the material presented in this paper draws heavily on that book. Some of the material in this paper was presented at the seminar "Mitigation of Environmental Impacts of Port Development in Developing Nations," held in Baltimore in November 1988.

1. F. W. Morgan, *Ports and Harbours* (London: Hutchinson, 1958).
2. Yehuda Hayuth, "Seaports: The Challenge of Technological and Functional Changes," *Ocean Yearbook 5,* ed. Elisabeth Mann Borgese and Norton Ginsburg (Chicago: University of Chicago Press, 1985), pp. 79–101.

TABLE 1.—WORLDWIDE EXPORTS OF COAL, 1986

Exporter	Metallurgical	Steaming	Total
U.S.A.	49.9	27.8	77.6
Australia	48.7	43.3	92.0
South Africa	5.2	38.8	44.0
Canada	22.4	4.8	27.2
Colombia	.6	5.0	5.6
Germany, Fed Rep	5.0	2.3	7.3
U.K.	.1	2.7	2.8
Poland	9.8	25.5	35.3
USSR	12.1	13.3	25.4
China	3.2	6.7	9.9
Others	3.2	5.8	9.0
Total	160.1	176.0	336.1
Seaborne	137.1	139.4	276.5

SOURCE.—Joint Coal Board, *Black Coal in Australia* (Sydney: Joint Coal Board, 1987).
NOTE.—Data are preliminary.

The drop in oil and gas prices after 1985 has once again slowed the growth in coal use. Further, rising concerns about the longer-term implications of the build-up of carbon dioxide in the earth's atmosphere may lead to future limitations on the use of all fossil fuels, including coal. Thus, the amount of coal that ports may be asked to handle in the decades ahead is inextricably linked to national energy and environmental policies.[3] Unlike oil, coal shipments will not be limited by the size of reserves.

Coal is now the second largest commodity in global international trade; 336.1 million tons (Mt) was traded in 1986.[4] Of this amount, 276.5 Mt moved by sea, roughly divided equally between metallurgical coal (137.1 Mt) and steam coal (139.4 Mt). The amounts exported by the major exporting countries are shown in table 1. Of the developing countries, China is the only one that exports substantial amounts of coal, but exports represent only about 1.5% of China's total coal production.

The amounts of coal imported by the different regions of the world are shown in table 2. Among the regions considered "developing," Asia other than Japan imported 45.1 Mt and Latin America 12.1 Mt. A large part of the Asian total is accounted for by South Korea, Taiwan, and Hong Kong, which have moved out of the "developing" category. Since the largest expansion in coal production and use has taken place in Asia, and this trend is expected to continue for at least the next decade, figure 1 shows some of the major coal ports on the continent.

3. Toufiq A. Siddiqi, "Environmental Standards and National Energy Policies," *Annual Review of Energy* 9 (1984): 81–104.
4. Joint Coal Board, *Black Coal in Australia* (Sydney: Joint Coal Board, 1987).

TABLE 2.—IMPORTS OF COAL BY REGION, 1986

Importer	Metallurgical	Steaming	Total
North America	5.8	9.3	15.1
Latin America	11.0	1.1	12.1
EEC	34.9	69.3	104.2
Other Western Europe	9.3	11.3	20.6
Eastern Europe	13.2	25.7	38.9
Japan	69.7	22.6	92.3
Other Asia	13.3	31.8	45.1
Others	2.9	4.9	7.8
Total	160.1	176.0	336.1

SOURCE.—Joint Coal Board, *Black Coal in Australia* (Sydney: Joint Coal Board, 1987).
NOTE.—Data are preliminary.

Not all the coal shipped from a port shows up in trade statistics. Countries are moving increasing amounts of coal domestically by sea, particularly when the existing rail system is overloaded, or when the coal is located on islands that are different from those that have the major demand centers. China and India illustrate the first case, Indonesia and the Philippines the second.

In China, for example, the largest coal-producing provinces are located in the north and northeast of the country, whereas many of the population centers are in the south (figure 1). The rail transportation system is already running at full capacity, so increasing amounts move by sea. Qinhuangdao has emerged as the largest coal port in China, handling over 50 Mt during 1986. A number of other ports handled over 10 Mt each, including Shijiusuo, Lianyungang, Qingdao, and Huangpu. The amount of coal handled in China's ports was well over 100 Mt, of which less than 10% was exported during 1986.

India is another country where the use of coal is being limited by the rail-based transportation system. The major producing provinces of Bihar and West Bengal, though convenient to Calcutta, are located 1,000 km or more from urban and industrial areas like Ahmadabad, Bombay, Bangalore, and Delhi (fig. 1). Haldia, downstream from Calcutta, has emerged as an important coal port for shipments to other parts of India.[5]

The large coal deposits in Indonesia are located on Sumatra and Kalimantan, whereas most of the population lives on Java. Thus, coal ports are being developed on all three islands to meet Indonesia's goal of relying on coal for much of its electric supply.

In the Philippines the main coal reserves are located at Semirara, Cebu, Surigao del Sur, and Cagayan.[6] Additional ports will need to be developed to

5. Editors' note.—For further information about India's ports see Satyesh Chakraborty, "Indian Ports: Problems Facing Management," in this volume.
6. Toufiq A. Siddiqi, H. H. Webber, and E. Winternitz-Russell, eds., *Coal Transportation in Asia and the Pacific: Infrastructure and Environmental Considerations* (Arlington, Va.: Pasha, 1985).

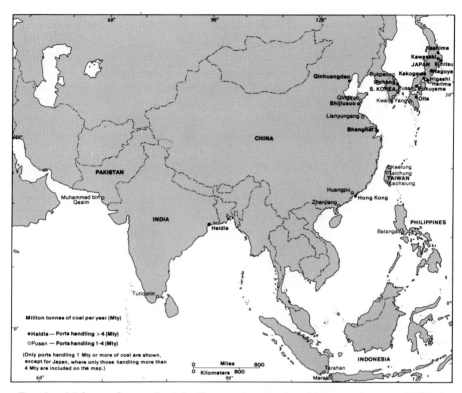

FIG. 1.—Major coal ports in Asia. Source: International Energy Agency, *Coal Information 1988* (Paris: Organization for Economic Cooperation and Development, 1988); and Toufiq A. Siddiqi, Herbert H. Webber, and Elizabeth Wintermitz-Russell, *Coal Transportation in Asia and the Pacific: Infrastructure and Environmental Considerations* (Arlington, Va.: Pasha, 1985).

move the coal to Manila and other urban and industrial areas in the Philippines.

The preceding discussion has attempted to illustrate that coal ports are of interest not only to countries exporting or importing coal, but also to countries that are planning to move increasing amounts of coal from one part of the country to another.

ENVIRONMENTAL CONSIDERATIONS IN PORT PLANNING

With two decades of experience in the industrialized countries, it has now come to be generally accepted that the most efficient way to include environmental considerations in almost all projects is to anticipate and deal with them at the earliest possible stage of planning. This is usually also the most cost-effective way. The opportunity for cost savings is high at the planning and preliminary design stages of most projects, and much lower when construction has already started.

The integration of environmental planning into total project planning is a broader concept than simple environmental impact assessment, which seeks to identify environmental impacts associated with project development. In many countries, environmental impact statements are often of limited use in environmental planning because they are prepared too late in project development and serve more as "public disclosure" documents than bases for planning.

Key elements of environmental planning for the development of a coal port include the policy framework (for instance, energy or trade) within which the coal port is to be built, the institutional framework necessary for environmental planning, the issues of environmental concern that must be addressed, the significance of these environmental issues, the information necessary to evaluate the project, the timing in the project cycle at which this information is required, and an environmental management plan.[7]

There are a number of steps during the project cycle at which environmental factors play an important role, including choice of alternative designs, site selection, and operational design. The integration of project information at each step is shown in figure 2. Each of the economic, engineering, and environmental analyses may proceed separately but should be integrated at the various decision-making steps. Decisions may be characterized as belonging to three types: what, where, and how. "What" decisions involve choosing the type of project to be developed. In the case of a coal port, "what" decisions involve the choice of alternatives (for instance, to modify an existing port, enlarge a port complex, or establish a new port). The second decision level, "where," consists essentially of site selection: choice of the area (for instance, a particular river or bay) and the choice of the specific site. The third level, "how," is concerned with procedures, methods, and technology required to mitigate expected adverse environmental impacts.

At each step it is important to have adequate environmental data to integrate the environmental factors with the economic and engineering factors. The environmental data required for each step will vary in specificity. At the "what" level, environmental analysis will generally rely on already pub-

---

7. A detailed discussion of these key elements is not possible in the limited space available here. The reader is referred to the following publication, prepared by an international team under the auspices of the Environment and Policy Institute of the East-West Center, for an extensive discussion of each of the key elements: John D. Wiebe, Herbert H. Webber, Peter Waterman, Wang Bao-Zhang, Tan Yongjie, Carl J. Sobremisana, Toufiq A. Siddiqi, Geraldine Knatz, and Suling Hum, *Environmental Guidelines for Coal Port Development* (Bangkok: United Nations Environment Programme [UNEP] on behalf of the Environment and Policy Institute, East-West Center, Honolulu, 1988). The discussion in the following sections relies heavily on the material included in that publication.

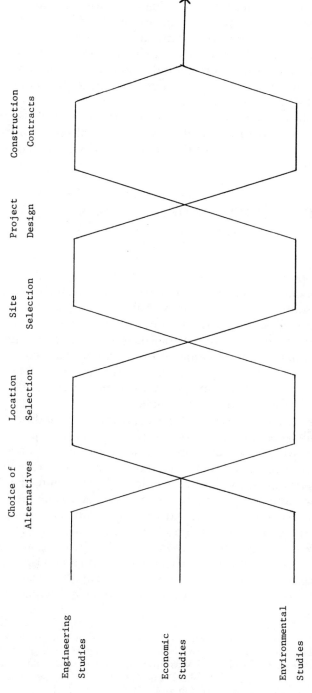

FIG. 2.—Input from environmental assessments at different stages of decision making. Source: John D. Wiebe, Herbert H. Webber, Peter Waterman, Wang Bao-Zhang, Tan Yongjie, Carl J. Sobremisana, Toufiq A. Siddiqi, Geraldine Knatz, and Suling Hum, *Environmental Guidelines for Coal Port Development* (Bangkok: United Nations Environmental Programme [UNEP] on behalf of the Environment and Policy Institute, East-West Center, Honolulu, 1988).

lished data, experience of professionals working in the area, and field surveys. "Where" decisions require site-specific information including inventories of natural resources, socioeconomic resources, and ecosystem structure and function. The "how" decision level requires baseline and monitoring data that provide quantitative information on natural resources, socioeconomic resources, and ecosystem structure and function relevant to project development. It may take 1–2 years to gather "how" data.

A number of possibly adverse environmental effects can be expected at different stages of the construction and operation of coal ports. These are summarized in figure 3. Only a few of the principal environmental concerns will be considered here.

THE ENVIRONMENTAL IMPACTS OF COAL PORTS

Overview

The nature of coal-port operations dictates siting along rivers and coastlines, often in estuaries or embayments that are of environmental significance either because of their resource value (e.g., commercial fishery) or because of their ecological sensitivity (e.g., sea-grass meadows or mangrove swamps). Estuaries and their surrounding wetlands are the spawning grounds of two-thirds of the world's entire fisheries harvest. Many of the most important commercial species spend all or an important part of their lives in estuarine waters. Estuaries and wetlands are also the home of waterfowl and other wildlife, many of which could not survive in other environments.

River shorelines and coastlines are also in great demand as industrial sites as well as residential, recreational, and commercial locations. Coal ports therefore often compete with these other uses for space or must accommodate these uses in their design. For example, in a recent case in North America in which a coal port was planned close to a pulp and paper mill, significant concerns were raised as to the potential for contamination of one product by another. Similarly, a coal terminal located near an urban community must take into account the potential pollution effects of its operation, including pollution of the water and air, visual effects, and noise. In addition, location on shorelines or at river mouths often means locating on a flood plain, with the attendant concerns for flooding and alteration of runoff characteristics.

Coal ports therefore may have a variety of environmental impacts associ-

FIG. 3.—Matrix of the adverse environmental impacts that should be expected from the construction and operation of a coal port. Source: Washington Public Ports Association, "Potential Coal Export Facilities in Washington: An Environmental Analysis," study commissioned by Washington Public Ports Association and the Washington Department of Ecology, in-house report, Olympia, Wash., October 1982.

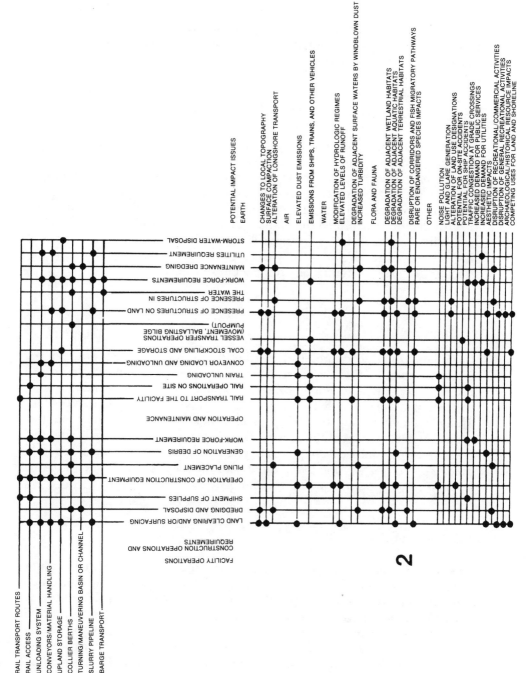

ated with their location, construction, and operation. The following section will highlight those impacts of significant concern. Succeeding sections will focus on mitigating these impacts by proper siting, design, and practices associated with port construction and operation.

Impact of Location

The location of the coal port is one of the more critical aspects determining the extent of environmental impact caused by the facility. The degree to which ecological or human resources are affected will depend on whether the port is located in an estuary or on an open coastline, and whether it is in an urban or rural area. In some cases, locations are fixed (that is, an existing port is expanded or refurbished to handle coal or as a receiving area for a power plant); in other cases, new port sites will be required. It is particularly in cases in which new port developments are being considered that attention to siting can alleviate many potential environmental problems.

Impacts of Construction

During the construction phase of a coal port, which generally covers a period of 18–24 months, there are a number of environmental impacts to be expected. These include, among others, dust and engine emissions from construction equipment, noise from construction activities, habitat loss and water-quality alterations due to dredging and shoreline modification, and the socioeconomic impacts of the construction labor force. Of these problems, the greatest adverse impacts are generally the ecological impacts associated with dredging and/or shoreline modification and the socioeconomic impact of the labor force on the surrounding community.

Ecological Impacts

The ecological impacts of port or harbor construction can be defined in three categories: (1) loss or alteration of a habitat due to dredging, filling, or other construction activities; (2) acute physical impacts on local populations due to entrainment by dredges, increased turbidity, noise disruption, and erection of physical barriers; and (3) temporary impacts due to increased air and water pollution. These impacts can generally be attributed to two major construction activities: (1) dredging and dredge spoil disposal and (2) shoreline modification (cutting and filling).

Filling and Shoreline Modification

Often, a coal port is situated where tidelands, marsh areas, shallow sea grasses, or mangroves have to be filled to produce the required backup lands. The loss of productive ecosystems reduces wildlife habitats and often adversely affects commercial and recreational fisheries by removing habitats crucial for larval and juvenile stages.

Where a coal port is sited on a coast with relatively high wave action (e.g., sandy or gravel beaches), the potential exists for interference with longshore transport of sediments. Generally, sand and gravel beaches are part of the transport process of sediments that moves particles along the beach. Any construction that interferes with the natural shoreline may cause a change in shoreline transport processes. The results of such changes include an accumulation of sediments in the immediate vicinity of the project, with the potential filling of berths, channels, and turning basins, and a reduction in sediment transport past the project, with the resulting erosion of beaches in the area. Hydraulic modeling of the area may be necessary to adequately forecast the changes in circulation and wave patterns that may occur.

The construction of causeways, groins, or seawalls often has the effect of intercepting longshore drift. Where causeways are constructed over tidal flats and in waters with normally high sediment load, they have been shown to be positive environmental features. One example is at Roberts Bank in Canada, where the construction of two causeways (one for the coal port, the other for a ferry terminal) has resulted in an area of increased water clarity and subsequent high productivity between the two causeways. Although still controversial, evidence appears to be accumulating that the construction of the two causeways has indeed enhanced the biological productivity of the area.

Impact of Construction Workforces

The construction of a coal port that will handle 10–15 million tons per year, for example, in North America, takes between 1.5 and 2 years. During this time the labor force will fluctuate, depending on the construction timetable. Construction jobs would be divided among carpenters, metalworkers, cement masons, electricians, pipe fitters, welders, and machine operators. Where coal ports are sited in large urban areas, it is likely that required personnel will be locally available for the work, and that adequate social services for the labor force will also be available.

However, if the coal port is sited in a rural area, a number of problems should be expected. The surrounding population will probably not have the specialized skills required for construction, and job-training programs should be anticipated. Even if there is adequate population in the area for the labor force, an influx of people looking for training and/or work should be expected. The increased population will create pressure for social services such as housing, schools, and fire and police protection. A construction work camp might have to be provided. Once the construction period is over, the labor force is likely to remain in the area. Since the number of jobs required for operation is much lower than that for the construction phase, the creation of a shanty town in the vicinity of the coal port might be anticipated.

Impacts of Operations

The major environmental concerns associated with coal-port operations are the emission of fugitive dust, the contamination of watercourses with leachate, and the disruption of local communities by port-generated noise. However, the range of environmental impacts of coal-port operations includes impacts on human health and safety, as well as on the biophysical and socioeconomic environments (see figure 3).

Studies from the United States (table 3) show annual fatalities for the mining, transportation, and shipping of coal through ports and harbors. In addition to these occupational hazards, there are other human-health impacts associated with coal-port operations.

Air Pollution

The transfer of coal inevitably results in the release of fugitive coal dust. Sources of these dust emissions include railcar unloading areas, conveyor belts, conveyor transfer points, stacker-reclaimers, ship loading points, and storage piles. The amount of fugitive dust and its dispersion characteristics will depend upon weather conditions (wind velocity, precipitation, humidity), topographical considerations (embankments, dikes, vegetation, coal-pile layout), coal characteristics (moisture content, particle size), and storage time.

Air-pollutant emissions can be estimated by direct testing, by using emission factors, or by some physical estimating procedure (such as relating capacity to emissions). When feasible, direct testing provides results of greatest accuracy. In most cases, particularly for fugitive emissions, estimation of emissions by the use of emission factors is satisfactory.

As an example, the predicted amount of fugitive coal dust expected from a coal port on the Columbia River is shown in table 4. In general, fugitive coal dust released during loading and unloading operations is considerable, with reported measurements of 200 tons of particulate per million tons of coal,[8] or

8. K. Bertram, P. Danzvardis, L. Fradkin, and T. Surles, *An Overview of the Environmental Concerns of Coal Transportation,* Argonne National Laboratory, Report no. ANL/EES-TM-99 (Argonne, Ill.: Argonne National Laboratory, 1980).

TABLE 3.—ESTIMATES OF HUMAN FATALITIES AND INJURIES
RESULTING FROM THE OPERATIONS OF COAL TERMINALS IN THE
UNITED STATES

	1975	1985	
		Low	High
Fatalities:			
Mining	18	14	15
Transportation:			
Inland	56	71	79
Ports and harbors	15	20	22
Total	89	105	116
Injuries:			
Mining	1,397	2,107	2,336
Transportaton:			
Inland	142	179	212
Ports and harbors	720	903	1,012
Total	2,259	3,189	3,560

SOURCE.—K. Bertram, P. Danzvardis, L. Fradkin, and T. Surles, *An Overview of the Environmental Concerns of Coal Transportation,* Argonne National Laboratory, Report no. ANL/EES-TM-99. (Argonne, Ill.: Argonne National Laboratory, 1980).

TABLE 4. ESTIMATES OF FUGITIVE DUST GENERATION FOR A
15-MILLION-TON-PER-YEAR COAL PORT PROPOSED FOR THE
COLUMBIA RIVER, KALAMA, WASHINGTON

	Emissions Factor[a]	Annual Capacity (million tons)	Annual Emissions (tons)
Railcar unloading	.4	15	3,000.00
Transfer points	.1	15	750.00
Loading into storage piles	.014	15	105.00
Wind erosion from storage piles	.020	1.5	15.00
Loading out from storage piles	.017	15	127.50
Loading into ships	.4	5	1,000.00
Total			4,997.50

SOURCE.—Portland District Corps of Engineers, "Final Environmental Impact Statement, Port of Kalama Marine Industrial Park and Bulk Handling Facility," report prepared for the Port of Kalama (Portland, Ore.: U.S. Army Corps of Engineers District, 1982).

[a]Expressed in lbs/ton, calculated from Environmental Protection Agency, Report no. EPA-450/3-77/010.

0.02% of the coal transferred.[9] A general value of 6,400 tons of fugitive dust per million tons of stored coal has been reported.[10] Lower-ranked coals (bituminous, subbituminous, and lignite) oxidize and slack (weather and disintegrate) more quickly than higher-ranked coals and may produce greater fugitive dust emissions when stored. Although this oxidation and weathering is known to result in the release of gaseous emissions (hydrocarbons, methane, carbon monoxide, and sulfur compounds), the concentrations are not detectable.

The dispersion of the fugitive coal dust is dependent on a number of factors, including particle size distribution. Fugitive dust settles out of the air relatively quickly, and with a wind velocity of 5 m/s, some 27% of the fugitive coal dust will travel a distance of 10 km.[11]

Coal dust particulates can have a number of deleterious effects.

A. Human Health

The smaller particles in fugitive coal dust (5–25 microns), which constitute 15%–40% of the total fugitive dust, persist in updrafts and turbulence and slowly settle out of the atmosphere. These fine particles can be inhaled and enter the lungs. A number of hazardous organic pollutants have been found in coal, including acenaphthene, benzidene, benzo(ghi)perylene, 2-chloronaphthalene, fluoranthene, and fluorene. Fugitive coal dust may affect human health in two ways: directly, and in combination with other materials. Coal dust can combine with other atmospheric pollutants and produce materials such as sulfate aerosols and peroxyacetyl nitrate (PAN) that may have deleterious effects on human health.[12] Recent studies have shown that coal dust can be genotoxic (that is, cause damage to the chromosomal material of cells).[13]

9. M. J. Chadwick and N. Lindman, eds., *Environmental Implications of Expanded Coal Utilization* (Oxford: Pergamon, 1982).

10. Interagency Coal Export Task Force, *The Environmental Effects of Increased Coal Exports* (Argonne, Ill.: Argonne National Laboratory and COWSAD Research Corporation, 1980).

11. L. Pelham, L. A. Abron-Robinson, M. Ramanthan, and D. Zimomra, "The Environmental Impact Statement, Port of Kalama Marine Industrial Port and Bulk Handling Facility," report prepared for the Port of Kalama, Washington, 1982.

12. A. J. Dvorak (project leader) et al., *Impacts of Coal-fired Power Plants on Fish, Wildlife and Their Habitats*, report prepared for the U.S. Department of Interior, Fish and Wildlife Service, by the Argonne National Laboratory, Report no. FWS/OBS-78/29 (Argonne, Ill.: Argonne National Laboratory, 1978).

13. Carcinogen Testing Laboratory, *Biological Testing of Canadian Coal Dust Samples for Genotoxic Activity*, report prepared for Environment Canada (Vancouver: British Columbia Cancer Research Center, 1982).

B. Biological Resources

The deposition of fugitive coal dust on vegetation (plant leaves) can reduce the light available for photosynthesis and thereby cause stress conditions in plants, making them more susceptible to diseases and parasites. Dust can also clog leaf stomata, reducing the flow of gases and further lowering photosynthesis. Finally, small dust particles (5–10 microns) can enter the plant tissue through the stomata and cause necrosis of the tissues.[14] This problem is usually most severe within 100 m of the storage pile but has been significant up to 1,200 m. Coal dust on leaves can also affect herbivores, causing allergic reactions and potentially toxosis.

Combustion emissions can also have a detrimental effect on wildlife and vegetation. Wildlife exposed to pollutants from combustion emissions may show responses similar to those of humans exposed to these pollutants. Vegetation is also affected by these air pollutants when gases enter the leaf tissues. In toxic concentrations this causes chlorophyll destruction and cell collapse.

The aquatic deposition of airborne coal dust can also have detrimental effects. Coal particulates may increase turbidity, which in turn reduces light penetration and photosynthetic activity. Changes in water chemistry, particularly pH, may occur along with the release of potentially toxic or bioaccumulated trace elements.

C. Socioeconomic Impacts

Coal-dust pollution is one of the issues raised most often by residents in the vicinity of coal terminals. In a community survey at Roberts Bank (British Columbia), blowing coal dust from the port's stockpiles or from moving trains, whether full or empty, was the most frequent complaint. Where fugitive coal dust settles on or near human settlements, the socioeconomic impact can be significant. Dust deposition on dwellings, clothing, and public buildings may create a public nuisance, degrade the quality of life, and lead to demands for mitigation by affected people. This soiling may be a mild nuisance, or it may result in significant economic ramifications such as the need for excessive painting, washing, or vacuuming in nearby houses. Where fugitive coal dust affects historic monuments or scenic areas, it can have an adverse impact on efforts to preserve national heritage and on industries such as tourism.

Dust emissions also have the potential to adversely affect neighboring industrial operations. A study carried out to assess the potential impact of a coal terminal on a pulp and paper operation 1.5 miles distant concluded that fine dust can travel as far as the mill and can downgrade the product a small amount.[15]

14. Dvorak et al.
15. Swan Wooster Engineering Co., Ltd., "A Study of Coal Dust Contamination

Water Pollution

Water in contact with coal will leach the soluble constituents of that coal. For coals with sulfur pyrite compounds, the production of acids (similar to acid mine drainage) will lead to a reduction of pH in the water in contact with the coal. The lowering of pH in turn promotes the solubility of metals occurring as contaminants in the coal. Further, a series of organic compounds may be leached from the coal. These soluble compounds, along with suspended fine coal particles, will contaminate water (from precipitation or wet suppression systems) that comes in contact with coal storage piles. The chemical composition of such coal-pile leachate is shown in table 5.

It is difficult to predict with accuracy the specific concentration of a given leachate, as the concentration and presence of a contaminant is dependent on a number of factors, including coal type and quality. Lower-ranked coals (lignite and bituminous) tend to oxidize and weather faster than higher-ranked coals (anthracite). Since pollutants consist mostly of oxidation products, lower-ranked coals might be expected to produce more contaminants.

There are several potential impacts on water quality of discharging such leachate.

1. The alteration of the pH of receiving streams.
2. The precipitation of metallic hydroxide in larger or higher-buffered receiving streams, which can result in flocculent coatings that cover the stream bottom and destroy benthic organisms.
3. Significant increases in the concentration of trace metals in receiving waters. Metals can be biomagnified in the food chain and may affect humans as well as other animals.
4. Percolation through soils and contamination of groundwater with heavy metals and depressed pH.
5. Reduced oxygen content of the water through chemical oxygen demand. Lower pH can affect fish species in aquatic bodies. The pH range of 4.5–4.7 will have adverse effects on many freshwater fish, including lake herring, yellow perch, lake chub, carp, and salmonid eggs and fry. A pH of 4.7–5.2 is harmful to brown bullhead, white sucker, and rock bass. A pH of 5.2–5.5 is harmful to lake trout and trout perch, and a pH of 5.5–6.0 is harmful to smallmouth bass, while a pH of between 6.0–9.0 is harmless to most fish.[16]

---

of Canadian Cellulose's Watson Island (Prince Rupert) Pulp Mill from the Operation of a Coal Terminal on Ridley Island and Coal Unit Train Access and Egress to the Proposed Terminal, Phase II," in-house report, 1980.

16. Dvorak et al.

TABLE 5.—EXPECTED POLLUTANT
CONCENTRATIONS IN COAL-PILE LEACHATE

Pollutant	Concentration (mg/l)
Total solids	500–3,000
Total dissolved solids	500–2,000
Total suspended solids	5–100
Total hardness (CaCO$_3$)	300–1,200
Alkalinity (CaCO$_3$)	100–600
Bicarbonate	100–160
Sodium	20–200
Boron	.7–.8
Potassium	5–30
Calcium	120–240
Sulfate	100–1,000
Chloride	2–12
Fluoride	.5–1.0
Silica	1.0–20.0
Manganese	30–150
Copper	.1–1.0
Zinc	.060–.020
Aluminum	.0–.03
Lead	.0–.1
Total iron	.09–.90
Ferrous iron	.0–.5
Nitrate	.3–2.3
TKN	.7–3.0
Total phosphate	.4–1.8
Ammonia	.4–1.8
Biochemical oxygen demand	1.0–3.0
Chemical oxygen demand	9–70
pH	6.0–8.3
Specific conductivity	30–500 microohms/cm
Dust suppressants	Unknown
Bilge discharge	Variable

SOURCE.—Washington Public Ports Association, "Potential Coal Export Facilities in Washington: An Environmental Analysis," report prepared for the Washington Public Ports Association and the Washington Department of Ecology, October 1982.

Discharge of untreated coal leachate into waterways will also lead to increased water turbidity. Increased turbidity can also be expected from dredging operations during both the operation and the construction of barge systems and coal transshipment and storage facilities. Increased turbidity, caused by the presence of coal fines in a body of water, reduces the depth of effective photosynthesis by rapidly absorbing radiant energy in the upper water layers, thus inhibiting growth of algae and other vegetation. Zooplankton populations could be secondarily affected owing to decreased amounts of vegetation.

Increased turbidity also delays the self-purification of water, thus allowing the transport of organic waste over long distances. While in high concentrations, usually greater than 40,000 mg/l, turbidity has been found to cause severe injury or death to many fish species.

Increased turbidity associated with coal-related activities may also damage or kill organisms in the benthos and the water column by clogging respiratory and feeding surfaces. Also, benthic organisms may be eliminated by toxic effects either directly through exposure to the toxic material or indirectly by limited food supplies.

An additional environmental impact originating with coal transportation systems may result from the use of herbicides used to control unwanted plant growth on rail, road, and conveyor rights-of-way. Some herbicides are persistent and can enter surface and groundwater and pose a potential health hazard to humans as well as becoming accumulated in food chains.

Noise

Noise is another impact of coal loading, unloading, and storage facilities. Table 6 lists noise levels for various activities in the transfer-storage sequence. To put this in some perspective, there is moderate sleep interference at a noise level of about 50 dBa, moderate hearing damage at 80 dBa, and severe hearing damage at 95 dBa, depending on the duration of the exposure to such noise.

In general, noise values for port operations are lower than those for rail and truck systems. However, noise from port operations is continual and is likely to be one of the more serious sources of community disturbance.[17] For the most part, noise associated with on-site terminal operations will be confined to the site itself. Noise and vibration associated with rail movements are generally of greater concern.

Socioeconomic Impacts

The operation of a modern coal port with an annual throughput of 8–15 million tons requires a labor force of between 80 and 120 in the industrialized countries. The number of persons employed may be higher in the developing countries. The payroll of this direct labor force will stimulate the local economy by some multiplier effect.

In already developed urban areas, the impact of these additional jobs on municipal services such as schools, hospitals, fire, and police services would

17. Port of Vancouver, *Environmental Impact Statement for Roberts Bank Coal Port Expansion* (Vancouver, British Columbia: Port of Vancouver, 1978).

TABLE 6.—ESTIMATED NOISE LEVEL OF OPERATIONS WITHIN A COAL
TRANSFER/TERMINAL FACILITY

Unit Operation	Noise Level (dBa)	Distance (m)
Locomotive and train noise:		
Locomotive engines	74	30
Train cars stop and start	18	30
Train cars wheel noise	14	30
Train movement in and out	93[a]	30
Unloading train:		
Bottom dump[b]	59	12
Rotary dump	60	12
Rotary dump	52	30
Loading ship:		
Shuttle conveyor[c]	75	15
Ship loader	69	30
Storage:		
Stockpile conveyor	65	15
Bulldozer activity and reclaim-		
ing process	75–95	15
Stacker-reclaimer	65	30
Reclamation equipment:		
Feeder	75	.9
Vibrator	110	.9
Hopper loading	78	15
Baghouse fans	85	1.0

SOURCES.—L. Pelham, L. A. Abron-Robinson, M. Ramanathan, and D. Zimomra, *The Environmental Impact of Coal Transfer and Terminal Operations,* report prepared for United States Environmental Protection Agency (Cincinnati, Ohio: Environmental Protection Agency, 1980), Environmental Protection Agency, Report no. EPA 600/7-80-169; Portland District Corps of Engineers, "Final Environmental Impact Statement, Port of Kalama Marine Industrial Park and Bulk Handling Facility," report prepared for the Port of Kalama (Portland, Ore.: U.S. Army Corps of Engineers District, 1982); and Washington Public Ports Association, *Potential Coal Export Facilities in Washington: An Environmental Analysis,* study commissioned by the Washington Public Ports Association and the Washington Department of Ecology, in-house report, Olympia, Wash., October 1982.

[a] Short-term peak noise level.

[b] This is an estimate for the bottom dump unloading facility based on the rotary dump facility. Both methods produce noise from coal impact, ventilation systems, and winches for car positioning.

[c] This is an estimate based on comparison with a traveling stacker.

probably not be noticed. Where the port is located in a rural area, however, provisions for these services should be anticipated.

Impacts of Ship Operations

Ships that transport coal to and from ports can have potentially adverse environmental impacts. Ships that load coal often carry ballast water in their tanks to maintain vessel trim. Although some bulk cargo carriers have separate ballast tanks, the majority do not. Since cargo tanks contain coal residue, ballast water will become contaminated in a manner similar to that of coal storage leachate and should be so treated. The cost of installing treatment facilities, however, may be prohibitive. At a proposed Long Beach coal port, it was estimated that 106 million liters of ballast water from between 150 and 200 ships per year will result from the shipment of 15 million tons of coal. Sedimentation tanks for this volume would require between 4 and 7 ha of land. In this case the treatment facility was judged prohibitively expensive. Instead, the requirement was made that ships clean the ballast water at sea by flushing the tanks at least twice and then discharge the resulting "clean" ballast water at dockside.

However, the most likely source of water pollution from ships at dockside is from bunkering activities. Bunkering involves the transfer of petroleum products to the ship, generally from barges tied up alongside. The most frequent source of oil spills in ports is bunkering activities; however, the spills are generally small (between 1 and 6 barrels) and can be readily cleaned up.

Berth areas, turning basins, and ship channels may require periodic dredging to maintain prescribed drafts. Such dredging can have adverse environmental impacts. These are discussed in the section on the dredging required during port construction.

Where ships travel through restricted channels, waves from ship movement and propeller rotation may wash up on the shore. Where the shoreline is shallow, juvenile fish are often found. The ship wash can carry these juvenile fish onto the shore, and since the wash subsides through the sediments, the fish may be stranded. Such stranding of juvenile fish has been a problem in the Columbia River estuary.

Shoreline erosion caused by ship-generated waves can also be a problem.

The Inland Transportation Link

In the discussion of the environmental impacts of coal ports, it is important to consider the question of boundaries. A coal port is only a part of the coal-transportation system. Except for those coal ports that are located at the sites of power plants and other industrial users, the coal loaded or unloaded at a port often moves some distance overland. Generally, this involves rail or truck systems. Both rail and truck systems have serious environmental impacts including fugitive coal dust and other emissions, increased accident rates, noise pollution, community disruption, and disruption of vehicular traffic. In the case of coal ports in North America,[18] important adverse impacts of transportation operations include community disruption and fugitive coal dust from the rail system passing through nearby communities.[19]

18. Roberts Bank in British Columbia (see n. 15 above); and Long Beach, Calif. (proposed; see Port of Long Beach, *Draft Environmental Impact Report for Long Beach International Coal Project*, Long Beach, Calif.: Port of Long Beach, 1983).

19. Port of Vancouver (n. 17 above); and Port of Long Beach (n. 18 above).

While related to the movement of coal through a port, these impacts are more specific to inland coal transportation. The environmental impacts resulting from the construction and operation of inland coal-transportation systems are described in detail in the companion volume to the coal-port guidelines.[20]

THE MITIGATION OF ENVIRONMENTAL IMPACTS

The lengthy discussion of environmental impacts in the previous section is not meant to discourage planners of coal ports. Solutions exist for almost all of the problems listed. As mentioned earlier, the costs of the available solutions decreases if environmental concerns are addressed fairly early in the planning process.

Only a few of the possible solutions are mentioned here. *Environmental Guidelines for Coal Port Development* provides much more detailed information.

Starting with the construction phase, there are a number of methods and procedures that can avoid or mitigate adverse environmental impacts. These environmentally preferred construction practices add little to total cost and yield a high measure of environmental protection. Procedures and methods (called here *environmental management plans*) to achieve these measures of environmental protection should be specifically established before construction starts. In fact, environmental management plans may be included as part of the construction contract itself.

Factors that should be considered in project design and environmental management plans include construction procedures, ancillary facilities, erosion control, noise-, air-, and water-pollution control, and waste-disposal and maintenance facilities.

Many of the above factors also need to be considered once the port becomes operational. For illustrative purposes, consider the mitigation of air-quality problems.

The greatest environmental concern at most coal transshipment facilities is the impact of terminal operations on air quality. Several techniques are available for controlling air pollutants during the operation of a coal-transfer terminal facility. These techniques will be most effective when terminal management is actively concerned with the reduction of dust pollution at the port. This can best be achieved when each shift superintendent is made responsible for dust control and where terminal operators are trained to recognize dust

20. H. H. Webber, Kenneth M. Bertram, Guo Kemu, Han Guangxu, Terrance P. Kearney, John H. Neate, Raj Kamar Sachdev, Jonathan P. Secter, Toufiq A. Siddiqi, C. Graeme Willcox, Peter Willing, Elizabeth L. Winternitz-Russel, and Wu Heping, *Environmental Guidelines for Overland Coal Transportation* (Bangkok: United Nations Environment Programme on behalf of the East-West Environment and Policy Institute, East-West Center, Honolulu, 1988).

sources and how to respond to them. A dust-control handbook should be developed which clearly describes how and when to operate the control systems and prescribes procedures for how to maintain a relatively clean terminal. Basically, two types of operations can be used: dust collection, as is the case in some enclosed systems using baghouses or scrubbers, and dust suppression using sprays and other techniques. Dust suppression is generally considered a better option where possible, because of lower maintenance cost and because the coal dust does not leave the product stream.

Operating techniques that could be adopted by all terminals without additional capital cost and with a minimal additional operating cost include the following:

Cessation of operations in very high winds (above 12 m/s)
Coordination of cleanup operations
Regular cleanup operations
Staking with a minimum of free-fall (less than 2 m)
Limiting reclaiming operations to low piles or behind piles during winds in excess of 8 m/s
Regular maintenance of handling equipment
Training staff to be pollution-conscious
Minimizing handling of coal
Enforcing speed limits on roadways in terminal area
Spraying roadways, stockpile work areas, and so on, with spray trucks
Keeping ships' loading spouts below hatch level

Additional control options are available to control dust emissions from coal-port operations. In general, there are two types of actions that can be taken: (1) dust collection and (2) dust suppression. In most terminals a combination of these are used. In cold climates, water-based dust-suppression operations are ineffective, and alternatives are required.

MONITORING AND SURVEILLANCE

Monitoring and surveillance are integral elements in the planning, assessment, and environmental management of coal-port facilities. Monitoring entails periodic or continuous collection of environmental data through all stages of project development in order to evaluate the effectiveness of protection measures and to suggest designs for more effective environmental management.

Surveillance is the process of inspection to evaluate the developer's implementation of environmental terms and conditions attached to a project.

Both monitoring and surveillance are important tasks in ensuring adequate environmental protection in the development of coal ports. Monitoring

TABLE 7.—ENVIRONMENTAL IMPACTS REQUIRING MONITORING OR
WHICH MAY REQUIRE MONITORING

Impacts that normally require monitoring:
  Fugitive coal dust from coal-storage stockpiles
  Water quality of leachate from treatment ponds
  Noise from trucks and rail traffic
  Community disruption from rail and truck traffic

Impacts that under certain circumstances require monitoring:
  Fugitive dust from railcar movement
  Fugitive dust from loading and unloading operations
  Livestock and animal kills on roads and railways
  Groundwater movement across road, rail, and pipeline rights-of-way

SOURCE.—John D. Wiebe, Herbert H. Webber, Peter Waterman, Wang Bao-Zhang, Tan Youngjie, Carl J. Sobremisana, Toufiq A. Siddiqi, Geraldine Knatz, and Suling Hum, *Environmental Guidelines for Coal Port Development* (Bangkok: United Nations Environment Programme [UNEP] on behalf of the Environment and Policy Institute, East-West Center, Honolulu, 1988), p. 185.

programs should be identified and developed during the design phase of the project, included in construction specifications, and, when necessary, continued as an integral part of project operations. Surveillance, on the other hand, is generally more important during construction but in some cases may also extend to operations.

The major task of surveillance is to ensure that environmental management plans developed during the project-design phase are successfully implemented during construction. Contractors and developers, when faced with inevitable on-site changes in construction design and procedure, generally rely on their experience and intuition to judge whether the new construction activity meets the intent of mitigation measures spelled out in the environmental assessment documents. This can result in disagreement among environmental agencies, developers, and contractors concerning the most economically appropriate and environmentally sound construction practices to follow. A successful surveillance program can influence changes in construction activities to avoid such conflicts or to expedite their resolution, in turn avoiding unwarranted delays. Surveillance helps achieve a balance between project integrity and environmental protection.

The environmental impacts that should have a monitoring program are shown in table 7. The general factors that should be included in monitoring plans are outlined in table 8.

CONCLUSION

The design, construction, and operation of any large project is a complex undertaking that requires input from the practitioners of a number of disciplines. As the awareness of environmental concerns in a number of countries

TABLE 8.—FACTORS TO BE INCLUDED IN MONITORING PLANS

Fugitive dust:
  The nature of fugitive dust as it relates to monitoring problems
  Methods required for measurement of fugitive dust
  Discussion of methods to sample that fraction (<5 microns) likely to enter the
    human respiratory system
  Usefulness of air-quality monitoring stations for measurement of fugitive dust
  Problems associated with the location of monitoring stations
  Threshold levels at which the action is warranted
  How the fraction measured relates to the total problem of fugitive dust
  Relationship of fugitive-dust monitoring to baseline data
Leachate:
  Procedure for monitoring treatment and discharge
  Procedures for monitoring untreated water runoff
  Importance of measuring pH and total dissolved solids
  Design of a monitoring program
  Threshold levels of treated water for compliance with discharge limits
  Meaning of monitored parameter (relationship between pH and metal solubility)
  Relationship of monitoring to baseline data
Marine habitat:
  Identification of critical species or valued ecosystem components (VECs)
  Cause-and-effect hypothesis of impact
  Appropriate scope and scale (geographical area)
  Relation to baseline data
Noise:
  Identification of situations likely to require noise monitoring
  Methods for noise measurement
  Design of noise-monitoring programs
  Threshold levels at which action should be taken
  Relationship of the monitoring program to baseline data
Community disruption:
  Locations likely to show community disruption
  Methods for measurement of community disruption
  Threshold levels at which action should be taken
  Relationship of the monitoring program to baseline investigations

Source.—John D. Wiebe, Herbert H. Webber, Peter Waterman, Wang Bao-Zhang, Tan Yongjie, Carl J. Sobremisana, Toufiq A. Siddiqi, Geraldine Knatz, and Suling Hum, *Environmental Guidelines for Coal Port Development* (Bangkok: United Nations Environmental Prgramme [UNEP] on behalf of the Environment and Policy Institute, East-West Center, Honolulu, 1988), pp. 186–87.

has increased, the need to incorporate such concerns in major projects such as ports has also become recognized. Although the discussion in this paper has focused on coal ports, the general approach, and many of the environmental problems cited, are common to most ports.

Transportation and Communication

Systems for Managing Ships and Cargoes in the Port of Rotterdam[1]

Jan van Ettinger
van Ettinger and Associates, The Netherlands

INTRODUCTION

A vessel traffic management system (VTMS) is a system to secure the safe and efficient flow of traffic within a large port. An international transport information system (INTIS) is a system to secure rapid and reliable flows of goods from producers to consumers. This article discusses the technological state of the art of these two systems, both of which are needed for the optimal utilization of a modern multimodal transport infrastructure.

The discussion uses the port of Rotterdam as a case study because, as the world's largest port, Rotterdam can illustrate how both systems function in combination with a multimodal transport infrastructure and in a global context.

Attention is given to the interrelated changes in global production, distribution, and consumption. These changes, since the 1970s and especially during the 1980s, have led to a profound revolution in shipping, in terms of quantities and qualities of goods to be shipped as well as modes and services to ship them.

The history of attempts to ensure the safety of shipping started just after 300 B.C. with the Pharos lighthouse and ended with the radar system to supply incoming and outgoing vessels with all necessary information. This shore-based radar marked the beginning of a limited VTMS.

Rotterdam has grown from a small shipping and fishing port, some 100 years ago, into the world's largest port. At present, annual traffic of goods totals 250 million tons, and the port has a multimodal (water, road, rail, pipeline, and air) infrastructure to handle it.

The VTMS, operational since 1987, will be completed by 1992, by which time the installed computer will combine radar tracking data with data on ships and port facilities.

In addition to a multimodal transport infrastructure and a VTMS, an

---

1. This paper was originally prepared for Pacem in Maribus XVI, Halifax, Nova Scotia, August 22–25, 1988.

INTIS secures rapid and reliable flows of goods throughout the transport chain from shipper to consignee. This system aims at supporting, with electronic data interchange (EDI), the transport information flows (paperless trading).

The article concludes with some comments about the VTMS and INTIS in the port of Rotterdam and the potential for this knowledge to be used by developing countries in the emerging global shipping system.

A REVOLUTION IN SHIPPING[2]

Since the 1970s, and especially during the 1980s, shipping has been going through a profound revolution stemming from interrelated changes in global production, distribution, and consumption.

Industrial production is no longer as centered in Western and Northern Europe and the East Coast of the United States as it was 40 years ago. This is due to the emerging industrial complex around the Pacific Basin, notably in Japan, Korea, Taiwan, China, Hong Kong, and Singapore, as well as the West Coast of the United States.

Within the United States, a shift of economic centers has occurred from the Northeast to the South and the West. Within Western Europe, the importance of the Ruhr area, Wallonia, and northeastern France can be observed, as well as the emergence of newer industrial centers in Baden-Würtemberg, Bavaria, northern Italy, and southern France.

The Chunnel (the proposed tunnel under the English Channel), which will transport 40 million passengers and 21 million tons of goods by 2002, and the high-speed railroad between the north German ports and southern Germany could make Rotterdam increasingly remote from the most important industrial centers and transport flows both in and outside Europe.

Industrial production is also changing. In the industrialized countries, a shift is taking place from labor-intensive, through capital-intensive, to knowledge-intensive production of high-value intermediate products. Basic industries such as iron, refining, and petrochemicals are losing relative importance.

A seller's market increasingly changes into a buyer's market. The demand for tailored products and a larger variety of goods increases. Maintenance of large stocks becomes risky and makes heavy demands on financial resources. Therefore, they are reduced by "just in time" production through flexible automation and robotization.

These developments in the industrialized countries have important consequences for the transport of goods. In addition, downstream operations are shifting to developing countries where raw materials, such as oil and ores, are

2. Rijksplanologische Dienst, *Ruimtelijke Verkenningen 1988* (Spatial explorations 1988) (The Hague: Staatsuitgeverij, 1988). See chap. 6 on international developments.

found. As to the flow of goods to and from Western Europe, the following can be expected: (1) a strong decrease in the massive flows of raw materials, (2) growth in transport of high-value (intermediate) products, (3) a more complex composition (types/qualities) of cargo, and (4) increased requirements as to the reliability of transport times.

The emergence of new economic centers lengthens supply and outlet lines between Rotterdam and these centers. Enterprises switch from self-transportation to professional transport firms, since the latter can work over longer distances with lower costs because they can take return cargo.

In addition, the chain between production and consumption, consisting of operations and movements, will be more fragmented. This is due to production for demand, decentralization of production units, "just in time" production, and reduced stockpiling. All of this leads to more frequent transport over shorter distances.

THE HISTORY OF ENSURING SAFETY[3]

Just after 300 B.C., a square, 130-m-tall lighthouse was built on the island of Pharos in the harbor of Alexandria. Its impressive appearance led the ancient Greeks to count it among the Seven Wonders of the World. Attempts to ensure the safety of shipping thus date from before the start of the Christian era.

The Romans were aware of the necessity of safe navigation. They built dozens of lighthouses along the Mediterranean shores, burning wood logs and fagots and later coal. Long afterward, following the invention of the Argand lamp in 1783, oil lighting came into general use.

The discovery of electricity and the incandescent lamp accelerated the development of shipping safety. At points of importance to navigation, on the shores of all the world's seas, towering lighthouses arose and cast strong beams of light seaward.

They were truly indispensable beacons to the seafarer, although their usefulness was poor in dense fog and low clouds. This shortcoming was largely solved in the years after World War II by the invention of radar.

Even before the war, the British had built a chain of radar stations to give early warning of approaching enemy aircraft. Originally, the echoes were received by large stationary antennae, but by the end of the war the much smaller revolving radar scanners had come into use.

The experience gained induced international shipping to adopt radar as a major aid, not only to prevent collisions, but also to facilitate navigation. The

---

3. Projectbureau Walradar Waterweg (Project Office Waterway Shore-based Radar), *Vessel Traffic Management in the Port of Rotterdam* (Rotterdam: Projectbureau Walradar Waterweg, 1984). See introduction.

realization that radar could be invaluable for ships entering ports, especially in the case of poor visibility, led to the birth of the shore-based or harbor radar.

The first Dutch port to benefit was IJmuiden, which, in 1951, started a radar system to "talk" seagoing vessels into and out of the port. Pilots were equipped with walkie-talkies to communicate with the radar operators, who kept them informed on the position of their ships and of nearby ships in relation to the shore.

In 1956, Rotterdam built a chain of seven shore-based radar stations along the New Waterway, reaching from the seashore into the heart of the city. On completion it was known as the most up-to-date system in the world. Its main objective was to assist ships on the river during times of poor visibility, but by the end of the 1960s the service was extended to 24 hours a day.

Over its long history, ensuring the safety of shipping evolved from only ensuring safety to facilitating navigation as well, from ensuring only the safety of the ship, the crew, and the cargo to that of the population and the environment, too, from a night service to one during day and night, and from a bad weather system to one for all weather. Through these developments a limited VTMS originated.

THE PORT OF ROTTERDAM[4]

It took Rotterdam just over 100 years to grow from a small shipping and fishing port into the biggest port in the world. Several factors contributed to this spectacular growth. The Industrial Revolution spread, around 1870, to western Germany and, early this century, to The Netherlands. Second, Rotterdam largely profited from the free-trade policy towards the end of the last century. Third, Rotterdam exploited its situation in the estuary of three wide western European rivers: the Rhine, the Meuse and the Scheldt. Fourth, the New Waterway was dug to the North Sea, which caused a scaleup of port activities.

Nowadays, Rotterdam's port activities are no longer centered in the town itself but near, and even in, the North Sea. Near the harbor mouth a canal is maintained by dredging to allow large bulk carriers, with a draft of 72 feet, to enter even at low tide (see fig. 1).

Owing to the arrival of ships of up to 300,000 dwt, Rotterdam has become not only a transit port for the continuously expanding hinterland, but also a distribution port for the foreland. A major part of the goods landed in

4. *Rotterdam Europoort Information 1985* (Havenkoerier BV: Rotterdam, 1984), *Rotterdam Europoort Information 1988* (Havenkoerier BV: Rotterdam, 1987); and TEMPO, *Consultancy in Port Related Matters* (Port of Rotterdam: Rotterdam, 1986).

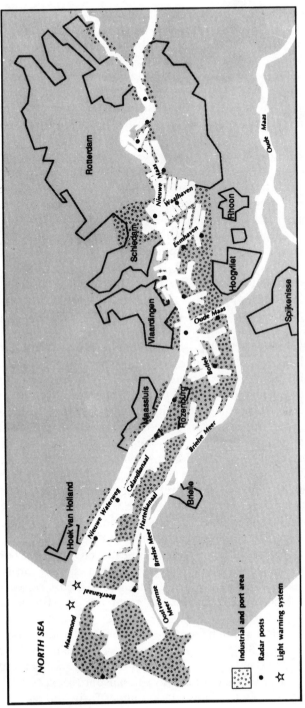

Fig. 1.—Port of Rotterdam

Rotterdam are being transshipped on smaller seagoing ships to ports in northern, western, and southern Europe.

Rotterdam's port traffic grew rapidly following World War II until, by 1972, 310 million tons were being handled, of which more than half consisted of oil products. Growth was then interrupted by several oil crises, and in 1983 traffic had dropped to 232 million tons. In 1987 it had climbed back to 250 million, and by 2000 it may amount to as much as 300 million tons.

Rotterdam occupies a special place in the following trade sectors:

1. Corn, with the closely linked trade in oil-seeds and fodder
2. Oil and oil products, which are also produced in the port area
3. Hides, with a part of their international market in Rotterdam
4. Nonferrous metals, with Rotterdam as the physical emporium of the London exchange
5. Citrus fruit, with a special harbor basin for this purpose
6. Coal, imported as well as exported from West Germany

Most port-related industries in Rotterdam need accessibility for large vessels. They include shipbuilding, oil refineries, blast furnaces, steel works, iron smelting, fertilizer manufactures, petrochemicals, electrochemicals, and engineering.

The international hinterland and foreland of Rotterdam's port comprise most of the highly industrialized areas of Europe, with a population of some 200 million. Including intermediate processing, Rotterdam's transit operations involve about 75% of the total traffic of the port.

In addition to Rotterdam's growing distribution traffic by sea, resulting from its main-port function, transit goods are carried to and from its international hinterland by inland shipping, rail, road, and pipeline. Air traffic, although largely limited to business travel, mail, samples, and spare parts, is also involved.

At present, the split of the purely national inland transport is 70% by road, 23% by inland shipping, and 7% by rail. In traffic crossing the inland frontiers, inland shipping holds a leading position with 54% of the total. This position is expected to be strengthened in the years ahead.

In the near future, all important European waterways will be adapted to the standard motorship of 1,500 tons. The construction of the Rhine-Main-Danube canal, extending Rotterdam's hinterland to the Black Sea and even to the Middle East, and the Rhine-Rhône canal will give further impetus to inland shipping in Europe.

Push-towing, at the Dutch frontier near Lobith, already accounts for 33% of all Rhine traffic. Six-barge push-tows, carrying altogether more than 12,000 tons of cargo, have recently proven to be technically feasible. A further development of this economic way of transport is expected.

Container traffic is the latest development in inland shipping. Special container vessels with a capacity of 90 20-foot containers are being built. Already, 20 specialized container terminals situated along the Rhine ensure the transport to and from nearby industries. Combinations of push-towing and container transport are possible.

Road haulage to and from the port of Rotterdam has continued to grow, especially in the past decade. It is estimated that far more than a million trucks per year pick up from or deliver to the docks. This includes trucks specialized in the transport of oil, perishables, heavy goods, and dangerous goods.

Rotterdam has direct access to a rail network covering all of Europe. For instance, important for the growing flow of chemicals, Interdelta connects the two major basins in Western Europe, those of the Rhine and the Rhône. In Europe's most sophisticated marshaling yard, south of Rotterdam, over 1,250 freight cars are sorted each day.

Rotterdam's accessibility to large vessels, and its refineries, have made it the starting point for a network of pipelines that pump oil and other products to destinations in The Netherlands, Belgium, and West Germany. One pipeline even goes as far as Ludwigshafen, a distance of 668 km. In the port area alone the network length is about 800 km.

Rotterdam airport has a capacity of two million passengers and 50,000 tons of goods per year. Due to its favorable climate (fog is the exception rather than the rule) it is often used as a port of refuge for Amsterdam or Brussels. It has excellent facilities for both freight and passengers.

The port of Rotterdam actively cooperates with its "sister" ports of Kobe and Seattle. Furthermore, its Technical and Managerial Port Assistance Office (TEMPO), assists developing countries in port-related matters through local consultancy and the training in Rotterdam of middle management personnel.

A multimodal (water, road, rail, pipeline, and air) transport infrastructure,[5] while important, is not sufficient on its own for the safe and efficient functioning of a port like Rotterdam. To this end a VTMS is vital.

VESSEL TRAFFIC MANAGEMENT SYSTEM[6]

A VTMS is both a navigational aid and a planning instrument, combined to form an integral traffic-management, information, and early warning system. To develop such a system, promote the safety of shipping and population, and further the efficiency of the port, the Projectbureau Walradar Waterweg (Project Office Waterway Shore-based Radar) was established in 1975.

---

5. Editors' note.—For further information on multimodal transport, see Bernhard J. Abrahamsson, "International Shipping: Developments, Prospects, and Policy Issues," in this volume.

6. Projectbureau Walradar Waterweg, *Vessel Traffic Management.*

In view of the extreme complexity of such a system, it was decided to divide its development into three phases: preparation (to survey port users and carry out studies), development (to describe the new system in broad terms), and realization (to actually construct the system).

A survey covering virtually all users of the existing system as well as a number of government departments was carried out. Nearly all respondents indicated that radar should be the primary sensor in the new VTMS. Two important requirements were that the system should permit traffic-management planning and play a vital part in dealing with catastrophes and accidents.

Several of the suggestions put forward by the users were that the system should be able to recognize the shape of the vessel, that it should be able to pinpoint objects ranging from 550,000-dwt tankers to buoys, that it should detect boats of wood or synthetic material, that observation should remain good at widely varying distances, that targets should be distinguishable even in confused situations, and that observation should not be dependent on weather conditions.

It was generally felt that guidance should be obligatory for very large crude carriers (VLCCs), vessels with dangerous cargo, and ships in distress. There was also a fairly general agreement that ships should be cleared for entry and departure. The government found, with a view to the well-being of the population, that vessel traffic control should be possible whenever required.[7]

Before realization of the system, the limits of modern technology had to be determined. Can a computer process raw radar signals obtained in a problem area—crammed with tower cranes, chimney stacks, buildings, and other tall structures—such as the port of Rotterdam? The technological feasibility of this was one of the most important conclusions from the two initial phases.

Automatic feeding of the computer with radar signals, from which the location, speed, overall measurements and other relevant data on incoming and outgoing vessels can be derived, is called automatic tracking. These data are processed, stored, and made available as systematic information by the computer.

Automatic tracking and registration of shipping movements by a computer will alleviate the task of the watchmen. Watching and interpreting radar images impose a strain on the operator in the long run. This is unacceptable when traffic is busy or visibility reduced. Furthermore, the operator should be

7. Projectbureau Walradar Waterweg, *Baseline for the Development of a Vessel Traffic Management System for the Port of Rotterdam* (Rotterdam: Projectbureau Walradar Waterweg, 1981). This document was written following study results and users' surveys. The comments and criticisms of the users were incorporated into the report. *Baseline* provided the guidelines for the Projectbureau Walradar Waterweg during the detailing and purchasing of the system.

able to spend his time on traffic-management work: decision making based on judgment.

The two initial phases started in 1977 and were finished in 1981. The last phase will take four years more than initially envisaged. Given the difficulties encountered with computer technology that combines radar tracking data with administrative data on vessels and port facilities, the system, though partially operational in 1987, will not be completed until 1990.

The VTMS will cost US$115 million and will consist of the following components:

1. A main traffic center, the Harbour Coordination Center
2. Three regional traffic centers linked to two local traffic centers
3. Twenty-six unmanned radar stations
4. A multi-radar tracking system
5. Eight locations for a closed-circuit television network
6. Three radio direction-finder systems
7. Devices measuring visibility, water level, wind direction, and wind speed
8. An extensive data-handling system
9. A communication system

The regional traffic centers (RTCs) collect data from surrounding radar towers. These will be simple masts supporting the radar scanners and containing an automatic tracking system. Microwave links will transmit the video to the RTCs while tracking data are carried to the centers by standard telephone lines. This will enable the RTCs to carry out traffic-management activities.

These localized functions are very important, because they allow the RTCs to give traffic information and advice directly to vessels within their region. Besides, there will be a constant exchange between these centers and the Harbour Coordination Center (HCC). By using several computers, the HCC may obtain a reliable picture of the traffic throughout its 40-km operational area.

The duties of the HCC include central traffic control and activities such as issuing the clearances for incoming seagoing vessels, supervising transport of dangerous goods, and directing operations in case of catastrophes. Preventing and dealing with catastrophes calls for a system that does more than just manage everyday shipping movements. Human skill and judgment remain indispensable. When exceptional circumstances arise, decisive action is completely dependent on human decisions.

Finally, in addition to a sophisticated VTMS for the safe and efficient use of a modern multimodal transport infrastructure, a port like Rotterdam needs an INTIS to secure rapid and reliable flows of goods from producers to intermediate consumers.

INTERNATIONAL TRANSPORT INFORMATION SYSTEM[8]

An INTIS was set up in 1985 to create and operate a communications network and information structure in Rotterdam. The INTIS's activities are aimed at supporting, with EDI, the information flows in the transport process—so-called paperless trading. Paper handling is still costing on average 7% of the value of international shipments.[9]

The information with which firms in the transport sector are working, such as bookings, forward instructions, and customs documents, can also be exchanged electronically. This is possible through a linking of the computer in one firm with that in another. Thus the need for rekeying will be reduced, and many sources of errors will be eliminated.

To have this information exchange take place efficiently, the firms linked to the INTIS network use standardized messages. These messages are being developed in phases for the total transport chain, on the basis of the requirements and wishes of the firms that will work with them. INTIS attunes these messages to the international standards for electronic information exchange (see fig. 2).

In this system every user has a "mailbox," which can be emptied at moments most convenient to the user. The entrance to the INTIS is guarded and the system provides a number of security levels for data protection. For linking to the INTIS network via a telephone line, a modem is needed.

The user can choose one of the following options: a personal computer (PC), a PC as an intermediary between the in-house computer and the INTIS, or a direct linking of the in-house computer to the INTIS network.

To users starting with a PC, the INTIS offers software with which messages can be drafted, changed, sent, and received.

All users have to pay an entrance fee of US$750. For this amount a mailbox is made available and the INTIS manual is received, as well as information on standardized messages. Further costs depend on the hardware with which the user starts as well as on the length of messages to be dispatched and the frequency of use of the network.

Since 1987, the INTIS has entered into international cooperation with the British and Scandinavian organizations to develop standards for electronic information exchange. This group has also been joined by a Belgian-based organization. As a result of this cooperation, the International Transport Message Scenario (ITMS) became available in September 1987.

8. INTIS BV, *INTIS: Network Facilities and Standardized Transport Message* (INTIS BV: Rotterdam, 1987); *INTIS Bulletin*, 1988, no. 1 (INTIS BV: Rotterdam, 1988). Editors' note.—INTIS BV is a limited company set up by the Port of Rotterdam to develop and manage INTIS. The *INTIS Bulletin* is the mimeographic series published by INTIS BV.

9. According to Smit International's president.

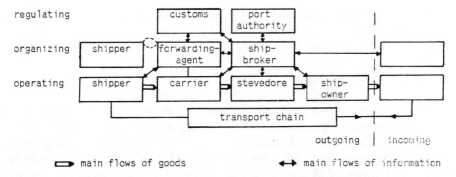

Fig. 2.—The international transport information system (INTIS) of the Port of Rotterdam: a network for information exchange. Source: INTIS BV (a limited company set up by the Port of Rotterdam to develop and manage INTIS).

The most important message from the ITMS is the International Transport Order, which has the great advantage that it is "intermodal." This means that it can be used for all modes of transport, not only by sea, but also by road, inland water, rail, and air. Thus, standardized messages are now available with which many firms can start electronic information exchange immediately.

New software programs for the PC have been developed to enable firms to start in a simple and safe manner with electronic data interchange in the transport sector. These programs are INTISFACE™ Shipping Instruction, for composing, sending, and receiving shipping instructions in a user-friendly manner, INTISFACE™ COCASYS, for exchanging information between ship broker and container terminal, and INTISFACE™ SAD, for composing and sending the Single Administrative Document (SAD), mandatory within the EC since 1988, to the customs office.

In September 1987, a project was initiated to explore how the Netherlands Railroads (NR) could best use the facilities of the INTIS network. In this project Holland Rail Container (HRC) and Netrail play an important role. HCR, a subsidiary of the NR, organizes container transport taking place partly by rail and partly by road. Netrail, a cooperative, integrates the demand side.

A new working group is being established for transport by road. Forwarding agents and shippers will determine which forms and messages lend themselves to electronic interchange and what the electronic messages should look like. The working group will occupy itself with the interchange between shippers and forwarding agents as well as among forwarding agents.

Since the beginning of 1988, the INTIS has offered its network users the possibility of forwarding electronic messages to firms linked to one of the two large international data communication networks, INS and GEIS. By linking to these networks, established by IBM and General Electric, respectively, users are able to interchange messages with firms and organizations in Asia and North America.

SOME CONCLUSIONS

Shipping has been undergoing a profound revolution since the 1970s. With respect to the transport to and from Western Europe, it is expected that flows of raw materials will decrease in relative importance, while flows of high-value products will grow, cargo will have a more complex composition, and requirements as to the speed of transport will increase. The total transport chain will be more fragmented due to production for demand, decentralization of production, "just in time" production, and reduced stockpiling.

The shipping infrastructure includes a network of waterways that will soon reach, through the Rhine-Main-Danube canal, to the Middle East and, through the Rhine-Rhône canal, to the Mediterranean. A road network of growing importance provides for national inland transport, and a network of pipelines reaches into Belgium and West Germany. Air traffic, so far largely limited to samples and spare parts, will gain importance.

A multimodal transport infrastructure constitutes an important but not sufficient condition for the safe and efficient functioning of a large port. To this end a VTMS is vital.

The developments in Rotterdam and other large ports throughout the world will lead to the gradual globalizing of shipping. The developing countries should play an active role in shaping a global shipping system. To this end, their actions should be channeled via the United Nations regional commissions and international organizations, such as UNCTAD and IMO.

The port of Rotterdam consists of a vast range of small elements, in the technological, organizational, and financial sense. Its know-how in port management could, therefore, be exploited by the developing countries and play a more effective role in the shaping of a global shipping system.

Marine Archaeology: A Trojan (Sea) Horse?

Philomène A. Verlaan
Environment and Policy Institute, East-West Center

PREFACE

In *Ocean Yearbook 2*, George Bass ended his article on marine archaeology by bemoaning the absence of international law protecting wrecks in international waters. He criticized Article 149 of what was then the Informal Composite Negotiating Text of the Third United Nations Conference on the Law of the Sea as follows: "Suggestions that antiquities found in international waters should belong to the country of historical or cultural origin are meaningless. . . . The public has yet to perceive antiquities beneath the sea as being . . . unique and precious to the people as a whole."[1] Eight years later, this suggestion is enshrined as a governing principle in Part XI, Article 149.[2] of the 1982 Law of the Sea Convention, (the LOS Convention). The LOS Convention was signed by 159 countries and now awaits 20 more ratifications of the required 60 to enter into effect. The pace of ratification has slowed considerably,[3] and as yet not a single Western industrialized nation has ratified. Three major Western industrialized nations with important marine capabilities are nonsignatories: the Federal Republic of Germany, Great Britain, and the United States. Although not yet officially in force, and despite the absence of the signatures of the above-mentioned countries, the LOS Convention is "acquiring normative powers."[4] This article will not discuss the extent to which the

1. George F. Bass, "Marine Archaeology: A Misunderstood Science," *Ocean Yearbook 2*, ed. Elisabeth Mann Borgese and Norton Ginsburg (Chicago: University of Chicago Press, 1980), pp. 137–54.
2. *The Law of the Sea: Official Text of the United Nations Convention on the Law of the Sea with Annexes and Index* (New York: United Nations, 1983), Sales No. E.83.v.5 (hereafter cited as LOS Convention), Article 149: "All objects of an archaeological and historical nature found in the Area shall be preserved or disposed of for the benefit of mankind as a whole, particular regard being paid to the preferential rights of the State or country of origin, or the State of cultural origin, or the State of historical and archaeological origin." The Area is defined in Article 1 as "the seabed and ocean floor and subsoil thereof, beyond the limits of national jurisdiction."
3. Uwe Jenisch, "The Future of the U.N. Law of the Sea Convention," *Aussenpolitik* (January 1988): 46–60.
4. Ibid.

LOS Convention is or is not customary international law. However, the LOS Convention is, by its very existence, profoundly influencing the way individuals and both signatory and nonsignatory nations view and treat the oceans, including the Area, which is the deep seabed beyond the bounds of national jurisdiction as defined in the convention.[5] The law governing objects of an archaeological and historical nature (OAHN) found in the Area has undergone a sea change since 1980. This is the result both of the LOS Convention and of advances in marine technology. The second part of this article (Sections III–V) describes the technological progress in marine scientific research that has rendered the deep sea accessible to man and thereby changed the status of OAHN from one of a largely theoretical to an immediately practical interest. The first part of this article (Sections I and II), describes the change in the legal status of OAHN under the LOS Convention. No longer is it true, as Bass deplored in 1980, that "wrecks found in truly international waters belong to their discoverers."[6] Since the LOS Convention, activities in the Area as regards OAHN are no longer clearly and freely open for any individual—legal or natural, person or nation—to engage in. Nor are OAHN, whether recovered or in situ, theirs to deal with as they see fit. These important changes in the accessibility and legal position of OAHN in the Area offer certain opportunities to those interested in the law and the science of the sea; the scope of these opportunities is explored in this article.

I. THE CONVENTIONAL CONTEXT

A. Part XI

The conventional context of OAHN is as follows. Part XI of the LOS Convention is entirely devoted to the Area and includes Article 149 on OAHN.[7] The first principle of Part XI governing the Area states that "the Area and its resources are the common heritage of mankind."[8] The second principle removes the Area and its resources or any part thereof from any claim or exercise of sovereignty by any state and from any appropriation by either states or persons.[9] For the purpose of Part XI, resources are defined as minerals.[10] Activities in the Area are defined as "all activities of, exploration for, and

---

5. LOS Convention, Art. 1.
6. Bass, p. 151.
7. LOS Convention, Pt. XI, Art. 134 (1).
8. LOS Convention, Pt. XI, Art. 136.
9. LOS Convention, Pt. XI, Art. 137 (1).
10. LOS Convention, Pt. XI, Art. 133. "For the purposes of this Part: (*a*) 'resources' means all solid, liquid, or gaseous mineral resources in situ in the Area at or beneath the sea-bed, including polymetallic nodules; (*b*) resources, when recovered from the Area, are referred to as 'minerals.' "

exploitation of, the resources of the Area,"[11] and are to be governed by Part XI.[12] Most of Part XI is devoted to regulating these activities. Part XI also governs marine scientific research (MSR) in the Area.[13] When carried out by states, MSR is *not* limited to resources, that is, it is not confined to the minerals of the Area.[14] Marine scientific research is not defined in the LOS Convention. Finally, Part XI establishes an International Sea-Bed Authority (the Authority), through which the parties to the LOS Convention "organize and control activities in the Area"[15] and with such powers and functions as are expressly conferred upon it by the Convention.[16] Incidental powers are limited to activities in the Area.[17]

B. OAHN—Ownership and Disposition

Objects of an archaeological and historical nature (OAHN) are part of the Area. They are otherwise undefined in the LOS Convention; what OAHN could encompass is a topic for another discussion. Because OAHN are part of the Area and because the LOS Convention defines the Area as the common heritage of mankind, it follows that OAHN are part of the common heritage of mankind. In consequence, OAHN may not be appropriated by anyone, nor may they be the subject of claims or exercises of state sovereignty. On the other hand, it is clear that OAHN are not "resources" as defined in the LOS Convention, nor do regulations governing "activities" in the Area apply to them. The OAHN must be preserved or disposed of according to certain rules. These rules are not especially clear, but they certainly bear little relation to those rules, in themselves also rather nebulous, obtaining before the LOS Convention for both jurisdiction over and ownership of OAHN.

Prior to the convention, the principle governing OAHN found on the deep seabed under international waters (what is now the Area) had been basically that of "finders, keepers."[18] No state had jurisdiction over activities by other states on the high seas, including searching for and recovering OAHN. After a recovered find, litigation over ownership usually ensued between various parties, including the original owners (or their heirs) both of the vessel and of the objects, their states of origin and their insurers, and the

11. LOS Convention, Pt. I, Art. 1 (2).
12. LOS Convention, Pt. XI, Art. 134 (2).
13. LOS Convention, Pt. XI, Art. 143, and in accordance with Part XIII.
14. LOS Convention, Pt. XI, Art. 143 (3).
15. LOS Convention, Pt. XI, Art. 157 (1).
16. LOS Convention, Pt. XI, Art. 157 (2).
17. Ibid.
18. Dean E. Cycon, "Legal and Regulatory Issues in Marine Archaeology," *Oceanus* 28 (Spring 1985): 78–89.

finders or salvors and their states of origin.[19] Under the LOS Convention, states still have no jurisdiction over the Area, but neither they nor individuals any longer have the right to appropriate any part of the Area, which the LOS Convention defines as including OAHN.[20] The issue of competing ownership presumably no longer arises, but it is not clear what is to take its place. Rights to OAHN are *not* "vested in mankind as a whole," contrary to what is specifically provided for resources, where rights are vested in mankind as a whole and the Authority is to act on its behalf.[21] The interpretation suggested here as most consistent with the LOS Convention is that the finder of OAHN holds them in a form of trust until the rules of Article 149 governing the use of OAHN are implemented. This was proposed in the context of the Area's resources by M. C. W. Pinto: "The 'common heritage' of these resources is not *res nullius* to be had for the taking; it is not *res communis* simply for enjoyment or use in common; it is more akin to property held in trust—held in trust for 'mankind as a whole,' for the public. It is therefore closest to *res publicae*, the property of the people to be administered by the people and for the people."[22] Article 149 restricts the use to which OAHN, once found, may be put: they must be "preserved or disposed of for the benefit of mankind as a whole."[23] In doing this, "particular regard" must be "paid to the preferential rights" of the following: "the state or country of origin, or the state of cultural origin, or the state of historical and archaeological origin."[24] The LOS Convention is silent on the relative weights, if any, to be attached to the rights, whatever they may be, of the listed types of states. It is not clear whether the use of the adjective "preferential" means that the rights of these states are to be subjugated to those of mankind as a whole. It is likely that in any event these states' rights include control over any rights remaining to traditional parties, such as finders and salvors. The continued existence of any such rights is doubtful at worst and highly diluted at best. Article 149 is not made subservient to "the rights of identifiable owners, the law of salvage or other rules of admiralty, or laws and practices with respect to cultural exchanges,"[25] nor is it described as being "without prejudice to other international agreements and rules of international law regarding the protection of OAHN."[26]

19. Susan J. Lindbloom, "Historic Shipwreck Legislation: Rescuing the *Titanic* from the Law of the Sea," *Journal of Legislation* 13 (1986): 92–111.

20. "Appropriate" is legally defined as "to make a thing one's own; to exercise dominion over an object to the extent, and for the purpose, of making it subserve one's own proper use or pleasure." Henry Campbell Black, *Black's Law Dictionary*, 5th ed. (St. Paul, Minn.: West, 1979), p. 93.

21. LOS Convention, Pt. XI, Art. 137 (2).

22. M. C. W. Pinto, "Statement," in *Alternatives in Deep Sea Mining*, ed. Scott Allen and John Craven (Honolulu: University of Hawaii Press, 1978), p. 13–19.

23. LOS Convention, Pt. XI, Art. 149.

24. Ibid.

25. LOS Convention, Pt. XVI, Art. 303 (3).

26. LOS Convention, Pt. XVI, Art. 303 (4).

These exceptions are clearly made in Article 303 of Part XVI on general provisions, entitled "Archaeological and historical objects found at sea." Under this article, states must protect such objects and must cooperate to that end, but they are constrained by the traditional rules as outlined above.[27] The absence of comparable language in Article 149 bolsters the interpretation suggested here that this article is intended to be a significant departure from the previous law governing OAHN under international waters (now the Area) that essentially gave priority rights to finders and salvors. Unfortunately, having made a good beginning, Article 149 does not follow through with clear guidelines for its implementation.

C. OAHN—Exploration and Investigation

The next issue is the extent to which, if any, exploration for and subsequent investigation in situ of OAHN in the Area is controlled by Part XI of the LOS Convention.[28] It is clear, as discussed above, that this is *not* part of "activities in the Area." Two conflicting motivations exist for this interest in OAHN: the pursuit of knowledge and the pursuit of financial gain.

1. OAHN and Marine Scientific Research (MSR)
The first question to be answered is, Can these exploration and investigation activities be considered MSR? The pursuit of knowledge is the essence of marine archaeology, whose nature is both marine and scientific. It is defined by Keith Muckelroy, a prominent marine archaeologist, as "the scientific study of the material remains of man and his activities upon the sea."[29] Even to remove any of these remains for further study from their submerged site without painstaking recording and investigation of their archaeological context in situ destroys the scientific value of the removed object and diminishes that value for the site as a whole for archaeologists, historians, oceanographers, and other scientists.[30] These remains are "submerged cultural resources . . . by their nature fragile, finite and non-renewable."[31] These remains also provide excellent opportunities for increasing our knowledge of the marine environment in general and benthic processes in particular. Examples include biological and chemical changes in submerged materials, the effect of sudden energy (food) inputs (of special interest in the oligotrophic deep sea), marine benthic community development and ecological succession

27. LOS Convention, Pt. XVI, Art. 303 (1).
28. Removal from the seabed, even for study, would arguably constitute "appropriation."
29. Edward M. Miller, "A Time for Decision on Submerged Cultural Resources," *Oceanus* 31 (Spring 1988): 25–34.
30. Ibid.
31. Ibid.

(wrecks can be viewed as a kind of artificial reef), local sedimentation and turbulence studies—in short, these sites are oases of intense fascination to the four major oceanographic disciplines as well. A persuasive case exists for the proposition that marine archaeology is clearly MSR within the meaning of the convention.

2. OAHN and Salvage

The other pursuit originates in the law of marine salvage and is essentially profit oriented. The concept of salvage is designed to promote the voluntary rescue of goods and vessels imperiled at sea. The reward, decided by the courts, is based largely on the value of the salvaged items. Thus the salvor is interested in the rapid and efficient retrieval of objects with as high a value as possible. The rest is often destroyed, albeit unintentionally, in the process or at best casually discarded.[32] This is irreconcilable with scientific interests and forms the crux of Bass's complaint, quoted at the opening of this article, about the legal situation of OAHN under the high seas. It is postulated here that Article 149 of the convention is intended to exclude the marine salvage interests from acquiring any significant economic advantage in exploring the OAHN in the Area, thereby substantially reducing the likelihood that exploration for salvage will occur.[33]

3. OAHN and the Authority

Interesting consequences derive from the conclusion that marine archaeology will be the primary activity as regards OAHN in the Area, and that this activity qualifies as MSR. In the Area, MSR "shall be carried out exclusively for peaceful purposes and for the benefit of mankind as a whole, in accordance with Part XIII."[34] States' parties may carry out MSR and so may the Authority. Marine scientific research is the only domain where the convention's grant of power to the Authority is not carefully restricted to minerals.[35] Thus, "the Authority may carry out MSR concerning the Area and its resources, and may enter into contracts for that purpose."[36] Furthermore, "the Authority shall promote and encourage the conduct of marine scientific research in the Area, and shall coordinate and disseminate the results of such research and analysis when available."[37] States' parties have obligations to

32. Ibid.
33. For those who doubt this effect by a convention that is not yet in force, the case of venture capital attitudes toward manganese nodule exploration and exploitation in the Area is an instructive example.
34. LOS Convention, Pt. XI, Art. 143 (1). Pt. XIII of the LOS Convention governs MSR, and does not change the conclusions in the text.
35. LOS Convention, Pt. XI, Art. 157.
36. LOS Convention, Pt. XI, Art. 143 (2).
37. Ibid.

promote international cooperation in MSR in the Area in a variety of ways and in cooperation with the Authority.[38]

In summary, Article 143 on MSR in the Area requires extensive involvement of the Authority with the conduct of MSR and, therefore, also with marine archaeology in the Area, both independently and together with states. The final version of Article 149 gives no such specific role to the Authority, nor does it specify dispute settlement procedures, although both items were contained in earlier drafts.[39] However, Part XV of the LOS Convention provides for a variety of ways by which disputes can be settled. More intriguing is the absence of a role for the Authority in the preservation or disposal of OAHN for the benefit of mankind as a whole. Nevertheless, it is proposed here that the MSR provisions of Part XI provide sufficient basis for assigning responsibilities to the Authority for OAHN found in the Area.

II. A ROLE FOR THE PREPARATORY COMMISSION[40]

A. Background

1. Prepcom
Prepcom was established pursuant to Resolution 1, adopted as an annex to the Final Act at the time of the signing of the LOS Convention in 1982. A second resolution, adopted at the same time, deals with investment protection for pioneer investors in deep seabed mining. While the convention awaits entry into force, Prepcom is to prepare for the Authority, its organs, and the International Law of the Sea Tribunal by issuing reports, proposals, and drafts on their organization, personnel, and operations. It also must process the exploration applications of the pioneer investors. It has no power to change the substance of the convention.[41] All full members of Prepcom are signatory states to the LOS Convention with voting rights. Countries that only signed the Final Act may attend meetings as observers. Prepcom has 159 full members and 15 observers. Of the three major nonsignatory countries, only the United States does not exercise its right as signatory of the Final Act to attend as an observer. Observers are given ample opportunity to participate: work proceeds quite informally by consensus procedure; only one vote has yet been

---

38. LOS Convention (see n. 14 above).

39. Anthony Clark Arend, "Archaeological and Historical Objects: The International Legal Implications of UNCLOS III," *Virginia Journal of International Law* 22 (1982): 777–803.

40. For further information about the Preparatory Commission, see Selected Documents, in this volume.

41. Jenisch (n. 3 above), p. 53.

taken since Prepcom started meeting semiannually in 1983.[42] Prepcom presently functions with a plenary, four special commissions, and one general committee.[43]

2. The United States

The reasons for U.S. opposition to the deep seabed mining regime and the U.S. government's consequent decision not to sign the LOS Convention and not to participate in Prepcom's current work are well known. On March 10, 1983, President Reagan issued a proclamation establishing an Exclusive Economic Zone (EEZ) for the United States and a statement on U.S. ocean policy. The latter addressed the LOS Convention specifically and stated that "several major problems in the Convention's deep seabed mining provisions" are the basis for the U.S. decision not to sign. The entirety of Part XI of the treaty was not rejected, either specifically or by implication. This caused a contributor to the professional correspondence series of the Law of the Sea Institute to query whether, in fact, the United States had rejected Article 149 on OAHN, as "the United States has repeatedly made it clear that its own objections to Part XI are due exclusively to the economic regime and machinery established for the natural resources of the Area. I have seen no reference to a United States objection to this particular Article [149] or its subject matter."[44] The United States does not appear to have any philosophical objections to the principles of protecting OAHN underlying Article 149, as the following examples indicate. Even the fact that OAHN lie in the Area should not be a stumbling block. The United States has supported the principle that the Area is the common heritage of mankind since 1970.[45] For application within U.S. waters, the U.S. Senate has passed a bill providing, inter alia, that shipwrecks designated as "historic" shall not be subject to the laws of salvage and finds.[46] The House has introduced similar legislation. As regards shipwrecks in international waters, the United States further showed support of the concept underlying Article 149 of the convention when the U.S. Congress established

---

42. Ibid., p. 54.

43. The plenary is concerned with general preparations for the Authority. The four special commissions are drafting the following, respectively: the compensation system for land-based producer states, the setup of the enterprise, the seabed mining code, and the organization of the Law of the Sea Tribunal. The general committee deals with pioneer investors under Resolution 2. Jenisch, p. 53.

44. Leigh Ratiner, "Unfinished Business," *LOS Lieder* 1 (July 1987): 1–2.

45. The United States voted in favor of the UN General Assembly Resolution 2749 (G.A. Res. 2749) embodying this principle in 1970 and has never repudiated it. G.A. Res. 2749 (25), Declaration of Principles Governing the Seabed and the Ocean Floor, and the Subsoil Thereof, Beyond the Limits of National Jurisdiction. 25 United Nations *General Assembly Official Records (UNGAOR)* Suppl. no. 28, at 24, UN doc. A/8028 (1970).

46. S. 858, The Abandoned Shipwreck Act of 1987, cited in Miller (n. 29 above), p. 34.

the *RMS Titanic* Maritime Memorial Act in 1986.[47] Congress found in relevant part that the *Titanic* "is of major national and international cultural and historical significance, and merits appropriate international protection." The secretary of state is instructed to enter into negotiations with Canada, France, Great Britain, and other interested nations for an international agreement to designate the *Titanic* as an international maritime memorial, and to protect its scientific, cultural, and historical significance. The U.S. Congress is also asking for a moratorium on any research or exploratory activities around the *Titanic* that could "physically alter, disturb or salvage" it until such an international agreement has been reached. Finally, the United States disclaims any extraterritorial sovereignty of any kind over "any marine areas or the *Titanic*."[48]

B. Proposal

First, the work of the Preparatory Commission (Prepcom) should be expanded to include this additional task of the Authority and define its scope and operation. A fifth special committee to deal specifically and solely with OAHN should be established. Finally, the United States should be invited to participate in this new special committee for OAHN.

C. Rationale

Why should Prepcom consider establishing the fifth special committee and why should the United States consider participating in it? As argued persuasively by many scholars almost from the moment of the Reagan proclamation, and most recently and eloquently stated by Professor Edward Miles in his banquet speech at the twenty-second annual conference of the Law of the Sea Institute in June 1988, it is absolutely necessary to "stabilize and facilitate the entry into force of the Convention,"[49] and the absence of the Federal Republic of Germany, Great Britain, and the United States "acts as a significant constraint on widespread ratification of the Convention."[50] Professor Miles proposes that the first step toward overcoming the hurdle of achieving accession to the convention by these three countries could be most constructively made by the United States agreeing to participate as an observer in Prepcom and to

47. R.M.S. Titanic Maritime Memorial Act of 1986, 16 *USCS* sec. 450rr., Editors' note.
48. Ibid.
49. Edward L. Miles, "Preparing For UNCLOS IV?" in *New Developments in Science and Technology: Economic, Legal and Political Aspects of Change: Proceedings of the Twenty-second Annual Conference of the Law of the Sea Institute,* ed. Lewis Alexander (Honolulu: University of Hawaii Press, 1989), in press.
50. Ibid.

commence the discussion of renegotiation of the controversial parts of the deep seabed provisions on an informal basis "through Prepcom, but not initially in it."[51] He points out that it is important to "find a way to revise parts of Part XI and Annex III in order to make the treaty as a whole acceptable to advanced industrial countries without triggering a major confrontation with a distrustful Group of 77, who nevertheless show willingness of being flexible."[52] The Group of 77, a coalition of developing countries, is skeptical of the practical (as opposed to the rhetorical) extent of the present commitment of the United States to the concept of the common heritage of mankind in the Area. The establishment of a fifth Prepcom special committee to deal with OAHN as a common heritage of mankind in the Area, with the active participation of the United States, may promote the creation of a constructive atmosphere for discussions in the other Prepcom meetings on the Area's mineral resources. A reexamination of mineral resource policies is a cardinal prerequisite for any U.S. reevaluation of its decision not to join the LOS Convention.

D. Timing

From a technological point of view, the ever-increasing accessibility to and improvement of working conditions in larger areas and deeper depths of the oceans make ratification of the LOS Convention a matter of growing urgency, if the OAHN in the deep sea are not to suffer the same fate as those located in shallower waters. The second part of this article will describe the most important of these developments.

III. THE TECHNOLOGY CONTEXT

Developments in marine science and technology have been and continue to be a driving force in the evolution of the law of the sea.[53] The assessment of the effects of these developments and the translation of these perceptions into law have spanned the spectrum from the truly visionary—see the famous Malta Resolution introduced to the United States General Assembly by Arvid Pardo in 1967—to the less than felicitous, as evidenced by the deep seabed mining portions of the LOS Convention. The discovery, in September 1985, by a joint U.S./ French scientific expedition of the *Titanic* lying at 4,000 m on the ocean floor outside (probably) any national jurisdiction has given an immediate

51. Ibid.
52. Ibid.
53. John P. Craven, "Technology and the Law of the Sea: The Effect of Prediction and Misprediction," *Louisiana Law Review* 45 (July 1985): 1143–59.

relevance to Article 149 unanticipated by the negotiators of the convention.[54] The status of OAHN on the deep seabed under Article 149 had always been relatively uncontroversial, "doubtless due, at least in part, to the relatively slight concern that many such objects would be found in areas seaward of the . . . EEZ and the continental margin."[55] It is important to remember that this essentially describes an assessment of marine technological capacity at the time rather than an estimate of quantities actually in situ.

A. Quantities and Condition of Deep-Sea OAHN

Archaeologists have never been in much doubt that the deep seabed harbors the fascinating remains of thousands of vessels. Even geologists have been moved to remark that "sunken statuary is probably the most valuable sea-floor resource per unit weight."[56] While it is true that most ships are wrecked close to land due to some shore-related accident, it has been estimated that at least 20% of all sinkings occur well away from the coast over deep water.[57] For example, in the 1860s the average annual loss rate of British ships far out to sea exceeded 250; Lloyd's insurance records show that of the ten thousand vessels on its registry, nearly a thousand were lost with nary a ripple between 1864–69 alone.[58] Deep-sea conditions also contribute to the preservation of nonwooden artifacts and parts of wrecks. The total absence of light excludes plant growth, the primary basis of the food web; consequently, animals are sparse at depth. The encrusting invertebrates such as hermatypic corals that flourish in the photic zones and rapidly cover and destroy vessels foundered in shallow water have only small, solitary relatives here.[59] Much of the deep (>500 m) ocean floor is covered with soft mud from a steady but slow sedimentation, estimated at about 2 cm per thousand years.[60] Despite the presence of bottom currents, there is little turbulence. Chemical reactions are

54. Lindbloom (n. 19 above), p. 106; but see also Ian Townsend Gault and David Vanderzwaag, "Now That Pandora's Box Is Open Could Canada Assume Responsibility for the Wreck of The *Titanic?*" *New Directions* (1987), pp. 6–7.

55. Bernard H. Oxman, "The Third United Nations Conference on the Law of the Sea: The Ninth Session (1980)," *American Journal of International Law* 75 (April 1981): 211–56.

56. K. O. Emery and Elazar Uchupi, eds. *Geology of the Atlantic Ocean* (New York: Springer, 1984), pp. 886–87.

57. Keith Muckelroy, *Maritime Archaeology* (Cambridge: Cambridge University Press, 1978), p. 150.

58. Peter Throckmorton, ed., *The Sea Remembers* (New York: Weidenfeld & Nicolson, 1987), pp. 222–23.

59. Robert D. Ballard, "Statement before the House Merchant Marine and Fisheries Committee, 29 October 1985," *Oceanus* 28 (Winter 1985): 47 (hereafter cited as "Statement").

60. Throckmorton, ed., p. 224; Ballard, "Statement," p. 47.

slowed by half for every 10-degrees-Centigrade (°C) drop in temperature; the deep-sea temperature hovers around a more or less uniform 4°C, although the bottom of some inland seas like the Mediterranean and the Aegean are rather warmer, about 13°C.[61] Low concentrations of dissolved oxygen and extreme pressures also slow biological activity. At the site of the *Titanic*, for example, the pressure is about 400 atmospheres or 6,000 pounds per square inch (psi).[62] The pH is well buffered at the slightly alkaline level of pH 8.2.[63] These deep-sea conditions all combine accordingly to retard the wreck's deterioration. The main natural enemy of deep-sea wrecks is the wood-boring mollusk, which seems to exist even at the depths of the *Titanic*—most of her woodwork had entirely disappeared.[64] As these organisms require oxygen to survive, only wood above the mudline is vulnerable.[65] In summary, the physical conditions in the deep sea are such that it is reasonable to expect to find both ships and their contents in generally good condition even after many centuries of submergence.[66] In fact, in the words of Charles Lyell, the "father" of the geological sciences, "It is probable that a greater number of monuments of the skill and industry of man will in the course of ages be collected together in the bed of the oceans, than will exist at any one time on the surface of the continents."[67]

B. Consequences of Accessibility for Marine Archaeology

The principal reason that these "capsules of life frozen in time"[68] still exist at all is largely due to their inaccessibility, until now, to the attentions of man. It is instructive to examine the history of marine archaeology as a function of developments in marine technology. The invention of the self-contained underwater breathing apparatus (scuba) by Jacques-Yves Cousteau and Emile Gagnan in June 1943 enabled the exploration of the ocean by amateurs and specialists alike to depths of about 100 m. In 1947 scuba gear was made available to the general public and by 1957 every single known wreck within those depths off the coasts of southern France and Liguria in Italy had been destroyed.[69] In the Caribbean, hundreds of sites are being ruined, some lit-

61. Throckmorton, ed., p. 224.
    62. Ballard, "Statement," p. 47. In the ocean, the pressure increases by 1 atmosphere for every 10 m of depth.
    63. Colin Pearson, ed., *Conservation of Marine Archaeological Objects* (London: Butterworths, 1987), p. 6.
    64. Throckmorton, ed., pp. 222–24.
    65. Pearson, p. 15.
    66. Ibid., Preface.
    67. Charles Lyell, *Principles of Geology* (1832) quoted in Throckmorton, ed., p. 226.
    68. Throckmorton, ed., p. 10.
    69. Ibid., p. 8.

erally dynamited by treasure hunters; a similar situation obtains off the coasts of many new nations.[70] Yet no one in the United States, at least, has ever made a financial profit from an ancient shipwreck. At most four wrecks might have had some commercial value, and all four "treasure" projects have lost millions of dollars.[71] The location and subsequent exploration of the *Titanic* dramatically illustrates the advances in deep-sea scientific abilities and technologies that are conceptually and practically accessible for the first time to the general public and its policymakers. More than 98% of the ocean bottom can be directly searched and reached by submersible and remotely operated vehicle (ROV) systems.[72] Robert Ballard, chief scientist of the expedition that found the *Titanic,* has stated that "this technology is now out of control."[73]

IV. OVERVIEW OF CURRENT MARINE TECHNOLOGY

The advances in this technology can be divided into search and site categories and further subdivided into remote and direct technology.

A. Remote Techniques

The three most common remote-search techniques use side-scan sonar, sub-bottom profiler, and magnetometer.[74] Responding as they do to different physical characteristics of a wreck, and having their own individual strengths and weaknesses, they form a complementary technical triumvirate. All three are portable, battery powered, and can be operated while being towed by small boats.[75]

Side-scan sonar examines the surface of the seafloor.[76] Shaped like a torpedo but called a "fish," it is an acoustic device that emits pulses of high frequency (50–500 kHz) sound in a narrow horizontal beam and a wide vertical beam. This combines good resolution with coverage that can extend to 300 m on both sides of the fish. The sound signals reflected back to the fish from the topographical differences of the seafloor are converted to visual

70. Ibid.
71. Ibid., p. 226.
72. Paul R. Ryan, "The *Titanic* Revisited," *Oceanus* 29 (Fall 1986): 4–15. The *Titanic* is no longer the deepest lost object found by a search at sea. The wreckage of South African Airways' flight 295 was found at 4,500 m depth in the Indian Ocean in mid-December 1987. Michael K. Kutzleb, "Anatomy of an Underwater Search," *Sea Technology* 29 (July 1988): 37–40.
73. Throckmorton, ed. (n. 58 above), p. 220.
74. Charles Mazel, "Technology for Marine Archaeology," *Oceanus* 28 (Spring 1985): 85–89.
75. Ibid.
76. Ibid.

images and recorded on paper. The results resemble detailed aerial photographs. Sonar signals from very rocky or otherwise highly irregular bottoms are difficult to interpret.

The sub-bottom profiler is an acoustic device used to peer into the sedimentary layers below the seafloor surface where the side-scan sonar is "blind."[77] It can locate sites that are completely buried. These sites are particularly interesting because, as mentioned earlier, burial under sediments increases the chances that the wreck is well preserved. The sub-bottom profiler aims low frequency (3.5–12 kHz) sound pulses vertically down at the sea floor. The sound partially reflects from and passes through each sedimentary layer below the surface until stopped by the bedrock. This produces an image in cross-section, with the wreck shown as a point discontinuity within the layers. Because the sub-bottom profiler cannot beam to both sides like side-scan sonar, it covers only a very narrow track just underneath the path of its tow boat and is, therefore, less useful in a general search over a wide area.

The magnetometer measures the strength of the local magnetic field.[78] It responds only to ferrous materials, (for example, bronze will not be detected) no matter at what depth they may be buried. Every location on the earth has its own "natural" level of magnetism. Concentrations of iron such as might be expected from wreck sites are registered as an anomaly of a particular shape and size over the background magnetism appropriate to a given area. Depending on the amount of iron in the wreck, the sensor must be towed relatively close to the site in order to detect it, as the strength of an object's magnetic field decreases rapidly with distance. Magnetometers work best in areas where the background magnetism is relatively uniform. Large quantities of ferrous rock on the ocean floor will register as anomalies in their own right and complicate the readings.

B. Direct Techniques

Having identified a target area and likely specific search points by these remote methods, the technology level then switches to direct-search techniques. These include tethered and untethered submersibles, ROVs, and, depending on depth, divers. All require surface support vessels. All are involved with site work as well. A distinction is made here between research submersibles and submarines, which are defined as military and will not be discussed here. Tourist submersibles, designed purely for viewing at shallow depth, are an actively growing separate industry and a special topic in their own right.[79]

77. Ibid.
78. Ibid.
79. An overview of the tourist submersible sector is given by Frank Busby, "A Promenade beneath the Sea," *Sea Technology* 28 (December 1987): 33–36; and, in the same issue, Robin Woodward Janca and Eddy J. Peters, "Bringing Submersibles to the Masses," pp. 19–22.

1. Submersibles

Twenty years have passed since George Bass launched the *Asherah,* named after a Phoenician sea goddess, the first research submersible designed for marine archaeology.[80] Much technological improvement has occurred since then. Submersibles are no longer limited by depth as far as marine archaeology is concerned, unless a foundered vessel rests in a deep ocean trench. France, the Soviet Union, and the United States all have submersibles capable of operating at depths of at least 6,000 m.[81] The Japanese will have a submersible with 6,500 m depth capacity on line in 1989.[82] There are a number of submersibles operating at depths up to 2,000 m.[83] Submersible crews are composed of the pilot and up to three observers in the 1-atmosphere (surface) pressure hull. Submersibles are battery powered, provided with one or two external manipulators, still and video cameras, and strong lights. Some are designed for "lockout," enabling exit and entry of "saturation" divers from the submersible at depth.[84] Their maximum speed is usually about 2 knots. They generally require a sizeable dedicated support vessel with stern-mounted A-frames and special cranes and winches to raise and lower the submersible into the water. Dives are therefore governed by sea state (none are undertaken past sea state 5),[85] and much time and money can be lost at sea waiting for reasonable weather.[86] The unique towed launch/retrieval/

---

80. George F. Bass and Donald M. Rosencrantz, "The Asherah—a Pioneer in Search of the Past," in *Submersibles and Their Use in Oceanography and Ocean Engineering,* ed. Richard A. Geyer (Amsterdam: Elsevier, 1977), p. 335.

81. France has Cyana (3,000 m) and Nautile (6,000 m) (Shinichi Takagawa, "Deep Submersible Project [6,500 m]," *Oceanus* 30 [Spring 1987]: 29–33); the Soviet Union has *Mir 1* and *Mir 2* (both at 6,000 m) (Professor Alexander Malahoff, University of Hawaii, personal communication, May, 1988); the United States has *Alvin* (4,000 m) and *Sea Cliff* (6,000 m) (Takagawa, p. 31).

82. Japan presently has *Shinkai* (2,000 m), and the new submersible will dive to 6,500 m. An important part of its mission is in fact to examine trenches off the coast of Japan (Takagawa, p. 30).

83. I am conducting marine scientific research experiments at these depths with D.S.V. (deep submergence vessel) *Pisces V,* one of two submersibles belonging to the Hawaii Undersea Research Laboratory (HURL) of the U.S. National Oceanic and Atmospheric Administration (NOAA) based at the University of Hawaii in Honolulu. D.S.V. *Makali'i,* (formerly *Star II*) the other submersible, can dive to 360 m and was originally used for the commercial harvest of precious coral off Hawaii, which grows beyond the reach of scuba divers.

84. Best known of the "lockout" type are the two submersibles of the Harbor Branch Oceanographic Institution, Johnson *Sea-Link I* and *II.* They have been involved in the search for the space shuttle *Challenger* and the exploration of the sunken battleship *USS Monitor.* Lynne Carter Hanson and Sylvia Earle, "Submersibles for Scientists," *Oceanus* 30 (Fall 1987): 31–38.

85. Naval Oceanography Command, *Manual of Manned Submersibles* (Bay St. Louis, Miss.: National Space Technology Laboratories [NSTL] Station, 1980), p. 3.

86. For example, in the summer of 1986 off Bermuda, scientists engaged in a series of marine biological studies, known as the Beebe Project, had 55 days of waiting at sea, due to weather conditions, to conduct five dives. Andreas B. Rechnitzer, "Beebe

transport (LRT) platform developed at the Hawaii Undersea Research Laboratory (HURL) for use in the perennially rough Hawaiian waters is attractive for marine archaeology as it does not require a large dedicated vessel and can be used in seas up to 5.5 m.[87] The HURL innovation illustrates the trend toward "deeper and cheaper" systems that are highly maneuverable, transportable, and easily deployed from small vessels to do in situ research.[88] One-person submersibles are being developed where the scientist is the pilot. The newest and very successful example of these is *Deep Rover*, introduced in 1984. The intention was to produce a submersible "so simple even a scientist can operate it."[89] It is essentially a lightweight (2,800 kg) acrylic sphere of 1,000 m depth capacity with thrusters and a 1.8 m long arm that can lift 90 kg. It has already been launched from 12 different types of platforms and operated by a variety of persons for whom it takes about half an hour to master the basic operating principles.[90] A one-person 1,500 m submersible, *Deep Flight*, is being designed to bridge the gap between towed search-and-survey systems and site-specific work vehicles. It weighs about 1,100 kg, cruises at four knots, has scanning sonar capacity, and can investigate, document, and sample targets encountered either visually or on sonar in situ.[91]

2. Tethered Submersibles
The submersible category also includes one-person tethered systems. These have come a long way since Alexander the Great allowed himself to be lowered, in what seems to have been a glass barrel, into the Aegean Sea circa 322 B.C.[92] More recently, this concept evolved into the Jims, named for Jim Jar-

Project—Looking for Sixgill Sharks . . . and More," *Sea Technology* 28 (December 1987): 10–17. Basically, each deep ocean dive requires a full day, while in shallower areas more than one dive can be conducted per day, weather permitting (Naval Oceanography Command, p. 27).

87. The LRT platform is controlled by its own pilot and two divers. It is neutrally buoyant and submerges with the submersible aboard to a depth of about 20 m, well away from the turbulent sea surface. The submersible is launched and retrieved at this depth. The LRT is attached to the surface vessel and can be towed with the submersible at 9–12 knots. Lynne Carter Hanson, ed. *The Marine Research Community and Low-Cost ROVs and Submersibles: Needs and Prospects: Summary of a Workshop* (Kingston, R.I.: Center for Ocean Management Studies, 1986), p. 43.

88. Another low-cost two-person submersible is the *Nekton/Delta*, which has become a highly effective scientific "workhorse" for benthic research to about 300 m depth (Hanson and Earle, p. 11).

89. Sylvia A. Earle, "Microsubmersibles: Putting More Scientists in Deep Water," *Sea Technology* 27 (December 1986): 14–21.

90. Ibid. Operators range from the designer's 13- and 14-year-old children to business executives, journalists, a retired British admiral, and, of course, many scientists.

91. Ibid.

92. Eugene Allmendinger, "Submersibles: Past-Present-Future," *Oceanus* 25 (Spring 1982): 18–29.

rett, the first person willing to try one in the 1930s. It is an anthropomorphic, articulated, 1 atmosphere pressure, armored diving suit that is tethered to the surface vessel for life support, communication, and deployment and has been used for benthic work to 500 m depth. At least 15 are still being used today.[93] More advanced forms of this tethered personal diving cocoon are the 18 Wasps and 30 Mantises, which can operate throughout the water column to 610 m and 700 m, respectively, with no decompression.[94] The Wasp was successfully used for the archaeological exploration of the three-masted bark HMS *Breadalbane*,[95] which sank 129 years ago in 110 m of water 600 miles north of the Arctic Circle and rests, perfectly preserved, under 1.8 m of solid ice.[96] A major problem with all tethered systems is keeping still in the water while surface vessel heaves about.[97] Dynamic positioning of the vessel can somewhat alleviate this, but adds considerably to operating expenses.[98] For archaeologists who work around structures there are also the problems of tether entanglement and breakage to be considered, and tether drag restricts maneuverability.[99] On the other hand, tethered systems have essentially unlimited power and life support supplied from the surface vessel, as opposed to autonomous systems, and are cheaper to use if the mission objectives permit.[100]

3. Remotely Operated Vehicles (ROVs)

ROVs started out large, expensive, and heavy for military and industrial (primarily offshore oil and gas) work. These were frequently less portable than the submersibles, required large support vessels and plenty of deck space.[101] This has until recently been true as well for the very few ROVs available to marine scientists.[102] The first low-cost, portable ROVs were only introduced in 1984 and have been so successful that several hundred small ROVs from over a dozen types of systems are now in use.[103] ROVs can either be towed or have their own propulsion system. The new small ROVs all have their own thrust capacity and have been described as "swimming eyeballs."[104]

93. Hanson and Earle, p. 33.
94. Ibid.
95. Ibid.
96. No author, no title, but see Rolex advertisement, on back cover of *Sea Technology* 28 (December 1987).
97. Hanson, ed., p. 44.
98. Ibid.
99. Allmendinger, p. 25.
100. Ibid.
101. Hanson, ed. (n. 87 above), p. 45.
102. Essentially, three were available with 6,000 m depth capacity: Angus from Woods Hole Oceanographic Institution, Deep Tow from Scripps Institute of Oceanography and Digitow from California Institute of Technology (Naval Oceanography Command [n. 85 above], pp. 218–19).
103. Hanson and Earle (n. 84 above), p. 38.
104. Ryan (n. 72 above), p. 4.

An imaginative combination of general search and site-specific ROVs can be seen in the Woods Hole Oceanographic Institution's Argo and Jason. Argo is towed and equipped with sonar and television cameras for a broad survey of the seafloor.[105] It also serves as a garage for Jason, which is tethered to Argo but has three-dimensional, self-propelled mobility, arms, and stereo color television.[106] When Argo spots an interesting area of the seafloor, Jason will be sent out for closer inspection and sampling. It was while testing this system that Robert Ballard found the *Titanic* in September 1985.[107] On Ballard's second voyage to the *Titanic*, the manned and unmanned elements of underwater research completed the convergence of complementarity toward which all these developments have been advancing. The autonomous manned submersible *Alvin* descended to the *Titanic* with Jason Jr., a tiny, tethered, mobile ROV carried in its own garage on the front.[108] *Alvin* supplies power, through the tether, for Jason Jr.'s 170 degree vertical scan cameras.[109] In the course of the explorations, *Alvin* settled on various parts of the *Titantic*'s outside decks and sent Jason Jr. to such places inside the ship as down the grand staircase, into staterooms, officers' quarters, and so on, where manned submersibles cannot go.[110] Untethered ROVs will be only briefly mentioned here as they are still primarily in the research and development stage, stimulated by the increasing sophistication of acoustic links, robotics, artificial intelligence, and miniaturization technology. At least 11 laboratories in the United States and Europe are focusing on creating an autonomous ROV that can perform inspection and information gathering tasks at any depth.[111]

4. Divers

Finally, there are the divers, working at ambient pressures on the site itself. The greatest depth reached to date is 460 m.[112] Working at these depths requires extensive support facilities designed to keep the diver at the local pressure. The facilities are placed on the seabed or on the surface vessel. This

---

105. Robert D. Ballard, "Argo and Jason," *Oceanus* 25 (Spring 1982): 32–35. Argo operates to 6,000 m, weighs 1,818 kgs, and flies at 20–40 m above the sea floor. When towed at 1 knot, Argo has a nearly vertical wire angle 100 m behind the ship. The sonar and camera systems complement each other to provide simultaneous, overlapping broad-swath optical and acoustical images. Stewart E. Harris & Katie Albers, "Argo: Capabilities for Deep Ocean Exploration," *Oceanus* 28 (Winter 1985): 99–101.
106. Ballard, "Argo and Jason," p. 33.
107. Ryan, p. 4.
108. Ibid.
109. Ibid., p. 6.
110. Ibid.
111. D. Richard Blidberg, "A Potential Untethered ROV for Ocean Science," in *Science Applications of Current Diving Technology on the US Continental Shelf*, ed. R. A. Cooper and A.N. Shepard, NOAA Symposium Series for Undersea Research (Rockville, Md.: NOAA Undersea Research Program, 1987), pp. 239–46.
112. Naval Oceanography Command, p. 7.

so-called saturation diving technique is useful for scientists who must spend a long time on the bottom as it improves significantly the cost-benefit ratio between what would otherwise be very short dives and very long decompression periods.[113] It is valuable to archaeological research where the finest and most delicate of detail work in situ is required. Saturation diving by archaeologists is only 10 years old but has already been extensively used on, for example, the Civil War ironclad *USS Monitor,* with the Johnson Sea Link lockout submersible.[114] The resting place of the *USS Monitor* has been designated a national marine sanctuary, and she herself has been designated a National Historic Landmark.[115] She sank in 1862 on her way to the siege of Charleston and lies in 70 m of water 20 miles southeast of Cape Hatteras, North Carolina. Her location in the midst of the shipping lanes, high currents, and very poor visibility water makes marine archaeological research there particularly challenging.[116]

5. Navigation and Positioning

Precision navigation and highly accurate and reliable positioning are essential to both archaeological search and site processes. For the general relocation of a site from the sea surface, accuracy to within a few meters is now available through a variety of systems out to 600 nm, at least off the coasts of the lower 48 U.S. States.[117] A network of satellites establishing the Global Positioning System (GPS) is also being put in place.

When it is complete, continuous and simultaneous fixes will be possible from any point on the globe.[118] This will be an appreciable improvement on the current situation far out at sea, where the normal navigation satellites pass within usable range only once every 5–6 hours or so. Only dead-reckoning navigation between satellite fixes is possible, which can be off by several nautical miles.[119] Even under optimum conditions, the resolution is no better than

113. James Woodell Miller, ed., *NOAA Diving Manual,* 2d ed. (Washington, D.C.: U.S. Government Printing Office, 1979), p. 12.

114. Saturation diving for archaeology was first used in 1978 for the excavation of the *Capistello,* lying at 92 m depth off the island of Lipari in the Aeolian Sea (Bass [n. 6 above], p. 149). For a discussion of work on the *Monitor,* see Edward M. Miller, "The Monitor National Marine Sanctuary," in R. A. Cooper and A.N. Shepard, eds., NOAA Symposium Series for Undersea Research (Rockville, Md.: NOAA Undersea Research Program, 1987), pp. 247–59.

115. E. M. Miller, "A Time for Decision on Submerged Cultural Resources" (n. 29 above), pp. 29–32.

116. Ibid.

117. U.S. Congress, Office of Technology Assessment, "Technologies for Underwater Archaeology and Maritime Preservation," Background Paper no. OTA-BP-E-37 (Washington, D.C.: U.S. Government Printing Office, September 1987), p. 40.

118. David M. Graham and Michael A. Champ, eds., "Ocean Space: Oceanographers at 'Ground Zero' in Seabed Exploration," *Sea Technology* 28 (June 1987): 23–29.

119. Ibid.

900 m, and more usually it is worse than 2 nautical miles.[120] Ballard was able to return to the site of the *Titanic* by using previously noted GPS fixes.[121]

6. Mapping

For on-site work, precision in positioning is "the key to all effective archaeology," and any tolerance greater than 2.0 cm is unacceptable.[122] It is probably in surveying and recording that the underwater environment is most restrictive: all intuitive, that is, land-based, conclusions are unreliable.[123]

Sizes are distorted, there are no natural horizons so that it is difficult to judge whether a plane is at all level, strong currents and poor visibility make it difficult to see the whole of an object at once, and a vessel "consists of an amazing series of planes and curved surfaces which are very difficult to comprehend outside the frame of reference of the vessel herself."[124] The most exciting development for shipwreck site mapping is the Sonic High Accuracy Ranging and Positioning System (SHARPS). The usual charting method can take months or years to complete and is virtually impossible at serious depths: it involves setting up a hand-placed grid of plastic lines or tubes, and then calculating thousands of reference points.[125] The SHARPS sets up three electronic transmitter/receivers around a site that accept signals from an electronic gun held by a diver, who pulls the trigger at each reference point. The points are then recorded on the shipboard computer.[126] With SHARPS vessels and artifacts can be outlined, labeled, and two- and three-dimensional maps made to accuracies of 1 cm.[127] As one example, the remains of an 1883 oyster boat were mapped in 1 hour with SHARPS, using over 1,200 reference points, where previous methods would have taken over 6 weeks.[128]

V. TECHNOLOGY TRANSFER

Excavation of submerged sites has been described as "an irreversible and unrepeatable scientific experiment."[129] These technologies have made it possible to extend the "non-destructive" approach to marine archaeology, where excavation and recovery of artifacts are undertaken, if at all, with the greatest reluctance and extreme care, because submerged sites are as irreplaceable

120. Ibid.
121. Ryan (n. 72 above), p. 6.
122. Hanson, ed. (n. 87 above), p. 32.
123. Muckelroy, p. 44.
124. Muckelroy, pp. 44–45.
125. U.S. Congress, pp. 41–42.
126. Ibid.
127. Ibid.
128. Ibid.
129. Blidberg (n. 111 above), p. 251.

and nonrenewable as land sites.[130] These technologies have also made archaeological sites throughout the world's oceans more accessible and more vulnerable to human depredation, as pointed out by Ballard. Their availability fuels the immediate need for implementation of the LOS Convention's provisions on OAHN on the deep seabed, and the development of protective legislation for submerged sites located within coastal states' jurisdiction.

A. Sources of Technology

Marine archaeology is extremely expensive and severely limited in funds. Because "archaeology's only reward is knowledge," little or nothing of economic value is produced.[131] On the other hand, as Bass points out, "underwater archaeology has such an undeniably romantic appeal that lesser funds are often needed than by other marine sciences for similar work."[132] Excavations receive more applications, from skilled volunteers of all disciplines, than can be accommodated.[133] Advances in the technology have usually first been stimulated by the needs of the military and the off-shore industries such as oil, gas, and mineral companies. These technologies are then tailored to the needs of marine archaeology and are frequently donated or made available at nominal cost together with the services of professionals in these fields, who are personally fascinated by marine archaeology and its technological challenges.[134] Marine archaeology seems to spontaneously generate a level of generosity and enthusiasm in much of the commercial (and defense) community that is quite unmatched in other marine sciences. This brings me to my final point, and back full circle to the LOS Convention.

B. Another Option under the Convention

I have suggested an approach whereby the presence of Article 149 in the LOS Convention might be constructively used to assist the coming into force of the convention and the eventual accession of the Federal Republic of Germany, Great Britain, and the United States. The existence and continuing development of technologies for marine archaeology under the unusually propitious and altruistic circumstances described above might also offer another positive contribution toward this end.

130. Hanson, ed., p. 30.
131. Bass and Rosencrantz (n. 80 above), p. 337.
132. Ibid.
133. Ibid.
134. U.S. Congress, pp. 36–37. The U.S. Navy, e.g., made its ROV Deep Drone available for work on the USS *Monitor* (E.M. Miller, "A Time for Decision on Submerged Cultural Resources [n. 29 above], p. 32).

1. The Use of Part XIV

Among the "problems" identified by President Reagan as contributing to the U.S. decision not to sign the LOS Convention were "stipulations relating to mandatory transfer of private technology."[135] Part XIV of the LOS Convention lays down the general principles on technology transfer."[136] It (1) covers all activities in the marine environment compatible with the LOS Convention, (2) sets out the objectives and (3) obligations of states parties, and (4) lists measures to achieve technology transfer.[137] The Authority is given certain powers and obligations with respect to transfer of technology in Part XIV, but as in Part XI, these are limited to activities in the Area, which, as noted above, do not incude marine archaeology.[138] On the other hand, in Part XIV states parties may work "directly or through competent international organizations and the Authority" to establish national and regional marine scientific and technological centers "in order to stimulate and advance the conduct of marine scientific research by developing coastal states" and these are not restricted to mineral resources activities as far as the Area is concerned.[139] The functions of the regional centers are nonexclusively and broadly described, encompassing MSR and related technologies.[140] As also noted above, the Authority does have certain powers with regard to MSR, which, it is argued, includes marine archaeology. It has been questioned whether or not the transfer of technology provisions "are really incompatible with the interests of the United States to the extent that they justify the decision to refrain from signing a Convention that includes, in the President's words, 'many positive and very significant accomplishments.' "[141]

The response has been persuasively argued in the negative, on the basis that "technology transfer is an established tool of promoting development, and the provisions in the Law of the Sea Convention are not unusual when compared to requirements now imposed on most investments by multinational enterprises in developing nations."[142] In that analysis it was concluded that Prepcom would be the appropriate forum within which to discuss further the ambiguities of the technology transfer provisions.[143]

135. As quoted by Jon Van Dyke and David L. Teichman, *Transfer of Seabed Mining Technology: A Stumbling Block to U.S. Ratification of the Law of the Sea Convention?* East-West Environment and Policy Institute Reprint no. 59 (Honolulu: East-West Center, 1984), p. 447.

136. LOS Convention, Pt. 14, Sec. 1.

137. For points 1, 2, 3, and 4, see LOS Convention, Pt. XIV, Art. 266 (2); Art. 268; Art. 266; and Art. 269; respectively.

138. For discrete differences, see LOS Convention, Pt. XIV, Arts. 273 and 274, and Pt. XIV, Art. 144.

139. LOS Convention, Part XIV, Arts. 275 and 276.

140. LOS Convention, Part XIV, Art. 277.

141. Van Dyke and Teichman, p. 428.

142. Ibid., p. 445.

143. Ibid., p. 446.

2. The Involvement of Prepcom

It is suggested here that transfer of technology with regard to marine archaeology for OAHN be included within the brief of the fifth special committee whose creation has been proposed above. These technologies are, of course, applicable to other marine activities, including commercial ones, whence they often originated. However, marine archaeology has been shown to be much less fraught with the preconceptions and imperatives derived from certain economic necessities and philosophies that have so far stymied discussions on mining technology transfer. In fact, a "creeping benevolence" could be said to exist toward this most impoverished and yet most intimately human of the marine sciences. If discussions were opened on technology transfer as well in the context of marine archaeology first, they could contribute toward the development of the constructive atmosphere that must obtain before discussions on deep seabed mining can begin. Discussions could take place within this fifth special committee on the possible functions of the Authority as regards OAHN and marine archaeological technologies. The actual operation of the Authority in this field, if and when the Convention enters into force, could be viewed as an experiment in international government and its own training ground for the very much more complex task of managing a truly commercial resource. In conclusion, our submerged past could contribute to our future. Article 149 could be the Trojan (Sea) Horse within the LOS Convention, loaded not with engines to destroy this treaty, but with seeds for its survival.

Mangrove Forests: An Undervalued Resource of the Land and of the Sea

Lawrence S. Hamilton
East-West Center, Honolulu

John A. Dixon
East-West Center, Honolulu

Glenys Owen Miller
East-West Center, Honolulu

INTRODUCTION

Mangrove forests have long been the "orphan forests," unvalued by anyone except groups of low-technology, indigenous coastal dwellers who have traditionally used the mangrove resource as an area of habitation and as a source of food and materials. In spite of scientific documentation that has appeared on the importance of mangrove communities, even in public media—particularly since the early seventies—probably most people today consider them foul, ill smelling, insect-ridden, difficult, and even dangerous. Because of the low esteem or outright hostility for mangroves on the part of the lay public, coastal-zone developers, many planners, and policymakers in government, mangrove forests are being degraded or destroyed globally on a scale that must surely match or exceed that of the tropical rainforests, which have finally captured media and public attention. A survey of the global status of mangrove forests published in 1983 found that vast areas were being destroyed either intentionally or as a secondary result of other activities.[1] This process has scarcely abated since that survey because decisions continue to be made—with regard to road construction, waste dumps, short-term wood exploitation, freshwater diversions, landfills for coastal structures, conversion to agriculture, mining, and the development of aquaculture ponds—that either ignore the value of the mangrove resource or place much higher value on the

---

1. P. Saenger, E. S. Hegerl, and J. D. S. Davie, eds., *Global Status of Mangrove Ecosystems*, International Union for the Conservation of Nature and Natural Resources (IUCN), Commission on Ecology, Paper No. 3 (Gland, Switzerland: IUCN, 1983), p. 33. This article was also published in the *Environmentalist* 3, no. 3, suppl. (1983): 40.

alternative land use. For instance, one important area of Thailand, which had 367,900 ha of mangrove forest in 1961, had lost 46% of this by 1987, a span of only 26 years.[2] If, in addition to these continued threats, mangroves are "squeezed" by a rise in sea level, they are in double jeopardy. It therefore seems appropriate to review their status, to look again at their values in terms of services and products, and to consider some of the current issues, particularly with respect to economic valuation and analysis of replacement alternatives.

THE NATURE OF THE MANGROVE ECOSYSTEM

Definition and Distribution

Mangroves are salt-tolerant forest communities that, together with their associated fauna, occur in tropical and subtropical intertidal areas of the world. Saenger et al. list 60 species of plants (trees, shrubs, palms, and ferns) that are exclusive and 23 species that are common or important (but not exclusive) to the mangrove habitat (table 1).[3]

Data on distribution and areal extent of mangroves are not very reliable; there has been no global assessment or inventory for them, as there has been for tropical rainforests under the GEMS program.[4] Data from a variety of published sources and from a questionnaire by the Scientific Committee on Oceanic Research were compiled by Saenger et al. in 1983 and are reproduced here with updating, wherever reliable subsequent information was available (table 2).[5] These data on areal extent are likely to be overestimates owing to the high rates of loss since the information was published.

General Types of Mangrove Ecosystems

Not only is there species variation in different habitats depending on salinity, period of inundation, nutrient and freshwater inputs, and local relief, but there is structural diversity as well. Mangrove forests range from tall closed forests of *Rhizophora mangle* and *Avicennia germinans*, which reach heights of

---

2. Sanit Aksornkoae, "Mangrove Habitat Degradation and Removal in Phauguga and Ban Don Bays, Thailand," *Tropical Coastal Area Management* 3, no. 1 (1988): 16.

3. Saenger et al., eds., pp. 9–16.

4. Global Environment Monitoring System (GEMS) is the collective effort to monitor the world environment in order to protect human health and preserve essential natural resources. The coordination center for GEMS was established within the United Nations Environment Programme (UNEP) in 1975.

5. Saenger et al., eds., pp. 11–12. Supplementary references are listed under table 2.

TABLE 1.—NUMBER OF SPECIES OF ASSOCIATED BIOTA RECORDED
FOR MANGROVES IN VARIOUS GEOGRAPHIC REGIONS WHERE DATA
ARE AVAILABLE

Taxonomic Group	Region*					
	1	2	3	4	5	6
Bacteria	10	...	...	...	...	...
Fungi	25	14	...	...	...	...
Algae	65	93	...	105	...	12
Bryophytes/ferns	35	5	...	2	...	2
Lichens	...	105	...	...	...	...
Monocotyledons	73	42	...	20	...	8
Dicotyledons	110	80	...	28	...	20
Protozoa	18	...	...	3	...	...
Sponges/Bryozoa	5	7	...	36	...	1
Coelenterata/Ctenophora	3	6	...	42	...	12
Nonpolychaete worms	13	74	...	13	...	3
Polychaetes	11	35	...	33	...	72
Crustaceans	229	128	...	87	...	163
Insects/arachnids	500	72	...	...	...	...
Mollusks	211	145	32	124	...	117
Echinoderms	1	10	...	29	...	23
Ascidians	0	8	...	30	...	13
Fish	283	156	...	212	...	114
Reptiles	22	3	...	3	...	...
Amphibians	2	...	...	2	...	...
Birds	177	244	...	138	...	...
Mammals	36	7	...	5	...	...

SOURCE.—P. Saenger, E. S. Hegerl, and J. D. S. Davie, eds., *Global Status of Mangrove Ecosystems,* International Union for the Conservation of Nature and Natural Resources (IUCN) Commission on Ecology, Paper No. 3 (Gland, Switzerland: IUCN, 1983), p. 15, and also published in the *Environmentalist,* Vol. 3, No. 3, Suppl. (1983).

  *Regions are as follows: 1 = Asia; 2 = Oceania; 3 = West Coast of the Americas; 4 = East Coast of the Americas; 5 = West Coast of Africa; 6 = East Coast of Africa and the Middle East.

40–50 m and diameters of 70 cm in parts of South American deltas and estuaries, to stunted shrubs less than 1 m tall forming open communities along arid coasts throughout the tropics. Grossly, the major climatic determinants of mangrove occurrence, structure, and species are air temperature (latitudinally) and rainfall (longitudinally). Structural variation occurs at a single location, in response to the local topography that characterizes the intertidal environment.[6] A typology that has proven useful in promoting

6. For an introduction to mangrove ecosystems, see Lawrence S. Hamilton, *Understanding Mangrove Ecosystems,* an audiovisual slide-cassette illustrated script program (Gland, Switzerland: IUCN, 1983). For geomorphological information, see B. Thom, "Mangrove Ecology—a Geomorphological Perspective," in *Mangrove Ecosystems in Australia: Structure, Functions, and Management,* ed. B. F. Clough (Canberra: Australia National University Press, 1982), pp. 3–17; and for mangrove ecology, see Samuel C. Snedaker and Jane G. Snedaker, eds., *The Mangrove Ecosystem: Research Methods* (Paris: Unesco, on behalf of the Unesco/SCOR Working Group 60 on Mangrove Ecology, 1984).

TABLE 2.—ESTIMATED AREAL COVERAGE OF MANGROVES BY COUNTRY

Country	Area (km²)	Country	Area (km²)
Asia:		East Coast of Americas:	
United Arab Emirates	30	United States	2,000
Pakistan	2,708	Cuba	4,000
India	3,500	Haiti	180
Sri Lanka	36	Dominican Republic	90
Bangladesh	4,050	Puerto Rico	65
Burma (Myanmar)	5,711	Guadeloupe	80
Thailand	2,686	Martinique	19
Kampuchea	100	Trinidad and Tobago	40
Vietnam	5,364	Jamaica	70
Malaysia	6,288	Guatemala	500
Brunei Darussalam	70	Belize	730
Singapore	5	Honduras	1,450
Indonesia	36,000	Venezuela	6,736
China	400	Guyana	800
Philippines	2,133	Suriname	1,150
Taiwan	2	French Guiana	55
Japan	4	Brazil	25,000
Total	69,087	Total	42,945
Oceania:		East and West Coasts of Africa:	
Papua New Guinea	4,116	Senegal	4,400
Australia	800	Gambia	600
Solomon Islands	642	Guinea-Bissau	2,430
Fiji	385	Guinea	2,600
New Caledonia	200	Sierra Leone	1,000
Tonga	10	Liberia	400
New Zealand	198	Benin	30
Micronesian Islands	88	Nigeria	9,730
		Cameroon	2,720
Total	6,439	Gabon	2,500
West Coast of Americas:	6,600	Zaire	200
Mexico	450	Angola	500
El Salvador	600	Kenya	450
Nicaragua	390	Tanzania	960
Costa Rica	4,860	Malagasy	3,207
Panama	3,570	Mozambique	850
Colombia	1,776	South Africa	11
Ecuador	25		
Peru		Total	32,588
		Estimated total world	
Total	18,271	mangrove coverage	169,330

Source.—P. Saenger, E. S. Hegerl, and J. D. S. Davie, eds., *Global Status of Mangrove Ecosystems*, International Union for the Conservation of Nature and Natural Resources (IUCN) Commission on Ecology, Paper No. 3 (Gland, Switzerland: IUCN, 1983). Additional information is updated from the following sources: S. Aksornkoae, "Mangrove Habitat Degradation in Phauguga and Ban Dan Bays, Thailand," *Tropical Coastal Area Management* 3, No. 1 (1988): 16; various articles in *Ambio*, Vol. 17, No. 3 (1988); various articles in *Bakawan* (1986–88); C. D. Field and A. J. Dartnall, eds., *Mangrove Ecosystems of Asia and the Pacific: Status, Exploitation and Management*, Proceedings of the Research for Development seminar, Australian Institute of Marine Science (AIMS), Townsville, Australia, May 18–25, 1985 (Townsville: AIMS on behalf of the Australian Committee for Mangrove Research, 1987); Lin Peng, "Ecological Notes on Mangroves of Southeast Coast of China Including Taiwan Province and Hainan Island," *Acta Ecologica Sinica* 1, No. 3 (September 1981): 283–90; Ong Jin-Eong, "Mangroves and Aquaculture," *Ambio* 11, No. 5 (1982): 252–57; *Siren*, No. 38 (October 1988); and United Nations Environment Programme (UNEP), *Co-operation for Environmental Protection in the Pacific*, UNEP Regional Seas Reports and Studies No. 97 (Nairobi: UNEP, Oceans and Coastal Areas Programme Activity Centre, 1988).

understanding of mangrove ecosystems was developed originally by Lugo and Snedaker[7] and is presented below, as slightly modified by Wharton et al.[8] See also figure 1.

1. *Overwash mangroves.*—Mangrove overwash forests are found around smaller, low-lying islands, along finger-like projections of larger land masses, and in shallow bays and estuaries. They are overwashed diurnally at high tides. There is an absence of vegetation below the canopy, and only small amounts of organic matter are exported on the tide. These islands are used as safe rookeries by many bird species. Tree regeneration is slow, as these forests are in low-nutrient, high-salinity areas exposed to high wave action.

2. *Fringe mangrove wetlands.*—Found in sheltered locations along fringes of protected shorelines and islands, fringe mangroves are best developed along shorelines where elevations are higher than mean high water. These mangroves are susceptible to strong winds, resulting in the deposition of large amounts of debris within the prop root system. If fringe mangroves receive terrestrial freshwater and nutrient runoff, they are taller and exhibit more biomass and higher productivity. A fringe mangrove zone on flat or gently sloping coastline may be extensive.

Fringe mangroves are refuges for fish and other wildlife, especially during stormy weather, and the calmer waters provide a valuable nursery area for fish. The prop roots abound with adhering flora and fauna. The well-developed prop root system traps much organic debris and sediment, and in still waters the sediments consolidate, allowing the mangroves to "advance" toward the water. High-energy waves and storm winds will, however, force the detritus inland and "recapture" land from the mangroves.

3. *Dwarf forests (scrub mangrove wetlands).*—Dwarf forests are found on flat coastal fringes. Many species may be represented, but they usually stand less than 1.5–2.0 m tall, even though the trees may be very old (40+ years). These mangroves grow under stressed conditions and represent the only flora that can grow in these marginal environments, which range from nutrient-poor marls and low-salinity waters to nutrient-rich soils and hypersaline soil/water conditions of dry environments. Scrubbiness characterizes those mangroves growing in colder climates. They are the least productive of mangrove types. These are not good fish habitats, but the wetlands support many varieties of wildlife, such as birds. Their role is beneficial to man in that they stabilize otherwise vulnerable soil, sand, and mud but are sensitive to further stress.

4. *Hammock mangrove wetlands.*—Growing on a slight rise (about 5–10 cm), and associated with reducing environments in a substrate of mangrove

7. A. E. Lugo and S. C. Snedaker, "The Ecology of Mangroves," *Annual Review of Ecology and Systematics* 5 (1974): 39–64.

8. C. H. Wharton, H. T. Odum, K. Ewel, M. Duever, A. Lugo, R. Boyt, J. Bartholomew, E. De Bellevue, S. Brown, M. Brown, and L. Duever, *Forested Wetlands of Florida—Their Management and Their Use,* Final Report to Division of State Planning (Gainesville, Fla.: Center for Wetlands, February 1976), fig. 102, p. 192.

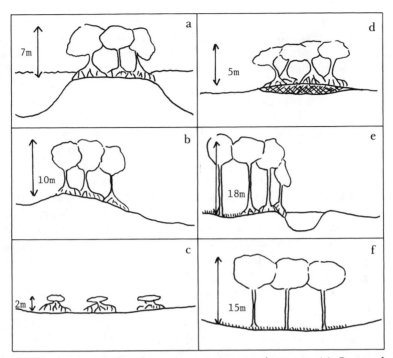

FIG. 1.—Physiognomy of major mangrove community types. (*a*) Overwash mangrove islands: overwashed by daily tides; high rate of organic exports; dominated by red mangroves, but all species may be present; south Florida, south coast of Puerto Rico; sensitive to ocean pollution. (*b*) Fringe mangrove wetlands: line water ways; high rate of organic exports; dominated by red mangrove; throughout south Florida, Puerto Rico, and Florida's east and west coast; sensitive to ocean pollution. (*c*) Scrub mangrove wetlands: on extreme environments; low organic exports; usually red or black mangroves; southeast Florida, south coast of Puerto Rico, high latitudes on west coast of Florida; sensitive to further stress. (*d*) Hammock mangrove wetlands: on land rises in south Florida; low export of organic matter; all mangrove species; south Florida Everglades; sensitive to fire and drainage; (*e*) Riverine mangrove wetlands: along flowing waters; high export of organic matter; all mangrove species, reds predominate; south Florida, north coast of Puerto Rico; sensitive to alterations of water flow. (*f*) Basin mangrove wetlands: in depressions or areas of slow water movement; high seasonal export of organic matter; black mangroves predominate; inland locations in south Florida and Puerto Rico; sensitive to alteration of sheet flow, seawater input, and prolonged high water. Source.—C. H. Wharton, H. T. Odum, K. Ewel, M. Duever, A. Lugo, R. Boyt, J. Bartholomew, E. De Bellevue, S. Brown, M. Brown, and L. Duever, *Forested Wetlands of Florida—Their Management and Their Use,* Final Report to Division of State Planning (Gainesville, Fla.: Center for Wetlands, February 1976), fig. 102, p. 192.

peat, these inland mangrove wetlands are molded in shape by upland water runoff, which flows around them toward the sea. The peat soil is highly saline, and growth of vegetation is limited by this factor and by the frequency of tidal inundation. Scrub mangroves and grasses are common. There is low export of organic material. In areas sensitive to fire, flood, and drainage, hammocks are safe refuges for wildlife, and these areas are often conserved as protected natural habitats for wildlife.

5. *Riverine mangrove wetlands.*—These occur on terminal floodplains of rivers and other freshwater drainages. Usually separated from the water by a berm but flushed with diurnal tides, the riverine mangroves extend as a floodplain forest behind the berm and are characterized by tall, relatively straight-trunked trees. They represent the most productive and spectacular mangrove forests. Freshwater nutrients and sediment enrichment from upland ecosystems and adequate ventilation of roots result in high productivity. The litter produced is the greatest of all mangrove systems, and the nutrients produced support a rich diversity of floral and faunal habitats during the alternating periods of low and high salinity. To maintain this healthy mangrove ecosystem, it is important to maintain the inflow of fresh water, nutrients, and sediments and a suitable depth of water in the wetlands to provide ventilation of roots. The water table is not far underground even if the surface soil dries out. Whereas water salinity varies widely, soil salinity remains more or less constant and about equal to that of seawater. These wetlands are particularly sensitive to alterations in water regime. They produce much organic detritus that will support food webs of finfish, mollusks, and crustaceans. Mangrove prop roots and pneumatophores ("breathing" roots that protrude above ground) slow down fast-moving water and precipitate out sediments. They are effective as buffers for hurricane-driven storms.

6. *Basin mangrove wetlands.*—These wetlands are found in drainage depressions channeling terrestrial runoff toward the coast and are located inland, behind fringe forests. They are inundated only by the highest of tides, although some closer to the coast may be influenced daily by tides. They have a tall canopy, with a high leaf-area index and a floor cover of pneumatophores. Inland, where tidal influence is minimal, various mangrove species are dominant. Basin mangrove wetlands are characterized by high soil salinity, anaerobic environment, water stagnation, and reducing soils. These mangroves typically excrete salt through their leaves (e.g., black mangroves) and have pneumatophores to allow the roots to breathe in the anaerobic mud or soil. The basin mangroves are efficient nutrient traps. They export high amounts of organic matter downstream.

Natural Functioning of Mangroves

Mangrove plant species are able to survive under saline conditions by processes of salt exclusion, as well as by storage and excretion of excess salt.

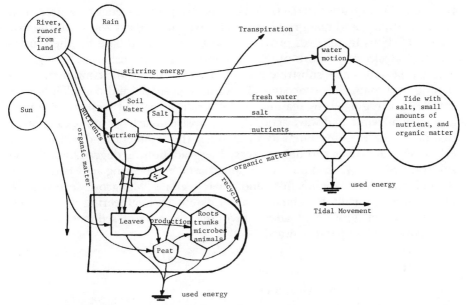

Fig. 2.—Mangrove swamp processes. Source.—C. H. Wharton, H. T. Odum, K. Ewel, M. Duever, A. Lugo, R. Boyt, J. Bartholomew, E. De Bellevue, S. Brown, M. Brown, and L. Duever, *Forested Wetlands of Florida—Their Management and Their Use,* Final Report to Division of State Planning (Gainesville, Fla.: Center for Wetlands, February 1976), fig. 127, p. 235.

Inundation may be handled by various root adaptations that allow the plants to obtain adequate oxygen, hence the prop roots, knee or ribbon roots, pneumatophores, and aerial roots that give most mangroves a distinctive appearance. Freshwater inputs into the ecosystem from rainfall, seepage from hinterland swamps, dilution of estuarine waters by upstream freshwater catchments, or freshwater flooding of the tidal zone are necessary to maintain appropriate salinity regimes by leaching or flushing and to maintain soil water above the wilting point in these hot environments where evapotranspiration rates can be high.

Nutrient inputs are also needed to maintain the growth of these ecosystems, which are *open* systems that export nutrients in the form of decomposable organic plant litter (see fig. 2). Nutrients come into the system from rainfall, wind, seawater, and by decay of organic material (assisted by benthic fauna, especially crabs). However, much of the most important nutrient supply comes from the freshwater inputs from the land. Both for salinity control and nutrient input, maintenance of the freshwater regime in the system is of paramount importance. This is a process that is often and easily disrupted by human activity either in or adjacent to the mangrove or even at some distance up the watershed.

The major process by which coastal waters are enriched is the export of decomposable organic material and its decomposition into protein-enriched fragments of detritus. In fact, detritus is the primary energy source in a tropical estuary, and mangroves generally are the primary source of the raw material.[9] Thus mangroves provide a most critical food base for the crustaceans, mollusks, and finfish of the tidal and near-shore fisheries.

A relatively stable substrate is also a prerequisite to the continued functioning of a mangrove ecosystem. Erosion, deposition, and consolidation of sediments is regulated by seasonal and episodic activity of freshwater runoff and by tidal, wind, and wave action.[10] Any upstream activity in the catchment that alters the flooding intensity or changes the sediment load will have an effect on mangroves through increased aggradation or degradation of the substrate. Similarly, on the seaward side, coastal engineering activity can alter the substrate by changing tidal and wave regimes. Mangrove forests have often unintentionally benefited in areal extent by engineering actions that reduced coastal erosion or added favorable deposits of sediments, so that mangrove areas expand seaward.

Products, Services, and Amenities

Direct products from mangrove ecosystems are important to the subsistence livelihood of many tropical coastal human communities, and they also provide a commercial base that is vital to many local and even national economies. Commercial and traditional products range from construction timbers to charcoal and pulpwood for paper, and from medicines to honey. One versatile species, the Nipa palm, is utilized as a source of alcohol fuel, sugar, vinegar, roof- and wall-thatching panels, hats, an intoxicating local beverage, a cure for herpes, a sweet confection, and a dozen other uses.[11] Appendix A lists direct products from mangrove forests.[12]

Mangrove ecosystems nourish adjacent fisheries in the coastal or estuarine environment. Odum calculated that 80%–90% of commercial fisheries in the Gulf of Mexico were dependent on mangroves at one or more stages of their life cycles.[13] Cintrón et al. stated that mangrove swamps may export over 30% of their net production to adjacent bays, thereby further

9. Lawrence S. Hamilton and Samuel C. Snedaker, eds., *Handbook for Mangrove Area Management* (Honolulu: UNEP, and East-West Center, Environment and Policy Institute, 1984), p. 26.

10. Saenger et al., eds. (n. 1 above), pp. 31–32.

11. Lawrence S. Hamilton and D. H. Murphy, "Use and Management of Nipa Palm (*Nypa Fruticans*, Arecaceae): A Review," *Journal of Economic Botany* 42, no. 2 (1988): 206–13.

12. Hamilton and Snedaker, eds., p. 2.

13. W. E. Odum, "Pathways of Energy Flow in a South Florida Estuary" (diss., University of Miami, Florida, 1970), as quoted by Cintrón et al. (n. 14 below).

contributing to coastal fisheries production.[14] These forests are also the habitat of a number of fauna that either are residents of the community (e.g., mangrove crabs) or find in these forests some critical element of their range of habitat requirements (such as a nursery area for juvenile fish or a safe roosting for herons or ibis). Though the data are incomplete, especially for Africa, 89 species of mangrove-associated animals that are "at risk" (threatened, vulnerable, endangered, or possibly extinct) have been listed.[15] Appendix B presents some of the products that are indirectly produced by mangroves through the linkages mentioned above.[16]

In addition to these direct and indirect products, mangrove forests provide various service and amenity functions. Coastal damage associated with cyclones and storm surges is reduced by the presence of mangroves. This has been the case for areas inland of mangroves on the Ganges-Brahmaputra Delta, in contrast to areas to the east or west in the Bay of Bengal.[17] Shoreline erosion is reduced by mangroves, and they have been planted in the United States, Australia, and China, for example, as an alternative to protective engineering works. As a coastal protection barrier, mangroves maintain themselves at no cost, and in the event of damage in a severe tropical storm they will self-repair without cost.

While many individuals regard mangrove areas with some distaste, mainly because of lack of information, these areas are attractive to many others, both naturalists and tourists who come to observe the interesting flora and fauna. Boardwalks into mangrove areas are appearing with increasing frequency to provide access, and boat trips have become very popular in some places. In these "swamps," one may see some of the interesting and often rare wildlife of the tropics. The most popular natural area visitor destination in Trinidad and Tobago is Caroni Swamp, where mangroves provide opportunities for bird watching, especially for viewing the scarlet ibis. In tropical Australia, mangrove boat cruises provide thousands of paying tourists with an unique experience "through ancient mangrove forests" and, in particular, encounters with the saltwater crocodile. Natal in South Africa, New Zealand, Florida, the Philippines, Trinidad and Tobago, Venezuela, Australia, and the United States are among the early pioneers in capitalizing on the recreational value of these interesting ecosystems.

Unfortunately, probably less than 1% of the total world mangrove resource has official protection status to safeguard the production of these

14. Gilberto Cintrón, A. E. Lugo, Ramon Martinez, Barbara B. Cintrón, and Luis Encarnacion, *Impact of Oil in the Tropical Marine Environment*, Technical Publication, Division of Marine Resources (Puerto Rico: Department of Natural Resources of Puerto Rico, 1981), p. 1.

15. Saenger et al., eds., pp. 47–49.

16. Hamilton and Snedaker, eds., p. 2.

17. Ibid., p. 1.

amenity and service functions.[18] When taken collectively, these services may outweigh the benefits from any other use man can imagine for these areas. Appendix C, taken from Wharton et al.,[19] lists a number of these services. The question of economic valuation of mangroves will be treated later in this article.

Of the many kinds of developments or impacts that affect the multiple-use production and maintenance of mangroves, three have been selected for discussion: conversion to aquaculture, wood harvesting, and oil pollution.

MANGROVE CONVERSION FOR AQUACULTURE

Worldwide, mangrove forests are being eliminated at an alarming rate, chiefly for conversion to brackish-water fishponds. There are an estimated 750,000 ha in brackish-water aquaculture ponds throughout the world.[20] In Asia alone, more than 400,000 ha of tidal land have been converted into brackish-water ponds for aquaculture.[21] Worldwide, Indonesia and the Philippines lead the conversion to brackish-water ponds, followed by India, Ecuador, Thailand, and Bangladesh.

Brackish-water ponds are often built on clear-felled mangrove lands. If cleared land is available, such as in salt ponds behind the mangroves, those areas will be used first because it will mean reduced pond-preparation costs. The ideal location for brackish-water ponds is landward of the mangrove belt, with channels cut through the mangroves to the open coastal waters for pond-water exchange. This siting causes minimal disturbance to the mangrove ecosystem.

In acid-sulfate prone soils, the water should never be entirely drained from the pond, and by keeping the soils covered with water the mud floor of the pond never gets exposed to air and, therefore, does not dry out. If the ponds do dry out, as in the more mechanically prepared ponds that are diked, drained, dried, and scraped flat with bulldozers, the sediments are susceptible to acid-sulfate conditions. The pH of the soil, which is normally near 7 before drainage, may drop to 4 or below. Good water quality is within the range of pH 6.5–8.5. Shrimp growth is slowed in acidic conditions, and below pH 4 the shrimp die.[22] Fish may die directly from the drop in pH or indirectly through

18. Ibid., p. 9.
19. Wharton et al. (n. 8 above), p. 234.
20. John Bardach, personal communication, 1987.
21. Bo Christensen, *Management and Utilization of Mangroves in Asia and the Pacific,* Food and Agriculture Organization (FAO) Environment Paper No. 3 (Rome: FAO, 1982).
22. Gilberto Cintrón and Yara Schaeffer-Novell, *Management of Stress in Mangrove Systems* (document prepared for the workshop, "Coral Reef, Seagrass Beds and Mangroves: Their Interaction in the Coastal Zones of the Caribbean Workshop," Dickinson University, St. Croix, West Indies, May 24–30, 1982, Working paper, UNESCO/ W.I.L.-F.D.U./IOCARIBE).

the toxicity of certain substances such as aluminum, iron, and manganese, which are mobilized at low pH.[23] Toxic leachate from pond runoff and drainage channels in acid-sulfate soils may also affect the mangrove ecosystem. While ponds can be treated with lime to neutralize the sulfuric acid, the amount of lime required may be prohibitively expensive.

Postlarval shrimp and juvenile fish targeted for aquaculture drift into the quiet nutrient-rich estuaries and mangrove waters and are recruited from the mangrove waters, their nursery grounds. This is the basis of the brackish-water pond culture. Seawater containing postlarval shrimp and juvenile fish is pumped into ponds through a filter system designed to limit entry to fauna the size of the targeted species and smaller. Traditionally, the screening of the incoming water is minimal, and it is not possible to separate shrimp from competitors or predators of the same size or smaller. If other species are caught, they may be discarded or live contentedly in the ponds, feeding on the shrimp until harvest time. In more controlled pond systems, the filtration of seawater may be such that only microorganisms that are food for the shrimp are admitted to the ponds, and the ponds are sometimes stocked by fry and larval shrimp that have been artificially raised. Grow-out ponds are also used to grow shrimp and juvenile fish to marketable size.

While the number of ponds continues to increase, the estuarine system deteriorates as the residual discharges from the ponds cannot be flushed away by the diurnal tidal flows. Ponds need frequent water exchange. For example, semiextensive operations need 3%–8% exchange, and requirements increase under more intensive farming. This can be illustrated with an example from the Gulf of Guayaquil region of Ecuador: at 30,000 ha of ponds, and a 5% pumping rate, the daily volume of water exchanged with the estuary is greater than the discharge of the Guayas River, Ecuador, during low tide flows.[24] Conditions will become worse if the cultures are intensified.

The free larvae live within the mangrove waters for 3–4 months[25] and grow up to 21 mm, at which stage they are called juveniles. They then swim to the deeper coastal waters to mature and reproduce. The *Penaeus merguiensis,* for instance, completes its entire life cycle within 12 km of the Malaysian coastline.[26]

After a few years, the shrimp ponds often become uneconomical enter-

23. Crispino A. Saclauso, "A Critique: Technical Consideration in the Use of Mangrove Areas for Aquaculture," *Bakawan* 5, no. 2 (1986): 6–8.

24. Lucia Colorzano, "Sources of Pollution and Principal Polluted Areas in the Pacific Coastal Waters of Ecuador," in *Cooperation for Environmental Protection in the Pacific,* UNEP Regional Seas Reports and Studies No. 97 (Nairobi: UNEP, Oceans and Coastal Areas Programme Activity Centre, 1988), pp. 87–97, esp. p. 92.

25. Roland C. W. Seow, "Mangrove Waters as Habitat of Different Species and Their Development Stages," *Bakawan* 5, no. 4 (December 1986): 6–8, esp. 8.

26. V. C. Chong, "Maturation and Breeding Ecology of the White Prawn *Penaeus merguiensis* (De Man)," in *Proceedings of the 5th International Symposium on Tropical Ecology* (Kuala Lumpur, Malaysia, April 16–21, 1979), pp. 1175–86.

prises because natural recruitment of larvae is no longer adequate or the ponds have become too acidic. It is more economical for fish farmers to abandon these ponds and find new areas to clear and rebuild. This shifting aquaculture is akin to terrestrial shifting agriculture.

Snedaker et al. predicted that the production of pond-raised shrimp can give per hectare returns four to five times larger than the economic value of product from irrigated agriculture.[27] The major limiting factor at the moment is the supply of larvae, which is an essential element in any intensification of production.

Examples of Brackish-Water Pond Conversion

The pond-development situation for most countries is very fluid. The Philippines and Ecuador are given as just two examples of particularly interesting cases.

Philippines.—In the Philippines, over 2,000 ha of mangroves are lost annually to fishponds;[28] by 1981, over 196,000 ha of mangrove forest had been converted to brackish-water aquaculture primarily for milkfish (*Chanos chanos*) and shrimp (*Penaeus*) (particularly *P. monodon*).[29]

In 1975, brackish-water fishponds covered an area of approximately 176,000 ha, 49% of which were leased from the government.[30] Production during 1975, primarily of milkfish and penaeid shrimp, contributed about 8%–10% of the total national fish production. In 1977 the brackish-water fishponds produced over 115,000 mt.[31] Production remains low, averaging about 600 kg/ha/year, and for some ponds it is as low as 100–200 kg/ha/year.

27. Samuel C. Snedaker, J. C. Dickinson III, M. S. Brown, and E. J. Lahmann, "Shrimp Pond Siting and Management Alternatives in Mangrove System in Ecuador," a USAID-funded project, published in 1986; excerpts from this report were published in *Bakawan* 6, no. 3 (September 1987): 2–3, and this quote is taken from the *Bakawan* reference.

28. Marta Vannucci, "The UNDP/UNESCO Mangrove Programme in Asia and the Pacific-Synopsis," *Ambio* 17, no. 3 (1988): 214–17, esp. 216. For figures of previous years, see Bureau of Forest Development, *Philippine Forestry Statistics* (Quezon City, Philippines: Diliman, 1977); and Natural Resources Management Center (NRMC), *Mangrove Inventory of the Philippines Using Landsat Data* (Quezon City, Philippines: Diliman, 1978).

29. Ricardo M. Umali, "Coastal Resources Development and Management: The Philippines Experience," Natural Resources Management Center report (Quezon City: NRMC, ca. 1982).

30. Miguel D. Fortes, "Mangrove and Seagrass Beds of East Asia: Habitats under Stress," *Ambio* 17, no. 3 (1988): 207–13, esp. 210.

31. Rogelio O. Juliano, James Anderson, and Aida R. Librero, "PHILIPPINES: Perception, Human Settlements, and Resource Use in the Coastal Zone," in *Man, Land and Sea*, ed. Chandra H. Soysa, Chia Lin Sien, and William L. Collier (Bangkok: Agricultural Development Council, 1982), pp. 219–40, esp. p. 232.

The poor yields are attributed in part to the fact that at least 60% of fishponds in the Philippines are affected by acid sulphate soil conditions.[32]

Ponds average 16 ha in size (ranging from a few hundred square meters to 500 ha). There are approximately 200,000 people who rely on the industry for all or part of their living.[33]

Ecuador.—Ecuador's total mangrove resource is about 180,000 ha. Estimates of percentages of mangrove cleared for brackish-water pond aquaculture vary from 5% to 29%, with as much as 50% in the Bahia de Caraquez region.[34] Investors and developers of ponds for penaeid shrimp production prefer semiarid climates and seek out salt flats, barren coastal areas, and former mangrove areas for the construction of ponds to reduce construction costs. However, the extremely rapid development of the aquaculture industry is forcing developers of new pond systems to convert productive mangrove forest and farmland.

The shrimp-grow-out pond mariculture industry of Ecuador began in 1969 with 600 ha of small ponds. Heavy investment began in 1977 once shrimp ponds proved successful. In 1980, 10,200 mt (worth US$66.4 million) of shrimp were exported, and a year later the harvest had almost doubled.[35] In 1982 the shrimp harvest from ponds accounted for 44% of Ecuador's fish exports and by 1984 was the largest industry based on renewable resources.[36] The industry ranks second behind petroleum as a major export commodity for Ecuador and has created at least 19,000 new jobs.[37] The mariculture ponds are so extensive that much of the coastline of southern Ecuador has been transformed,[38] and there are now over 89,000 ha in shrimp ponds (1984 figures).[39] The only limitation is the availability of naturally occurring shrimp larvae for pond recruitment seeding.[40]

Prior to 1980, Ecuador was the only South American state that had made a significant investment in shrimp aquaculture. The perceived financial suc-

32. Saenger et al., eds. (n. 1 above), p. 40.

33. Scott E. Siddall, Joseph A. Atchue III, and Robert L. Murray, Jr., "Mariculture Development in Mangroves: A Case Study of the Philippines, Ecuador, and Panama" (1984), p. 43.

34. Ibid., p. 17.

35. Ibid., p. 31.

36. James M. Kapetsky, "Development of the Mangrove Ecosystem for Forestry, Fisheries and Aquaculture" (paper presented at the symposium "Ecosystem Redevelopments: Ecological, Economic and Social Aspects," Budapest, Hungary, April 5–10, 1987).

37. Siddall, p. 16.

38. Snedaker et al. (n. 27 above), p. 2.

39. Colorzano (n. 24 above), p. 92; however, Kapetsky states that there were 61,000 ha in shrimp ponds in Ecuador as of March 1986 (p. 13).

40. Kapetsky, p. 13. During March 1986, Kapetsky states that one-half of the shrimp ponds in Ecuador were empty because of seed-stock shortage. Hatchery-reared seed-stock production is increasing to substitute for the diminished wild-caught postlarva availability.

cess, however, has resulted in most Central and South American countries encouraging or planning to stimulate mariculture investment in their countries.[41]

Impact of Brackish-Water Pond Construction on the Estuarine Habitat

Brackish-water pond conversion has had a devastating effect on the stocks of shrimp and fish found naturally in mangrove-influenced waters. Conversion has been profitable only to investors who are in the enterprise for short-term gain. Unless managed wisely, with knowledge of the interrelationships involved, brackish-water pond culture is a destructive practice.

> [Even] prior to the "El Nino" event of 1982/83, the shrimp mariculture industry [in Ecuador] began to experience increasingly severe and unpredictable shortages in the wild post-larval shrimp that are harvested in shallow coastal waters to stock commercial growout ponds. For several years, the demand for PL [post-larval shrimp] had exceeded their natural availability and led the shrimp producers . . . to begin to seek a solution to the continuing shortage. . . . It was, therefore, locally concluded that the extensive conversion of the coastal mangrove forests to shrimp growout ponds . . . was somehow implicated in the reduction of the availability of larval and juvenile shrimp.[42]

As postlarval and juvenile shrimp and juvenile fish are found where there is a source of nutrient-enriched organic food matter, there is less chance that they will occur once this food resource has been depleted.[43]

The Socioeconomic Impact of Brackish-Water Pond Conversion

The impact of conversion on people who dwell in and rely on the mangrove forests for their homes, food, and jobs is being given increased attention.[44]

41. Samuel C. Snedaker, "Some Traditional Uses," *Bakawan* 6, no. 2 (June 1987): 10–12; and IFREMER-France Aquaculture, "The Preselection of Sites Favourable for Tropical Shrimp Farming," Bulletin of Commerce Series (Paris: IFREMER, 1987).
42. Snedaker et al., p. 2.
43. In another article on Ecuador, the same authors state that "although all mangroves have not been eliminated in the major production region, we hypothesize that the disproportionate elimination of sources of dissolved organic matter may be the dominant cause of the reduction in wild shrimp post larvae stocks" (see Enrique J. Lahmann, Samuel C. Snedaker, and Melvin S. Brown, "Structural Comparisons of Mangrove Forests near Shrimp Ponds in Southern Ecuador," *Interciencia* 12, no. 5 [September/October 1987]: 240–43, quoted at 242).
44. Two papers that raise this issue in the Philippines, e.g., are Richard L. Ed-

In a socioeconomic study of the barrio people of Bataan, the Philippines, it was concluded that those in the lowest socioeconomic class (who often live in the mangrove areas) were adversely affected by the conversion of mangrove to fishpond.[45] Whereas middle and upper classes participated actively in the economic aspects of fishpond enterprises if they owned or leased the land or were the managers, the barrio people were only marginally involved and were paid minimum wages for working in the ponds. For them, their food supply largely came from what they harvested from the mangroves and adjacent public waters. Once the mangroves were cleared for ponds, there was little left to harvest. As a result, family and community structures disintegrated. It was suggested that, instead of converting entire mangrove swamps into ponds, a section of the swamp should be left for the barrio inhabitants and for maintaining ecological resources for the fish and other marine life that are dependent on the mangrove swamps.

A Malaysian study by Foo and Wong illustrated the social and economic importance to fisheries of conservation of mangrove forests.[46] Sabah has about 12,000 fishermen, and more than half of them either live among the mangrove swamps or fish in and near the swamps. The crab *Scylla serrata*, which lives in the mangrove swamps, brings these small-scale fishermen about Malaysian $1.2 million/year. This source of income is reduced or eliminated if the mangrove swamps become ponds.

Current Management Policy

The fact that fish catches decline as mangroves are destroyed has been confirmed in many places in Asia.[47] However, current demand and high prices paid for shrimp in the Japanese and American markets encourages further expansion of shrimp ponds, and in Asia this is being facilitated by loans from the Asian Development Bank and the World Bank. "All fishery uses of the mangrove ecosystem can be sustainable. . . . Nearly all aquaculture uses of mangroves are sustainable. The exception is pond aquaculture, mainly

wards, "Socio-economic Implications of Fishpond Development in the Mangrove Swamp Barrio of the Philippine Islands" (Humboldt State University, Arcata, Calif., December 16, 1982); and A. B. Velasco, "Socio-cultural Factors Influencing the Utilization of Mangrove Resources in the Philippines: *Fishpond vs. Other Uses*," *Tropical Ecology and Development* (1980), pp. 1185–93.

45. Edwards, pp. 41–42.

46. H. T. Foo and J. T. S. Wong, "Mangrove Swamp and Fisheries in Sabah," *Tropical Ecology and Development* (1980), pp. 1157–61.

47. For example, P. Martosubroto and N. Naamin, "Relationship between Tidal Forests (Mangroves) and Commercial Shrimp Production in Indonesia," *Marine Research in Indonesia*, No. 18 (1979), pp. 81–86; and Marta Vannucci, "Overall Productivity of Mangroves with Special Reference to Indus Delta," *Bakawan* 5, no. 3 (September 1986): 9–11.

of shrimps and milkfish, which uses about 0.6 million hectares in countries with mangroves. At present, there are no economically viable sustitutes for pond culture of shrimp. Therefore, remedies for the conversion of mangroves for pond aqaculture are required. Intensification of shrimp culture instead of extensification is one remedy."[48]

A greater awareness of the importance of healthy mangrove ecosystems, coupled with ongoing research and legislative regulation, is beginning to slow the rate of destruction of mangrove forests, for example, in the Philippines, where 45% of the original mangrove forest is now occupied by fish ponds.

Intensification must be promoted, but pond operators need to know how to change their management practices. Nongovernmental agencies, such as shrimp and fish farmers' associations, may be able to assist with personnel and training programs for pond operators. In the meantime, destructive practices will continue through ignorance or lack of education about alternatives. Periodic coastal aerial and ground surveillance can help monitor future pond construction, particularly those constructed without authority.

WOOD HARVESTING

One of the major disturbances to mangrove ecosystems throughout the world is the cutting of the trees and the removal of wood for products such as poles, fuelwood, charcoal wood, pulpwood, and chips for rayon, fibreboard, or paper. In terms of area affected, it was suggested in Saenger et al. that the scale of impact for wood harvesting in commercial enterprises (often clear felling) was on the order of 10,000–500,000 ha for a single decision about allocating mangroves.[49] This would compare with conversion to salt ponds (100–1,000 ha) and mining or mineral extraction (10–100 ha). For instance, currently underway in the Bintuni Bay area, West Irian, Indonesia, is a large mangrove chipping project undertaken by a Japanese corporation that, through a consortium, will cut 135,000 ha over a 10-year period. This involves yearly production of 200,000 tons of chips (initially) to 400,000 tons (final stages).[50] In Sabah, Malaysia, beginning in 1970, 40% (122,750 ha) of the total mangrove area of that state was allotted for woodchip export to Japan,[51] and, since then, additional areas, including a state mangrove reserve, have been allocated to wood harvesting.

48. James M. Kapetsky, "Convention of Mangroves for Pond Aquaculture: Some Short-Term and Long-Term Remedies," *Bakawan* 5, no. 3 (September 1986): 13 (abstract only).

49. Saenger et al., eds. (n. 1 above), p. 33.

50. Japan Tropical Forest Action Network, "Bintuni Bay Mangroves for Woodchipping," *Asian Wetland News* 1, no. 2 (1988): 3.

51. H. T. Chan, "Malaysia," in *Mangroves of Asia and the Pacific: Status and Management* (Bangkok: UNDP/Unesco, 1986), pp. 131–50.

The impact of traditional woodcutting by individuals from local communities may be very small. However, in view of the importance of mangroves as a local source of fuelwood and charcoal wood, and the increasing demand for cooking and heating fuel worldwide, the cumulative impact per year is very large.

Whether wood-harvesting disturbances are light in impact and the functioning of the system quickly restored or heavy and with serious long-term impairment depends on the degree to which sound forest management procedures are implemented. These practices involve principally a minimum disturbance to the water regime (both saline and fresh), prompt regeneration, and, if the area is clear felled, limited size and a controlled pattern of the area cleared.

There is a long and valuable experience in mangrove forest management, particularly well developed in Peninsular Malaysia, but also existing in Bangladesh and India.[52] Based on excellent applied research programs dating back to the 1920s, several state forests in Malaysia at Klang, Matang, and in Johore have been on a controlled management regime of patch clear felling (plus thinnings in Matang) that involves rotations of from 20 to 30 years, with artificial planting employed where reproduction is inadequate.[53]

More recently, in Venezuela's Orinoco delta area, a successful harvest system that results in good natural regeneration has been based on strip clear felling at right angles to the waterway channel.[54] The strips vary in width from 20 to 50 m, depending on tree height. Thailand has adopted the strip clear felling system but places the strips at 45 degrees to the waterway. In both countries, the forest adjacent to the clear-cut strip remains unharvested until well after the open area has been fully regenerated. The uncut forest and older age classes of regenerated forest serve to maintain the important organic detritus/nutrient supply function of the ecosystem and to provide habitat for mangrove fauna that require tall closed forest.

The "selection" system of cutting that is based on a minimum size of stem has also been used effectively in some places. When fully implemented, it produces a more or less continuous forest of different sizes (and ages) with scattered openings, rather than a mosaic of even-aged stand or strips. It therefore maintains most of the functions and services that have previously been described for mangrove ecosystems. This is the cutting method that may be commonly used by individual woodcutters, though they may also be selecting for preferred species, and this could have a negative impact on future forest composition. On forests under state control in Bangladesh's Sunder-

52. See, e.g., R. G. Dixon, *A Working Plan for the Matang Mangrove Forest Reserve, Perak* (Malaysia: Perak State Forestry Department, 1959).
53. Hamilton and Snedaker, eds. (n. 9 above).
54. A. Luna-Lugo, *Manejo de manglares en Venezuela* (Merida: Latin American Forestry Institute, 1976).

baans, this system is employed, with lower limits of from 11.8 cm to 56.3 cm depending on species. These delta forests are an important source of commercial pulpwood, as well as subsistence and commercial fuelwood. In addition, they provide storm surge protection, are the home of the endangered Bengal tiger and other wildlife, and nourish a valuable fishery within the delta and offshore.

In general, however, many past and some recent harvesting concessions of large size have been granted where the volumes removed are obviously unsustainable and where the other products and services of mangrove forests are being lost or impaired for a long time. One current large harvesting project involves leaving an uncut buffer strip along the coast but does not restrict the size of the area cut as long as isolated, scattered "mother trees" are left to provide for regeneration of the area. The buffer strip will certainly look more aesthetic from a viewpoint on the sea and will be effective in reducing erosion, but a fringing mangrove forest does not produce as high a quality nor as great a volume of nutrient detritus as does the interior forest, nor is it as biologically rich in habitats and species. Moreover, although the scattered seed trees may assist in obtaining regeneration of the stand, these isolated trees over large areas do not provide refuges for other components of the mangrove ecosystem, particularly birds and mammals.

Of more serious consequences than wood harvesting is the handmaiden of "conversion," which all too often follows logging. With the forest biomass removed, these logged areas appear to be viable opportunities for the development of aquaculture ponds or an expansion of agriculture from adjacent "fresh-soil" coastal fields. With the high market price of shrimp in particular, the conversion is being carried out by both government-backed schemes and private initiative. In this respect, the situation is akin to that of the tropical rainforests, where it is conversion to cattle ranching and sedentary or shifting agriculture that is responsible for the high rate of loss, not logging per se. Similar to the situation with rainforest, much of this conversion is ill advised and will be only short-lived. There is indeed a "shifting aquaculture" that parallels "shifting agriculture" as fertility declines and management problems increase. Conversion to aquaculture is one of the major issues in mangrove conservation, which has already been treated separately in this article.

The annual wood increment in mangrove stands is not high enough to make wood harvesting by itself economically attractive as a sustainable enterprise for the land owner.[55] Therefore, when comparing only the net annual revenue from wood alone with most other alternatives such as clearing off the forest and going into aquaculture, sustained-yield forestry comes out on the short end. It is the multiple products and services to linked resources such as the near-shore fishery that need to be assessed and valued, and this combined revenue should be compared with the alternatives. It is important to recog-

55. Growth rates are given by Christensen (n. 21 above), pp. 15–16.

nize that when *properly* carried out, wood harvesting may either not reduce these linked values or may reduce them slightly or only briefly. In contrast, wood harvesting that removes seed sources, destroys habitats, disturbs sediments, and alters water regimes can greatly impair these other values.

It is important also to recognize that not all wood harvesting is of an organized, mechanized, commercial nature. In many areas where subsistence landowners or fishermen are engaged in low-technology practices, many wood products are harvested informally. These products, such as fuelwood, poles for fish traps, bark for medicines, and many others listed in Appendix A, are extremely important both economically and culturally to local communities. Since they do not enter the market, they are seldom assigned a monetary value in any accounting procedure when commercial use of mangroves is being considered. They are usually harvested on a sustainable basis because of local dependency, and along with them many nonwood products may be harvested in a multiple-use system, for example, mangrove crabs. Of particular interest, and representing a potential for increased local products, are those many and unusual products available from management of the nipa palm (sugar, toddy, thatch, and alcohol).[56]

OIL POLLUTION IN MANGROVE ECOSYSTEMS

One particular environmental threat to mangrove areas is water pollution, especially from oil spills. The mangrove environment—low-energy, sheltered shores with permeable, sticky sediments and a tangled mass of aerial roots—entraps and restricts the degradation and flushing out of oil.

There are neither techniques nor suitable equipment that can safely and effectively clean up an oil spill in mangroves. Most often it is left to nature to do the job, and regeneration may take up to 20 years and much longer in less than ideal habitats. Should the mangrove forests be subjected to further environmental stresses in the meantime, the regeneration period will be even longer. Persistence of oil in the mangrove swamp is probably more serious to the ecosystem than the absolute toxicity of the oil.

Recovery may be accelerated by cutting and removing dead standing wood (1–2 years after a spill) after the bulk of the damage is evident, flushing the area with clean seawater, or reseeding and periodic thinning. Forest regeneration to a prespill productive state is often delayed because of less-than-ideal conditions. The delay is exacerbated by several factors, including chronic seepage and bleed water and long-term effects due to associated leaf loss.

In mangrove regions damaged by oil spills following the *Peck Slip* and *Zoe*

56. Hamilton and Murphy (n. 11 above).

Colocotroni (Puerto Rico) accidents,[57] leaf fall was dramatic, and the results devastating, for 50% of the canopy was lost 43 days after the event; 75% was lost after 61 days and 90% after 85 days.[58] Defoliation in more lightly oiled sites of the same area occurred more gradually. In areas of oil pollution, where residual oil remained, growth was stunted and saplings seemed chlorotic and were heavily damaged by insects. Saplings die off before maturity if the residual oil is not sufficiently degraded or leached out. At Cilacap, Indonesia, a mangrove mortality of 98% has been associated with oil-refinery effluents.[59]

Oil damage in biological systems includes (*a*) direct mortality due to fouling, coating, and asphyxiation, contact poisoning, and absorption of soluble toxic fractions in the water column; (*b*) indirect mortality caused by destruction of food sources (organisms lower in the food chain); (*c*) destruction of sensitive juvenile forms; (*d*) incorporation of sublethal amounts of oils and other hydrocarbon fractions, which may lower tolerance to infection and other stressors; (*e*) destruction of food value of fishery resources due to incorporation of hydrocarbons into the flesh and "tainting" (development of flavor unacceptable to consumers); and (*f*) incorporation of mutagenic and carcinogenic agents into the marine food web.[60]

Risk and Damage Prevention

Baseline studies are usually unavailable to assist in an assessment of the damage following an oil spill. Unfortunately, it is usually only as a result of an oil

57. Cintrón et al. (n. 14 above). Authors' note.—The *Zoe Colocotroni* spill was caused when the ship grounded and the captain lightened the vessel by pumping 5.7 million liters (1.5 million gallons) of crude oil overboard. The oil was carried by currents and winds into Bahia Sucia, a semienclosed bay where much of the oil was stranded in mangrove forests. Litigation ensued, and the case, *Commonwealth of Puerto Rico v the S.S. Zoe Colocotroni*, 456 F.Supp.1327, 1978, set several precedents for damages. Damages for approximately 92 million organisms, killed because of the oil pollution, were set at US$6,086,083.20 (US$751,368.30/ha of damaged mangrove forest). However, on appeal, the U.S. Court of Appeals (1) ruled on August 12, 1981, that the Commonwealth of Puerto Rico could not collect damages for the loss of the 92 million organisms since there was no plan to actually purchase the organisms and use them to restore the area; and (2) rejected the claim of US$559,500.00 for replanting mangroves in the oil-polluted areas because of previously mentioned conflicts concerning exactly how large an area of mangroves was damaged and based on the fact that replanting mangroves in oiled sediments seemed "pointless." See Roy R. Lewis III, "Impact of Oil Spills on Mangrove Forests," research report (Mangrove Systems, Inc., Tampa, Fla., July 1981), p. 15.

58. Cintrón et al.

59. Ibid., p. 22.

60. M. Blumer, "Scientific Aspects of the Oil Spill Problem," *Environmental Affairs* 1 (1971): 54–73.

spill and the ensuing legal cases that studies of oil pollution in mangroves have been done. Research into mortality thresholds has been carried out,[61] and a review of the literature on the effects of oil on mangroves, including estimates of the amounts of oil damage that cause death, has been made by Odum and Johannes.[62] Vulnerability of forest types to oil pollution is not uniform.[63]

Riverine forests.—These are the least vulnerable to oil pollution because surface freshwater flows tend to move oil slicks away from river mouths. Oil slicks near river mouths aggregate between salt and freshwater masses. In highly stratified waters, seawater enters the estuary as a wedge and mixes through vertical advection with the freshwater. The greatest vulnerability to pollution is during dry seasons.

Basin mangroves.—These are most vulnerable to stressors originating inland and are normally less vulnerable to seaborne pollution. Often a berm separates the mangroves from the fringe or the sea. These basin types are most vulnerable to inland spills from pipelines and storage facility leakage.

Fringe and overwash forests.—These are the most vulnerable mangrove ecosystems to seaborne oil pollution. Damage is a function of the flushing rate and the wave energy available. Tide and wave energy can mitigate the impact of oil in the outer fringe. The inner fringe suffers the highest tree mortality, and often complete defoliation will result in sheltered locations; the outer fringe will suffer, too, but not as severely.

Sometimes, dispersants may be used on an oil spill. Dispersants do not remove oil from the environment; they simply enhance incorporation into living systems by absorption and adsorption. However, dispersant-oil mixtures can be more toxic than the dispersant itself. Small dispersant-oil particles are easily ingested by filter feeders. Dispersant-enhanced biodegradability

61. See, e.g., Lai Hoi-Chaw and Feng Meow-Chan, eds., *Fate and Effects of Oil in the Mangrove Environment* (Pulau Pinang: Universiti Sains Malaysia, 1984); Charles D. Getter and Bart J. Baca, "A Laboratory Approach for Determining the Effect of Oils and Dispersants on Mangroves," (in review) Proceedings Symposium on Oil Spill Dispersants American Society of Testing and Materials (Philadelphia, 1982); Joseph Yuska, "Microbial Degradation of Oil in Marine Environments," appendix to report by Samuel C. Snedaker and Melvin S. Brown, "Effects of the Port Sutton Oil Spill (October 1978) on the Mangrove Community" (in-house publication of the Division of Biology and Living Resources, Resenstiel School of Marine and Atmospheric Science, University of Miami, 1980); and see also Samuel C. Snedaker, "Oil Spills in Mangroves," working paper (East-West Environment and Policy Institute, Honolulu, Hawaii, 1981).

62. William E. Odum and R. E. Johannes, "The Response of Mangroves to Man-induced Environmental Stress," in *Tropical Marine Pollution,* ed. E. J. Ferguson Wood and R. E. Johannes, Elsevier Oceanography Series No. 12 (Amsterdam: Elsevier Scientific Publishing Co., 1975), chap. 3, pp. 52–181.

63. A. E. Lugo, G. Cintrón, and C. Goenaga, "Mangrove Ecosystems under Stress," In *Stress Effects on Natural Ecosystems,* ed. G. W. Barrett and R. Roenberg (New York: Wiley, 1981), pp. 129–53.

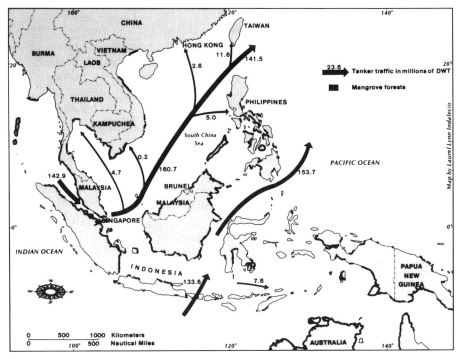

FIG. 3.—Transport of crude oil and mangrove distribution in Southeast Asia. Source.—Drawn from several maps in J. R. Morgan and M. J. Valencia, eds., *Atlas for Marine Policy in Southeast Asian Seas* (Berkeley and Los Angeles: University of California Press, 1983). Reproduced with permission.

may increase biological oxygen demand, particularly in sheltered areas with limited water circulation that are already susceptible to eutrophication.

The primary strategy for an oil spill heading toward a vulnerable mangrove coastline should be to protect the mangroves by using booms and absorbent barriers. To do this, a contingency plan *must* be formulated so that abatement and mitigation strategies can be planned *before* a crisis and *implemented immediately* when required.

Figure 3 shows the correlation between oil industry activities and mangrove areas in Southeast Asia. In a region where conflict over resources is severe, regional and national marine pollution regulations and contingency plans have been established in order to alleviate environmental damage from an oil spill to the greatest degree possible.[64]

64. For an overview of the ASEAN legislative response to marine pollution, see Amado S. Tolentino, Jr., "Legislative Response to Marine Threats in the ASEAN Subregion," *Ambio* 17, no. 3 (1988): 238–42.

VALUATION OF MANGROVES

For most commodities, values are determined by the market. This is also true for some of the products that can be extracted from a mangrove forest: so many dollars per hundred poles or per kilogram of crabs. Unfortunately, the directly marketed goods (or services) produced by a mangrove ecosystem represent only a fraction of the total array of goods and services that yield benefits to individuals and societies. As a result, mangroves are commonly considered as low-value ecosystems and are, therefore, prime candidates for conversion to other uses such as aquaculture ponds, infrastructure (e.g., ports, marinas, and coastal roads), agriculture, housing, and even garbage dumps.

Two questions need to be asked when assessing the "development" alternatives being proposed for mangrove areas. First, are the expected benefits from conversion large and sustainable, and, second, what are the true costs (direct and indirect) of losing the mangrove ecosystem?

The magnitudes of expected benefits from mangrove conversion can be calculated. For some uses, particularly infrastructure such as ports, industrial sites or housing, the conversion is permanent, and the economic benefit from use of the mangrove area is the difference between the costs of conversion (clearing, filling, and drainage) and the next less expensive alternative site. In many cases there are compelling reasons for building near the water (e.g., ports): the value of such uses is very large, and the conversion is clearly justified. In Singapore, for example, the major industrial estate and port of Jurong is partially built on reclaimed mangrove areas; in Malaysia, the new port at Kelang is also built on reclaimed mangrove forest. The justification for other infrastructure developments may not always be as clear. One should note that in some areas the final costs of creating dry land from mangroves has been considerably higher than initially projected because of extra costs for foundations and soil-stabilization measures, as well as construction and maintenance of seawalls. In Kelang, for example, a large area of mangrove adjacent to the new port also was converted for industrial development and housing. Due to a variety of economic, locational, and engineering factors, this land, reclaimed at considerable expense, lies largely unused; traditional mangrove ecosystem products have also been lost, thereby compounding the economic costs of the project.

For other types of conversion, however, the benefits may be both smaller than anticipated and less permanent. Mangrove soils have proved difficult to manage in some areas on a sustained basis, both for agriculture and aquaculture. In some places, this has led to what was earlier referred to as "shifting aquaculture." Acid-sulfate soils are not uncommon in mangrove areas and are a major cause of these problems. For man-made structures, property damage may be caused by storm surges or typhoons. In either case, the net benefits from conversion are smaller than initially expected.

The question of expected benefits from conversion, both their magnitude and their permanence, is not discussed further here. It is sufficient to say that this is a crucial piece of information when assessing whether or not the net benefits of conversion are positive or negative. Expected benefits are usually documented "up-front" in any proposal for an alteration or conversion of mangroves. What are often not documented at all, or documented imperfectly, are the opportunity costs of achieving those benefits. To answer this question, therefore, one has to assess the full range of mangrove-derived goods and services that would be lost by conversion. This is the true "opportunity cost" of conversion, that is, what society must give up to use the mangrove for a port, an aquaculture pond, or a housing estate. Measuring this opportunity cost, however, has proved very difficult.

The difficulty in measuring the true economic value of a mangrove stems from two factors—some products or services do not have market prices, and the goods and services produced occur both within the mangrove and outside of it. To a large extent, the within-mangrove, on-site goods and services are land based, while the off-site effects are usually aquatic or coastal, reflecting the mangroves' role as a bridge between the sea and the land. Other sections of this article illustrate these points; Appendixes A, B, and C list the wide variety of goods and services produced by a mangrove ecosystem, from the tangible production of poles or crabs to less-tangible aspects like nutrient flows or breeding habitat. The production of goods and services and the interaction of both fresh and salt water result in dispersed locations for mangrove products. Figure 4 illustrates how the relationship between *location* and *valuation* can be expressed in a simple 2 × 2 matrix.

When the value of an existing mangrove forest is assessed, the analysis traditionally has included only those items found in quadrant 1 of figure 4: goods and services that have *market prices* and are found within the mangrove. The value of these quadrant 1 products is frequently small compared to the expected benefits from conversion; consequently, extensive mangrove areas are converted each year in the name of "economic efficiency." As knowledge has been gained of the ecosystem interactions in mangroves, greater attention has been paid to quadrant 2 products, those that occur off site but are marketed. These largely consist of fish and shellfish caught in adjacent waters.

Quadrants 3 and 4, those goods and services without market prices that are found both within and outside of the mangrove, are usually ignored. For some of these goods and services, the valuation problem is due to its subsistence role in the local economy (e.g., traditional medicine, minor mangrove forest products). For others, identification and quantification of the effect are sometimes a problem (e.g., storm-surge protection), and placing monetary values may therefore be difficult. Nevertheless, these contributions of the mangrove may be quite important in terms of the total benefits produced by the mangrove system.

	Location of Goods and Services	
	On-site	Off-site
Marketed	1 Usually included in an economic analysis (e.g., poles, charcoal, woodchips, mangrove crabs)	2 May be included (e.g., fish or shellfish caught in adjacent waters)
Nonmarketed	3 Seldom included (e.g., medicinal uses of mangrove, domestic fuelwood, food in times of famine, nursery area for juvenile fish, feeding ground for estuarine fish and shrimp, viewing and studying wildlife)	4 Usually ignored (e.g., nutrient flows to estuaries, buffer to storm damage)

FIG. 4.—Relation between location and type of mangrove goods and services and economic analysis. Source.—Lawrence S. Hamilton and Samuel C. Snedaker, eds., *Handbook for Mangrove Area Management* (Honolulu: East-West Center, 1984), fig. 42, p. 110. (Matrix developed by J. A. Dixon and P. R. Burbridge.)

Past Valuation Efforts

Limited data are available on the actual yearly value of mangroves. Table 3 presents some examples gleaned from the literature and reported in Hamilton and Snedaker.[65] The reported values (in US$/ha/year) range from estimates for the complete ecosystem (quadrants 1–4) to estimates of forestry products (quadrant 1 goods: on site and marketed) to fishery products (quadrant 2: off site and marketed). The range of annual values reported is great, from as low as $25/ha/year for forestry products in Malaysia to over $1,000/ha/year in Thailand when fishery and forestry benefits are included. Most of these values are for gross financial benefits using market prices to assign values. Net financial benefits, after costs of production are subtracted, would be less.

Thailand.—The Thai figures summarized in Table 3 are an interesting case. A carefully done case study in Chanthaburi Province in southeast Thailand examined a traditional mangrove-based economy and presented annual values for a variety of products including forest products, nipa thatch, fisheries, oyster culture, shrimp farming, and agriculture.[66] (Note that a mangrove is a naturally sustainable ecosystem, while the alternative uses after conversion require contiuous maintenance and management.)

Values were estimated for each of the mangrove-dependent products,

65. Hamilton and Snedaker (n. 9 above), p. 115.
66. Christensen (n. 21 above).

TABLE 3.—EXAMPLES OF VALUES PLACED ON MANGROVE SYSTEMS
AND MANGROVE ECOSYSTEM PRODUCTS

Type of Resource or Product and Location	Date	Value Placed on Resource (US$/ha/year)
Complete mangrove ecosystem:		
Trinidad	1974	500
Fiji	1976	950–1,250
Puerto Rico	1973	1,550
Forestry products:		
Trinidad	1974	70
Indonesia	1978	10–20 (charcoal and wood chips)
Malaysia	1980	25
Thailand	1982	30–400
Fishery products:		
Trinidad	1974	125
Indonesia	1978	50
Fiji	1976	640
Queensland	1976	1,975
Thailand	1982	30–100 (fish); 200–2,000 (shrimp)
Recreation, tourism:		
Trinidad	1974	200

Source.—Lawrence S. Hamilton and Samuel C. Snedaker, eds., *Handbook for Mangrove Area Management* (Honolulu: United Nations Environment Programme (UNEP) and East-West Center, Environment and Policy Institute, 1984), table 20, p. 115.
Note.—All of these estimates are approximate and are presented to give a range of values placed on mangroves and mangrove ecosystem products. The values in each locale will vary.

both for present productivity levels and for potential levels with improved management. These values are presented in table 4. The shrimp-farming values are large, ranging from $200 to $2,000/ha/year. These values are for commercial shrimp farms on converted land that are partially dependent on the remaining mangrove as a source of shrimp fry. Still, the total value per hectare from forest and fishery products from an intact mangrove ecosystem is substantial, from $160/ha at present to a potential of over $500/ha. When compared to one alternative use, rice farming, one sees why a broader economic analysis is needed. The expected return from agriculture ($165/ha/year) is large compared to the annual per hectare value from charcoal production ($30). However, when one adds in the present value of fishery products caught within and outside the estuary, the two uses are equivalent in value. When future increases in potential income are considered, and when the contribution of mangroves to shrimp farming is also included, the in situ value of mangroves becomes very substantial. In addition, forestry and fishery production are fairly labor intensive and have important employment generation potential. In the Thai case, one potential production system, based on intensified management of the natural mangrove (particularly utilizing nipa palm), was about equal to shrimp cultivation in creating jobs and better than rice farming.

TABLE 4.—TENTATIVE ECONOMIC COMPARISON OF VARIOUS FORMS
OF LAND USE IN THE MANGROVE AREA

	Gross Income	
	Present $/ha/year	Potential $/ha/year
"Forestry":		
Official charcoal production*	30	400
Fishery inside estuary	30	30
Mangrove-dependent fishery outside	100	100
Oyster culture	. . .	60
Total forestry†	160	590
Shrimp farming	206	2,106
Rice farming	165	. . .

SOURCE.—Bo Christensen, *Management and Utilization of Mangroves in Asia and the Pacific,* Food and Agriculture Organizaton (FAO) Environment Paper No. 3 (Rome: FAO, 1982).
NOTE.—In terms of employment opportunities, charcoal production ranks rather low but forestry plus associated fisheries is comparable to other land uses.
*Nipa is estimated at $230/ha/year.
†At present, 1 ha of the remaining mangrove forest is estimated to contribute US$15 to shrimp farming in the form of fry.

Note that even these estimates only include a limited number of the goods and services included in Appendixes A, B, and C. As such, they are *minimum estimates* of the yearly value of mangrove products. And yet, the decision as to whether mangroves should be converted (or destroyed) is made by comparing this minimum, partial estimate with the total expected benefit from conversion. No wonder mangroves are being lost at such a rapid pace!

Indonesia.—Issues similar to those in Thailand were raised in a recent report on coastal resources management in Indonesia.[67] Indonesia has the largest mangrove area of any country in the world, and conversion, both to dryland agriculture and to *tambak,* or fishpond culture, is widespread. Indonesian mangroves are highly productive and diverse and under severe pressure from overharvesting of forest products and conversion. In spite of an official policy against further *tambak* extensification, Burbridge and Maragos observed this occurring throughout the country. They noted that the potential for sustainable production from mangrove areas was largely ignored and that "a prevailing but erroneous view among many advocates of mangrove conversion is that the swamps are wastelands with little value in comparison to 'higher' uses such as *tambak* and rice culture."[68]

Ecuador.—Large areas of mangrove forest have been converted to shrimp grow-out ponds in Ecuador. This development has been particularly

67. Peter W. Burbridge and James E. Maragos, *Coastal Resources Management and Environmental Assessment Needs for Aquatic Resources Development in Indonesia,* Report No. 33 (Washington: International Institute for Environment and Development, 1985).
68. Ibid., p. 24.

rapid in the southern Gulf of Guayaquil—16% of the mangrove forest has been lost between 1966 and 1982. In addition to the physical loss of mangrove forest, researchers believe that this has also resulted in the measured decline in abundance of shrimp postlarvae in Ecuadorean estuaries.[69] This is a major concern of shrimp pond owners who rely on these wild postlarvae as a stock source for their grow-out ponds. As a result, productivity falls, and shrimp farmers may have to use more expensive hatchery operations to produce shrimp larvae.

Fiji.—A recent study on mangroves in Fiji examined the trade-offs between conservation and reclamation in a number of mangrove areas.[70] Although the alternatives considered in each site were different, the combined economic and ecological analysis yielded useful results and policy guidance.

Reclamation sites for agricultural production (both with and without irrigation) were examined. The analysis demonstrated that the net benefits of the agricultural production itself were *negative* (due to low yields, soil-related problems, and high capital costs) and that sizable fishery and forestry benefits were also lost after reclamation. (These values were determined using an economic social welfare approach, as opposed to the financial analysis reported in the other studies.) In one site, for example, the per hectare annual "development benefit" after conversion for agriculture and aquaculture uses was a negative Fijian $F 516. In addition, one must add fishery losses ($F 150) and forgone forestry production ($F 9) for a total economic "loss" of $F 675/ha/year. Conversely, if the land had not been reclaimed, annual production of fishery and forestry products worth about $F 160/ha would be expected. In this case, conversion clearly made no sense, even if only quadrant 1 outputs were considered. As in the Thai case, the fishery benefits are considerably larger than forestry benefits.

Further analysis revealed that, in most cases, indigenous Fijians with recognized coastal rights received only a small portion of the actual economic loss to society from decreased forestry and fishery production.[71] If the future fishery and forestry benefits that will be lost due to conversion are valued over 50 years at a 5% discount rate, they average $F 2,734/ha for fishery products and $F 164–$F 217/ha for forestry products, or a total of over $F 2,900/ha. In contrast, landowners were paid only about $F 520/ha for nonindustrial uses, or less than 20% of forgone fishery and forestry benefits (but roughly three times the estimate of forgone forestry benefits alone). Again, one sees that by considering only the on-site, terrestrial component, one clearly undervalues the economic importance of mangroves.

69. Lahmann et al. (n. 43 above).

70. Padma N. Lal, "Conservation or Reclamation: Economic and Ecological Interactions within the Mangrove Ecosystem in Fiji" (Ph.D. diss., Department of Agricultural and Resource Economics, University of Hawaii, 1989).

71. Ibid.

Conclusions

Although much more detailed analysis of the economics of mangrove ecosystems is needed, the following general conclusions can be drawn.

1. Decisions on whether to convert mangroves to other uses are frequently based on the value of marketable forestry products produced by the natural unmanaged mangrove. These values may be quite low. Fishery and marine products, both within the mangrove and in nearby waters, are frequently much more valuable than forest products.

2. Whereas a natural mangrove is a self-sustaining, productive ecosystem, many conversion-based alternative uses have proved to be expensive to construct and maintain or have produced disappointing economic results due to low and declining productivity. Nevertheless, normal market forces will almost always favor conversion of mangroves to other uses. This is a direct result of the dispersed nature of the products of, or those dependent on, the mangrove and the problem of assigning monetary values to some goods and services. Because of this "market failure," government intervention is essential if mangroves are to be used in a socially optimal manner.

3. The linked land-ocean system of a mangrove forest creates complicated and far-reaching ecosystem linkages affecting the production of a wide range of socially valuable goods and services.

4. Subsistence production of various nonmarketed goods and services may be very important in some areas but is rarely reflected in any economic analysis of a mangrove ecosystem. These products may be an important part of a cultural tradition. New uses for mangroves, through tourism development and as wildlife habitats, may also become increasingly valuable in the future.

In sum, a more comprehensive economic analysis of mangrove forests and the alternative uses being proposed may well demonstrate that many mangroves yield greater social net benefits as natural ecosystems. In cases where conversion is clearly necessary or justified, sound physical/social/economic analysis can help plan conversions that reduce the loss of mangrove forest benefits to a minimum.

MANAGEMENT AND FUTURE DIRECTIONS

Mangrove ecosystems are best viewed as multiple-resource systems that provide, when allowed to function properly, an array of products and services. Any proposed conversion or management for a single purpose should be most carefully analyzed, using the best biophysical and socioeconomic knowledge and methodologies. Too often, the net social benefits projected from mangrove conversion have only looked attractive when based on a very short-term analysis. Moreover, because these systems are linked hydrologically to current and future land- and water-use activities upstream, transboundary impacts that impair the functioning of these intertidal systems need to enter any analysis. This open system of linked resources involving the uplands, lowlands, intertidal area, and marine environment argues for comprehensive coastal zone planning and management.

Although true joint use (production) occurs in many areas in a complementary fashion (e.g., harvesting mangrove crabs and sustainable fuelwood cutting), it is often appropriate to assign priority functions to "zones" of mangrove forests in order to decide on proposed new uses. This planning, or zoning, can be done nationally but will in the last analysis have to be done from detailed information at a local level. A possible priority use assignment might include the following.

1. *Preservation zone.*—The objective here would be to maintain mangrove communities without human disturbance to protect genetic, species, and community diversity and to provide areas for scientific research and education. There might be limited recreational use of a nonconsumptive kind, such as for bird watching. Apiaries might well be permitted. Such zones would at the same time provide such other services as shoreline protection, breeding grounds and shelter for aquatic and terrestrial fauna, and a source of nutrients for near-shore fisheries.

2. *Sustained-yield wood and nonwood products.*—Controlled harvesting with provision for replenishment or regeneration would be the priority use in this zone. Much of this might be of a hunting/gathering nature or subsistence wood harvesting. When nipa palm is present, a sustainable harvesting system for the sap or for the leaves as shingles could be carried out. Commercial wood harvesting that employs appropriate methods to minimize site damage and to obtain adequate regeneration of desirable species would be permissible in some areas where local communities were not dependent on the resource.[72] Appropriate harvesting of wood, mangrove oysters or crabs, sap, leaves, medicines, and other direct products should not greatly impair the contribution to coastal protection or to the near shore fisheries. Fish exploitation in mangrove waters should be controlled through capture regulations, as in Caroni Swamp Reserve in Trinidad.

3. *Conversion areas.*—In recognition of the imperative for coastal zone development, it is prudent to identify and classify mangrove areas whose elimination would not greatly impair the goods and services that most areas are producing. These might be areas of exceptionally high salinity and very low productivity or isolated areas too small to be maintainable as viable ecosystems. It is important to have some "throwaway" areas that can be sacrificed as alternatives to having development occur in key or critical areas.

4. *"Hold" areas.*—Given the aforementioned "orphan" nature of mangroves, many countries will not have sufficient information about all of their mangrove forests. Until enough information is gathered to permit rational

72. See the forest-management guidelines in Hamilton and Snedaker, chap. 9.

allocation to some use, such areas should be zoned as a reserve or bank, pending classification. Such reserve areas should not be regarded as fair game for intervention, as is often true for unclassified areas of public domain. They should be strictly protected in a holding status until adequate assessment and evaluation can be made.

5. *Restoration areas.*—The technology for reforestation with several mangrove species is known well enough to permit cleared or degraded areas to be restored. Former aquaculture areas that have been abandoned may often recolonize naturally through breaking open the banks and allowing a natural freshwater-tidewater regime to reestablish. Other areas may require planting of seedlings or propagules. Through these actions, restoration of some, or eventually all, of the products and services produced by mangroves may be realized.

Suggestions for the development of national mangrove plans may be found in Hamilton and Snedaker.[73] It has been recognized that lack of familiarity with multiple-purpose mangrove management and lack of planning skills for integrating mangrove management into broader issues of coastal-zone planning has limited this activity. The United Nations Development Programme and Unesco have combined forces under the Major International Project on Research and Training Leading to the Integrated Management of Coastal Systems (COMAR)[74] to implement a Mangrove Programme in Asia and the Pacific, the aim of which is to understand the structure and dynamics of mangrove systems in order to implement adequate management practices. The first project was "Research and Training Pilot Programme on Mangrove Ecosystems in Asia and the Pacific." In 1987, the second UNDP/Unesco project, "Research and Its Application to the Management of the Mangroves of Asia and the Pacific" was initiated. As part of the second project, a major survey of the mangroves of Ranong Province (Thailand) has been completed, and a meeting of the South Pacific states' mangrove representatives was held at the Unesco office in Apia, Western Samoa, in February 1988. Similar internationally sponsored programs are being implemented in other regions of the world.

If adequate planning and management are to be achieved, there must concomitantly be programs to raise the level of awareness of citizens and politicians at all levels in government about the important role mangroves fulfill as part of the coastal complex.

Global climate change, with its possible attendant change in sea level,

73. Hamilton and Snedaker, sec. 5.
74. For a synopsis of the UNDP and Unesco involvement in mangroves in Asia and the Pacific, see Vannucci, "The UNDF/UNESCO Mangrove Program" (n. 28 above). See Also C. D. Field and A. J. Dartnall, eds., *Mangrove Ecosystems of Asia and the Pacific: Status, Exploitation and Management,* Proceedings of the Research for Development Seminar, Australian Institute of Marine Science (AIMS), Townsville, Australia, May 18–25, 1985 (Townsville: AIMS on behalf of the Australian Committee for Mangrove Research, 1987).

may pose more serious threats to mangrove forests than to any other of the world's forest biomes or major communities. Since mangroves are the predominant vegetation of tropical and subtropical sheltered, low-relief, coastlines and estuaries under tidal influence, even a modest increase in sea level would alter the freshwater/seawater flushing regime and the edaphic conditions that affect the growth and reproduction of mangrove vegetation. While gradual change would permit some natural migration landward, much of the current terrain (except where there are broad tropical coastal plains or low-relief deltas) would not be conducive to mangrove establishment until long-term erosion and sedimentation had again created more suitable substrate. Most small island mangrove resources would be particularly vulnerable. This "squeeze" occurs because mangrove vegetation is adapted to saline conditions of various degrees, but as salinity decreases landward, more aggressive species that adapt to "fresh" environments out-compete mangroves.

For management purposes, tropical coastal zones must always be regarded as complex areas with many fragile but economically and culturally important ecosystems, such as coral reefs and mangroves. The limited nature of these resources and their keystone function in long-term sustainable development must be recognized. Development should seek to maximize compatibilities and optimize the multiple use of these resources that are of the land and of the sea.

APPENDIX A

DIRECT PRODUCTS FROM MANGROVE FORESTS

Uses	*Products*
Fuel	Firewood for cooking, heating
	Firewood for smoking fish
	Firewood for smoking sheet rubber
	Firewood for burning bricks
	Charcoal
	Alcohol
Construction	Timber for scaffolds
	Timber for heavy construction (e.g., bridges)
	Railroad ties
	Mining pit props
	Deck pilings
	Beams and poles for buildings
	Flooring, paneling
	Boat-building materials
	Fence posts
	Water pipes
	Chipboards
	Glues

APPENDIX A—(*Continued*)

Fishing	Poles for fish traps
	Fishing floats
	Fish poison
	Tannins for net preservation
	Fish attracting shelters
Agriculture	Fodder
	Green manure
Paper production	Paper of various kinds
Foods, drugs, and beverages	Sugar
	Alcohol
	Cooking oil
	Vinegar
	Tea substitutes
	Fermented drinks
	Dessert topping
	Condiments from bark
	Sweetmeats from propagules
	Vegetables from propagules, fruits, or leaves
	Cigarette wrappers
	Medicines from bark, leaves, and fruits
Household Items	Furniture
	Glue
	Hairdressing oil
	Tool handles
	Rice mortar
	Toys
	Matchsticks
	Incense
Textile and leather production	Synthetic fibers
	Dye for cloth
	Tannins for leather preservation
Other	Packing boxes

SOURCE.—Lawrence S. Hamilton and Samuel C. Snedaker, eds., *Handbook for Mangrove Area Management* (Honolulu; United Nations Environment Programme and East-West Center, Environment and Policy Institute, 1984), table 1, p. 2.

APPENDIX B

INDIRECT PRODUCTS FROM MANGROVE FORESTS

Source	*Product*
Finfish (many species)	Food
	Fertilizer
Crustaceans (prawns, shrimp, crabs)	Food
Mollusks (oysters, mussels, cockles)	Food
Bees	Honey
	Wax

APPENDIX B—(*Continued*)

Birds	Food
	Feathers
	Recreation (watching, hunting)
Mammals	Food
	Fur
	Recreation (watching, hunting)
Reptiles	Skins
	Food
	Recreation
Other fauna (e.g., amphibians, insects)	Food
	Recreation

Source.—Lawrence S. Hamilton and Samuel C. Snedaker, eds., *Handbook for Mangrove Area Management* (Honolulu: United Nations Environment Programme and East-West Center, Environment and Policy Institute, 1984), table 2, p. 2.

APPENDIX C

SERVICES PERFORMED BY MANGROVE WETLANDS (IN VARYING INTENSITIES) THAT ARE USEFUL TO HUMANITY AND TO THE REGION TO WHICH THEY BELONG

1. Production of high organic matter, which may be consumed locally, accumulated (in peats), or exported to other ecosystems
2. Channelization of local productivity into wood production for human use
3. Creation of peats, which are valuable as fuels or as substrate for certain activities
4. Export of nutrients, which may increase fish or other marine organism production (such as shell fish), which are also useful to people
5. Excellent protection of shore erosion. Mangrove forests are notorious for their prevention of erosion and reversal of the process (by accelerating deposition)
6. First line of defense against such catastrophies as hurricanes, tidal waves, or periods of high seas
7. Preferred habitat for nesting of birds. In many areas, mangrove ecosystems are the only remote areas left for many species of wildlife
8. Spawning waters for a multiplicity of marine and estuarine fish and other aquatic species (manaties, turtles, etc.)
9. Feeding nurseries for many of these species
10. Protective cover to these species and many more during periods of stress
11. Retention basin for flood waters (fresh and marine)
12. Removal of nutrients, heavy metals, and other substances dissolved or suspended in flood waters by scrubbers
13. Overall dampening of wild environmental oscillations, particularly those close to cities (noise, water, dust, water pollution, etc.)

Source.—C. H. Wharton, H. T. Odum, K. Ewel, M. Duever, A. Lugo, R. Boyt, J. Bartholomew, E. De Bellevue, S. Brown, M. Brown, and L. Duever, "Forested Wetlands of Florida—Their Management and Their Use," Final Report to the Division of State Planning (Gainesville, Fla.: Center for Wetlands, February 1976), table 3, p. 236.

"Soft" Beach Protection and Restoration*

Roger H. Charlier, Charles De Meyer, and Daniel Decroo
HAECON N.V., Harbour and Engineering Consultants, Ghent

INTRODUCTION

Coastal erosion is a worldwide occurrence principally along seashores but noticed also along many a lake shoreline. It has been reported in the literature for several decades.[1] It encompasses considerable economic consequences, endangering, for instance, valuable properties. The rise in water level is a major cause of the phenomenon but anthropic actions have also played an important part.

Stopping coastal erosion and halting shoreline retreat is considered impossible. Reducing the rate of erosion and restoring endangered areas, on the other hand, falls within the reach of coastal engineers. Various approaches have been tried, with unequal success and, occasionally, with undesired if not unexpected consequences.

The traditional approach has been to construct "hard" coastal defense structures such as groins, breakwaters, seawalls, revetments, tetrapods, and the like. Often they have been destroyed, all have required maintenance, and many have merely displaced the problem from one site to another. In more recent years another approach has been tried, either by itself or in conjunction with the conventional measures—beach nourishment by artificial means.

EDITORS' NOTE.—This study was funded by a grant from HAECON N.V., Harbour and Engineering Consultants, of Ghent, Belgium. Credit for figs. 2 and 7–9 is due the U.S. Army Corps of Engineers. This is the third article Professor Charlier has written for *Ocean Yearbook*. The earlier two articles are "Other Ocean Resources," *Ocean Yearbook 1*, ed. Elisabeth Mann Borgese and Norton Ginsburg (Chicago: University of Chicago Press, 1978), pp. 160–210, and "Water, Energy, and Nonliving Ocean Resources," *Ocean Yearbook 4*, ed. Elisabeth Mann Borgese and Norton Ginsburg (Chicago: University of Chicago Press, 1983), pp. 75–120.

1. Roger H. Charlier, *Professional Geographer* 7, no. 2 (1955): 10–13; and U.S. Department of the Interior, *Our Vanishing Shoreline* (Washington, D.C.: Government Printing Office, 1957).

The "soft" approach is not new, since records mention beach nourishment immediately after the end of World War I near Santa Barbara, California.[2] However, techniques have been refined and source material varies more in origin; the U.S. Army Corps of Engineers has used dredgings in beach restoration works.

Artificial beach rebuilding has, in some instances, even become beach creation, as has been the case, for instance, in Monte Carlo.[3]

Perhaps the largest successful beach restoration project has been undertaken along the eastern coast of Belgium, involving some 8 million m³ (11 million cu. yd.) of material.[4] The core of the beach-nourishment approach is discussed herewith.[5]

BEACH DYNAMIC PROCESSES

Unconsolidated beaches result from accumulation processes, and appear as sandy beaches (less than one-third of world beaches) or muddy beaches (either colonized by mangroves or in sheltered areas). The principal types are dunal systems, estuarine environments, littoral spits, lagoons, and ponds. Today most beaches, due to natural and/or anthropic action, are thinning or retreating. On the French Atlantic coast, for example, some 850 km (527 miles) of littoral shoreline recedes approximately 1 m (3.28 feet) a year as sea level increases worldwide, on average 1.2–1.5 mm/year.[6]

2. Charlier et al., "Hard Structures and Beach Erosion," *International Journal of Environmental Studies*, vol. 21, no. 1 (1989): 29–44.

3. Ibid.

4. Charlier et al., "Hard Structures and Beach Erosion;" G. De Moor, "Premiers effets du rehaussement artificiel d'une plage sableuse le long de la côte belge," *Les côtes atlantiques d'Europe, évolution, aménagement, protection*, vol. 9, ed. A. Guilcher (Brest: Centre National pour l'Exploitation des Oceans [CNEXO], 1979), pp. 97–114; G. De Moor, "Recent Beach Erosion along the Belgian North Sea Coast," *Société de Géologie de Belgique, Annales* 88 (1979): 143–57; G. De Moor, "Erosie aan de belgische kust," *De Aardrijkskunde*, n.s., no. 4 (1980): 279–94; P. Kerckaert, P. P. Roovers, and A. Noordam, "Artificial Beach Renourishment on the Belgian East Coast," paper presented at the 18th International Conference on Coastal Engineering, Cape Town, 1982; and P. Kerckaert, P. P. Roovers, A. Noordam, and P. De Candt, "Artificial Beach Renourishment on the Belgian Coast," *Journal of Waterways, Ports, Coastal and Ocean Engineering* (January 1985).

5. P. P. Roovers, P. Kerckaert, A. Burgers, A. Noordam, and P. De Candt, "Beach Protection as a Part of the Harbour Extension at Zeebrugge, Belgium," in *Proceedings of the 25th International Navigation Congress*, Permanent International Association of Navigation Congresses (PIANC), May 10–16, 1981, Brussels and Oxford (Edinburgh: Pergamon, 1981), vol. 2, pt. 5, pp. 755–69; and both articles by Kerckaert et al. (n. 4 above).

6. M. Auzel and J. Bourcart, "Erosion des plages," *Bulletin d'Information de la Commission Oceanique et d'Etudes des Côtes* (1950), pp. 379–82; E. C. F. Bird, *Coastline Changes* (Somerset, N.J.: Wiley, 1985); R. H. Charlier and M. Auzel, "Géomorphologie

Human action and ignorance of shoreline processes often necessitate decisive measures, particularly where economic values are important. Selection of erosion-control measures then depends on the type of shore and land use. Alternatives include construction of hard structures, artificial and induced natural beach replenishment, passage or traffic restriction, vegetation protection, and outright abandonment.

The Belgian coast, for example, has been subjected to severe erosion, and human coastal settlements have disappeared during historical times.[7] The Dutch coast has been similarly affected. So has the American shoreline, with barrier islands particularly endangered. Even such famed beaches as those of Copacabana (Rio de Janeiro) and Monte Carlo need attention (fig. 1, top).

In Italy, massive settlement of coastal regions, tourism and industrial use, enlargement of harbors, construction of coastal defenses, pumping of underground water, and reclamation works have accelerated natural subsidence. Inland modifications for a variety of purposes have compounded the problems of Italian beaches.[8]

Natural causes bring about coastal erosion, and undoubtedly sea-level rise is leading to the "drowning of beaches;" anthropic action also plays a major role in shoreline retreat.[9] Man may accelerate the process by creating a

---

côtière: Migration des sables sur la côte belge," *Zeitschrift für Geomorphologie* 5, no. 3 (1961): 181–84; R. H. Charlier, "North Sea Beach Erosion in Belgium," in *Proceedings of the 23rd International Geological Congress,* Prague, August 1968 (International Geological Union/UNESCO affiliated, 1968), pp. 167–71; R. H. Charlier, "Beach Erosion and Salvage along the Belgian Shore," in *Proceedings of the 7th International Sedimentological Congress,* University of Reading, England, 1967, paper C-1, pp. 1–3; S. Y. Chew et al., "Beach Development between Headland Breakwaters," in *Proceedings 14th Coastal Engineering Conference,* Copenhagen, June 24–28, 1974, ed. Billy Edge (New York: American Society of Civil Engineers, 1975), vol. 2, pp. 1309–1418; H. Howa, "L'érosion du littoral du Nord-Médoc (Gironde)," *Bulletin de l' Institut de Géologie du Bassin d'Aquitaine* 38 (1985): 57–68; J. Gribbin, "The World's Beaches Are Vanishing," *New Scientist* 102 (May 10, 1984): 30–32; F. P. Shepard and H. R. Wanless, *Our Changing Coastlines* (New York: McGraw-Hill, 1971); J. E. L. Verschave, "La défense et le maintien des plages belges entre Zeebrugge et la frontière Néerlandaise," *Bulletin Technique de l'Association des Ingénieurs issus de l'Universite de Louvain* 89, no. 1 (1961): 19–29; and P. P. Wong, "Beach Evolution between Headland Breakwaters," *Shore and Beach* 49 (1981): 3–12.

7. De Moor, "Premiers effets," "Recent Beach Erosion," and "Erosie aan de belgische kust."

8. M. Zunica, "Human Influence on the Evolution of the Italian Coastal Areas," in *Proceedings 23rd International Geographical Congress* (1968), pp. 87–93; and M. C. Barth and G. J. Titus, eds., *Greenhouse Effect and Sea Level Rise: A Challenge For This Generation* (New York: Van Nostrand Reinhold, 1984).

9. P. Bruun, "Sea Level Rise as Cause of Beach Erosion," *Journal of Waterways, Ports and Harbors* (ASCE) 88 (1962): 117–30; J. M. Darling, "Seasonal Changes in Beaches of the North Atlantic Coast of the United States," in *Proceedings 9th Conference of Coastal Engineering,* September 14–18, 1964, Lisbon, Portugal, ed. Billy Edge (New York: American Society of Civil Engineers, 1964), pp. 236–48; W. Kaufman and O. H.

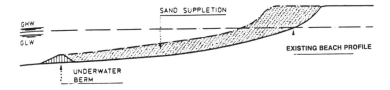

Suspended beach (Monte-Carlo, Monaco)

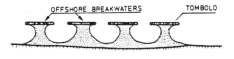

Offshore breakwaters (Tel Aviv, Israel)

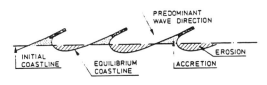

Headland system (Singapore)

FIG. 1.—Some beach-restoration possibilities. Top, suspended beach (Monte Carlo, Monaco); Center, offshore breakwaters (Tel Aviv, Israel); Bottom, headland system (Singapore). Source: HAECON N.V., Ghent, Harbour and Engineering Consultants, Belgium.

greenhouse effect on earth and additionally by local actions. Roads, board-walks, and buildings stand in the path of shifting sands.

In Tunisia, port construction at Tabarka has disturbed natural sand accretion and caused substantial transformations. Similar consequences resulted from port works at Bizerte. Though no loss of beach space has been reported so far along the Moroccan coast, removal of beach material for construction purposes between Rabat and Casablanca may create future problems, and new alternative sources of sand and gravel are sought.[10]

Off the coasts of Lebanon and Israel, dredging for construction material too close to shore has led to severe loss of beach territory. Efforts have been

Pilkey, *The Beaches Are Moving: The Drowning of America's Shoreline* (Durham, N.C.: Duke University Press, 1983); Khelila, "Nos plages vont-elles disparaître?" *La Gazette Touristique de Tunisie* 14 (1985): 23–25; M. L. Schwartz, "The Scale of Shore Erosion," *Journal of Geology* 16 (1968): 508–17; and J. E. Webb, *The Erosion of Victoria Beach: Its Causes and Cure* (Ibadan: University Press, 1960).

10. Personal communication with the Ministry of Public Works, General Directorate for Geology, Rabat, Morocco, 1986.

made, here and elsewhere, to counteract severe coastal erosion, for example, with the construction of offshore breakwaters or headlands (fig. 1, center and 1, bottom). Similar situations arose on Brunei, and Indonesia is alarmed over intensive coastal erosion. While, on Bali, wave attack and natural causes are the culprits in the shrinking of the beaches of Uluwatu and Tanah Lot, shoreline retreat at Siyut and Lebih can be ascribed to river mouth shifting, and losses of beach width on southern Bali are due to man's activities. Tourism has made spectacular advances on this island. The lengthening of the runway at Ngurah Rai airport appeared advisable, but the 800 additional meters of runway jutting out into the sea has resulted in serious damage at Kuta Beach (fig. 2). Traditional coral extraction, and ensuing gradual destruction of the reef, have brought about beach regression at Batumadeg, Sanur, Nusa Dua, and intensified the phenomenon at the already damaged Kuta Beach. Both coral and sand are used for ornamental and building materials in local construction; removal of river material is the principal cause of erosion on Gumbrich Beach. The tourist demand for hotel rooms "on the beach" has led to the siting of such facilities too close to the sea, and construction interferes with natural beach maintenance. Extraction of coral poses equally severe problems on the Maldive Islands.

A typical example of coastal erosion problems in developing countries is provided by Bénin. Geophysical factors and, more importantly, the construction of the Akossombo (Ghana) barrage and the development of the harbors of Lomé (Togo) and Cotonou (Bénin), are responsible for the spectacular retreat of beaches of the Gulf of Bénin. Here the coastline migrated inland by at least 20 m (66 feet) between 1892 and 1900. A dramatic acceleration of beach loss has occurred since 1960, when the Ghana barrage was constructed and important harbor improvements were undertaken at both Lomé and Cotonou. The barrage is a sediment trap, and longshore currents now remove littoral sand and erode the nearshore ocean floor. At Lomé, the harbor jetty impedes sand transport, sediments accumulate on its western flank, and there is no beach nourishment east of it. Only fronting the Tropicana Hotel, Lomé, has erosion abated, apparently due to the outcropping of calcareous cemented sandstone (beach rock) which, once denuded, acts as a natural breakwater.

Where groins were built at Lagos, Nigeria, to counter habor silting, sand accumulated west of Lighthouse Beach groin, and the sand bank enlarged about 400 m (1,320 feet) seaward, while to the east the shoreline retreated 1,250 m (4,063 feet) over about 60 years. Wave energy has increased in some areas as a result of the greater depths which developed near groins. Along this segment of the Nigerian coast, beach retreat has ranged from 5 m to 9 m (16 to 30 feet) each year since 1970.

Accumulation of removed material takes place at relatively shallow depths, and a natural breakwater may build up and stabilize the beach because of reduced wave power. Erosion varies in time and space—along the

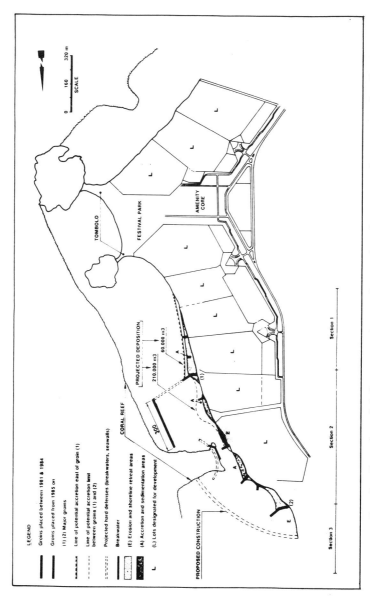

FIG. 2.—Location of defenses, erosion, and accretion along some shore areas of Bali, Indonesia. After Indonesian Institute of Hydraulic Engineering.

American Atlantic shore it is intense in winter, but during the summer accretion may take place.[11] Quantification must take into account the volume of sediment moved in tidal and seasonal cycles and on long-term scales, as well as the transport capacity of waves, currents, and winds.[12]

Gently sloping sandy beaches are far more vulnerable than gravel beaches, and rocky coasts are more resistant to the waves' onslaught. Seasonal changes are marked, and the damage caused by storm waves in the winter may be compensated during the calm summers if no other factors intervene. This is due to long-period waves in the summer. However, too often the coastal dune barrier is tampered with, breached, or, even more frequently, in industrialized countries, flattened by bulldozers. The recent popularizing of dune buggies and other vehicles has increased the damage.[13] Pathways for storms are thus created, and the dune can no longer protect the beach.

Surging storm waves riding on an abnormally high sea level can cause erosion, sedimentation, and short-term flooding. Storm waves erode foredunes and bluffs and carry sediment offshore and alongshore to beachways. High surges from large coastal storms transport sediments from the beach and foredune onto the back of the barrier, resulting in upward and inland gain in these areas as the ocean front is eroded. Damage to buildings is considerable.

Dunes play the role of stabilizers, sometimes "feeding the beach," sometimes receiving beach sand. Overwash may actually start new dune building.

During quiet periods, sand is returned onshore to the beaches, but not to the dunes; storms cause landward progress of dune erosion. Counteractions include the building of seawalls, but these are expensive and interfere with the sand supply of adjacent areas. Another measure calls for replenishment of the beach with new sand, and even more drastic measures require the relocation of buildings from the foredune crest.[14]

BEACH APPEAL

Beaches have steadily gained importance as recreational areas since the 1920s, and the democratization of travel has increased their accessibility to ever larger numbers of people, particularly since World War II. Though facilities were already well developed, economic growth in the 1960s increased the

11. See n. 9 above.

12. R. H. Charlier and M. Vigneaux, "Study of Management and Conflicts in the Coastal Zone, Part 2," *International Journal of Environmental Studies* 26 (1986): 206–15.

13. H. R. Mahoney, "Dune Busting: How Much Can Our Beaches Bear?" *Sea Frontiers* 26 (November–December 1980): 323–30; M. Primack, "Battling the Beach Buggies," *Sanctuary* 21 (July–August 1982): 7–10.

14. V. K. Tippie, "Coastal Canals Can Damage the Environment," *Maritimes* 20, no. 1 (1976): 12–14.

demand for space on adjoining land and use of the beach to the very edge of the sea. Although this phenomenon was particularly evident in the industrialized countries, it also manifested itself very rapidly in developing countries, even in quite remote areas of the world.[15]

Vacationers cause serious local damage to intertidal habitats along the British coast, particularly at Dorset, where comparison is possible between exploited coasts and adjacent coasts controlled by the Ministry of Defence and closed to the public. On the latter there is an abundance of species rare or absent on adjacent beaches.[16] The Isles of Lundy, Skamer, and Farne have been proposed as nature reserves, a measure also recommended for parts of Tor Bay and the Isles of Scilly.

However, demand for ocean-side space did not originate only with tourists; it came with equal force from energy and manufacturing industries, commercial enterprises, and from the traditional vocations of agriculture, animal husbandry, and fishing.[17]

During planning, little attention has been paid to the possible negative consequences for beach evolution and marine erosion. Harbor extensions or construction, whether for industrial purposes or for marinas, have led to extensive dredging of access channels, construction of dams on rivers, particularly in estuaries, and large-scale retrieval of bottom material for construc-

15. R. H. Charlier and A. Haulot, "Coastal Belt Touristic Occupance and Ecological Impact," in *Proceedings of the 3rd International Conference on Ocean Management,* August 13–17, 1975, Tokyo (Tokyo: Japan Management Association, 1975), pp. 277–84.

16. The Nature Environmental Research Council, *Marine Wildlife Conservation,* publications ser. B, no. 5 (January 1973); R. H. Charlier, A. Haulot, et al., "Coastal Belt Tourism, Economic Development and Environmental Impact," *International Journal of Environmental Studies* 10, no. 4 (1978): 161–72; R. H. Charlier, A. Haulot, and L. Verheyden, "Coastal Environmental Dilemma: Economic Development versus Tourism," in *Proceedings of the International Conference on Ocean Development,* Tokyo, 1978, vol. 5, paper E-3, pp. 72–78; J. Clarck, *Coastal Ecosystems: Ecological Considerations for the Management of the Coastal Zone* (Washington, D.C.: Conservation Foundation, 1974); C. N. Ehler and D. J. Basta, "Strategic Assessment of Multiple Use Conflict in the Exclusive Economic Zone," *Oceans '84,* Preprints of the Joint Annual Meeting of Marine Technology Society/Institute of Electrical and Electromechanical Engineers (MTS/IEEE), Washington, D.C., September 10–12, 1984 (Washington, D.C.: MTS/IEEE, 1984); R. Healy and J. Zinn, "Environment and Development Conflicts in Coastal Zone Management," *Journal of the American Planning Association* 51 (Summer 1985): 263–336; I. P. Joliffe, "Man's impact on the coastal environment," *Geographica Polonica* 34 (1976): 73–90; A. R. Orme, et al., eds., *Coasts under Stress* (Berlin: Börnträger, 1980); and R. Thompson, "America's Threatened Shorelines," *Congressional Quarterly-Editor Congressional Research, House of Representatives* 2, no. 17 (1984): 819–36.

17. J. M. Bryden, *Tourism and Development* (Cambridge: Cambridge University Press, 1973); and R. H. Charlier and M. Vigneaux, "Study of Management and Economic Conflict in the Coastal Zone, Part 1," *International Journal of Environmental Studies* 26 (1986): 177–89.

tion works, with the concomitant risk of important reductions in bed-load material from the natural supply of sand needed to resupply the beach.[18]

Beaches are in constant evolution; continued surveillance is required. Defense against erosion and beach area reduction has heretofore induced engineers to build protective structures. In several countries such counter-measures mean building of groins, detached breakwaters, or similar construction works.[19] In the United States, due to the absence of eminent domain on several coasts, governmental help until recently was excluded. This resulted in the building of makeshift breakwaters which were often inefficient, highly temporary, and eyesores.[20]

In recent years beach nourishment, namely artificial accretion, has been called upon as a coastal protection method—soft in contrast to hard medicine. Such an approach has been followed in Belgium, for example, in the beach-replenishment project at Knokke-Heist, near the Dutch border (fig. 3). A beach-renourishment project involving the sluicing of about 8.4 million m³ (11.2 million cu. yd.) of sand was executed along the eastern part of the Belgian coast between Zeebrugge and the Dutch border.[21] With further works now undertaken, the volume exceeds 10 million m³ (14.25 million cu. yd.). Further nourishment plans were undertaken on the west coast (Oostduim Kerke and the Coxyde [Koksÿde] vicinity) and carried out during 1989. Completed in 1979, the east coast works covered a stretch of 8 km (5 miles). The beach was widened by about 100 m (325 feet), and in some places it required the construction of an entirely new beach.

A few kilometers to the east, extensive works were undertaken to expand the harbor of Zeebrugge. The neighboring resort of Heist lost its beach,

18. W. Harrison, "Environmental Effects of Dredging and Spoil Deposition," *World Dredging Conference*, 1967, Technical Papers, pp. 535–59; G. Govatos and I. Sandi, "Beach Nourishment from Offshore Sources," *Shore and Beach* 27 (1969): 40–49; and S. J. Williams and D. B. Duane, *Construction in the Coastal Zone: A Potential Use of Waste Material* (Washington, D.C.: U.S. Army Corps of Engineers, 1975).

19. For example, Belgium, Holland, Great Britain, Italy, and the United States; "Strategy for Beach Preservation," *Geotimes* 30, no. 12 (1986): 15–19; P. Bruun and J. S. Purpura, *Emergency Measures to Combat Beach Erosion* (Gainesville, Fla.: Florida Engineering and Experiment Station, 1963); A. Führböter, "A Refraction Groyne Built by Sand," in *Proceedings of the 11th International Conference on Coastal Engineering*, June 24–28, 1974, Copenhagen, ed. Billy Edge (New York: American Society of Civil Engineers, 1975), vol. 2, pp. 1451–69; M. Petersen, "German Experience on Coastal Protection by Groins: A Review," *Bulletin of the Beach Erosion Board* 17 (1963): 38–54; and R. Silvester, "Natural Headland Control of Beaches," *Continental Shelf Research* 4, no. 5 (1985): 581–96.

20. R. H. Charlier, "Shifting Sands," *Long Island Business* 4, no. 3 (1956): 1–16; and James K. Mitchell, *Community Response to Coastal Erosion* (Chicago: University of Chicago Press, 1974).

21. Roovers et al. (n. 5 above); De Moor, "Premiers effets" (n. 4 above); and F. Depuydt, "Het Belgisch Strand," *Verhandelingen Koninklÿke Akademie voor de Wetenschappen* (1980): 122.

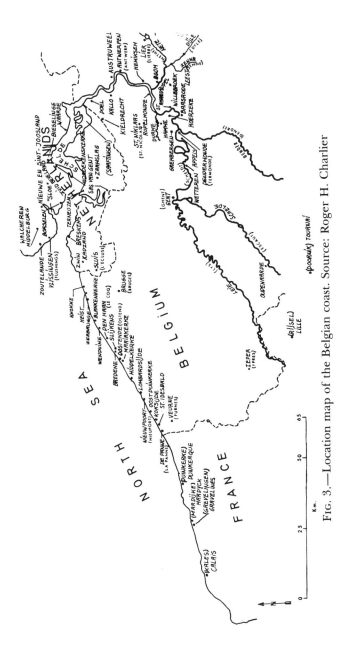

Fig. 3.—Location map of the Belgian coast. Source: Roger H. Charlier

meaning no dry sand area was left at high tide; next to it, the so-called beaches of Duinbergen and Albert-Strand were actually remnants of the dune barrier, and Knokke-Zoute itself had only a few square yards of dry beach left at high tide. Heavy storms were carrying the meager supply of sand seaward. Once completed, the beach-renourishment follow-up required a comprehensive survey program to observe the beaches' further evolution, to ascertain the accuracy of the study prognosis, and to recommend new measures to be taken.[22]

HARD COASTAL DEFENSE WORKS

Common hard defenses are groins (set perpendicular to the coast), detached breakwaters (built parallel to the coast), sometimes submerged, and seawalls. Each of these protective structures has a specific function.

The most commonly used materials along the North Sea shores are quarrystone or rubble. Smaller-size stones form the structure's core, which is armored by the larger-size material.

Scrap tires, often used for floating breakwaters, can also serve as construction material for fixed breakwaters. Treated timber piles are driven into the sea bottom and tires are piled on them (fig. 4, top left). In areas of moderate tidal range and low-gradient bottom slopes, concrete-filled burlap bags can be used to build a permeable breakwater. Filter material to prevent settling of the structure must be placed underneath (fig. 4, bottom left).

Groins can be built of treated timber, metal pipes of steel or aluminum, fuel drums, or concrete-filled bags (fig. 4, top right and 4, bottom right).

A flat concrete block, the HARO℠[23] was recently developed for protection of at-sea structures against wave action. It has a large central opening, and both of its short sides are widened at the base. It thus achieves a high degree of porosity when used in two layers, and good stability. Its high porosity and lesser weight allow a 30% reduction of the amount of required concrete. The decreased wave run-up permits a reduction in breakwater crest height. The new units have been used in the Zeebrugge, Belgium, harbor extension works.

Structures

Groins
The term groin is usually applied to fingerlike structures placed perpendicular to the shoreline. Groins are commonly built as a field along an eroding

22. See n. 4 above.
23. "The HARO block" is the registered trademark of HAECON N.V. Harbour and Engineering Consultants, Inc., Ghent, Belgium.

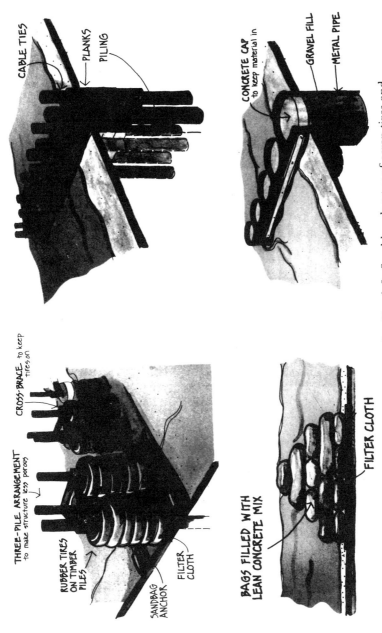

FIG. 4.—Hard coastal defence works. Top left, fixed breakwater of scrap tires and treated timber piles; Top right, groin built of timber; Bottom left, breakwater of lean concrete–filled bags; Bottom right, groin built of pipes. Source: U.S. Army Corps of Engineers.

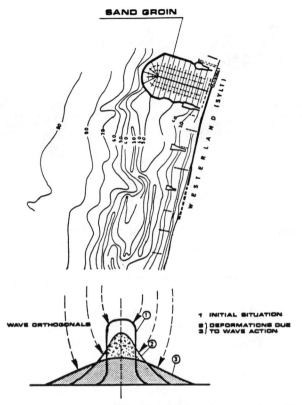

SAND GROIN

WAVE ORTHOGONALS

↑ INITIAL SITUATION
2) DEFORMATIONS DUE
3) TO WAVE ACTION

Fig. 5.—Construction of sand groin as coastal protection, Sylt, West Germany. Source: HAECON N.V., Ghent, Harbour and Engineering Consultants.

coast (fig. 5). They are short, squat structures jutting out to sea, built to reduce the longshore sediment transport in a certain beach area and to capture passing sand (fig. 6, left). When placed too close together, groins cause waves to stir up material, and currents will carry material offshore (fig. 6, right). If not well designed they can cause severe erosion, extending leeward along several kilometers of the coastline to seaward. A field of groins will maintain a stable beach when filled to capacity with sand, but will not provide full protection against stormdriven waves. However, the groin field usually triggers downdrift erosion which can be somewhat mitigated by using beach fill, which is in fact a variant, on a small scale, of beach nourishment.

To counteract longitudinal sand transport which resulted in silting of the entrance to Lagos harbor, Nigeria, two groins were constructed between 1907 and 1969 (fig. 7). The consequence was sediment accumulation west of Lighthouse Beach's groin, but east of the harbor near Victoria Beach's groin, erosion took a considerable toll, with the shoreline retreating 1.25 km (0.8 miles). To redress the problem, groins were lengthened, and artificial nourishment

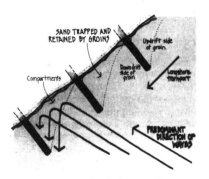 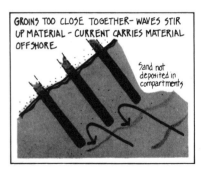

Fig. 6.—Groin fields. Left, groin field. Right, groins placed too close together. Source: HAECON N.V., Ghent, Harbour and Engineering Consultants.

of 1 million tons (1.102 million U.S. tons) of sand/year was carried out. However, shoreline retreat has continued, depending on the site, at a rate of 5 to 9 m (16 to 30 feet) a year (fig. 8).

Despite some unfortunate effects, the investment made over the years in groins must not be viewed as entirely lost. Wave action will transform an existing groin, if left untouched, into a sand *hoft*. Wave direction being variable, the sand is carried in different directions and eventually is distributed around the hoft. The groin-become-hoft hence exerts a stabilizing effect on the beach.

Apparently, unfavorable effects of groins need not occur. A recently built groin at Masirah Island, Oman, has been reported by Everts to have created a slightly more stable beach rather than an accumulation of sediment or increased erosion.[24]

Jetties

More massive than groins, their principal function is to redirect tidal flow to protect a beach or a navigation channel. However, construction of jetties results in groinlike effects by redirecting littoral drift. They are occasionally used to cause bypassing. Bypassing consists of dredging sand from behind the upstream jetty and bringing it via pipeline to behind the downstream jetty.[25]

Such port structures could obstruct sediment drift, and sediment may accumulate updrift, sometimes causing siltation of the port inlet. In the United States, Mexico, and India, for example, sediment bypassing by means of pumping is practiced.

---

24. C. H. Everts et al., "Sedimentation inversion at Masirah Island, Oman," Miscellaneous Papers (Washington, D.C.: U.S. Army Corps of Engineers, Engineering Research Center, 1983), available through National Institute of Science and Technology, Fort Belvoir, Va.

25. M. Yajinia et al., "Application of Sand By-passing to Amanohashidate Beach," *Coastal Engineering in Japan* 26 (1983): 151–62.

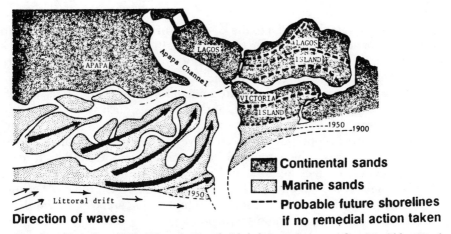

Continental sands

Marine sands

**- - - Probable future shorelines
if no remedial action taken**

FIG. 7.—Shoreline shift, Victoria Beach Lighthouse, Lagos. After M. Akle, "Problèmes d'érosion côtière dans le Golfe du Bénin," *Siren* 29 (September 1985): 20–31.

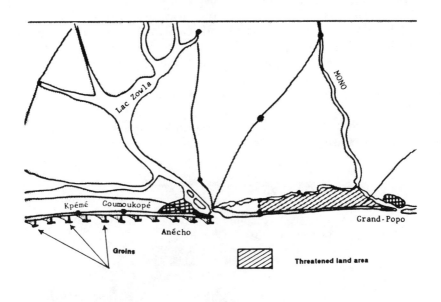

FIG. 8.—Proposed shoreline protection scheme, Victoria Island Beach. After M. Akle, "Problèmes d'érosion côtiere dans le Golfe du Bénin," *Siren* 29 (September 1985): 20–31.

Breakwaters

Detached breakwaters (parallel to the coast) are offshore barriers intended to reduce wave action on the existing coastline (fig. 9). They absorb wave energy before waves reach the shore. This will not only reduce onshore/offshore transport (especially during storm conditions) but will, in certain circumstances, reduce littoral drift and create tombolos due to wave-diffraction effects. However, in regions with significant tidal ranges, such as the North Sea, these structures would normally have their crest at mean or low water level. Under severe storm conditions the water layer above the crest will no longer stop incoming waves, which continue to attack and erode the beach area. Furthermore, wave attack on the structure is heaviest under these conditions.

The height of a breakwater poses serious problems if it is too low, because inadequate protection will ensue; if it is too high, interference with the shore processes will result in generating new problems.

Breakwaters come in great varieties. Besides the traditional construction, submerged breakwaters with openings for outflowing surge water, steel pilings, and other configurations exist. Permeable breakwaters offer some advantages, but effectiveness may be impaired where they are too porous. Floating breakwaters may not be suitable in areas of high wave energy, but they do have the advantage of being able to adjust their direction with changing winds or currents. Discarded tires are commonly used as construction material for floating breakwaters. Easily installed and readily adaptable, they have a high energy-absorption power.

Tying materials can be made of steel, rubber, or plastic material in the form of wires, ropes, chains or belts. They are subject to various wear and tear factors: fatigue, abrasion, corrosion, ultraviolet radiation, and biodegradation.

The low cost of floating breakwaters and their suitability where water level differences are great or bottom conditions poor have made them attrac-

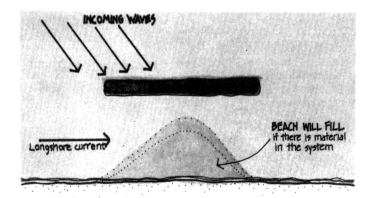

FIG. 9.—Breakwater, overhead view. Source: U.S. Army Corps of Engineers

tive wave-energy dissipators when funds are limited. Little energy is re-flected.[26] They have been used, for example, along England's southeast coast. Interestingly, a painting attributed to John Constable titled "Floating Break-waters at Brighton," which is displayed in Brighton's art gallery, shows open breakwaters installed long ago.

A secondary role, as wave-energy dissipators of certain structures whose function is energy production, has been praised. At the forefront of these are several wave-energy conversion systems.[27] Not only do such devices "absorb" wave energy to convert it into electrical or mechanical power, but they also fulfill the auxiliary role of de facto breakwaters.

Leonard and Huspeth have proposed membranes which could be filled with seawater.[28] Once rigid, they could be used as temporary breakwaters at low initial cost. However, only mathematical design models have thus far been developed.

Seawalls

Seawalls are structures separating land and water areas and are built to pre-vent further coastline retreat, whereas a bulkhead is placed on a bank or bluff to halt land sliding and protect an inland area against wave action. Venetian *murazzi* are examples of seawalls.

26. P. H. Adee, "A Review of Developments in Using Floating Breakwaters," in *Proceedings of the Offshore Technical Conference, May 3–6, 1976*, pt. 2 (Houston: Gulf, 1976), pp. 225–36; R. Arafa, "Mechanics of Wave Breaking over Floating Breakwaters in Erodible Bed," in *Proceedings of the 7th Miami International Conference on Alternative Engineering Sources, 1985*, December 9–11, 1985, Coral Gables, Fla., ed. T. N. Veziroglu (Coral Gables, Fla.: University of Miami, Florida Clean Energy Research Institute, 1986), p. 273; V. W. Harms, "Floating Breakwater Performance Compari-son," in *Proceedings of the 17th Coastal Engineering Conference (ASCE)*, March 23–28, 1980, Sydney, ed. Billy Edge (New York: American Society of Civil Engineers, 1980), pp. 2137–58; A. J. Harris, "Les brise-lames flottants," *Expomat—Actualites* 74, no. 4 (1986): 44–45; N. C. Kraus, "Shoreline Change behind Segmented Detached Break-waters at Holly Beach, Louisiana: Prototype Behavior and Model Prediction," in *Pro-ceedings of the 7th Symposium on Coastal and Ocean Management* (in press); E. P. Richy and R. E. Nece, "Floating Breakwaters: State of the Art," paper presented at the Confer-ence on Floating Breakwaters, Kingston, R.I., 1974, pp. 1–9; and T. Uda and Y. Murai, "Development of a Self-sustained Permeable Offshore Breakwater," in *Proceed-ings of the Techno-Ocean '88 Symposium*, November 16–18, 1988, Kobe, vol. 1, pp. 146–53.

27. P. A. Kakuris, *Surgebreaker Offshore Reef Systems* (Chicago: Great Lakes Eviron-mental Marine Ltd., 1983); C. W. Shabica, "Sediment Accretion Associated with Surgebreaker Offshore Reef, a Low Cost Shore Protection Device: An Update on the Corps of Engineers Section 54 Program," in *Oceans '85: Ocean Engineering and the Environment*, Proceedings of the Joint Annual Meeting of Marine Technology Society/ Institute of Electrical and Electromechanical Engineers (MTS/IEEE), San Diego, Calif., November 12–14, 1985, part 1 (Washington, D.C.: MTS/IEEE, 1985), pp. 576–80; and H. Kondo, et al., *Wave Energy Dissipating Structure* (in Japanese) (Tokyo: Morikita, 1983).

28. J. Leonard and R. Huspeth, Engineers at Oregon State University.

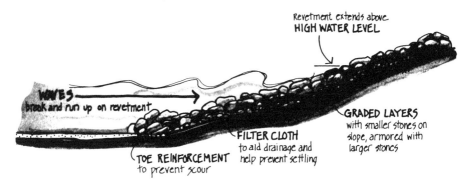

FIG. 10.—Revetment. Source: U.S. Army Corps of Engineers

Seawalls prevent erosion of land behind them; they do not protect the shore in front of them. They are most appropriate for boating and fishing ports. Revetments and bulkheads play the same role. Revetment designates a facing placed on a bank or bluff or other stone structure (fig. 10). Revetments are similar, for a sloping coast, to seawalls; bulkheads are used to prevent backshore property from sliding into the surf zone. Revetments are layers or blankets of strong, nonerodable material placed on the shore. Revetments built on relatively gentle slopes are stable; they do not greatly interfere with littoral drift. Nevertheless, they may induce local erosion elsewhere, particularly if they are steep.

In The Netherlands, masonry has traditionally been used for revetments, although asphalt revetments are not unusual. Asphalt may well become the more common material not only because of a progressive shortage of paving stones but because it can be installed more quickly. Asphalt concrete on sand is an approach that is followed on a large scale along Dutch shores.

Seawalls and bulkheads must be high enough to prevent overtopping, have a toe protected against scour, have weep holes to relieve pressure, be securely anchored and tied so as not to tip, and be defended against percolation and runoff.

On eroding beaches with predominant longshore transport, such defenses will not alter the longshore transport, and beach erosion in front of the seawall will continue. The erosion increases wave attack on the seawall and offshore transport. As a consequence, the stability of the structure is undermined, and it may fall over, a not exceptional occurrence. The beach may even disappear as wave force is reflected seaward instead of being spent across the beach. Sand, when disturbed, may be carried off seawards or along the shore away from the wall protected area.

Seawalls were already being built in the Venice region in the eighteenth century. These *murazzi* were but one of the anthropic interventions into natural process which can be traced back to the thirteenth century. Additionally, stone jetties protecting harbor entrances acted as groins.

Occasionally, natural formations may play the role of temporary or long-term seawalls. This is, for instance, the case with *kurkar* ridges (more or less carbonate cemented sands, loams, and sandstones) on the shallow shelf north of the Nile delta (Gaza, Israel, Lebanon), which parallel the coastline. Their effects are the same as those of man-made seawalls.

Gabions and Mattresses

Attempts have been made to counteract beach erosion and to encourage accretion by using flexible concrete mattresses, or by synthetic seaweed grids (The Netherlands).

Gabions offer a related approach. They are rectangular metallic cages filled with various-sized pebbles and spread out; one gabion type resembles a mattress.[29] They are usually 0.5 m (1.64 feet) high and as long as 8 m (26.2 feet), with a mesh size of 8 to 12 cm (3.15–4.72 inches). The mattress is laid upon a slope of appropriate gradient. However, the materials must be present in situ, and one has to cope with corrosion.

Gabions also can be used for revetments. Rocks filling the gabions must be at least 10 cm (4 inches) in diameter lest they slip through the mesh (fig. 11). This type of revetment is permeable, takes on the soft terrain's shape, and absorbs wave energy while successfully resisting the strongest currents. Several layers can be placed on top of one another and tied together with wire. However, gabions are poor protection for the beach itself as they may reduce or prohibit accretion by sand of terrestrial origin.

Gabion breakwaters have been constructed along East Coast Park on the southeastern coast of Singapore.

Artificial Reefs

Denmark and England both claim the earliest development of erosion control through use of artificial seaweed. In England, bundles were placed along 275 m (900 feet) of shoreline, 245 m (800 feet) offshore, and in 4.5-m-deep (15-foot-deep) water, making a 27.5-m-wide (90-foot-wide) barrier. No conclusions could be drawn on the effectiveness. Later experiments were conducted in the same countries and in The Netherlands, Germany, Norway, and France.[30] In The Netherlands and France, positive results were achieved in

29. A. Q. White, "Impact of Beach Renourishment on a Sandy Beach Ecosystem," *Estuaries* 4, no. 3 (1981): 259.

30. K. A. Atherley, "Seascape (R) Synthetic Seaweed: A Failed Solution to Erosion in Barbados," in *Proceedings of the 7th Symposium on Coastal and Ocean Management;* S. Benton et al., *Cape Hatteras Shoreline Erosion Workshop: Summary Report* (Columbia, N.C.: North Carolina Department of Natural Resources and Community Development, 1983); R. L. Brashears and J. S. Darnell, "Development of the Artificial Seaweed Concept," *Shore and Beach* 35, no. 2 (1967): 35–41; H. H. Dette, "Effectiveness of Beach Deposit Nourishment," in *Coastal Sediments '77: Proceedings of a Specialty Conference*, Charleston, S.C., 1977 (New York: American Society of Civil Engineers, 1977),

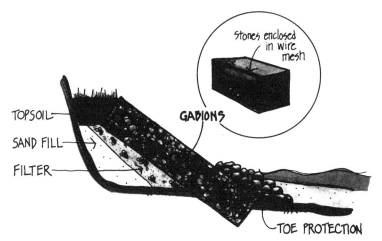

FIG. 11.—Gabion as revetment: rock fill. Source: U.S. Army Corps of Engineers

reducing scouring near man-made structures by lengthening their density in the bundles.[31]

The effects of seaweeds, whether natural or artificial, are greatest at and close to the surface. Seaweed "barriers" can be effective by reducing incoming wave height; such a barrier acts as a floating porous breakwater. But artificial seaweed fields present two disadvantages: an effective field, even if placed so that it would not be a navigational or recreational hazard, would be more expensive than other beach protection methods, and it has unfavorable effects upon beaches and shorelines further downdrift.

Beavis and Charlier[32] and others have also proposed the use of artificial reefs made of *living* plant material as "natural breakwaters" (fig. 12). Algae—for instance, kelp and other brown seaweeds such as *Macrocystis pyrifera* and

pp. 211–227; Geomidi Consultants, "Protection against Coastal Erosion (Scour), Saintes-Maries-de-la-Mer, France," Geomidi Consultants, Company Report, 1985; Linear Composites, "Erosion Control Systems," ICI Fibres, Company Report, North Yorkshire, England, 1986; S. M. Rogers, Jr., "Artificial Seaweed for Erosion Control," *Shore and Beach* 55, no. 1 (1987): 19–29; Seascape Technology Inc., "Seascape at Cape Hatteras," Seascape Technology Inc., Company Report, Greenville, Del., 1984; and H. G. K. ten Hoopen, "Recent Applications of Artificial Seaweeds in the Netherlands," *Coastal Engineering* 166 (1976): 2905–15.

31. *Non-structural Beach Erosion Protection* (Long Beach, Calif.: Department of Public Works and Tidal Agency, 1983); J. Hall, et al., "Creative Shoreline Management through Community Partnership," paper presented at the Shore and Beach Protection Association Annual Meeting, Santa Cruz, Calif., 1986; and Rogers (n. 30 above).

32. Papers presented at the International Seaweed Symposium, São Paulo, July–August, 1986. Date is of preprints published by the organizing committee. A special volume was later published as *Hydrobiologia* 151/152 (1988).

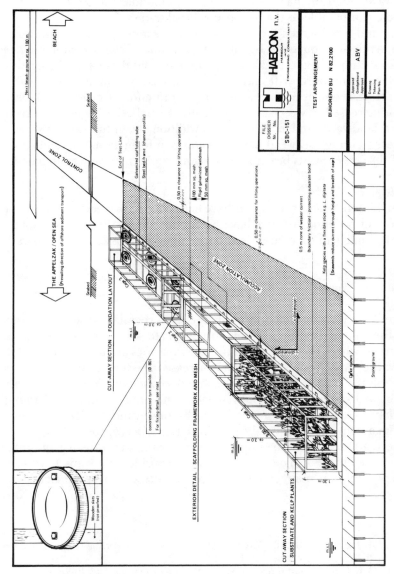

FIG. 12.—Cages array for algae artificial reef. Source: HAECON N.V., Ghent, Harbour and Engineering Consultants.

Laminaria species—appear particularly suited for the purpose. Wave attenuation can be achieved by large kelp beds. Beds over 0.8 km (0.5 miles) wide and several kilometers long, in water 15.25 m (50 feet) deep, could theoretically reduce wave height by half, as far as 0.8 km (0.5 miles) into the field. They play a role comparable to that of porous breakwaters.[33]

A natural-weed screen has been proposed as a means of stabilizing sand losses of artificially nourished beaches. A short line of cages, containing *Laminaria,* for example, could break up rip currents that develop at the heads of groins where longshore drift may carry a bed load of sediments seaward.

As artificial nourishment buries the bottom ends of groins, back-beach erosion is increased. A weed screen could break up the confluence of currents and favor landward sand accumulation. Additionally, grains carried seaward are deposited on the seaweed barrier, leading to formation of a sandbank in front of the screen; thus beach nourishment can be naturally maintained.

Tetrapods

In Japan, 50-ton concrete tetrapods have been used to protect the east coast of Honshu, and tetrapod barriers have been built at Sylt, Federal Republic of Germany. Similar to them are the tribars, tripods, dolosses, and hanbars.

Environmental Aspects

Groins and breakwaters will affect only the wave regime and sediment transport in their immediate area. They do not create solutions for the adjacent coastal areas, where the sediment transport capacities are unchanged. On the contrary, they may cause a spectacular increase in sediment transport on the lee side of the structure, causing renewed coastal erosion. In that case, the problem has merely been moved from one area to another. Although reports have overwhelmingly faulted seawalls for causing beach erosion, some authors maintain that in several cases this has not been proven.[34]

If recreational aspects are important, any coastal structure (e.g., groin) on the beach should be avoided. Furthermore, considerable sums have to be spent repeatedly on maintenance and repair due to storm damage to the structures.

33. R. A. Dalrymple, et al., *Physical and Environmental Aspects of Ocean Kelp Farming*, no. 111 (Chicago: Gas Research Institute, 1982).

34. E. C. F. Bird, "Coastal processes," in *Man and Environmental Processes*, ed. K. J. Gregory and D. E. Walling (Folkestone, Kent: Dawson, 1979), pp. 82–101; O. H. Pilkey and T. D. Clayton, "Beach Replenishment: The National Solution?" in *Proceedings of the 5th Symposium on Coastal and Ocean Management*, Seattle, 1987, organized by the U.S. Environmental Protection Agency (in press); and R. L. Wiegel, "Trends in Coastal Erosion Management," *Shore and Beach* 55, no. 1 (1987): 3–11.

BEACH NOURISHMENT AS COASTAL PROTECTION

More appropriate coastal protection solutions are often provided by a durable defense system and by modifying the beach material itself. With more extensive knowledge of coastal dynamics and the availability of physical and complex mathematical models, morphological processes and modifications can be predicted to a considerable extent. Additionally, technical advances in dredging equipment provide new and more economical possibilities for beach restoration (fig. 13).

Beach nourishment has been considered a potentially desirable form of shore protection, since wave action spreads the material along the shoreline. Costs have been lowered by the use of booster pumps, allowing piping of material over long distances, and the use of hopper dredges. Dredged material from navigation-channel maintenance can be used for beach restoration. Hence both deposition of sand and beach fill are suitable techniques. No substantial biotic effects have been observed.

The earliest beach-nourishment project may well have been the dredging, in 1920, of Santa Barbara Harbor, California, and deposition of the material on the eroded beach downcoast from the breakwater.[35] White reported that with beach nourishment involving over 2,500,000 m³ (3,280,000 cu. yd.) of material deposited over 3 years, the original beach fauna was destroyed, but new fauna have developed.[36] Assessments of artificial nourishment of beaches have been made for over 30 years.[37]

35. W. C. Penfield, "The Oldest Periodic Beach Nourishment Project," *Shore and Beach* 28, no. 1 (1960): 9–15.

36. See n. 29 above.

37. P. Bruun, "Cost-Effectiveness of Coastal Protection with Reference to Florida and the Carolinas, U.S.A.," *Journal of Coastal Research* 1, no. 1 (1985): 47–55; D. W. Berg, *Factors Affecting Beach Nourishment Requirements At Presque Isle Peninsula, Erie, Pennsylvania* (Ann Arbor: University of Michigan, Great Lakes Research Division, 1965), publication 13, pp. 214–21; H. H. Dette, "Effectiveness of Beach Deposit Nourishment;" J. R. Giardino et al., "Nourishment at San Luis Beach, Galveston, TX: An Assessment of the Impact," pp. 1145–57; and R. S. Grove et al., "Fate of Massive Sediment Injection on a Smooth Shoreline at San Onofre, California," pp. 531–38; both in *Coastal Sediments 87: Proceedings of a Specialty Conference on Advances in Understanding of Coastal Sediment Processes*, May 12–14, 1987, New Orleans, La., ed. N. C. Kraus (New York: American Society of Civil Engineers, 1987); J. V. Hall, Jr., "Artificially Nourished and Constructed Beaches," Beach Erosion Board, technical memo 29, 1952; N. C. Kraus, "Shoreline Change" (n. 26 above); J. R. Lesnik, *An Annotated Bibliography on Detached Breakwaters and Artificial Headlands*, no. 79 (Fort Belvoir, Va.: Coastal Engineering Research Center, 1979, mimeographed); R. D. Norby, *Evaluation of Lake Michigan Nearshore Sediments For Nourishment of Illinois Beaches* (Champaign, Ill.: Illinois State Geological Survey, 1981); K. F. Nordstrom, J. R. Allen, D. J. Sherman, and N. P. Psuty, "Management Considerations for Beach Nourishment at Sandy Hook, NJ, USA," *Coastal Engineering* 2 (1979): 215–36; J. Pope, "Segmented Offshore Breakwaters: An Alternative for Beach Erosion Control," *Shore and Beach* 54, no. 4 (October 1986): 3–6; C. L. Thompson, "Beach Nourishment Monitoring on the

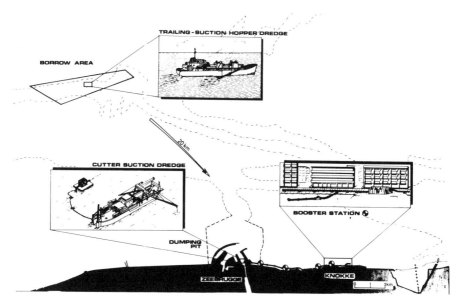

FIG. 13.—Knokke: beach restoration process. Source: HAECON N.V., Ghent, Harbour and Engineering Consultants.

Segmented offshore breakwaters were built, based on the premise that they would attenuate direct wave pounding on the beach, cause a transformation of the incoming waves, and induce deposition of drifting sediment in the lee of the structures (fig. 14). The ensuing beach salient may even become connected to the breakwater segment and form a tombolo.

Wave energy is reduced in the spaces between the successive segments as well. A beach is thus constructed, but a certain amount of sand is removed thereby from the longshore transport system. To balance this loss, artificial nourishment equaling the amount removed is desirable. This approach counteracts the negative impact the structures would have outside the zone where they have been placed. However, local conditions may be such that artificial nourishment does not eradicate adverse impact.

However, breakwater structures are expensive, and the lack of examples of segmented systems acts as a deterrent to funding. Yet they have been in use in Japan for more than 30 years, generally built rather close to shore.[38] Others

Great Lakes," in Kraus, ed., *Coastal Sediments '87*, pp. 1203–15; and T. L. Walton and J. A. Purpura, *Beach Nourishment Along the Southeast Atlantic and Gulf Coasts* (Tallahassee, Fla.: University of Florida, Florida Sea Grant Program, 1979), pp. 11–24.

38. Refer to references in note 37. Also, see L. D. Nakashima et al., "Initial Response of a Segmented Breakwater System, Hetty Beach, Louisiana," pp. 1399–1416, C. S. Sonu and J. F. Warwar, "Evolution of Sediment Budget in the Lee of a Detached Breakwater," pp. 1361–68, both in Kraus, ed., *Coastal Sediments '87*.

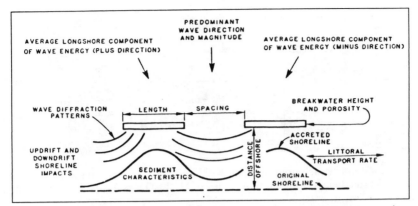

FIG. 14.—Segmented offshore breakwater design considerations. Source: HAECON N.V., Ghent, Harbour and Engineering Consultants.

are found in Denmark, France, Italy, Israel, and Singapore. In the United States several segmented breakwaters have been constructed in locations from Boston to Virginia and on the Great Lakes.

Many projects combine sand addition with construction of hard defense structures. This was done in Monaco, Israel, and Singapore (fig. 1), and is a commonly recognized approach especially where suitable material is available close by.

To counter recession of weak outcrop cliffs by building and maintaining a broad beach, the construction of groins that intercept longshore drifting can be supplemented by beach nourishment on the foreshore or updrift.[39] However, when such groins intercept longshore drifting, erosion may be accentuated, because sediment is withheld from the downdrift area. With sea walls, fronting beaches are rapidly depleted by scour due to the reflection of storm waves from the wall. In some cases, for instance, in Belgium, depending on predominant wave direction governing the littoral drift, simple sand supply

39. J. C. Boothroyd, "Geologic Processes Pose Problems for the Rhode Island Shore," *Maritimes* 29, no. 2 (1985): 1–3; Y.-H. Chu and E. B. Hands, "Shoreline Erosion Protection and Beach Nourishment," in *Proceedings of the 5th Symposium on Coastal and Ocean Management* (n. 36 above); P. J. Godfrey, "Barrier Beaches of the East Coast," *Oceanus* 19, no. 5 (1976): 27–40; L. P. Johnson, "Natural Shore and Beach Restoration Enhancement and Preservation Systems," in *Proceedings of the 5th Symposium on Coastal and Ocean Management;* W. Kaufman and O. H. Pilkey, *The Beaches Are Moving* (n. 9 above); Laboratoire Central d'Hydraulique de France, *Plage et littoraux artificiels* (Paris: Ministère de l'Equipement du Logement et de l'Aménagement du Territoire, 1972); F. Lowenstein, "Beaches or Bedrooms: The Choice as Sea Level Rises," *Oceanus* 28, no. 3 (1985): 20–29; M. W. Mugler, "A Problem Can Be a Resource: Beach Nourishment with Dredged Material," *Water Spectrum* (Spring 1983), pp. 38–45; and U.S. Department of the Interior, *Coastal Barrier Resources Systems*, draft report to Congress (Washington, D.C.: U.S. Department of the Interior, 1985).

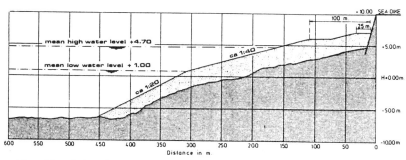

F<small>IG</small>. 15.—Simple beach nourishment, Knokke-Heist, Belgium: typical cross sections before and after beach nourishment. Source: HAECON N.V., Ghent, Harbour and Engineering Consultants.

alone can provide adequate coastal protection (fig. 15). In Westhampton, New York, however, when sand addition was stopped but more groins were added, storms stripped the sand beyond the groin field in the downdrift direction.

By adding new sand to a beach but not to the immediate offshore area, the beach slope may become out of equilibrium with the adjacent underwater area, and the shoreface profile may steepen. During storms, waves remove sand, and the "shoreline" is again rolled back landward. This must be carefully considered when an artificial accretion project is undertaken. The slope of the "filled beach" should match as far as possible the slope of the original beach.

One advantage of the nourishment approach is that it can be easily adapted to changes in coastal patterns. So-called permanent structures are not flexible and may become harmful in a new situation, necessitating removal. In the long run, artificial accretion is less costly.

Material to be used, carefully selected as to nature and source site, may be brought to the beach to be restored and dumped offshore, leaving it to the sea to carry it to the shore and determine its distribution. It may also be deposited at a single location on the beach which acts as a distribution center or feeder beach. Still another approach is to spread the material directly on the endangered beach, restore it immediately, and take simultaneous appropriate soft measures to protect it. Finally, the continuous-supply method involves creation of a fixed or semifixed plant which first intercepts and then passes on the littoral drift to the downdrift side. In some cases, a combined approach has been taken—for example, offshore dumping and onshore nourishment. Beach nourishment has been applied to lake beach restoration in the Province of Quebec at Lac St. Jean.

While the technique is the same for lake and ocean beaches, there are some environmental differences to be considered. Grain size distribution varies daily due to wave climate, storms, and water height, but changes are less pronounced on lakes than on ocean beaches because of water level variations on lakes with ensuing base level readjustments. A vivid example is provided by Lake Michigan, particularly in the greater Chicago area and the Indiana Dunes region. Water level variations have caused, at numerous sites, catastrophic beach erosion. On the Indiana shoreline, the beach has often regressed to such an extent that waves assault the foot of the dune, houses are due to topple into the lake, a youth summer camp has been abandoned, and property values have taken a dive. Along the metropolitan Chicago shorefront, emergency measures had to be taken during and after the 1987 winter. Indecision, conflicting interests, disagreements among various agencies, short supply of funds, hesitation on the type of measures to take, local rather than overall and long-term action have been the prevailing attitudes. Though Lake Shore Drive occasionally had to be closed to traffic, and waves licked the base of buildings, action seems again delayed (and still is) as immediate dangers receded due to a 1988 lowering of lake water level. It may prove, in the long run, a costly procrastination.

Dune nourishment and artificial building may provide additional beach protection. The dune blocks wave attack and provides sand to the beach. Such an approach is being tested at Mantoloking, New York, where a 6-m-high (19.5-foot-high) dune has been erected.

Fill should resemble as much as possible the original material. Coarser fill will erode slower, and finer fill will erode faster than the original material. The source area must be selected so no significant deepening of nearshore areas occurs; otherwise wave energy may be intensified and cause or exacerbate coastal erosion. The same applies when dredging material for beach nourishment. Somewhat coarser and better sorted sand than beach material is preferable. Dredging up to 60 m (195 feet) poses no problem, and it is possible to pipe it over an 8-km (5-mile) distance.

Beach nourishment has been considered by the U.S. Army Corps of Engineers as a method to dispose of the 265.5 million m^3 (300 million cu. yd.) of dredged material it picks up annually. This material is usually dumped at sea, damaging benthic life. Under proper circumstances, the use of dredged material for beach nourishment not only contributes to the development and maintenance of waterway transportation, but also provides hazard reduction and recreational benefits consistent with environmental mandates and the material laws governing coastal systems. However, indiscriminate ocean dumping of dredgings should not be tolerated.

Beach nourishment is the least costly, most environmentally acceptable disposal alternative. Despite its current prevalence, beach nourishment cannot be considered a panacea for dredged-material-disposal ills. For 75% of navigation projects the dredged material is not a practical, cost-effective, permissible source for beach-nourishment supply. Most frequently, the material is too fine-grained to be effective for shore protection. Other constraints are excessive costs, detrimental environmental effects, and technical limitations.[40]

40. Erosion Technology Systems Inc., *Sea Grid* (Wilmette, Ill.: Erosion Technology Systems Inc., 1985).

Unintentional beach nourishment has also yielded good results. Near Wokington (Cumbrian coast, England) mining wastes and industrial slag are used as beach nourishment material. In Denmark, at Hoed, flint, gravel, and limestone waste products are regularly dumped. Quarry wastes feed the Cornwall beaches at Porthoustock, and mine wastes and slag at Porthallow Coal fulfill the same role.

The following steps are suggested in the establishment and management of artificial beaches: determination of various conditions affecting the beach under consideration and choice of the size and methods for the project, assessment of the environmental conditions, detailed facilities design, implementation, and, after completion, beach administration.

The survey program related to a beach-nourishment scheme involves bathymetric soundings, studies of beach evolution (eventually by remote sensing), and monthly terrestrial measurements (beach profiling). At least once a year sampling should be carried out on the beach, the nearshore, and the seabed. Winds, littoral drift, and tides affect sediment transport, and effects must be measured. Hence the survey requires the gathering of tide, wave, current, and wind data.

On the Strait of Georgia (Vancouver, Canada), in Monte Carlo (Monaco), at Hook (Hoek) van Holland (The Netherlands), and on Bora Bora (Society Islands, near Tahiti), artificial beaches have been built. In Monaco dolomite chips were used instead of sand (fig. 1). In Vancouver large cobbles were used to stem sand cliff erosion, and such nourishment has apparently had the desired effect. The use of calcium carbonate sands is under study.[41]

Beach-Nourishment Schemes

Several schemes have been undertaken at various sites (see table 1). In the United States, it cost US$65 million to restore 25 km (15.5 miles) of beach in

41. R. T. Cunningham, "Evaluation of Bahamian Oolithic Aragonite Sand for Florida Beach Nourishment," *Shore and Beach* 34, no. 1 (1966): 18–19; K. A. Downie and H. Saaltink, "An Artificial Beach for Erosion Control," *Proceedings of the Conference on Coastal Structures* (New York: American Society of Civil Engineers, 1983), pp. 846–59; W. C. Eiser and C. P. Jones, "Performance Evaluation of a Beach Nourishment Project in Myrtle Beach, South Carolina;" T. E. Lankfort and B. J. Baca, "Comparative Environmental Impacts of Various Forms of Beach Nourishment;" and D. Roellig, "Shoreline Response to Beach Nourishment;" all in *Proceedings of the 7th Symposium on Coastal and Ocean Management* (n. 10 above); M. N. Nichols and C. Cerco, "Coastal Dunes and Sand Resources," *Proceedings of the 5th Symposium on Coastal and Ocean Management;* A. Roos, "Artificial Beach at Hook of Holland," *Shore and Beach* 45, no. 2 (1977): 19–23; and L. Tourmen, "The Creation of an Artificial Beach in Larvoatto Bay, Monte Carlo, Principality of Monaco," *Proceedings of the 11th Conference on Coastal Engineering,* June 24–28, 1974, Copenhagen, ed. Billy Edge (New York: American Society of Civil Engineers, 1975), vol. 1, pp. 558–69.

TABLE 1.—LOCATIONS OF SOME BEACH NOURISHMENT SCHEMES

Year	Location	Quantity of Material m³	cu. yd.
1919–78	Southern California coast	108,000,000[a]	114,000,000
1961–62	La Croisette, Cannes	100,000	130,000
1967–68	Redondo Beach, Calif.	1,100,000	1,450,000
1969–70	Copacabana Beach, Brazil	3,500,000	4,100,000
1971	Goeree, Holland	600,000	780,000
1972	Sylt, West Germany	900,000	1,750,000
1974–75	Bournemouth, England	650,000	850,000
1975–87	Lagos, Nigeria	1,000,000[b]	1,300,000
1977	Rockaway Beach, N.Y.	48,000,000	62,100,000
1977–79	Knokke-Heist, Belgium	8,400,000	11,000,000
1980	Gold Coast, Queensland	2,400,000	3,150,000
1980	Miami Beach, Fla.	87,000,000	117,000,000
1982–83	Isle of Langeoog, West Germany	200,000	260,000
1983	Pet Foreshore, England	19,000[b,c]	24,750
. . .	Walland, England	31,000[b]	41,000

NOTE.—Compiled by Roger Charlier from many sources and personal information.
[a]Total amount over a 60-year period and covering 60 projects.
[b]Amount is annual recharge.
[c]"Shingle."

Miami Beach; at Redondo Beach 1,100,000 m³ of material was used to rebuild the beach.[42]

Knokke-Heist, Belgium
Offshore transport of beach material resulted in almost total beach submergence at high tide east of Zeebrugge (fig. 16). The Appelzak Channel gradually extended itself to within 500 m (550 yards) of the harbor's seawall, thus favoring beach erosion, while the seawall, itself threatened by storm waves, hampered sand transport to and accretion on the beach (fig. 17). While dune area had expanded during the 1980s, the fore- and the backshore lost ground due to deflection of longshore currents (Duinbergen, Knokke-Zoute; figs. 18, 19) and sand withdrawal (Heist; fig. 16). Beach regression due to the effects of tidal currents affected by the new Zeebrugge harbor is to be expected.

In the case of Knokke-Heist, predicted sand loss and coastline regression was weighed against sand-suppletion and groin-building solutions.[43] Total capital and maintenance costs of beach nourishment, with or without groin construction, were not significantly different. This, and the fact that beach restoration was at the time a necessity, led to the choice of beach renourish-

42. See nn. 37 and 39 above.
43. Kerckaert et al., "Artificial Beach Renourishment on the Belgian Coast" (n. 4 above).

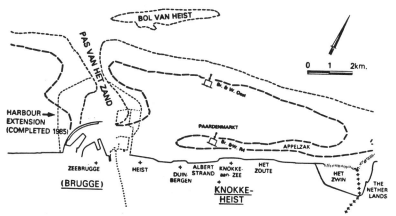

Fig. 16.—East coast, Belgium: general situation, 1976. Source: HAECON N.V., Ghent, Harbour and Engineering Consultants.

ment as a coastal protection measure. The entire project was accompanied by a well-defined coastal observation program which is ongoing. It permits monitoring of coastal changes due to the harbor-extension works at Zeebrugge. A detailed description of the project has been presented by Rovers, Kerckaert, and others, and the recent beach evolutions have been the subject of several reports (fig. 20).[44]

Along this eastern shoreline segment of Belgium, it has been noted that the Appelzak gully, which acted as a trap for sediments and materials removed from the beach, has been substantially reduced in size, and checks carried out indicate that it is filling up. It is noteworthy that the "filler" material is mainly silt. The westerly segment of the Appelzak is doomed to disappearance.

Due to the silting of a suitable material area at depths somewhat exceeding 10 m (32.5 feet), large seagoing trailing-suction hopper dredgers were used. The dredged material was dumped into a pit from which a cutter-suction dredge and booster stations pumped them onto the beach.

An economic assessment was made based only upon the economic rent derived from beach-based tourism. Income from visitors was estimated in 1984 at US$142 million/year (1984: 1 US$ = 66 BEF; 1988: 1 US$ = 35 BEF). Beach maintenance costs US$1,128,000, or about 20¢/tourist/day. However, the amortization of the investment in the original project is not accounted for in the calculation, and in this particular instance, the dredging can be written off as a no-cost item since it was required to provide a navigation channel.

44. Roovers et al., "Beach Protection" (n. 5 above); "Artificial Beach Renourishment" (1982) and Kerckaert et al., "Artificial Beach Renourishment" (1982) and "Artificial Beach Renourishment" (1985) (n. 4 above).

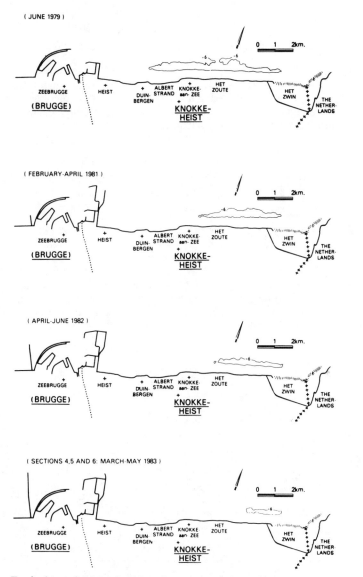

FIG. 17.—Evolution of "Appelzak" gully, east coast, Belgium. Source: HAECON N.V., Ghent, Harbour and Engineering Consultants.

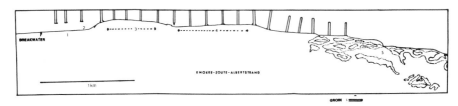

FIG. 18.—Knokke-Zoute, Albert Plage, Belgium. (1) Casino built on dune; (2) area used as "beach" is actually dune; (3) beach completely covered at high tide, except for a few square yards; (4) beach reduced in most locations to a strip of a few meters of sand not normally covered at high tide; and (5) Zwin area where dunes still exist and protect marshes inland. Source: Roger H. Charlier.

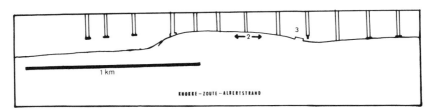

FIG. 19.—Knokke-Zoute, Belgium. (1) Where imposing dunes once stood; (2) bathing facilities once stood here on the beach but were destroyed decades ago because of sand removal; and (3) areas where the beach has been either destroyed or reduced to a string of a few meters. Source: Roger H. Charlier.

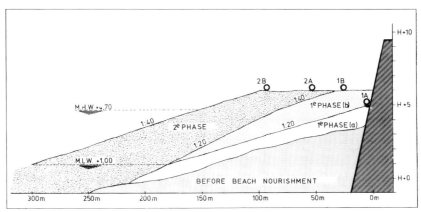

FIG. 20.—First- and second-phase execution plan, east coast, Belgium. Source: HAECON N.V., Ghent, Harbour and Engineering Consultants.

Ostend

The soft approach may not be applicable in all instances. A particular case that comes to mind is Ostend, Belgium's most famous coastal resort. Construction of a fort, at the turn of the century, seriously disturbed normal sand transport. Building of groins did not solve the problem. Strong erosion, often the consequence of storm waves caused by refraction-induced wave-energy concentration, is endangering the beach, the dike, and the structures beyond it.

Beach nourishment alone is not suited because Ostend is also a cross-Channel port, and the navigation channel must be kept open. Large-scale beach nourishment would extend beach slopes too far seaward. Yet the toe of the boardwalk-dike is presently almost unprotected against sapping by the sea.

Requirements considered for the Ostend project were (1) reduction of wave overtopping of the existing seawall (flood prevention), (2) prevention of increased sedimentation in the existing port access channel, as a result of the coastal defense works, (3) taking into account future recreational beach activities, and (4) limitation of the project within the existing groin structure extension. As sand suppletion alone would not effectively contribute to these requirements, the mixed use of sand and gravel is envisioned. This approach considers the coastal dynamics without any hard defense structure. The gravel layer thickness, the equilibrium profile, and the overtopping frequencies of the seawall were tested in a wave flume. Wave attack and deformation of the gravel core were successfully tested in a wave tank at the Hydraulic Research Laboratory of the Belgian Ministry of Public Works at Borgerhout (near Antwerp). The mixed gravel-sand option apparently provides for feasible and dynamically stable coastal protection.

Four alternative solutions were tested, based on the idea of confining the sand in a gravel enclosure with steeper slopes. The volume of materials needed thereby became considerably reduced, as were expenditures, by some 40%. Through continuous monitoring, subsequent shore behavior has been controlled, and, in addition, the hydrometeorological data thus gathered have built up a valuable information bank for future projects.

West Coast, Belgium

The largest beaches are located west of Ostend in the direction of the French border; until recently they had not been affected by erosion. Lately, however, some concern has been voiced about the area between Coxyde (Koksÿde) and Nieuport (Nieuwpoort). The decision was made to provide nourishment to the area (fig. 13).

Bournemouth, Dungeness, and Portobello, United Kindgom

In Bournemouth, construction of a seawall cut off the sand supply from neighboring cliffs. Beach nourishment was decided upon to remedy the situation.[45] Beach rebuilding material was pumped ashore by trailing-suction hopper dredgers. Some 110,000 m³ (144,100 cu. yd.) of sand was placed within 200 m (656 feet) of the shore.

At Dungeness, Kent, beach material is dredged in surplus areas and spread where the beach is eroding, thereby balancing the sedimentary budget.

At Portobello, near Edinburgh, replenishment of the beach occurs in 10-year spans. The results are satisfactory, but there is concern whether this is sufficient.

Sylt and Langeoog, Federal Republic of Germany
The projects at Sylt and Langeoog considered the construction of a sand spit (refraction groin) as adequate coastal protection (fig. 5). Under wave action the sand groin moves shoreward, feeding the adjacent beach. The orientation of the sand groin has to correspond to the predominant wave action. Seafloor sand was dredged to rebuild beaches on Sylt which had earlier been given a protective system of jetties and tetrapod barriers.

Elsewhere in Germany, the century-old tradition of groin construction and maintenance has been set aside in favor of nourishment. The method has been followed on the shores of the Frisian Islands. The first attempt was made on Norderney during 1951 and 1952, and followed by a second larger intervention in 1957.[46] From then on, artificial accretion was used on several beaches: Fohr (1963), Baltrum (1968), Borkum (1969 and 1970), Langeoog (1971), and Westerland-Sylt (1972). None, however, involved a volume of material as impressive as in Belgium.

France
In the Huttes area of the Medoc (southwest France), a groin built in 1854 was rapidly broken and carried out to sea. With an average retreat of the beach of 10 m/year (10.93 yards/year), the beach requires artificial nourishing and is reprofiled each year (fig. 21). One of the earliest beach-nourishment schemes was undertaken near Cannes (Mediterranean coast) on the fashionable La

45. R. H. Willmington, "The Renourishment of Bournemouth Beaches, 1974–1975," in *Conference on Shoreline Protection* (London: The Institution of Civil Engineers [ICE] and Thomas Telford Ltd., 1983), pp. 157–62.

46. Dette (n. 30 above); H. H. Dette and J. Gartner, "Time History of a Seawall on the Island of Sylt," in Kraus, ed., *Coastal Sediments '87*, pp. 1006–22; J. Kramer, "Artificial Beach Nourishment on the German North Sea Coast," in *Proceedings of the 11th Conference on Coastal Engineering,* June 24–28, 1974, Copenhagen, ed. Billy Edge (New York: American Society of Civil Engineers, 1974), vol. 2, pp. 1465–83; H. Kunz, "History of Seawalls and Revetments on the Island of Norderney," in Kraus, ed., *Coastal Sediments '87*, pp. 974–89; and H. Kunz, "Shoreline Protection of the East Frisian Islands of Norderney and Langeoog," in *Proceedings of the 5th Symposium on Coastal and Ocean Management* (n. 34 above).

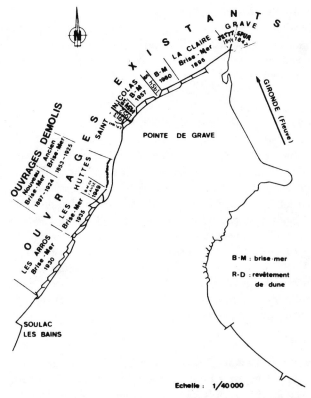

Fig. 21.—Shoreline protection, Pointe de Grave, France. After Laboratoire Central d'Hydraulique de France, 1979. Redrawn and modified by R. H. Charlier.

Croisette Beach, where 100,000 m³ (130,000 cu. yd.) of sand was placed during 1961 and 1962.

Scheveningen, The Netherlands

Westerly gales reduced the width of Scheveningen, the most celebrated beach in The Netherlands (near The Hague) to the extent that at high tide and under bad weather conditions, the sea washed over and against the boulevard. Beach widening became a necessity for both economic and safety reasons.

Here sand is dredged 20 km (12.5 miles) inland and taken by ship to a pontoon which connects a pipeline to the beach. The sand is pumped from the ship through the pipe to the beach. At Goeree, beach nourishment in 1971 involved 600,000 m³ (787,500 cu. yd.) of sand.

Soviet Union

Various beach-nourishment schemes have been implemented in the European part of the Soviet Union, for instance, at Sochi, Sukhumi Bay,

Gelendzhik, and Planerskoye (Black Sea), and on the Sea of Azov. Beaches south of Sochi have narrowed or disappeared altogether due to port construction. Remedial action consisted of building groins, submarine breakwaters, and artificial replenishment. At Odessa, 52% of the nourishment material was lost over a 7-year span.[47]

Australia and New Zealand

Restoration of the Melbourne coastal beaches was carried out by piping sand from the floor of Port Phillip Bay. On the Gold Coast, Queensland, beach nourishment was used to reconstruct beaches south of Brisbane.[48]

In New Zealand, severe beach erosion at Balaena Beach, Wellington, demanded protective action and restoration. The choice was made to rebuild the beach artificially in 1981 and 1982, by using dredged sandy granular gravel. The beach was monitored between 1982 and 1984. Nearly all the nourished material was retained in the littoral zone during this period, notwithstanding considerable sediment mobility in response to locally generated wind waves. Long-term beach erosion affects, for instance, Waihi Beach and Ohiwa Spit (both on Bay of Plenty coast), while anthropic action is responsible for having enhanced progressive erosion at Omaha Beach. Protective and corrective action appears mandatory.[49]

Rio de Janeiro

Famed Copacabana Beach, Rio de Janeiro, has been subject to erosion for some time. Charlier reported early signs of an accelerating problem in 1967. In response to a study of a physical model which considered the characteris-

47. V. V. Dodin and V. V. Ponomarenko, "Dynamics of Artificial Beaches for Conditions in Odessa," in *Geologiya poberezhya i dna chernogo morya*, no. 6 (Kiev: Kiev University Press, 1972), pp. 145–54; E. K. Grechishchev, L. A. Morozov, and Y. S. Shul'gin, "Artificial Sand Beaches and Organization of Their Aggradation," in *Ukreplenie morskikh beregov* (Moscow: Moskva Transport Publishers, 1972), pp. 60–68; V. P. Zenkovich and A. G. Kiknadze, "The Sea Coast Investigations in Georgia," in *Man and Nature in Geographical Science* (Tbilisi: Georgian Soviet Socialist Republic Academy of Science, 1981); and V. P. Zenkovich, "Geomorphological Problems of Protecting the Caucasian Coast," *Journal of Geography* 139 (1973): 460–66.

48. W. S. Andrew, "Entrance management and beach restoration," *Shore and Beach* 53 (1981); and J. A. Quilty and A. H. Wearne, "Evaluation of a Phase of Extreme Coastline Erosion," *Journal of the Soil Conservation Service of New South Wales* 31 (1982): 179–92.

49. L. Carter and J. S. Mitchel, "Stability of an Artificially Nourished Beach, Balaena Beach, Wellington, N. Z.," *New Zealand Journal of Marine and Freshwater Research* 19, no. 4 (1985): 535–52; K. G. Harray and T. Healy, "Beach Erosion at Waihi Beach, Bay of Plenty, New Zealand," *New Zealand Journal of Marine and Freshwater Research* 12, no. 2 (1978): 99–107; and T. Healy, "Conservation and Management of Coastal Resources: The Earth Science Basis," chap. 14 in *The Land, Our Future: Essays on Land Use and Conservation in New Zealand*, ed. A. G. Anderson (Wellington: Longman Paul/New Zealand Geographical Society Inc., 1980), pp. 239–60.

tics of wind, waves, tides, sand, and bathymetric conditions, it was considered that artificial nourishment would be the most appropriate action to take.[50] A mixed approach was decided upon, and 2 million m^3 (2,625,000 cu. yd.) of sand was dumped offshore while another 1.5 million m^3 (1,967,000 cu. yd.) was deposited on Copacabana Beach itself over a distance of 4.2 km (2.6 miles). This allowed the beach to widen from its remaining 55 m (179 feet) to about 140 m (455 feet).

United States

A project to restore 25 km (15.5 miles) of coast in the Miami Beach area has been priced at US$65 million. At Westhampton Beach, New York, an original proposal to install a very limited number of groins and deposit a large amount of sand on the beach met with local resistance. Numerous groins were built and successive storms over a 5-year period removed the sand from areas downdrift of the groins. Houses have toppled into the sea as the high tide pushes its way under the shoreline buildings. This has also occurred in other nearby regions.

In Virginia Beach, Virginia, beach nourishment is required each year and 28,999 m^3 (38,000 cu. yd.) of sand is placed on the beach at a cost of US$1.5 million. A similar solution has been suggested which would double the width of the beach, but it requires huge expenditures to truck in 2 million m^3 (2,625,000 cu. yd.) of sand.

Seawalls have been prohibited in the state of Maine, and so have similar bulkheads. The prohibition is an effort to force better planning in coastal development and to halt interference with the natural systems. Apparently other states will follow suit. In some cases, however, seawalls have saved lives, as in Galveston, Texas, but they do not solve the coastal erosion problem because seawalls reflect waves and they are, in fact, self-destructing.

"The beach is inherently a dynamic system and any attempt to make it stable by . . . construction of groins, seawalls, bulkheads, and other engineering devices . . . is ultimately self-defeating."[51]

Other areas in the United States where beach nourishment was selected as a means of remedying coastal erosion include the coast of Washington state (Ediz Hook), Coronado Strand (San Diego), and downcoast from the Channel Islands Harbor in California, Sandy Hook (New Jersey), Presque Isle Peninsula (on Lake Erie), the famed Waikiki Beach (Hawaii), and Cape Hatteras (North Carolina).

Successful restoration of the Sandy Hook Beach by nourishment has been repeated at several other locations on the southeast Atlantic and the

50. D. Vera-Cruz, "Artificial Nourishment of Copacabana Beach," in Edge, ed. (n. 19 above), vol. 2, pp. 1451–61.

51. J. Craig Potter, Assistant Secretary of the Interior, at the Cities on the Beach conference, Washington, D.C., 1986.

Gulf of Mexico coasts (Huntington Island, S.C., Cape Canaveral Beach, Fla., Harrison County, Miss.). Spectacular results were achieved at Key West, Florida, where a 203-mm (8-inch) blanket of sand was deposited over a 405-mm (16-inch) layer of crushed limestone.[52] Virginia Beach, Virginia, fared less well, and the problem has been ascribed to the fact that the nourishment ("borrowed") material was finer and not as well sorted as the original ("native") beach material.

On Key West, gravel 8.35 mm (3/8 inch) in diameter, so-called pea sized, was placed on the beach and gradually intermixed with fine sand during wave storms. The area between the existing groins was filled to capacity immediately; this allowed natural sand to bypass and continue deposition processes downdrift. The coarser sand is more likely to resist storm-induced displacement.

In the state of Georgia, a combination of hard and soft counter-erosion measures was taken. Tybee Island has a history of shifting shorelines and severe coastal erosion.[53] Groins, seawalls, and revetments have been built over the years. However, more recently, beach nourishment was carried out as an alternative and complementary approach. Beach nourishment was also carried out further south at Hallandale, Florida. There, the project was subjected to a thorough ecological impact study which showed that 7 years after the operation had been completed no ill effects on the fish life were evident.

Close to 100 sites in the United States have been the beneficiaries of federal government beach-rebuilding projects with use of dredged material. It has been shown that such use of dredged material has the least negative environmental impact.

Projects under Study

India
Beach nourishment has been recommended for 300 km (186 miles) of severely eroding coastline in southwest India for the past 15 years; a short segment has been restored in this manner at Purakkad.

Lazio Coast, Italy
A feasibility study for coastal protection works on the Lazio Coast showed that beach nourishment could be considered as an effective and preferable alternative to the traditional building of detached breakwaters. Appropriate sand

52. See n. 37 above.
53. J. F. Oertelm, J. E. Fowler, and J. Pope, *History of Erosion and Erosion Control Efforts at Tybee Island, Ga.,* U.S. Army Coastal Engineering Research Center, report CERC 85-1, 1985.

qualities are required to contribute to predominant longshore transport. A future maintenance program of the beach based on a 5-year cycle has been suggested.

The Lido di Ostia, near Rome, is an important coastal resort for Italian vacationers, but coastal erosion has severely reduced the beach area. A feasibility study conducted by HAECON N.V. (Ghent) recommended the monitoring of the dredging carried out on the nearby coastal zone, and a beach-nourishment scheme which included the use of 2.6 million m^3 (3.4 million cu. yd.) of sand. In its *Progetto Littorale 1983* study, the City of Rome drew up plans for new developments along the "Roman beach area."[54] The plans consider the establishment of a marina at Isola Sacra. Material removed from there, and the additional sand from marine sources, could be used for a beach-nourishment scheme to restore the damaged Fiumicino beaches.

Other Locations

Examples of other sites for which beach nourishment has been recommended as a deterrent for coastal erosion are Amanohashidate Beach in Japan[55] and Folly Beach in South Carolina.[56] In the latter location, beach nourishment over a distance of 5,140 m (3,187 yd.) involving over 522,000 m^3 (686,840 cu. yd.) of material is proposed to restore 9.5 ha (23 acres) of beach. The feasibility study points to the following impacts: temporary disturbance of the benthos, increase in water turbidity, decline of aesthetic values, increase of automobile emissions and noise levels. All are temporary, however.

In Japan, 66 beaches have artificial nourishment plans, and 21 others have already commenced restoration by this method.[57]

In Indonesia, the severely threatened coast of south Bali is the subject of proposed restoration, maintenance, and protection of its beaches. Erosion results from wave attack, airport and hotel construction, and coral removal for decorative or construction purposes. Existing protection structures are to be rehabilitated and integrated in the new scheme, and the project combines traditional hard structures with extensive beach nourishment.

54. City of Rome, *Progetto Littorale (Spiaggia di Roma)* (study re "Roman Beach Area") Roma, Citta di Roma (an in-house report), 1983.

55. Yajinia et al. (n. 25 above).

56. *Beach Erosion Control and Hurricane Protection* (Charleston, S.C.: Army Engineers District, 1985), app. 2; T. W. Kana and M. L. Williams, "Beach Nourishment at Myrtle Beach, South Carolina: An Overview," *Proceedings of the 5th Symposium on Coastal and Ocean Management* (n. 34 above); and K. F. Nordstrom and J. R. Allen, "Geomorphologically Compatible Solutions to Beach Erosion," *Zeitschrift für Geomorphologie*, suppl. 34 (1980): 142–45.

57. Y. Nakayama et al., "Construction of Artificial Beaches in Japan," *Civil Engineering in Japan* 21 (1982): 100–13.

SOCIAL ACTION

Soft action against coastal erosion can be buttressed by a series of socioeconomic measures. Their type and extent should be determined by a social impact assessment (SIA). They include minimizing the danger of damage by allowing buildings only back of the foredune crest, denying flood and disaster insurance and utilities service in a delineated area, and rezoning areas which contribute to the erosion problem.

The issue of "disappearing beaches" has been the subject of considerable concern for a long time. Charlier among many others discussed Long Island shore problems and drew parallels more than 3 decades ago, but it is during the last 10 years that concern has escalated.[58] The vulnerability of U.S. barrier islands and beaches was emblazoned at an Annapolis workshop.[59] Increased demand for and use of beaches and dunes triggered a reappraisal of laissez-faire attitudes,[60] and measures once thought in the United States to be in the best interests of both public and property owners were discontinued.[61]

A recent study commissioned by the International Geographical Union concluded that further research is needed to formulate an appropriate overall plan of action for erosion control.[62] Three-quarters of the world's sandy coastline is in retreat, some at a rate of a meter a year. A combination of factors is now destroying beaches that built up in the thousands of years following the end of the latest ice age.

In conclusion, coastal erosion progresses inexorably, and nature will have its way. A sea-level rise ranging from 0.6–3.3 m (2–11 feet) is predicted for the year 2010.[63] Man, however, can mitigate the effects of the process and to a considerable extent help natural factors preserve the beaches.

58. Charlier, in *Professional Geographer* (n. 1 above), and "Shifting Sands" (n. 20 above).

59. Barrier Islands Workshop, Annapolis, *Barrier Islands and Beaches* (Washington, D.C.: Conservation Foundation, 1976).

60. E. J. Kahn III, "The Seashore's Secrets," *Boston Magazine* 77 (June 1985): 140–41, 150–52, 154; and articles referred to in note 13 above.

61. R. D. Behn and M. A. Clark, "The Termination of Beach Erosion Control at Cape Hatteras," *Public Policy* 27 (Winter 1979): 99–127.

62. International Geographic Union, Commission on Marine Geography. Secretary, H. D. Smith, University of Wales, Institute of Science and Technology, Colum Drive, Cardiff CF1 3EU, United Kingdom.

63. See Lynne Rodgers-Miller and John E. Bardach, "In Face of a Rising Sea," *Ocean Yearbook 7*, ed. Elisabeth Mann Borgese, Norton Ginsburg, and Joseph R. Morgan (Chicago: University of Chicago Press, 1988), pp. 177–90.

The Peaceful Purposes Reservation of the UN Convention on the Law of the Sea*

Boleslaw Adam Boczek
Kent State University

I. INTRODUCTION

Historically, the oceans have not only played a crucial role in facilitating international communication and providing a source of living and nonliving resources but have also served as a battleground for innumerable armed conflicts and an area for projecting naval power as a tool for achieving political goals. The oceans have traditionally been used for peaceful as well as nonpeaceful purposes. In the more recent period, the emergence of the bipolar balance of terror, combined with the rapidly advancing naval warfare technology, has favored increasing utilization of the oceans for military purposes.

This inquiry focuses on the peaceful purposes clauses of the 1982 United Nations Convention on the Law of the Sea.[1] It is useful to bear in mind the following list of major categories of military activities in the oceans:[2] (1) navi-

---

* EDITORS' NOTE.—This is a condensed version of the paper prepared for Pacem in Maribus XV, Malta, September 7–11, 1987. An earlier article to which readers are referred is Boleslaw A. Boczek, "The Concept of Regime and the Protection and Preservation of the Marine Environment," *Ocean Yearbook 6,* ed. Elisabeth Mann Borgese and Norton Ginsburg (Chicago: University of Chicago Press, 1986), pp. 271–97.

1. *The Law of the Sea: Official Text of the United Nations Convention on the Law of the Sea with Annexes and Index* (New York: United Nations, 1983), Sales n. E. 83.v.5. (hereafter referred to as "the Convention").

2. This list is based on the following literature: for a useful chronology of naval nuclear developments in the years 1946–85, see William M. Arkin et al., "Naval Nuclear Weapons Developments: A Selective Chronology," in *The Denuclearisation of the Oceans,* ed. R. B. Byers (New York: St. Martin's Press, 1986), pp. 237–63. The geostrategic environment of the nuclearization of the oceans is reviewed in William M. Arkin et al., "Ocean Space and Nuclear Weapons: The Geostrategic Environment," in Byers, ed., pp. 21–40. The extent of nuclearization in terms of nuclear capabilities, roles, and missions of the nuclear powers is presented in Williams M. Arkin et al., "The Nuclearisation of the Oceans: Roles, Missions and Capabilities," in Byers, ed., pp. 41–74. For some other analyses see, e.g., Jonathan Alford, "Some Reflections on Technol-

---

gation on the surface and in the water column (and overflight), including routine cruises, naval maneuvers, and other exercises with or without weapons tests and use of explosives, and projecting "naval presence" as an instrument of foreign policy ("gunboat diplomacy");[3] (2) providing strategic deterrence in the form of nuclear ballistic missile submarines; (3) surveillance of the potential adversaries' naval and other military activities, of which anti-submarine warfare (ASW) forms an essential part (with the use of various seabed-based devices such as sonars and other acoustic detection systems); (4) emplacement of navigation and communication devices in the sea and on the seabed; (5) emplacement of conventional weapons such as mines; (6) military research; and (7) logistical support, including maintaining naval bases.

The progressing militarization, and especially nuclearization, of the oceans raises the question of whether and to what extent international law imposes any limitations on the military use of the sea in peacetime; in other words, What is the permissible scope of such activities from the international legal point of view? The matter attracted the attention of the international community in connection with nuclear tests on the high seas and then, after 1967, was closely related to the debates in the UN Sea-Bed Committee and the negotiations for the conclusion of the Sea-Bed Arms Control Treaty in 1971.[4] Consequently, much of the pertinent literature of that period focused on the legal aspects of the military uses of the seabed.[5]

---

ogy and Sea Power," 38 *International Journal* (1983): 397–408; Neville Brown, "Military Uses of the Ocean Floor," in *Pacem in Maribus I* (Malta: Royal University of Malta Press, 1971), pp. 115–21; Roger Villar, "Weapon Developments in the 1980s: Sea," in *1982 RUSI & Brassey Defence Yearbook* (London: Royal United Services Institute, 1982), pp. 18–193; O. Wilkes, "Ocean-based Nuclear Deterrent and Anti-submarine Warfare," *Ocean Yearbook 2*, ed. Elisabeth Mann Borgese and Norton Ginsburg (Chicago: University of Chicago Press, 1980), pp. 226–49. Military uses of the seabed are summarized, with rich references to further literature, in Patrizio Merciai, "La démilitarisation des fonds marins," 88 *Revue générale de droit internationale public (RGDIP)* (1984): 46–113. See also Rex J. Zedalis, " 'Peaceful Purposes' and Other Relevant Provisions of the Revised Composite Negotiating Text: A Comparative Analysis of the Existing and Proposed Military Regime for the High Seas," 7 *Syracuse Journal of International Law and Commerce* (1979): 1–35; and Rudiger Wolfrum, "Restricting the Use of the Sea to Peaceful Purposes: Demilitarization in Being?" 24 *German Yearbook of International Law (GYIL)* (1982): 200–41.

3. The latter aspect of the use of the oceans has been widely discussed in politico-strategic literature. For a recent analysis of naval strategy within the context of the Convention see Ken Booth, *Law, Force, and Diplomacy at Sea* (London: Allen & Unwin, 1985), esp. bibliog.

4. Treaty on the Prohibition of the Emplacement of Nuclear Weapons and Others Weapons of Mass Destruction on the Sea-Bed and the Ocean Floor and in the Subsoil Thereof (hereafter cited as the Sea-Bed Arms Control Treaty), February 11, 1971, 955 *United Nations Treaty Series (UNTS)*, p. 115; also in *International Legal Materials (ILM)* 10 (1971): 145.

5. See, e.g., Jens Evensen, "The Military Uses of the Deep Ocean Floor and Its Subsoil," in *Proceedings of the Symposium on the International Legal Regime of the Sea-Bed,*

At the Third United Nations Conference on the Law of the Sea (1973–82) (UNCLOS III), attention shifted to the resource aspects of the deep seabed and other areas of the ocean, as well as to the navigational rights of the maritime nations concerned about the mobility of their navies. In general, military aspects of the sea were a neglected issue at UNCLOS III, and yet its final product does include a series of very concise provisions on the use of the ocean for peaceful purposes. This article will analyze and interpret the meaning of these clauses of the Convention. The crucial question is whether they add anything to existing general and conventional international law governing the military uses of the oceans in time of peace. Do these sweeping clauses demilitarize the oceans in the way that the Antarctic Treaty demilitarized the Southern continent or do they otherwise impose any limitations on the traditional military uses of the sea?[6]

II. ANTECEDENTS OF THE PEACEFUL PURPOSES RESERVATION

In order to shed light on the meaning of the term "peaceful purposes" in the Convention it is logical to examine those international legal instruments that use this term and can provide some guidance in the interpretation of this rather vague notion. One has to remember, however, that the term must be interpreted separately within the context of each specific document and analogies must not be rashly drawn.

A. The Antarctic Treaty

The term "peaceful purposes" appeared first in the Antarctic Treaty, whose preamble recognizes "the interest of all mankind that Antarctica shall continue forever to be used exclusively for peaceful purposes," and refers to the treaty as a whole as "ensuring the use of Antarctica for peaceful purposes only."[7] In the dispositive part of the treaty this principle is followed by specific prohibition of "any measures of a military nature" some of which are listed, inter alia, as examples, namely establishment of military bases and fortifications, the carrying out of military maneuvers, as well as testing of any

---

ed. L. Stucki (1970), pp. 535–56; R. W. Gehring, "Legal Rules Affecting Military Uses of the Sea," *Military Review* (1971), pp. 168–224; F. Krüger-Sprengel, "Militärische Aspekte der Nutzung des Meeresbodens," in *Die Nutzung des Meeresgrundes ausserhalb des Festlandsockels (Tiefsee)* (Leyden, 1970); W.Kühne, *Das Völkerrecht und die militärische Nutzung des Meeresbodens* (Leyden, 1975); and P. S. Rao, "Legal Regulation of Maritime Military Uses," 13 *Indian Journal of International Law* (1973): 425–54.

6. See Sec. II A of this article. This is also discussed by Joseph R. Morgan, "Naval Operations in the Antarctic Region: A Possibility?" in this volume, pp. 362–377.

7. The Antarctic Treaty, December 1, 1959, 402 *UNTS* 71.

type of weapons.[8] However, the treaty specifically permits the use of military personnel or equipment for scientific research or for any other peaceful purposes.[9] The treaty also prohibits nuclear explosions and disposal of radioactive waste,[10] and it provides for a system of verification.[11]

Although the treaty does not give any clear and explicit definition of peaceful purposes, its language, and especially the exceptions from the general prohibition of military activities, demonstrates that under the Antarctic Treaty the phrase means complete demilitarization of the Antarctic.[12] However, even though the peaceful purposes reservation defines and governs the treaty area, that is, south of 60° South Latitude,[13] the legal status of the waters surrounding the continent is subordinated to the preservation of the High Seas freedoms.[14] While the matter is not entirely free of controversy, this would mean that demilitarization did not apply to the High Seas area of the Antarctic waters (whatever that area is). Hence, for example, one could argue that military maneuvers in these waters, carried out by a treaty party, would not contravene the treaty.[15] However, such an interpretation would run counter to the spirit of the Antarctic Treaty and must be rejected.

B. The Outer Space and Moon Treaties

The demilitarization of Antarctica evidently influenced the legal nature of outer space as stipulated by the Outer Space Treaty of 1967,[16] but the language of this treaty does not necessarily decree the demilitarization of outer space as such, as does the Antarctic model. It must be borne in mind that two

8. Antarctic Treaty, Art. 1(1). For a comprehensive examination of this clause see F. M. Auburn, *Antarctic Law and Politics* (Bloomington: Indiana University Press, 1982), pp. 94–98.

9. Antarctic Treaty, Art. I(2).

10. Ibid., Art. 5(1).

11. Ibid., Art. 7.

12. Earlier examples of demilitarization of certain areas include the Treaty Relating to Spitzbergen, February 9, 1920, 2 *League of Nations Treaty Series (LNTS)* 7; and the Convention Concerning the Non-fortification and Neutralization of Aaland Islands, October 20, 1921, 9 *LNTS* 217.

13. Antarctic Treaty, Art. 6.

14. Ibid.

15. W. M. Bush, *Antarctica and International Law: A Collection of Inter-State and National Documents* (New York: Oceania, 1982), p. 67.

16. Treaty on Principles Governing the Activities of States in the Exploration and Use of Outer Space, Including the Moon and Other Celestial Bodies, January 27, 1967, 610 *UNTS* 205 (hereafter referred to as Outer Space Treaty). Readers are referred to an article by Nicholas Mateesco Matte, "The Law of the Sea and Outer Space: A Comparative Survey of Specific Issues," *Ocean Yearbook 3*, ed. Elisabeth Mann Borgese and Norton Ginsburg (Chicago: University of Chicago Press, 1982), pp. 13–37.

different legal regimes apply to two different zones in outer space: the moon and other celestial bodies on the one hand and outer space around the earth as such, on the other. Specifically, the peaceful purposes reservation governs only the moon and other celestial bodies and not outer space as such.[17] As in the Antarctic Treaty,[18] the reservation is followed by specific prohibitions: the establishment of military bases, installations, and fortifications; the testing of any type of weapons; and the conduct of military maneuvers. But, unlike the Antarctic Treaty, this specification appears to be exhaustive. The use of military personnel and equipment for peaceful purposes is, after the Antarctic pattern, allowed.[19] It can be argued, that, although unlike the Antarctic Treaty, the peaceful purposes clause relating to the moon and other celestial bodies does not broadly prohibit "any measures of a military nature" but only specific military activities, it still has to be interpreted, in accordance with the underlying policy objectives, as denoting complete demilitarization of the moon and other celestial bodies.[20]

Outer space proper is not demilitarized as such since the Outer Space Treaty only prohibits placing in orbit around the Earth any objects carrying nuclear weapons or any other kinds of weapons of mass destruction, installing such weapons on celestial bodies, or stationing them in outer space in any other manner.[21] Yet despite the limited scope of the treaty's language it can be argued that, on the basis of the phrases referring to outer space as a "province of all mankind" and exploration and use of outer space "for the benefit and in the interests of all countries,"[22] any military activity anywhere in outer space contravenes the treaty.[23] On the other hand, the view maintaining complete demilitarization of outer space conflicts with the position of the United States and the Soviet Union in the drafting stages of the treaty and international practice allowing placing military reconnaissance satellites in orbit.[24]

Consistent with the regime of the moon and other celestial bodies under the Outer Space Treaty, the Moon Treaty of 1979 dedicates these areas exclusively for peaceful purposes.[25]

17. Outer Space Treaty, Art. 4(2).

18. See Sec. II A of this article.

19. Outer Space Treaty, Art. 4(2).

20. Carl Q. Christol, *The Modern International Law of Outer Space* pp. 25–26; Eric Stein, "Legal Restraints in Modern Arms Control Agreements," 66 *American Journal of International Law (AJIL)* (1972): 255, 260–61.

21. Outer Space Treaty, Art. 4(1).

22. Outer Space Treaty, Art. 1.

23. See Motte (ed.), *Space Activities and Emerging International Law* (Montreal: McGill University, Institute and Centre of Air and Space Law, 1984), p. 290.

24. See Stein, pp. 262–63.

25. Agreement Concerning the Activities of States on the Moon and Other Celestial Bodies (hereafter referred to as the Moon Treaty), Art. 3 in 18 *International Legal Materials (ILM)* (1979): 1434. The Moon Treaty entered into effect on July 11, 1984. See Carl Q. Christol, "The Moon Treaty Enters into Force," 79 *AJIL* (1985): 163.

C. The Sea-Bed Arms Control Treaty

The adoption of the Outer Space Treaty in 1967 coincided with the onset of a great debate in the United Nations and other bodies on the future legal regulation of the seabed and ocean floor beyond the limits of national jurisdiction. The debate was precipitated not only by concern about competition over the resources of the seabed but also by the hope of preventing or at least limiting the use for military purposes of vast and important areas of the ocean. It was perhaps not a coincidence that the peaceful purposes reservation appeared at that time in the Maltese proposal for a "declaration and treaty concerning the reservation exclusively for peaceful purposes of the seabed and the ocean floor, underlying the seas beyond the limits of present national jurisdiction, and the use of their resources in the interests of mankind,"[26] subsequently so eloquently championed by Ambassador Arvid Pardo in his famous "common heritage of mankind" address before the first committee of the United Nations General Assembly.[27] Two conflicting approaches to the construction of the term "peaceful purposes" crystallized in the General Assembly's Sea-Bed Committee which adopted the Declaration of Principles in 1970,[28] and subsequently in the negotiations on the Sea-Bed Arms Control Treaty of 1971.[29]

In the Sea-Bed Committee, numerous states, and in particular the developing countries and the Soviet Union and its allies, favored a total demilitarization of the seabed and ocean floor after the Antarctic model. In their view, no military use whatever, even for clearly defensive purposes, could be regarded as peaceful, a position considered consistent with the declaration of the seabed as a common heritage of mankind excluded from appropriation by any states or persons, according to the argument that something owned and managed in common cannot be used for military purposes by one party alone.[30] On the other hand, most Western nations, and especially the United States and the United Kingdom, insisted on the interpretation of the use for exclusively peaceful purposes as meaning the use that is consistent with the

---

26. See the accompanying Memorandum of the Maltese Government, UN document A/6695, August 18, 1967.

27. UN document A/C.1/P.V. 1515 and P.V. 1516. Readers are also referred to an article by Arvid Pardo, "The Evolving Law of the Sea: A Critique of the Informal Composite Negotiating Text (1977)," *Ocean Yearbook 1,* ed. Elisabeth Mann Borgese and Norton Ginsburg (Chicago: University of Chicago Press, 1978), pp. 9–37.

28. Declaration of Principles Governing the Sea-Bed and the Ocean Floor, and the Subsoil Thereof, beyond the Limits of National Jurisdiction (hereafter cited as Declaration of Principles). UN General Assembly Resolution 2794 (XXV) (1970).

29. See n. 4 above.

30. See, e.g., statements by: the Soviet Union, UN document A/AC.125/SR.16 (1968); Bulgaria, UN document A/AC.135/SR.16 (1968); Peru, UN document A/AC.135/SR.17 (1968); and Malta, A/AC.1/PV.1582 (1968).

law of the UN Charter; consequently military activities in furtherance of the right of self-defense (in whatever way this right is construed) were peaceful.[31] The "nonmilitary" versus "nonaggressive" interpretations continued to divide the negotiating states at the Eighteen States Disarmament Conference convened to draft a treaty banning nuclear weapons from the seabed.[32] On the other hand, a certain relaxation of tension and maybe even a rapprochement of the superpowers' views set the stage for two important developments. First, the UN General Assembly adopted the famous Declaration of Principles.[33] The concept of peaceful purposes was not defined or clarified, however. Second, the negotiations at the Disarmaments Conference produced the Sea-Bed Arms Control Treaty in 1971.[34]

While recognizing in its preamble the common interest of mankind in the use of the seabed for peaceful purposes, the treaty does not clarify this term. Neither does it contain any reference to it in its dispositive provisions, calling only for continued "negotiations in good faith concerning further measures in the field of disarmament for the prevention of an arms race on the sea-bed, the ocean floor and the subsoil thereof."[35] This programmatic statement,

31. For the United States, the United Kingdom, and Norway, see UN document A/AC.135/SR.17 (1968) (n. 30 above).

32. For the Soviet position see the Soviet draft treaty prohibiting the use of the seabed for military purposes. UN document Endc/240 (1969). For Soviet comments on this issue at that time see, e.g., G. F. Kalinkin, "Ob ispol'zovanii morskogo dna iskluchitelno v mirnykh tsel'akh" (On the use of the seabed exclusively for peaceful purposes), *Sovetskoe gosudarstvo i pravo*, no. 10 (1969), pp. 117–22; G. M. Melkov, "Yuridicheskoe znachenie termina 'iskluchitel'no v mirnykh tsel'akh'" (The legal meaning of the term "exclusively for peaceful purposes"), *Sovetskiy Ezhegodnik Mezhdunarodnogo Prava* 1971 (1973): 153–60; V. I. Vaneyev, "Demilitarizatsia dna morey i okeanov" (Demilitarization of sea and ocean bed), in *Strategia imperializma i bor'ba SSSR za mir i razoruzhenie*, ed. V. Ya. Aboltin et al. (1974):258–72. As discussed below in Sec. IV A of this article ("The Specific Reservations: Debate in the Plenary"), the Soviet Union changed its position at UNCLOS III, virtually joining the United States against the Third World demands for demilitarization of the oceans. For the U.S. draft of the seabed treaty see UN document Endc/249 (1969).

33. The Declaration of Principles (n. 28 above) proclaimed the area of "the seabed and ocean floor, and the subsoil thereof, beyond the limits of national jurisdiction, ... as well as the resources of the area" as the common heritage of mankind, reserved exclusively for peaceful purposes.

34. Sea-Bed Arms Control Treaty (n. 4 above). The early drafts of the treaty are reprinted in *ILM* 8 (1969): 659 and *ILM* 9 (1970): 392. For a summary review of the negotiations leading to the conclusion of the treaty see Merciai (n. 2 above), pp. 68–72; Wolfrum (n. 2 above), pp. 220–23; and, generally, L. Migliorino, *Fondi marini e armi di distruzione di massa* (Milan: Giuffré, 1980). Only El Salvador and Peru voted against the treaty. Ecuador and France abstained and 19 other countries were absent (*ILM* 10 [1971]: 151). Documents leading to the treaty are collected in U.S. Arms Control and Disarmament Agency, *International Negotiations on the Seabed Arms Control Treaty* (Washington, D.C.: Government Printing Office, 1973).

35. Sea-Bed Arms Control Treaty, Art. 5. It has been emphasized that this clause introduces into the treaty a dynamic aspect of time dimension indicating clearly that

combined with the conviction of the drafters in the preamble that the treaty constitutes only "a step towards the exclusion of the sea-bed, the ocean floor and the subsoil thereof, from the arms race," must lead to the logical conclusion that the use in the preamble of the phrase "peaceful purposes" cannot refer to an actually imposed legal obligation to use the seas for nonmilitary purposes but only projects a policy goal for the future. Moreover, the Sea-Bed Arms Control Treaty bans only certain specific military activities, does not cover the offshore marine zone, and raises some problems of verification and enforcement.

D. General Assessment

The general conclusion to be drawn from examination of the peaceful purposes clauses in international instruments must be that no agreed understanding of the notion of peaceful purposes emerged prior to the convening of UNCLOS III, and that the term "peaceful purposes" must be construed within the context and circumstances of each specific instrument in which it is employed. It might be added that no clarification can be obtained from other treaties that use the term, such as the Nonproliferation Treaty (1968) which, inter alia, reserves the parties' right to develop research, production, and use of nuclear energy "for peaceful purposes."[36]

As far as the application of the concept of peaceful purposes to the oceans is concerned, no mention of it can be found in the Geneva Conventions of 1958. The Sea-Bed Arms Control Treaty seems to be the only purely marine-oriented international agreement in force that uses the term "for peaceful purposes" (and that only in the preamble).

Yet, one must not overlook the fact that the ideological origins of the Sea-Bed Arms Control Treaty as well as the roots of UNCLOS III can be traced to the early concepts of the common heritage of mankind and the use of the seabed for the benefit of mankind as a whole, principles which can be used in support of the thesis that at least bona fide attempts should be made for excluding this area from both offensive and defensive national military activities.

The contrast between the "nonmilitary" interpretation and the one that excludes only those military activities which are in conflict with the UN Char-

---

the ultimate objective of the process, of which the treaty is only one stage, is demilitarization of the seabed. See Elisabeth Mann Borgese, "The Sea-Bed Treaty and the Law of the Sea," in Byers, ed. (n. 2 above), pp. 88, 91.

36. Treaty on the Non-proliferation of Nuclear Weapons (hereafter cited as the Nonproliferation Treaty), July 1, 1968, Art. 4(1), 729 *UNTS* 161; *ILM* 7 (1968): 811. Cf. the Statute of the International Atomic Energy Agency, October 26, 1956, 276 *UNTS* 3.

ter prescribing threat or use of force, reemerged, this time within a wider marine context, at UNCLOS III.

III. BINDING LEGAL RESTRAINTS UPON MILITARY USES OF THE SEA

A. In General

All peacetime military uses of the seas are subject to the general UN Charter ban on the threat or use of force in international relations "against the territorial integrity or political independence of any State, or in any other manner inconsistent with the Purposes of the United Nations."[37] There is, therefore, universal agreement that any military activity at sea that is incompatible with this fundamental principle of contemporary international law is not considered to be for peaceful purposes. Although not explicitly stated in any of the 1958 Geneva Conventions, the peaceful purposes reservation in this sense has, since 1945, been implicitly recognized in its application to the activities of states in the oceans. The question, however, is whether beyond this obvious, very general, ambiguous, and controversial principle there are any further legal restraints on military activities at sea under today's international law.[38]

The answer is that there are some restraints on states' peacetime military activities at sea under international law. They are of two categories: conventional and customary (general) international law.

B. Conventional Restraints

Conventional restraints may be either global, regional, or bilateral. There may be treaties imposing quantitative limitations on the naval arsenals of the parties concerned. At this time there are no such treaties in force. The Washington Naval Armaments Treaty of 1922 and the US-USSR SALT I agreement on offensive missiles are only historical examples.[39] However, naval disarmament has been on the agenda of the UN Disarmament Commission, more recently at its 1987 session, and a study on the naval arms race has been prepared for its review by the United Nations.[40] Another category of interna-

---

37. Nonproliferation Treaty, Art. 2(4).

38. Further elaboration of this fundamental problem is, of course, far beyond the scope of the present inquiry.

39. Treaty for the Limitation of Naval Armament, February 6, 1922, 25 *LNTS* 202. For the other examples see generally R. A. Hoover, *Arms Control: The Interwar Naval Limitation Agreements*, Monographs in World Affairs, vol. 17 (Denver, Colo.: University of Denver, 1980).

40. See "Verification, Naval Disarmament Discussed by 1987 Disarmament Commission," 24 *United Nations Chronicle*, no. 3 (August 1987), p. 16.

tional agreement concerns efforts, belonging to the general category of "confidence-building measures," designed to prevent incidents at sea that might escalate into open warfare. The bilateral U.S.-Soviet Agreement of 1972 with its 1973 protocol and similar U.K.-Soviet Agreement of 1986 belong to this category.[41] Finally, there are a number of global and regional arms control agreements that prohibit testing and deployment of weapons of mass destruction and directly or indirectly apply to the oceans.

1. *Global Agreements: Restraints under the Sea-Bed Arms Control Treaty*

The Sea-Bed Arms Control Treaty of 1971 is the most important global instrument limiting military activities in the ocean. As already noted, this agreement is not a general demilitarization measure with regard to the seabed, but only a compromise first step toward such demilitarization.

Without defining the weapons concerned, the treaty prohibits emplantment or emplacement on the seabed, beyond the 12-mile limit from the coastline, of "nuclear weapons or any other type of weapons of mass destruction as well as structures, launching installations or any other facilities specifically designed for storing, testing or using such weapons."[42] It is clear that the treaty suffers from a number of gaps and ambiguities. Apart from the fact that its preamble reference to peaceful purposes cannot be interpreted to mean "nonmilitary purposes," the treaty leaves the 12-mile coastal zone, an obvious area for emplacing nuclear weapons, free for nuclearization by the coastal state or an invited ally;[43] it does not ban conventional weapons; and it even seems to allow weapons of mass destruction that are mobile and capable of navigation, such as nuclear missile submarines temporarily resting on the seabed. Actually, the treaty bans a weapon system that did not exist at the time of its conclusion and is still not likely to be developed. On the other hand, it does nothing to limit the arms race within the context of strategic deterrence at sea by means of submarines, let alone acoustic and other defensive devices used in antisubmarine warfare. Also, strictly speaking, facilities that are not "specifically designed" for storing, testing, or using weapons of mass destruction are not prohibited.[44] The treaty does not provide for an adequate system of verification,[45] and an additional handicap is that France and China are not parties.

---

41. Agreement on the Prevention of Incidents on and over the High Seas, May 25, 1972, 852 *UNTS* 151; Protocol, May 22, 1973, *ILM* 12 (1973): 1108; and U.K.-U.S.S.R., Agreement of 15 July 1986 Concerning the Prevention of Incidents at Sea beyond the Territorial Sea, *Naval Forces* 8, no. 1 (1987): 14–15.

42. Sea-Bed Arms Control Treaty, Arts. 1 and 2.

43. This rule was maintained by the Second Review Conference of the Sea-Bed Arms Control Treaty. See Eric P. J. Myjer, "The Law of Disarmament and Arms Control: Implications for the Law of the Sea," in Byers, ed. (n. 2 above), pp. 104, 108.

44. This results from Sea-Bed Arms Control Treaty, Art. 1(1).

45. For the existing system see Sea-Bed Arms Control Treaty, Art. 3.

For these reasons the Sea-Bed Arms Control Treaty does not make a significant contribution to making the oceans more peaceful. Its importance lies in the fact that it was intended as a step in negotiations leading to a more complete disarmament under effective international control. However, no such negotiations on the global level have taken place. It must be added that the conclusion in 1982 of the Convention on the Law of the Sea requires action designed to adjust the provisions of the Sea-Bed Arms Control Treaty to the new legal context created by the Convention.[46]

2. *Other Treaties*

The modest restraints under the Sea-Bed Arms Control Treaty are complemented by the previously negotiated ban on testing nuclear weapons under the Partial Test Ban Treaty of 1963.[47] This treaty bans nuclear tests, irrespective of their purpose, not only in the atmosphere and outer space but also under water in all the ocean zones. Underground tests and tests conducted "in any other environment" are banned only if the explosion "causes radioactive debris to be present outside the territorial limits of the State under whose jurisdiction or control such explosion is conducted."[48] In the prevalent opinion nuclear tests in the subsoil of the High Seas are prohibited.[49]

For reasons of comprehensiveness one should finally mention the bilateral U.S.-Soviet Treaty on Limitation of Anti-Ballistic Missile Systems of 1972 (ABM treaty) under which the two countries have undertaken not to develop, test, or deploy antiballistic missile systems that are sea-, air-, or space-based or mobile land-based.[50]

3. *Regional Agreements*

Military uses of the sea can also be limited under regional agreements. As reserved in the Sea-Bed Arms Control Treaty itself, this treaty "shall in no way affect the obligations assumed . . . under international instruments establishing zones free from nuclear weapons."[51] At the time the treaty was negoti-

---

46. This issue is analyzed in Borgese (n. 35 above).

47. Treaty Banning Nuclear Weapons Tests in the Atmosphere, in Outer Space and under Water (hereafter cited as the Partial Test Ban Treaty), August 8, 1963, 48 *UNTS* 43. See also Egon Schwelb, "The Nuclear Test Ban Treaty and International Law," 58 *AJIL* (1964): 642.

48. Partial Test Ban Treaty, Art. 1(1)(a)(b).

49. See Brown (n. 2 above), pp. 60, 73. See also Tullio Treves, "Military Installations, Structures, and Devices on the Sea-bed" (hereafter cited as "Military Installations"), 74 *AJIL* 808, 820–2.

50. Treaty on Limitation of Anti-Ballistic Missile Systems (hereafter cited as ABM Treaty), May 26, 1972, Art. V(1), *ILM* 11 (1972): 784.

51. Sea-Bed Arms Control Treaty, Art 9. Cf. the Treaty on the Non-Proliferation of Nuclear Weapons, July 1, 1968, 729 *UNTS* 161.

ated, the Tlatelolco Treaty of 1967 was one such instrument.[52] It was followed by the Rarotonga Treaty of 1985.[53]

There have been other, less successful, attempts at demilitarizing certain ocean areas, especially as sponsored by the UN General Assembly, which has endorsed the idea of establishing nuclear-free zones in various parts of the Third World; the General Assembly even designated the Indian Ocean as a zone of peace. Conflicting national strategic and political interests as well as the ambiguity of the term "zone of peace" have so far prevented the UN General Assembly resolution on the Indian Ocean from acquiring legal validity in a treaty form.[54]

52. Treaty for the Prohibition of Nuclear Weapons in Latin America (hereafter referred to as the Tlatelolco Treaty), February 14, 1967, 634 *UNTS* 281; *ILM* 6 (1967): 521. For a detailed analysis see Alfonso García Robles, "Mesure de désarmement dans les zones particulières: le Traité visant l'interdiction des armes nucléaires en Amérique latine," 133 *Recueil des Cours, Academie de Droit International (RCADI)*, no. 2 (1971): 43–134; J. R. Redick, "Regional Nuclear Arms Control in Latin America," 29 *International Organization* (1975): 415–45; Jozef Goldblat and Victor Millan, "Militarization and Arms Control in Latin America," *Stockholm International Peace Research Institute (SIPRI) Yearbook* (Cambridge, Mass.: MIT Press, 1982), pp. 393–425.

53. South Pacific Nuclear-Free Zone Treaty (hereafter referred to as the Rarotonga Treaty), with three Protocols, August 6, 1985, in *Law of the Sea Bulletin*, no. 6 (October 1985), p. 24, and reproduced in *Ocean Yearbook 6*, ed. Elisabeth Mann Borgese and Norton Ginsburg (Chicago: University of Chicago Press, 1986), pp. 594–605. The publication of the entry into force of this treaty appears in *Ocean Yearbook 7*, eds. Elisabeth Mann Borgese, Norton Ginsburg, and Joseph R. Morgan (Chicago: University of Chicago Press, 1988), p. 529. See also, United Nations, *The Law of the Sea: Current Developments in State Practice* (New York: United Nations, 1987), pp. 192–207.

54. The concept of the zone of peace is analyzed in M. Malita, "The Concept of 'Zone of Peace' in International Politics," 10 *Revue roumaine d'Etudes internationales* (1976): 680–91. Nuclear-free zones in general are reviewed in G. Delcoigne, *An Overview of Nuclear Weapon Free Zones* (Bologna: Johns Hopkins University Center, 1982). Regional approaches to denuclearization are discussed in contributions to Part 4 of Byers, ed. (n. 2 above). Zones of peace and nuclear-free zones are examined in great detail in Sandra Szurek, "Zones exemptes d'armes nucléaires et zones de paix dans le tiers-monde," 88 *RGDIP* (1984): 114–203. The UN General Assembly declaration on the Indian Ocean as a zone of peace, A/Res. 2832 (XXVI) Dec. 16, 1971, was adopted by 61 votes, with 55 abstentions (including the United States and the Soviet Union). The declaration and its follow-up are discussed in Barry Buzan, "Naval Power, the Law of the Sea, and the Indian Ocean as a Zone of Peace," 5 *Marine Policy* (1981):194–204; M. C. W. Pinto, "The Indian Ocean as a Zone of Peace," in Byers, ed. (n. 3 above), pp. 145–56.

In 1979, the UN General Assembly decided to hold a conference on the Indian Ocean. Originally scheduled for 1981, it was postponed and was then prepared by a special ad hoc committee for the year 1988. See "Ad Hoc Committee on Indian Ocean Concludes First Session in 1987," 24 *UN Chronicle*, no. 2 (1987):53. For a possibility of creating a nuclear-free zone in the Black Sea, see Dumitru Mazilu, "The Black Sea and Demilitarisation," in Byers, ed. (n. 2 above), pp. 157–63. For further information on the Indian Ocean as a zone of peace, see Stanley D. Brunn and Gerald L. Ingalls, "Voting Patterns in the UN General Assembly on Uses of the Seas," *Ocean Yearbook 7*, eds. Elisabeth Mann Borgese, Norton Ginsburg, and Joseph R. Morgan (Chicago: University of Chicago Press, 1988), pp. 52–64.

The Tlatelolco Treaty.—This Latin American treaty's ban on testing, production, acquisition or any form of possession of nuclear weapons in Latin America applies, in addition to land area, to the internal waters, the territorial sea and in general to "any other space over which the State exercises sovereignty in accordance with its legislation."[55] Unlike the Sea-Bed Arms Control Treaty, the Latin American treaty refers only to nuclear weapons, but not to weapons of mass destruction, and includes a definition of nuclear weapons.[56] Nuclear explosions for peaceful purposes are allowed under certain circumstances.[57] There is no ban on the emplacement or emplantment on the seabed of structures and installations designed to store, test or use nuclear weapons.

The Tlatelolco Treaty incorporates two protocols. Protocol I is intended to commit states having territories in Latin America.[58] Among them the Netherlands, the United Kingdom, and the United States have ratified the Protocol, but France has not yet done so. Protocol II commits nuclear weapon states, whether or not situated in the Americas.[59] China, France, the Soviet Union, the United Kingdom, and the United States have ratified it, each with a statement or a declaration.

The treaty currently applies to the waters over which the parties have sovereignty, but, projecting into a more peaceful future, the treaty envisages the time when its nuclear-free zone will extend to a much wider ocean zone, going even beyond the 200-mile claim of some Latin American countries.[60] However, for this to happen it is necessary that the treaty be ratified by all states situated in their entirety south of 35° North Latitude, with Protocol II ratified by all nuclear powers.[61] These conditions have not yet been fulfilled.[62]

The Rarotonga Treaty.—The other regional arrangement for a nuclear-free zone is the South Pacific Nuclear-Free Zone Treaty (the Rarotonga Treaty) of 1985, negotiated under the auspices of the 13 countries of the South Pacific Forum.[63] The treaty declares a nuclear-free zone covering most

55. Tlatelolco Treaty, Art. 3. The phrase "in accordance with its legislation" means that for some Latin American countries (Argentina, Brazil, Ecuador, El Salvador, Nicaragua, Panama, Peru and Uruguay) the nuclear-free zone extends 200 miles from the base lines. This, however, is without prejudice to the right of not recognizing such extravagant claims to the territorial sea. See Treves, "Military Installations," pp. 808, 826.

56. Tlatelolco Treaty, Art. 5.

57. Ibid., Art. 18.

58. Ibid., Additional Protocol I, February 14, 1967, 634 *UNTS* 362.

59. Ibid., Additional Protocol II, Feburary 14, 1967, 634 *UNTS* 364.

60. Ibid., Art. 4(2), indicating the geographical limits of the future zone. For the 200-mile claims see n. 55 above.

61. Tlatelolco Treaty, Arts. 4(2) and 28(1).

62. Cuba, Guyana, and some other Caribbean countries are not yet parties to the treaty. Argentina has not ratified it and Brazil and Chile have ratified it but, unlike the other parties, have not agreed to be bound by the treaty before the conditions of Art. 28(1) are fulfilled. As noted, France has not yet ratified Protocol I. See Merciai (n. 2 above), p. 102, n. 93.

63. Rarotonga Treaty; See n. 53 above. This treaty is analyzed in detail in

of the Pacific Ocean south of the equator and prohibiting ownership, stationing, manufacturing, or testing of nuclear explosive devices in the territories of the parties which, in addition to land, include the internal waters, the territorial sea and archipelagic waters, and the seabed and subsoil beneath.[64] Dumping of radioactive waste is also prohibited.[65] However, it is left to each individual state to decide for itself to allow visits by foreign nuclear-powered ships and ships carrying nuclear weapons to its ports, and navigation by such ships in its territorial sea and archipelagic waters. The same applies to admission of aircraft carrying nuclear weapons and their transit of the airspace of the state concerned.[66] Like the Tlatelolco Treaty, the Rarotonga Treaty incorporates protocols: Protocol I intending to commit the states having territories situated within what it designates as the South Pacific Nuclear-Free Zone (France, the United Kingdom, and the United States); and Protocols II and III designed to commit France, China, the Soviet Union, the United Kingdom, and the United States not to use or threaten to use any nuclear weapons against the parties and any territory within the zone, and not to test any nuclear weapons anywhere within the zone.[67]

C. Customary (General) International Law Restraints

In the absence of any treaty restriction, military use of the sea is governed by the principles and rules of customary (general) international law of the sea, as codified and partially developed by the 1958 Geneva Conference on the Law of the Sea, especially in the Convention on the High Seas.[68] The freedoms of navigation and overflight are among the most important freedoms of the high seas specifically listed in the 1958 convention.[69] It has always been recognized

---

Georges Fischer, "La zone dénucléarisée du Pacifique Sud," 1985 *Annuaire Français du Droit International (AFDI)* (1986): 23–57; and Elizabeth L. Gibbs, "In Furtherance of a Nuclear-Free Zone Precedent: The South Pacific Nuclear-Free Zone Treaty," 4 *Boston University International Law Journal* (1986): 387–431.

64. Rarotonga Treaty, Art. 1. This Art. also gives a definition of the "nuclear explosive device."

65. Rarotonga Treaty, Art. 7.

66. Rarotonga Treaty, Art. 5. State practice varies among the parties to the treaty. For example, New Zealand, Solomon Islands, and Vanuatu firmly refuse visits by warships carrying nuclear weapons. Western Samoa bars visits by nuclear-powered ships. On this and on the controversy surrounding the antinuclear stand of New Zealand's Labour Government of Prime Minister Lange, see Fischer, pp. 37–39.

67. For the texts of these Protocols see United Nations *The Law of the Sea Bulletin* (n. 53 above), pp. 205–07. While China and the Soviet Union view favorably the establishment of the South Pacific Nuclear-Free Zone, the United States decided not to sign the Protocols. See 87 *Dep't State Bull.* (no. 2121), (April 1987): 53. The position of France, whose nuclear tests prompted the negotiations for the treaty, is also negative.

68. Convention on the High Seas, April 29, 1958, 450 *UNTS* 82.

69. Ibid., Art. 1.

in customary international law that navigational uses of the High Seas by warships are part of the freedom of navigation, subject only to the general requirement of "reasonable regard" to the interests of other states.[70] As demonstrated by a decisive rejection in 1958 of a Soviet-sponsored proposal designed to ban naval and air exercises of long duration near foreign coasts or on international sea routes,[71] military maneuvers are not considered as an activity exercised without reasonable regard to the interests of other states in their exercise of the freedoms of the High Seas, even though they necessarily cause temporary exclusion of a part of the sea from use by other states. By themselves, military exercises cannot be presumed to conflict with the "reasonable regard" principle; nor can they be regarded as contravening the prohibition of subjecting any part of the High Seas to national sovereignty. Nonnavigational military activities are also subject to the reasonable regard principle as codified in 1958.[72] The controversial issue of nuclear tests on the High Seas, though raised in Geneva in 1958, was not brought to vote, and was eventually regulated by the Partial Test Ban Treaty.[73] However, even under customary international law it would be difficult to argue that a nuclear test on the High Seas was an activity showing reasonable regard to the interests of other states.

The issue of the military uses of the High Seas has recently been complicated by the emergence of the Exclusive Economic Zone (EEZ), as an institution of customary law of the sea, prior to the entry into force of the 1982 Convention on the Law of the Sea. In this transitional period in the history of the law of the sea, when the rules governing the EEZ are undefined and in flux, the issue of military activities in a foreign EEZ is highly controversial. In the prevailing opinion of publicists and maritime nations, all lawful uses of the sea, including military maneuvers, apply to the EEZ as part of the freedoms of the sea. On the other hand, some Third World countries contend that they have the right to restrain military activities of foreign countries in their EEZs.[74]

70. *Ibid.* See also Bernard H. Oxman, "The Regime of Warships under the United Nations Convention on the Law of the Sea," 24 *Virginia Journal of International Law* (1984): 809. And see, generally, Willem Riphagen, "La navigation dans le nouveau droit de la mer," 84 *RGDIP* (1980): 144; D. Momtaz, "Les forces navales et l'impératif de securité dans la Convention des Nations Unies sur le droit de la mer," in *Essays on the New Law of the Sea*, ed. Budislav Vukas (Zagreb: Sveučilišna naklada Liber, 1985), pp. 230–43.

71. See the proposal of Albania, Bulgaria, and the Soviet Union, in fourth session UN Conference on the Law of the Sea, UN document A/Conf.13/C.2/L. 32. The proposal was rejected by 43 votes to 13, with 9 abstentions.

72. This problem is extensively discussed in Uwe Jenisch, *Das Recht zur Vornahme' militärischer Übungen und Versuche auf Hoher See in Friedenszeiten* (Hamburg, 1970).

73. See *UNTS* 48, (n. 47 above).

74. This problem is analyzed at length in Boleslaw A. Boczek, "Peace-Time Military Activities in the Exclusive Economic Zone of Third Countries," 19 *Ocean Development and International Law* (1988): 445–68.

As far as the military use of the deep seabed, beyond the claimed Continental Shelf, is concerned, traditional international law has no specific rules.[75] The matter is controversial, but applying by analogy the principles of the traditional regime of the High Seas, one can argue that, subject only to the reasonable regard principle and any treaty prohibitions such as the Sea-Bed Arms Control Treaty and, of course, the law of the UN Charter, international law does not disallow military use of the deep seabed.[76]

IV. AT UNCLOS III

A. The Specific Reservations: Debate in the Plenary

The debate on the peaceful use of the oceans, initiated by the Maltese delegation in 1967 within the context of the deep seabed, reemerged at UNCLOS III, but this time in a wider application to the ocean space as a whole. However, it must be remembered that, while securing naval mobility for their fleets was perceived by the maritime powers as an essential condition for accepting the Convention, very little, if anything, was stated openly by the delegates of these powers about military activities at sea. As a result, the final text of the Convention stipulates peaceful use for the following marine zones and activities. First, the High Seas are "reserved" for peaceful purposes.[77] Second, through the cross-reference of Article 58(2), this principle applies also to the EEZ as being clearly compatible with, and even mandatory for, the legal status of the zone. Third, since under the Convention's regime the EEZ also includes the seabed and its subsoil for up to 200 miles offshore, the peaceful purposes reservation applies equally to the continental shelf up to such distance. In the Continental Shelf beyond 200 miles the coastal state has sovereign rights over natural resources,[78] but the shelf, if any, is not part of the coastal state's EEZ; nor does it form part of the deep seabed Area. Still, the peaceful purposes reservation must apply to this part of the seabed, which

75. The 1958 Geneva Conventions on the High Seas (450 *UNTS* 82) and on the Continental Shelf (499 *UNTS* 311) do not deal with this subject.

76. "Military Installations" (n. 49 above), pp. 851–52. Among the arguments cited in support of this position are: (1) the legislative history of Art. 2 of Geneva Convention on the High Seas (rejection of the Brazilian proposal to use the term "waters of the high seas" instead of "high seas"); (2) analogy to the right of laying submarine cables and pipelines on the deep seabed as one of the High Seas freedoms; (3) the fact that even relatively permanent uses of the High Seas are not perceived as conflicting with the prohibition of claims to sovereignty; and (4) the fact that if military activities on the seabed violated its *res communis* nature under general international law, there would have been no need to conclude the Sea-Bed Arms Control Treaty. See Zedalis (n. 2 above), pp. 23–24; and see, generally, Jenisch.

77. Convention, Art. 88. (in the French text: "used exclusively"). See discussion of the deep seabed Area reservation in Sec. V C below.

78. Convention, Art. 77(1).

for purposes other than exploring and exploiting its natural resources must be considered High Seas and thereby subject to the peaceful purposes reservation of Article 88.[79] Fourth, the Area is open to use exclusively for peaceful purposes.[80] Consequently, marine scientific research there must be carried out exclusively for such purposes;[81] installations used for carrying out activities there must also be used in this way;[82] and the Review Conference will have to ensure the maintenance of all these principles.[83] Fifth, the Convention has no rules specifically providing for peaceful use in so far as the internal waters, the Territorial Sea and Archipelagic Waters are concerned. However, assuming, for the time being, that the definition in Article 301 entitled "Peaceful uses of the sea" (added in 1980), which as a general provision governs the entire ocean space, has the same meaning as the specifically directed Articles 88 and 141, then the whole ocean, without any geographical limitation, is covered by the peaceful purposes reservation. Finally, the Convention stipulates, also without any such limitation, the principles of conducting marine scientific research exclusively for peaceful purposes and of promoting international cooperation in such research for peaceful purposes.[84] Furthermore, a coastal state's consent to marine scientific research to be carried out in its EEZ or Continental Shelf by another state or by an international organization shall be granted if such research is for exclusively peaceful purposes.[85]

The 1976 debate in the Plenary Committee elicited a wide spectrum of views.[86] In the extreme view of Ecuador the peaceful purposes reservation meant complete prohibition of all military activities at sea, both offensive and defensive. Its delegate insisted that "the use of the ocean space for exclusively peaceful purposes must mean complete demilitarization and the exclusion from it of all military activities."[87] The opposite view interpreted the reservation as simply meaning use of the ocean consistent with the Charter of the United Nations and other rules of international law. As argued by the U.S. delegate:

> The term "peaceful purposes" did not, of course, preclude military activity generally. . . . [T]he conduct of military activities for peaceful purposes was in full accord with the Charter of the United Nations and with

79. Treves, "Military Installations" (n. 49 above), p. 817, and "La notion d'utilisation des espaces marins à des fins pacifiques dans le nouveau droit de la mer [1980] (hereafter cited as "La notion"), 26 *AFDI* (1981): 687–99.
80. Convention, Art. 141(1).
81. Ibid., Art. 143(1).
82. Ibid., Art. 147(2) (d).
83. Ibid., Art. 155(2).
84. Ibid., Art. 240(a) discusses scientific research for peaceful purposes; Convention, Art. 242(2) addresses international cooperation in pursuit of such research.
85. Convention, Art. 246(3).
86. This debate is ably summed up in the intervention by the delegate of Iran. 5 *Official Records* (1976): pp. 65–66.
87. Ibid., p. 56.

the principles of international law. Any specific limitation on military activities would require the negotiation of a detailed arms control agreement. The Conference was not charged with such a purpose and was not prepared for such negotiations. Any attempt to turn the Conference's attention to such a complex task could quickly bring to an end current efforts to negotiate a law of the sea convention.[88]

Except for the above-quoted exceptionally radical Ecuadorian statement, Third World delegates realistically admitted that the reservation could not be construed to exclude all military activities from the oceans.[89] Emphasis was rather on the need for agreeing upon a definition of the vague concept of peaceful use; establishment and gradual expansion of zones of peace and nuclear-free zones; banning military exercises on the High Seas and especially in the EEZs of foreign countries; and continued negotiations on further demilitarization of the seabed as directed by the Sea-Bed Arms Control Treaty.[90] One new, but not entirely unexpected, feature of the debate was the change in the attitude of the Soviet Union. Having by the time of the debate achieved the status of a world naval power, the Soviet Union considered it expedient not to support efforts at demilitarizing the oceans. While paying lip service to the principle of peaceful use of the oceans, the Soviet delegate—like his U.S. colleague—argued that such complex problems as establishing zones of peace and security, eliminating naval bases, and so on, were "beyond the scope of the work facing the Conference on the Law of the Sea" and that a "complete and constructive solution of those issues would be possible only within the framework of the appropriate United Nations bodies or at other international conferences and forums dealing with the problems of disarmament, international security and world peace."[91]

Ultimately the debate at the fourth session of UNCLOS III in 1976 produced no clarification of the peaceful purposes reservation. It remains in the Convention as vague and ambiguous as in its original formulations. Unlike the Antarctic and outer space treaties, the Convention clauses do not

88. Ibid., p. 62.

89. See, e.g., the Philippines, in 5 *Official Records* (1976): 65; Tunisia, ibid., p. 67.

90. See statements by the delegates of Madagascar, 5 *Official Records* (1976): 57–58; Iraq, ibid., p. 59; United Arab Emirates, ibid., pp. 63–64; the Philippines, ibid., pp. 64–65; Iran, ibid., pp. 65–66; Somalia, ibid., p. 66; Pakistan, ibid., pp. 66–67; and Tunisia, ibid., pp. 67–68. The position of Romania came closer to that of the Third World countries than to the Soviet stand on the matter. See Romania, ibid., p. 57.

91. 5 *Official Records* (1976): 59. This position coincided with that of the United States and was also supported by Bulgaria, ibid., pp. 60–61, and Cuba, ibid., p. 61. The evolution in the Soviet attitude is traced in Anthony P. Allison, "The Soviet Union and UNCLOS III: Pragmatism and Policy Evolution," 16 *ODIL* (1985): 109–36.

elaborate on the specifics of the peaceful purposes principle, *in themselves* providing no clues for interpretation.

B. The General Provision on "The Peaceful Uses of the Seas"

Article 301 of the Convention, clearly inspired by Article 2(4) of the UN Charter,[92] ostensibly provides the answer to the question of what the Parties meant by peaceful purposes in the various provisions to this effect. It is significant that the origins of Article 301 can be traced primarily to developing countries, some of which, at least in 1976, had interpreted peaceful uses as nonmilitary uses, but the definition of this article follows the point of view of the maritime powers, namely that the peaceful purposes formula proscribes only those military activities which are incompatible with the law of the UN Charter. However, even if Article 301 does define peaceful purposes in this manner, the definition is still very vague and general, incorporating all the ambiguities of the Charter into the law of the sea. All the controversial issues concerning the threat or use of force, aggression, self-defense, and so on, will have to be faced in the interpretation of the Convention. The concept of threat will be especially vexing since, being a subjective notion, easily used or abused for political purposes, it can be invoked by a state against any military activity of another country. In this sense, there is no need to resort to the interpretation of "peaceful" as "nonmilitary."

Article 301 adds nothing to the obligations of states since they are already bound by the UN Charter whose rules in any case prevail over any conflicting obligations of states under international agreements.[93] Moreover, within the context of the Convention itself, if the peaceful purposes formulas of, for instance, Articles 88 and 141 mean the same as the peaceful purposes or uses under Article 301,[94] then they are redundant since Article 301 is a provision of general application.

92. Art. 2(4) of the UN Charter requires all members to "refrain in their international relations from the threat or use of force against the territorial integrity or political independence of any State, or in any manner inconsistent with the Purposes of the United Nations." However (unlike the original proposal), Art. 301 of the Convention uses the phrase "principles of international law embodied in the Charter of the United Nations" instead of "purposes," in order to make reference not just to Chap. 1 of the Charter (Purposes and Principles) but to all international law of the Charter. Another reason may have been the desire to avoid any subjective considerations of "purposes" of the Charter being included in the Convention. See Treves, "La notion" (n. 79 above), pp. 688. Provisions almost identical to Art. 301 are found in Arts. 19(2)(a), 39(1)(b), also applying to the context of Arts. 52 and 54.

93. UN Charter, Art. 103.

94. In the English text there is a difference between the wording of the title of Art. 301, which reads "peaceful uses of the seas" and the wording of the specific reservations using the phrase "peaceful purposes." But both mean the same. As a matter of fact, there is no such difference in the French text, using "for peaceful purposes" (*à des fins pacifiques*) in both.

V. AN ANALYSIS OF THE PEACEFUL PURPOSES CLAUSES

The concise peaceful purposes clauses of the Convention seem, at first sight, to impose far-reaching limitations upon military activities at sea. Whether they really do so must be elucidated by analyzing each clause in accordance with the principles and rules of interpretation as developed in the practice of international law. The main purpose of this process is to ascertain, in good faith, the intention of the Parties in accordance with the ordinary meaning of the terms used in the Convention in their context and in the light of the Convention's objective and purpose.[95]

Although, as exemplified by the Antarctic and outer space treaties, the peaceful purposes formula is not a novelty in "international legislation," any interpretation of its meaning in the 1982 Convention cannot be influenced by arguments used for purposes of those earlier treaties which are (including also the Antarctic Treaty) basically arms control agreements. This cannot be said of the 1982 Convention. Its preamble, while recognizing the desirability of "establishing through the Convention . . . a legal order for the seas and oceans which . . . will promote the peaceful uses of the seas and oceans," does not make any references that would clearly point to the Convention as an international arms control agreement. If the specific peaceful purposes clauses were to be arms control provisions, they would have to be followed by fairly long series of rules applying in detail the general idea of various situations and activities in the respective marine zone. No such provisions can be found in the Convention.

A. The High Seas

Article 88 states that "The high seas shall be reserved for peaceful purposes."[96] Does this mean that all the navies of the contracting Parties would have to be scrapped? However such complete naval disarmament were "devoutly to be wished," the context of the Convention clearly contradicts it. The Convention reaffirms the traditional freedoms of the High Seas which,[97] inter alia, comprise freedom of navigation and overflight, without any ban on navigation by warships. Moreover, the context of the Convention indicates that the instrument takes at least some military activities on the High Seas for granted as one of the routine uses of the sea. The existence of warships is

95. Vienna Convention on the Law of Treaties, UN document A/Conf.39/27 (1969), reprinted in 63 *AJIL* (1969): 875, Art. 31(1).
96. In the French text "reserved" is rendered by "used exclusively" (*utilisée exclusivement*).
97. Convention, Art. 87.

considered a normal phenomenon and such vessels even enjoy a privileged status.[98] Among the activities incompatible with the meaning of innocent passage in the Territorial Sea, the Convention lists some military activities such as exercise or practice with weapons of any kind, and the launching, landing, or taking on board of any aircraft or military device.[99] This special exclusion clause implies that such activities are permissible outside of the Territorial Sea, in the zones which are covered by the peaceful reservation of Article 88, that is, the High Seas and the EEZ. Finally, the optional exclusion of disputes concerning military activities from compulsory judicial settlement would make no sense if they were illegal under the reservation.[100] Furthermore, the Convention explicitly includes in the concept "military activities by government vessels and aircraft engaged in non-commercial service."[101] Nowhere in the context of the Convention can one find any suggestion that Article 88 bans military activities on the High Seas.[102] However, in the same way as in traditional international law, the exercise of the freedoms of the High Seas,[103] including freedom of navigation, is subject to "due regard for the interests of other states . . . and also with due regard . . . with respect to

98. Convention, Arts. 32, 95, 236. Also in the Convention, see references to warships in Arts. 29, 30, 31, 102, 107, 110, 111, 224, and submarines in Art. 20. See, generally, Oxman (n. 70 above).

99. Convention, Art. 19(2)(b)(e)(f).

100. Convention, Art. 298(1)(b). See also Sec. VI of this article.

101. Ibid.

102. There is general consensus on this among the commentators. See, among others, Pyotr Barabolya, "Changes in the Legal Regime of the Sea and Their Influence on Navigation," in Vukas, ed. (n. 70 above), p. 189; Isaak I. Dore, "International Law and the Preservation of the Ocean Space and Outer Space as Zones of Peace," 15 *Cornell International Law Journal* (1982): 1–61; Jens Evensen, "The Law of the Sea Regime," in Byers, ed. (n. 2 above), pp. 77–87; Francesco Francioni, "Peacetime Use of Force, Military Activities and the Law of the Sea," 18 *Cornell International Law Journal* (1985): 203–26; Stepan V. Molodtsov, "The Exclusive Economic Zone: Legal Status and Regime of Navigation," *Ocean Yearbook 6*, ed. Elisabeth Mann Borgese and Norton Ginsburg (Chicago: University of Chicago Press, 1986), pp. 203–16, esp. p. 214 (not a "complete" ban); Momtaz (n. 70 above); R. W. G. de Muralt, "The Military Aspects of the UN Law of the Sea Convention," 32 *Netherlands International Law Review* (1985): 78–99; Oxman (n. 70 above); Jean-Pierre Quéneudec, "Zone économique exclusive et forces aéronavales," in René-Jean Dupuy, ed., *La gestion des ressources pour l'humanité: Le droit de la mer—The Management of Humanity's Resources: The Law of the Sea* (Boston: Martinus Nijhoff, 1982); Elmar Rauch, "Militarische Aspekte der Seerechtsentwicklung," in Wolfgang Graf Vitzthum, ed., *Aspekte der Seerechtsentwicklung* (Munich: Hochschule der Bundes Wehr, 1980), pp. 75–121; Elliot L. Richardson, "Law of the Sea: Navigation and Other Traditional Security Considerations," 19 *San Diego Law Review* (1982): 553, and "Power, Mobility and the Law of the Sea," 58 *Foreign Affairs* (1980): 902–19; Riphagen (n. 70 above); Treves, "Military Installations" (n. 49 above), and "Le nouveau regime des espaces marins et la circulation des navires," in Vukas, ed., pp. 202–21 (n. 70 above); and Zedalis (n. 2 above).

103. Convention on the High Seas, Art. 2 (n. 68 above).

activities in the Area."[104] Under these conditions, military exercises are, in the prevailing view, permissible.[105] Whether the existence of the EEZ has resulted in special restrictions on grounds of "peaceful purposes" is another question.

B. The Exclusive Economic Zone (EEZ)

Through the cross-reference of Article 58(2), the High Seas peaceful purposes reservation applies also to the EEZ in so far as it is not incompatible with the part of the Convention governing this new, sui generis, marine zone. The Convention guarantees that all states may, within Exclusive Economic Zones, have freedom of navigation and overflight, lay submarine cables and pipelines, "and other internationally lawful uses of the sea related to these freedoms, such as those associated with the operation of ships, aircraft and submarine cables and pipelines, and compatible with the other provisions of [the] Convention."[106] In the prevailing interpretation,[107] this provision allows states to conduct naval maneuvers in a foreign EEZ, subject to the interests of other states,[108] including the coastal state,[109] and the rules of the Convention and other rules of international law governing navigation on the High Seas.[110] On the other hand, the language of Article 58 is ambiguous enough for the promoters of the coastal states' extensive jurisdiction in the EEZ to claim that foreign military activities cannot be accommodated under the adopted formula, as a use of the sea that is not lawful internationally or incompatible with the reservation of the High Seas for peaceful purposes. Thus some states, such as Brazil, Cape Verde, and Uruguay, insisted in their declarations upon the signing of the Convention that the provision in question did not authorize other states to carry out, in the EEZ, military exercises and maneuvers or other military activities without the coastal state's consent.[111] The Cape Verde

104. "Reasonable regard" of the 1958 Convention is replaced in the 1982 Convention by "due regard," but this change does not seem to have any substantive meaning.

105. This is admitted by Evensen in "The Law of the Sea Regime" (n. 102 above), pp. 85–86.

106. Convention, Art. 58(1).

107. See, e.g., Evensen, "The Law of the Sea Regime" (n. 102 above), pp. 85–86.

108. Convention, Art. 87(2) cross-reference in Art. 58(1).

109. Convention, Art. 58(3).

110. Convention, Art. 58(2).

111. See the interpretative declarations of Brazil, Cape Verde, and Uruguay, made under Art. 310 of the Convention in, *Law of the Sea Bulletin*, no. 5 (July 1985), p. 45. "Declarations" are possible under Art. 310 although reservations and exceptions are not allowed (Art. 309). Declarations or statements can be made, with a view, inter alia, to the harmonization of the declaring state's laws and regulations with the provisions of the Convention, "provided that such declarations or statements do not pur-

and Uruguay declarations even explicitly state that military exercises are not a peaceful use of the oceans. This is apparently also the position of Brazil.[112] All this was strongly opposed by the maritime nations, as evidenced by Italy's declaration upon the signing of the Convention: "[a]ccording to the Convention, the coastal State does not enjoy residual rights in the Exclusive Economic Zone. In particular, the rights and jurisdiction of the coastal State in such zone do not include the right to obtain notification of military exercises or manoeuvres or to authorize them."[113]

In addition to the issue of navigation by foreign warships in the EEZ, a controversy arose at UNCLOS III over nonnavigational uses of the zone for such purposes as emplacing military installations, structures, and devices in the Continental Shelf of a foreign EEZ.[114] The Convention is silent on the legality of military installations and devices. It approaches the issue in general terms, granting the coastal state exclusive jurisdiction to construct and use artificial islands irrespective of their purpose.[115] At least in the interpretation of the maritime powers, third states have the right to place military installations and structures in the EEZ. This would, in principle, include emplacement and emplantment of weapons. However, such activities are limited by the "due regard" clause of the Convention and applicable arms control treaties.[116] Of direct relevance here is the possibility that the coastal state may argue that placing military devices by a third state, and especially offensive weapons, has nothing in common with the exercise of the freedom of the High Seas and can be regarded as a nonpeaceful use and a threat of force in violation of Article 301 and the UN Charter. The coastal state may also contend that the rights not specifically attributed by the Convention, such as the right of third states to conduct maneuvers or place military devices in the EEZ fall within the competence of the coastal state.[117] In general, the Convention leaves enough latitude for the coastal state to contend that a foreign military activity in its EEZ is not "for peaceful purposes."

---

port to exclude or to modify the legal effects of the provisions of [the] Convention in their application to that State" (Art. 310).

112. *Law of the Sea Bull.,* no. 5 (July 1985), p. 45. For Brazil, see also Declaration of December 22, 1988, made upon ratification of the 1982 Law of the Sea Convention, in *Law of the Sea Bull.,* no. 12 (December 1988), p. 8.

113. *Law of the Sea Bull.,* no. 5 (July 1985), p. 45; and *Law of the Sea Bull.,* no. 4 (February 1985), p. 13. Similar considerations apply to the problem of installations and structures in the EEZ.

114. On "due regard," see Boczek, "Peace-Time Military Activities" (n. 74 above); and generally Treves, "Military Installations" (n. 49 above").

115. Convention, Arts. 56(1)(b)(i); 60(1)(a). See Treves, "Military Installations" (n. 49 above), pp. 840–48; Merciai (n. 2 above), pp. 102–06.

116. Convention, Art. 58(3).

117. This issue, generally regulated in Art. 59, is not directly related to the topic of the present inquiry.

C. The Area

The other major reservation for peaceful purposes is stipulated in Part XI of the Convention among the provisions governing the Area, that is, "the seabed and ocean floor and subsoil thereof, beyond the limits of national jurisdiction."[118] Article 141 states that "the Area shall be open to use exclusively for peaceful purposes by all States, whether coastal or land-locked, without discrimination and without prejudice to the other provisions of this Part."[119]

The interpretation of Article 141 raises some serious questions that do not arise with regard to the parallel provision in Article 88. One problem is that, whereas the UN Charter–based (and Art. 301–based) construction of Article 88 enjoys ample explicit and implicit support and corroboration in the provisions of the Convention, and especially those that recognize the legality of warships in all zones of the ocean, there exists no analogous military use of the deep seabed expressly allowed by the Convention in the Area.[120]

Then, when reading the English text of the Convention, it appears at first sight notable that the texts of the two major peaceful purposes clauses (Arts. 88 and 141) differ in that under Article 88 the High Seas shall be "reserved" for peaceful purposes, whereas Article 141 (like the other relevant provisions of Pt. II) does not reserve the Area for peaceful purposes, but stipulates that it shall be open for use "exclusively" for peaceful purposes.

However tempting it is to conclude otherwise, the use of "exclusively" does not make Article 141 any different from Article 88. The problem is purely semantic and related to the drafting process of the Convention. First, to be "reserved" (Art. 88) means, in English, to be "used exclusively" (Art. 141). Second, the phrasing of Article 141 simply follows the terminology that has been traditionally and almost automatically employed with regard to the deep seabed ever since 1967 and even before in the Antarctic Treaty. The committee drafting Article 141 copied the linguistic model of the "classical" period of the debates on denuclearization of the deep seabed in the late 1960s and early 1970s, and subsequent drafting of the final English text did not apparently harmonize the drafts that originated in two different committees. Third, the problem does not exist for those who use the French text of the Convention.[121] The French text uses the phrase "exclusively for peaceful purposes" (*exclusivement à des fins pacifiques*) both in Article 88 and in Article 141. Under the Convention all the six versions of the text (Arabic, Chinese, English, French, Russian, and Spanish) are equally authentic.[122] According to

---

118. Convention, Art. 1(1).

119. Convention, Art. 141. This Article has virtually the same wording as para. 5 of the Declaration of Principles of 1970 (n. 34 above).

120. Zedalis (n. 2 above), pp. 26–27.

121. For example, Treves does not even raise this problem. See Treves, "La notion" (n. 79 above).

122. Convention, Art. 320.

international law principles of interpretation, the terms of the treaty are presumed to have the same meaning in each authentic text, but if a comparison of the authentic texts discloses a difference of meaning, the difference should first be removed by reference to the context and in the light of the object and purpose of the treaty and its *travaux préparatoires*. If the differences cannot be removed, "the meaning which best reconciles the texts, having regard to the object and purpose of the treaty, shall be adopted."[123] The application of these principles to the peaceful purposes reservation clauses in the 1982 Convention makes one conclude that the adverb "exclusively" adds nothing to the construction of Article 141 in the English version of the Convention.

Other arguments that can be raised in support of an extensive (nonmilitary meaning) interpretation of Article 141 are based on the intrinsic connection between the peaceful purposes reservation and the principle declaring the Area and its resources "the common heritage of mankind,"[124] combined with the principle that "[a]ctivities in the Area shall . . . be carried out for the benefit of mankind as a whole."[125] There is no doubt that the original, Maltese proposal of 1967 to reserve the deep seabed for peaceful purposes as a common heritage of mankind implied the demilitarization of the seabed.[126] However, the interpretation of Article 141 does not lead to this conclusion within the context of Part XI. If the drafters of this Part meant to completely demilitarize the Area, they certainly did not produce a text that consistently and clearly stated and developed this proposition.

The provisions of Part XI govern "activities in the Area,"[127] by which is meant only "all activities of exploration for, and exploitation of, the resources of the Area."[128] Much of Part XI deals with activities thus defined, although the Part does envisage "other activities in the marine environment." But does the principle of common heritage of mankind allow military activities, a category of "other activities" that certainly is not resource oriented? The Convention does not give any definitive answer to this question. For those who start from the premise of the common heritage principle, total demilitarization of the deep seabed is a logical consequence. On the other hand, for those who follow a strict contextual interpretation, the answer is that the Convention does not prohibit military activities as such since the principles of the common heritage of mankind and the benefit of mankind as a whole apply only to the resource-oriented activities in the Area.

Yet, even if one accepts that Article 141 cannot be construed to proscribe all military activities in the Area, there are some provisions in Part XI which

123. Vienna Convention on the Law of Treaties (n. 95 above), Art. 33 in connection with Arts. 31 and 32.
124. Convention, Art. 136.
125. Convention, Art. 140(1).
126. See nn. 27 and 28 above and the accompanying text.
127. Convention, Art. 134(2).
128. Convention, Art. 1(1)(3).

limit the scope of any such activities. First, they must not amount to, or result in, any claim or exercise of sovereignty or sovereign rights by any state over any part of the Area or its resources, and second, no state (or any other person) may appropriate any part of the Area or its resources.[129] The precise meaning of "appropriation" is not clear, but it ostensibly constitutes a form of exclusive possession and control falling short of sovereign rights.[130] Therefore, any exclusive use by a state party of any portion of the deep seabed for military purposes that indicates permanent intent would be prohibited under Article 137.[131]

Competing uses of the Area would raise the question of accommodation between resource-oriented and "other" activities in the deep seabed. One of the principles governing the Area is that accommodation in such cases is to be based on reciprocal reasonable regard for the other kind of activity.[132] As in the case of "due regard" in the exercise in the freedoms of the High Seas,[133] there is no further elaboration of this general clause except for the rules concerning installations used for carrying out activities in the Area, generally designed—like the rules concerning installations in the Continental Shelf—to safeguard the freedom of navigation.[134] In addition, such installations must—as already noted—be used exclusively for peaceful purposes.[135] The Convention does not assign priority to one kind of activity over the other. In practice, the chances are that the exclusive rights of the entity exploring and exploiting the Area "on behalf of mankind as a whole" will have an advantage over any conflicting activity.[136]

The general conclusion concerning the meaning of Article 141 is that it does not proscribe all military activities. But to be for peaceful purposes such activities must be conducted with reasonable regard for "the activities in the Area" and, of course, meet the requirement of Article 301. Again, as with the High Seas, the concept of threat will be crucial. It is difficult to articulate a precise test, but if a military activity in the deep seabed can reasonably be

129. Convention, Art. 137(1).

130. Cf. the Canadian paper submitted to the UN Seabed Committee in 1971, UN document A/AC.138/59, in 26, United Nations General Assembly *Official Records* (*UNGAOR*) suppl. 21, 205 UN document A/8421 (1971), cited in Zedalis (n. 2 above), pp. 31–32, n. 107.

131. Zedalis (n. 2 above), p. 32.

132. Convention, Art. 147 (1)(3).

133. Convention, Art. 87(2).

134. Convention, Art. 147(2).

135. Art. Convention 147(2)(d).

136. Treves, "Military Installations" (n. 49 above), p. 855. Moreover, as noted by Treves, one must rule out the possibility that activities in the Area, especially those of the Enterprise, might be conducted not "for peaceful purposes" (perhaps there is even a legal presumption of their peaceful nature) whereas one cannot take this for granted in the case of a conflicting activity. Ibid; See also Treves, "La notion" (n. 79 above), p. 695.

regarded as going beyond the criteria of "peacefulness" established in international practice, then such activity should be prohibited as a threat to international peace. It can be argued, for example, that as a result of the Sea-Bed Arms Control Treaty, whose prohibitions have been recognized even by nonparties, the activity of emplacing nuclear weapons on the seabed cannot be regarded as a use for peaceful purposes. This might develop into a legal presumption of nonpeaceful use. On the other hand, in view of the fact that international practice seems to have acquiesced in listening and similar defensive devices on the seabed, the activity of placing them there could be regarded as legal under the peaceful purposes reservation of Article 141.[137] Specific determination would have to be made in each case according to the procedures provided for by the Convention.[138]

D. Marine Scientific Research

The peaceful purposes reservation applies also, as a general principle, to marine scientific research. Such research "shall be conducted exclusively for peaceful purposes."[139] States and competent international organizations are required to promote international cooperation in marine scientific research "for peaceful purposes."[140] Finally, as far as marine scientific research in the EEZ and on the Continental Shelf is concerned,[141] coastal states, whose consent is needed for the conduct of research by other states and by international organizations, shall, in normal circumstances, grant such consent for projects to be carried out exclusively for peaceful purposes.[142]

The concept of research for exclusively peaceful purposes is not defined, but—like the other reservations of this kind—it cannot be construed as research for "nonmilitary" purposes, for at least two reasons. First, such a construal would be unacceptable to maritime powers who find research essential for defense purposes, and second, it is not possible to draw a clear dividing line between "peaceful" and "military" or "defensive" and "offensive" research. While it is true that states asking permission to conduct research in a

137. See Treves, "Military Installations" (n. 49 above), p. 819, and "La notion" (n. 79 above), pp. 693–94.

138. See Sec. VI of this article.

139. Convention, Art. 240(a). Readers are referred to an article by Alexander Yankov, "A General Review of the Law of the Sea: Marine Science and Its Application," *Ocean Yearbook 4*, ed. Elisabeth Mann Borgese and Norton Ginsburg (Chicago: University of Chicago Press, 1983), pp. 150–75.

140. Convention Art. 242(2). There is no adverb "exclusively" here.

141. The coastal state has jurisdiction with regard to marine scientific research in the EEZ; the Convention, Art. 56(1)(b)(ii) and 246(1)(2), and the Continental Shelf; the Convention, Art. 246(1)(2).

142. The Convention, Art. 246(2)(3).

foreign EEZ or on the Continental Shelf must indicate the nature and objectives of the project,[143] this rule can, in practice, be circumvented.

VI. INTERPRETATIVE MECHANISMS UNDER THE CONVENTION

The analysis of the peaceful purposes reservation in the Convention has revealed that no all-embracing definition of this clause can gain consensus or even is possible at this time. In its "constructive ambiguity,"[144] the concise peaceful purposes provisions are acceptable both to the maritime powers interested in military uses of the sea and those states that aspire to restrict and eliminate all such uses. Either group can find it possible to interpret the reservation in its own way. The question is whether the Convention itself envisages any procedures for a legally binding interpretation of the clause in cases of specific disputes over its meaning. Before turning to this problem, it should be noted that as far as activities carried out by the Enterprise are concerned, it is reasonable to accept the legal presumption of peaceful use of the Area, something that cannot in principle be taken for granted for activities by national contractors.[145]

In general, two types of situations may occur under the Convention where an interpretation of the peaceful purposes reservation can have binding force for the parties concerned.[146] First, there are situations where a Party has, under a rule of the Convention, the power directly to impose an interpretation upon the other Party. Thus under the rules on marine scientific research in the EEZ and on the Continental Shelf, the coastal state has the discretion to refuse consent to research by other states or international organizations if it decides that the research is not for exclusively peaceful purposes.[147] Under the Convention,[148] the researching state cannot challenge this decision by means of third party settlement, because the coastal state is not obliged to accept submission to settlement of any disputes arising out of its exercise of the right or discretion to grant consent.[149] Of course, in any case

143. The Convention, Art. 248(a).
144. Booth (n. 3 above), p. 88.
145. Treves, "La notion" (n. 79 above), p. 698.
146. See Treves, "Military Installations" (n. 49 above), p. 818, and "La notion" (n. 79 above), pp. 694–98.
147. See Sec. V D of this article.
148. Convention, Art. 297(2)(a)(i).
149. The same applies to the coastal state's decision to order suspension or cessation of a research project if it subsequently decides that it is not exclusively for peaceful purposes. This results from the Convention, Art. 253(1)(a). On the other hand, if one assumes that the dispute over the exclusively peaceful nature of a research project concerns not the "exercise" of the right to give consent but the presupposition of such exercise, then the dispute might be submitted to compulsory settlement. See Tullio Treves, "Principe du consentement et recherche scientifique dans le nouveau droit de la mer," 84 *RGDIP* (1980):253, esp. pp. 264–65.

there exists the possibility to take advantage of the military activities optional exemption under Article 298.

In another situation of the first type, the International Sea-Bed Authority is the party that in performance of its limited inspection powers has the right to determine that, contrary to Article 147(2)(d), an installation used for carrying out activities in the Area is not used exclusively for peaceful purposes.[150] The chances are that the Authority will adopt an interpretation tending toward the nonmilitary point of view. The Authority may warn the contractor state that emplaced military installations and, under the circumstances, even decide to suspend or terminate the contractor's rights under the contract.[151] A contractor that does not observe the peaceful purposes clauses may subsequently be disqualified by the Authority from obtaining another contract,[152] but has the right to appeal to the Sea-Bed Disputes Chamber of the International Tribunal for the Law of the Sea.[153]

The other type of situation includes cases where the Convention provides for a third-party binding dispute settlement. While most disputes over whether a certain activity, for example weapons practice, is for peaceful purposes can be submitted to compulsory procedures entailing binding decisions, the optional exception to the applicability of such procedures of "disputes concerning military activities, including military activities by government vessels and aircraft engaged in non-commercial service" is likely to reduce the possibility of generating case law interpreting the peaceful purposes reservation.[154]

VII. GENERAL CONCLUSIONS—FROM SOFT TO HARD LAW?—AGENDA FOR IMPLEMENTATION

Much has been written here and elsewhere about the undefined nature and ambiguity of the peaceful purposes reservation in the 1982 Convention on the

150. The International Sea-Bed Authority has the right to inspect all installations in the Area used in connection with activities there. Convention, Art. 153(5). There have been proposals to grant the Authority verification powers with regard to all kinds of installations. See proposal of Canada, March 2, 1972, in the Seabed Committee. UN document A/AC.138/SC.1/SR. 33. See also intervention of the Iranian delegate at UNCLOS III, 5 *Official Records* 66; and see further Borgese "The Sea-Bed Treaty and the Law of the Sea" (n. 35 above), pp. 97–99.

151. The Convention, Annex III, Art. 18. However, the Authority may not execute such decision until the contractor has been accorded a reasonable opportunity to exhaust judicial remedies available to him pursuant to the provisions of Part II dealing with the settlement of disputes (Sec. 5).

152. This results from Convention, Annex III, Art. 4(2).

153. Convention, Art. 187(d).

154. Convention Art. 298(1)(b). See also Mark W. Janis, "Dispute Settlement in the Law of the Sea Convention: The Military Activities Exemption," 4 *ODIL* (1977): 51–65.

Law of the Sea. Yet, the very fact that the text of the Convention includes such reservation is a turning point in the history of the law of the sea. Peace, the most cherished value in the relations of nations, has found a place in the codification of international law governing the oceans.

Important principles in international legislation are often deliberately couched in general and intentionally ambiguous terms in order to allow conflicting political, strategic, and economic interests enough latitude of interpretation and thus acceptance of the instrument. This is the case with the peaceful purposes reservation in the Convention.

However, the context of the Convention proves beyond any doubt that, at least in its High Seas reference, the reservation does not ban military activities as such. But it is controversial whether it transcends the general obligation of states to refrain from the threat or use of force prohibited by the principles of international law embodied in the Charter of the United Nations. In other words, are there any military uses of the sea that, while not necessarily incompatible with these principles, are still illegal by virtue of contravening the peaceful purposes reservation? The adoption of Article 301, which defines peaceful uses of the seas at the minimum level of compatibility with the law of the UN Charter, suggests a negative answer to this question. In that case, what is the purpose of the specific clauses? The drafters of the Convention leave this matter open, making it possible to interpret the reservation as restricting or even eliminating all military activities from the oceans, while at the same time accommodating the interests of those who argue that any military activity in the oceans that is not incompatible with the principles of the Charter can be considered activity for peaceful purposes. As with a number of other issues at UNCLOS III, this compromise confirmed that "[t]he essence of the emerging law of the sea would be largely a political compromise between perceived national interests, couched in legal language. Since there were fundamental divergences with regard to perceived national interests among States on a number of important questions, such compromises could sometimes be reached only by deliberate ambiguity of language."[155]

It is significant, however, that the Convention does not have any specific reservation for marine zones under national sovereignty, which might suggest that the High Seas and the Area were subject to more stringent requirements. Furthermore, in accordance with the principle of effectiveness, it could be argued that the peaceful purposes clauses must be allowed some legal effect other than meant by Article 301; otherwise they would be redundant.

It will take many years of international practice for some clarification of the reservation to emerge. Also, it is only after the Convention enters into

---

155. Arvid Pardo (International Ocean Institute) at the sixty-third Plenary Meeting, Fourth Session of UNCLOS III, New York, 1976, in 5 *Official Records*, p. 46. And see also generally Jean-Pierre Queneudec, "Les incertitudes de la nouvelle Convention sur le droit de la mer," in Vukas, ed. (n. 70 above), pp. 47, 49–52.

effect that a possibility will exist, in specific cases of international disputes, of an authoritative interpretation of the reservation by judicial organs envisaged by the Convention. Unfortunately, the prospects for developing a body of case law on this issue will be reduced by the optional exception of disputes concerning military activities from compulsory procedures for settling disputes.

The chances are, therefore, that military uses of the sea will remain a source of international friction and conflict, with some coastal states challenging the legality of foreign naval operations in their EEZ and on their Continental Shelf. As already demonstrated by declarations made when signing or ratifying the Convention and by domestic legislation, naval maneuvers in the EEZ, placement of ASW devices on the Continental Shelf, and weapons testing on vast areas of the High Seas may come for questioning on grounds of being incompatible with the principle of the peaceful use of the sea. In response, maritime states (which need not necessarily include only the present naval powers but also newcomers from the Third World) will assert their own interpretation of the peaceful purposes reservation, trying to prove that the Convention in no way restricts their naval mobility as long as they do not infringe on the UN Charter's proscription of force and the definition under Article 301 of the Convention. A naval power can rationalize its military activity as self-defense, totally compatible with "the principles of international law embodied in the Charter of the United Nations," specifically Article 51, while a coastal state, in turn, can, especially in a situation of high political tension, consider foreign maneuvers in its EEZ as a hostile act that can qualify as illegal threat of force.[156]

Does all this mean that a treaty provision that, like the peaceful purposes reservation, is susceptible to any interpretation that is mandated by perceived national interest, has no legal value for the regulation of the military uses of the sea? Is it just rhetoric, and are the repeated references to peaceful purposes in the Convention nothing but ritual incantations?[157] Even worse, some believe that the vagueness and ambiguity of the reservation "foster controversy and contribute to international tension."[158]

This undefined nature of the reservation has prompted some commentators to note that rather than being a firmly posited law it belongs to the category of "soft law" (*droit programmatoire*), which only expresses aspirations of the drafters and their policy goal for the future without living up to the

---

156. But again, it can be argued on behalf of the naval power that such a "low threshold of anxiety" of the coastal state cannot, apart from the due regard clause, diminish the rights of the flag state. Richardson, "Law of the Sea," (n. 102 above), p. 574.

157. This expression, *rite conjuratoire*, is used by Treves in "La notion." See Treves (n. 79 above), p. 698.

158. Arvid Pardo, "Foreword," in Byers, ed. (n. 2 above).

realities of the current circumstance.[159] The peaceful purposes reservation of the Convention certainly fits very well in this concept, but this does not yet mean that it has no legal impact or potential. It is noteworthy that while fitting the concept of soft law, the peaceful purposes clauses in the 1982 Convention are not couched in less binding verbal phrasing such as "may" or "should" or "endeavor to"; all are introduced by the mandatory "shall" formula.

It has been increasingly recognized that over time a soft law provision can, in a process of gradual accretion, mature into a hard or firm rule of law. The reservation in questiton may be such a provision. It may seem to be mere rhetoric, but it may also appear to some as "a useful first step in setting the agenda for negotiation of the future law."[160]

Thus, however "soft" the reservation may be, it will provide a legal base for efforts at restricting military uses of the oceans. Combined with the growing realization of the need for genuine arms control, these efforts may eventually succeed in operationalizing the soft concept of peaceful purposes of the Convention into a firmer rule of law.[161] Briefly, the concept of the peaceful use of the Area should be developed in concrete hard-law terms within the framework of the linkages between the 1971 Sea-Bed Arms Control Treaty and the 1982 Convention. In geographical terms, the scope of the seabed treaty should be extended to include also the waters under coastal state sovereignty, including archipelagic waters. Functionally, the gaps of the 1971 treaty should be closed, by banning all types of weapons on the seabed. In terms of verification, a major step would be to grant the International Sea-Bed Authority the right to inspect not only resource-related installations but also other installations on the entire seabed area.

However difficult the implementation of the principle of the peaceful uses of the Area and the seabed in general may be, Article 88 presents an even more serious challenge. Realistically, the reservation of the High Seas for peaceful purposes cannot be operationalized in terms of complete demilitarization. However, in view of the seemingly brighter prospects for arms control today, a mutual balanced reduction of the nuclear strategic submarine forces is within the reach of possibility. Also, as demonstrated by the Rarotonga Treaty, the establishment of nuclear-free zones is also possible. The translation of the High Seas reservation into concrete legal obligations, in so far as

159. This aspect is raised in Treves, "La notion" (n. 79 above), pp. 698–99.

160. Oxman (n. 70 above), p. 831. As noted by O'Connell with regard to the peaceful purposes reservation in the Declaration of Principles (n. 28 above), "reiteration of the formula in the Sea-Bed Committee and persistent lip service which has been given to it, have created a political milieu in which no credit is likely to accrue to any assailant on the sea bed, however defensible its motives," See D. P. O'Connell, *The Influence of Law on Sea Power* (Manchester: Manchester University Press, 1975), p. 159.

161. Some general ideas on this theme can be found in the debate on the peaceful reservation at UNCLOS III. See, e.g., Madagascar, 5 *Official Records* (1976), p. 58; Iran, ibid., p. 66.

conventional naval forces are concerned,[162] will be a protracted process that can only bring about some restrictions on the military uses of the sea but cannot be expected to demilitarize the oceans. Negotiating various confidence-building measures, such as limitations on the number and scale of naval exercises in specific regions and exchange of information on naval matters, would be the first stepping stones; these could be followed by prohibition of naval exercises and weapons tests in certain areas of the ocean, especially those with the busiest sea lanes, and expansion of such areas into peace zones where no military activity would be permissible. A reduction of conventional and nuclear naval forces would proceed parallel with all these measures and within the general arms control framework. Finally, any real progress in developing the concept of the peaceful uses of the seas under the Convention would require elimination from its text of the optional exception of military activities from compulsory dispute settlement.

Most of the items on the above-outlined agenda for converting the soft law of the peaceful purposes reservation into firm rules of law seem highly unrealistic at this time. Yet, in the longer term, the trend in the law of the sea, initiated by the innovative but vague reservation, may eventually reverse the process of militarization of the oceans and thus contribute to making them more peaceful.

162. See the UN Study on the Naval Armament Race, UN document A/140/535 (1985).

Naval Operations in the Antarctic Region: A Possibility?

Joseph R. Morgan
Environment and Policy Institute, East-West Center

INTRODUCTION

Interest in the Antarctic region is increasing, partially due to the fact that in 1991 the Antarctic Treaty will be open for possible revision.[1] If the Treaty, which currently governs the continent and the waters of the Southern Ocean south of 60°S latitude, is to be revised in any substantial way, military strategic considerations might play a part. While claims to Antarctic land have thus far not yielded economic benefits to the seven countries claiming sovereignty over various parts of the continent, there is considerable interest in the resources of the surrounding seas, including living resources in the south polar ocean and mineral resources from the ocean floor.[2]

Moreover, Antarctica is considered to be militarily strategic by some nations, chiefly because they fear that Southern Ocean sea lanes could be controlled by naval vessels operating from bases on the continent. The possibility that strategic considerations played a part in both early exploratory efforts and current policies among nations with strong interests in the regions should not be discounted. In some cases strategic concerns could be strong enough to be a dominant factor in the decisions of nations to maintain claims on the continent.

Elsewhere in this volume Boleslaw A. Boczek discusses the peaceful pur-

---

1. Article 12 of the Antarctic Treaty permits modification or amendment at any time by unanimous agreement of the contracting parties and further states, "If after the expiration of thirty years from the date of entry into force of the present Treaty, any of the Contracting Parties . . . so requests . . . a Conference of all the Contracting Parties shall be held as soon as practicable to review the operation of the Treaty." Since the treaty came into force in 1961, such a conference could be held in 1991 or thereafter.

2. See John E. Bardach, "Fish Far Away: Comments on Antarctic Fisheries," *Ocean Yearbook 6*, ed. Elisabeth Mann Borgese and Norton Ginsburg (Chicago: University of Chicago Press, 1986), pp. 38–54, for a discussion of living resources. Mineral resources in Antarctica are discussed in Sir Anthony Parsons, *Antarctica: The Next Decade* (Cambridge: Cambridge University Press, 1987), pp. 76–97.

poses resolution in the Convention on the Law of the Sea. In a masterful analysis he concludes that the resolution that the oceans be used for peaceful purposes only is not likely to have any substantial effect on the size of navies and their operations in the seas of the world.[3] The Antarctic Treaty of 1961, which is effective for both the Antarctic continent and the waters of the Southern Ocean, has a similar peaceful purposes resolution. The purpose of this article is, in part, to compare the effectiveness of the two international law documents in influencing the use of naval force to achieve political ends in the world ocean and in the more limited Antarctic waters.

MILITARY PROVISIONS OF THE ANTARCTIC TREATY

The Preamble to the Antarctic Treaty of 1961 states, in part, that "it is in the interest of all mankind that Antarctica shall continue forever to be used exclusively for peaceful purposes and shall not become the scene or object of international discord." These words are strikingly similar to statements in the Convention that "The high seas shall be reserved for peaceful purposes," "The Area shall be open to use exclusively for peaceful purposes by all States," and "Marine scientific research in the Area shall be carried out exclusively for peaceful purposes and for the benefit of mankind as a whole."[4] However, the Antarctic Treaty goes on to state specific provisions prohibiting in various ways the use of military force in Antarctica, while the Convention provides nothing to amplify the hortatory statements quoted above.

In the Antarctic Treaty, Article 1 elaborates on the peaceful purposes resolution by stating, "Antarctica shall be used for peaceful purposes only. There shall be prohibited, *inter alia*, any measure of a military nature, such as the establishment of military bases and fortifications, the carrying out of military manoeuvers, as well as the testing of any type of weapon." Other Articles of the Treaty serve to put teeth into the idea that the region must be used for peaceful purposes only. For instance, Article 5 prohibits any nuclear explosions in the Antarctic, and Article 7 requires advance notice to be given of "any military personnel or equipment intended to be introduced . . . into Antarctica."

However, Article 6 of the Treaty states, "The provisions of the present Treaty shall apply to the area south of 60° South latitude . . . but nothing in the present Treaty shall prejudice or in any way affect the rights . . . of any State under international law with regard to the high seas within that area."

---

3. Editors' note.—See Boleslaw Boczek, "The Peaceful Purposes Reservation of the United Nations Convention on the Law of the Sea," in this volume.

4. *The Law of the Sea: Official Text of the United Nations Convention on the Law of the Sea with Annexes and Index* (New York: United Nations, 1983), sales no. E.83.v.5., Arts. 88, 141, and 143 (1).

Since naval vessels have traditionally had the right to operate freely on the high seas, the Treaty provides no more safeguards than does the Convention regarding use of the oceans for peaceful purposes only. Hence, any differences between the waters south of 60°S latitude and other parts of the oceans regarding the use of navies must be sought not in international treaty law but in an analysis of the general strategic situation of the Antarctic region.

GEOSTRATEGY AND THE ANTARCTIC REGION

Antarctica and its surrounding seas might be considered a geostrategic region[5] for one or more of the following reasons: there are valuable economic resources on the continent and in the seas around it, nations have overlapping claims to the continent which they might be willing to defend with military force, and the location of the Antarctic continent could provide bases for naval ships and aircraft, which in turn could control shipping in the Southern Ocean.

Economic Resources

Thus far there have been no demonstrated economic benefits to exploiting resources in the Antarctic region. Although the continent certainly has coal and may have other minerals of potential value, the difficulties of extraction and the distance to markets make the prospects for profitable exploitation very questionable.[6] The Antarctic continent could possibly be used for commercial or military airfields, since air operations in the region have been successfully conducted since 1928.[7] It is now commonplace to land large planes on the Antarctic ice. However, there is really no need for Antarctica as a site for commercial or military airports, since the continent does not lie close to established air routes and there is little foreseeable prospect that air routes in the Southern Hemisphere will increase in number in the near future.[8] Moreover, the increased range of commercial and military jet aircraft now enables planes to overfly the Antarctic region without the necessity of refueling.

---

5. Geostrategic regions are areas where geopolitical and strategic factors are combined in such a way that one or more countries might consider the region important enough to undertake military action to protect or enhance their interests.

6. Parsons, p. 93.

7. J. F. Lovering and J. R. V. Prescott, *Last of Lands: Antarctica* (Carlton South, Victoria: Melbourne University Press, 1979), p. 139.

8. Central Intelligence Agency, *Polar Regions Atlas* (Washington: CIA, 1978), p. 5.

National Claims

There are seven national claimants to territory in Antarctica: Argentina, Australia, Chile, France, New Zealand, Norway, and the United Kingdom (fig. 1). Argentina and Chile base their claims on the division of the world between Spain and Portugal by the Pope in 1495, both insisting that they are legitimate heirs to old Spanish claims.[9] The other countries use the more common ideas of discovery, exploration, and effective occupation as the basis for claims. All but Norway have adopted the sector principle in their claims, laying claim to all of the continent between certain specified meridians of longitude, which of course meet at the South Pole. Norway, because it has objected to the sector principle in the Arctic region, since it would gain far less territory than either the USSR or Canada, has been careful not to recognize the principle in the Antarctic. Consequently, its claim is to a coastal region extending an unspecified distance inland.

The sector from 90°W to 150°W is thus far unclaimed, although it is generally understood that if the U.S. were to make a claim to it other nations would not object.[10] The U.S., however, while reserving the right to make future claims, thus far is not a claimant nation, despite a long record of Antarctic exploration, discovery, and scientific research.

United States activity in Antarctica has a long history; in 1819 Nathaniel Palmer, then in command of a small sealing vessel, may have been the first to discover the continent. Most of the U.S. activities in the nineteenth century were in what has now become the Australian sector. The Wilkes expedition mapped the coast of what is now known as Wilkes Land. American aviators, particularly Richard Byrd and Lincoln Ellsworth, were very active in the early decades of the twentieth century in the still unclaimed sector west of the Antarctic Peninsula. Byrd Land and Ellsworth Land appear as prominent place names in the region.

Australia claims the largest area of the continent (approximately 42%), while France claims only a narrow slice, between 136°W and 142°W longitude. The French claim is sandwiched in between Australian claims, and it is clear that the two countries have been careful to respect each other's sovereign rights on the continent. This is certainly not the case with the U.K., Argentina, and Chile, all three of which have overlapping claims to parts of the Antarctic Peninsula. The three countries maintain bases on Deception Island, off the west coast of the Peninsula, and, although there have been no actual hostilities over the overlapping claims, exchanges of diplomatic notes have been frequent.

9. John Hanessian, Jr., "National Interests in Antarctica," in *Antarctica*, ed. Trevor Hatherton (New York: Praeger, 1965), pp. 12, 15.
10. Central Intelligence Agency, p. 43.

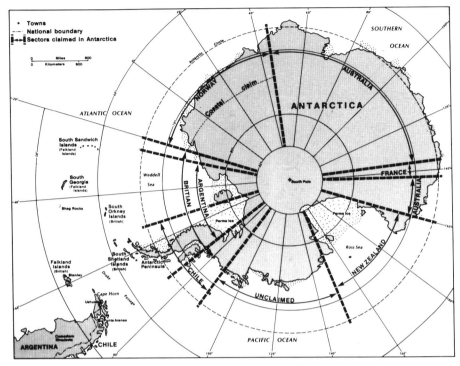

Fig. 1.—The Antarctic region

The USSR does not make any Antarctic claims but, like the U.S., points out to claimant nations that it demands to be consulted in Antarctic matters.[11] Russian exploratory and scientific activity in the Antarctic region dates back to the work of Bellingshausen in 1819–21, and the USSR has been particularly active during the IGY, with several bases on the continent and large annual resupply and scientific expeditions.

Norway's coastal claim is based on considerable exploration activity, including, of course, the work of Roald Amundsen and his party in reaching the South Pole in 1911. Moreover, Norwegians have been quite active in Antarctic whaling; the coastal claim was undoubtedly for the primary purpose of establishing bases for Norway's whaling fleet on both Antarctic continental land and subantarctic islands. Claims to inland territory were not necessary, and, as has been previously mentioned, any reference to inland claims based on the sector principle would have hurt Norway's position in the Arctic.[12]

Neither Argentina nor Chile was very active in Antarctic exploration prior to making its claims, but both countries have since established a number

11. Ibid.
12. Lovering and Prescott, pp. 182, 183.

of stations on the continent and offshore islands. Consequently, both countries announced their claims rather late in the game, although Argentina stated that it really had no need at all to make a formal claim since the Argentine sector of the Antarctic has always belonged to Argentina as the inheritor of Spanish territory. Chile announced its claim on November 6, 1940 in the following terms: "All lands, islets, reefs of rocks, glaciers (pack ice) already known or to be discovered, and their respective territorial waters, in the sector between longitude 53° and 90° West, constitute the Chilean Antarctic or Chilean Antarctic Territory."[13]

The wording of the claim is interesting in several respects. It follows the sector principle, but neither of the longitude limits corresponds to a limit of Chilean territory. The 90°W limit lies between the island of Juan Fernandez, at about 80°W, and Easter Island, in the vicinity of 110°W. It is possible that the 90°W boundary was chosen to avoid overlap with the Norwegian claim to Peter I Island, which lies 10 nm west of the 90°W meridian.[14] The eastern limit of the Chilean claim lies just 25 nm east of Clarence Island, the easternmost of the South Shetland islands.[15]

In 1941, Argentina claimed the sector between 25°W and 68°34'W; in 1957 the western boundary was changed to 74°W, just west of the most westerly point of Argentina. The 68°34' boundary had corresponded to the meridian designated as the boundary between Chile and Argentina on the island of Tierra del Fuego by the treaty between the two countries signed in 1881.[16] There is overlap between the Argentine and Chilean claims from 53°W to 74°W, which includes all of the Antarctic Peninsula.

On July 21, 1908, the United Kingdom laid claim to the area south of 60°S between longitudes 20°W and 80°W. Thus, there is an overlap with the Chilean claim between 53°W and 80°W, and with the Argentine claim between 25°W and 74°W. Most of the important Antarctic Peninsula and the offshore islands are claimed by all three countries.

In 1923 the British laid claim to the area south of 60°S between 160°E and 150°W. The area was designated as the Ross Dependency with administration to be vested in the Governor of New Zealand.[17]

The current Australian Antarctic territory was originally claimed by the U.K. in 1933 and accepted by Australia in the same year. It includes all the land between 45°E and 160°E, except for the French Adelie Land claim between 136°E and 142°E, which had been made on March 29, 1924.[18]

Norway, despite a long record of exploration and whaling activity in the

13. *Polar Record* 8, no. 53 (1956): 147.
14. Lovering and Prescott, p. 174.
15. Ibid.
16. Ibid., p. 175.
17. Ibid., p. 181.
18. Ibid., p. 182.

South Polar region, made no claims to territory on the continent itself until January 14, 1939, when it claimed "the mainland coast of the Antarctic between the British and Australian claims, that is, between meridians 20°W and 45°E."[19] Prior to that date Norway claimed only Peter I Island and Bouvet Island, presumably as potential whaling stations. The German expedition of 1938–39 and the numerous flights from the seaplane tender *Schwabenland* undoubtedly convinced the Norwegians that a German claim to Antarctic continental territory was forthcoming; the Norwegian claim was presumably made to forestall a German claim, particularly since the claimed area was one in which Norwegian explorers had been most active in the years 1926–37.[20]

The Antarctic Treaty, signed by all the claimant nations as well as the U.S., USSR, Belgium, Japan, and South Africa (which *does* claim some subantarctic islands), has managed to avoid any attempt to settle problems of overlapping claims. Article 4 of the Treaty reads:

1. Nothing contained in the present Treaty shall be interpreted as:
   (*a*) a renunciation by any Contracting Party of previously asserted rights of or claims to territorial sovereignty in Antarctica;
   (*b*) a renunciation or diminution by any Contracting Party of any basis of claim to territorial sovereignty in Antarctica which it may have whether as a result of its activities or those of its nationals in Antarctica, or otherwise;
   (*c*) prejudicing the position of any Contracting Party as regards recognition or non-recognition of any other State's right of or claim or basis of claim to territorial sovereignty in Antarctica.
2. No acts or activities taking place while the present Treaty is in force shall constitute a basis for asserting, supporting or denying a claim to territorial sovereignty in Antarctica or create any rights of sovereignty in Antarctica. No new claim, or enlargement of an existing claim, to territorial sovereignty in Antarctica shall be asserted while the present treaty is in force.

The Treaty, therefore, permits signatories to maintain their present claims while at the same time recognizing or not recognizing the claims of others. Most important, it forbids additional claims or extensions of current claims by any of the signers as long as the Treaty remains in force. Since other articles of the Treaty require free exchange of information and permit scientists of all nations to work in the region, the issue of sovereignty is greatly downplayed. Moreover, military activity is forbidden on the continent; hence, national claims specifically for the purpose of maintaining naval and air bases are precluded. In short, the Treaty thus far has made the issue of overlapping

19. Ibid., p. 183.
20. Ibid.

sovereign claims and the idea of national territories in Antarctica relatively inconsequential. Whether these issues will remain inconsequential remains to be seen. Certainly, if Antarctica is considered to be strategically important to one or more nations, the problems of sovereign claims will become more critical.

STRATEGIC CONSIDERATIONS

As previously noted, Article 1 of the Antarctic Treaty states that Antarctica shall be used for peaceful purposes only and that establishment of military bases in the Treaty area is forbidden. Likewise forbidden are military maneuvers and the testing of weapons. However, all of the signatory nations carried out work in Antarctica prior to the Treaty, and it is certainly possible that strategic considerations have played a part in national policies of countries that have been active in the region. Claimants to Antarctic territory may some day wish to use coastal areas over which they claim jurisdiction as sites for naval or air bases; it is likely that some claims were initially made to forestall the activities of other possible claimants.

Since few of the claimants operate large navies with worldwide base facilities—only France and the U.K. have overseas bases, and the British in recent years have shown a tendency to pull back from worldwide commitments and concentrate on defense of the home islands—it seems improbable that current Antarctic claims, if they have any strategic value at all, are for other than defensive purposes. It is always possible, however, that the claims themselves, particularly if they are contested or overlapping, such as in the Antarctic Peninsula area, may lead to military action. Since the Antarctic Treaty forbids the establishment of military bases on the continent it can be argued that signatory nations with Antarctic claims are effectively preventing strategic influence from becoming important, since signers of the Treaty presumably are adhering strictly to the Treaty's provisions, and are being careful to avoid military activities.

Bertram gives five motives for Antarctic activities: adventurous, economic, scientific, political, and strategic.[21] On the surface it appears that all of the currently active Antarctic Treaty nations are pursuing scientific objectives, but political, economic, and strategic objectives might also play a part in their continued activities on the continent. The adventurous motive peaked with the attainment of the South Pole and is now no longer an important factor. Antarctic research activities are now almost always sponsored by governments, and the logistic support provided enhances safety and efficiency. Clearly, scientific rather than adventurous motives are now preeminent.

21. G. C. L. Bertram, *Antarctica Today and Tomorrow* (Dunedin: University of Otago, 1958), p. 3.

The fact that seven nations claim parts of the continent and an eighth claims some subantarctic islands indicates that political motives are still important. But a political motive for Antarctic activity could be in effect either economic or strategic. The real question to be asked is, Why do the eight countries claim Antarctic or subantarctic territory? The Antarctic continent almost certainly has mineral resources of value; coal in considerable quantities has been discovered, and the offshore ocean floor may contain oil. But are the minerals valuable enough to make their exploitation feasible in view of the formidable environmental obstacles that must be overcome? Antarctic waters likewise contain valuable resources in the form of the abundant large zooplankton, krill. Thus far logistic and processing problems have precluded successful economic exploitation of the resource.

But what of strictly military considerations? Is the continent valuable as a site for naval bases? Probably not, since modern warships require sophisticated support facilities and need supplies of fuel, food, spare parts, and ammunition. They also require extensive repair facilities—shipyards, for example. The Antarctic cannot provide for these needs, and moreover the presence of sea ice for much of the year makes operations dangerous for both surface ships and submarines. Icebreakers, of course, are the obvious exception, but this type of vessel really has little in the way of military capability; its chief use is to provide support for resupply vessels and a platform for certain scientific investigations.

According to Hanessian, "The U.S. Department of Defense has consistently taken the position that the Antarctic area presents little strategic value to the United States."[22]

The U.S. analysis is in agreement with that of most other students of the problem, who generally base their thinking about military strategy on the actions and motives of industrialized Northern Hemisphere nations. Of the Northern Hemisphere industrialized countries only the U.K. has been truly concerned with the strategic importance of the Antarctic. The recognition by the British of the strategic location of the Drake Passage, between the southern tip of South America and the northern extremity of the Antarctic Peninsula, led the U.K. to voice strong objections to the idea of establishing a UN trusteeship in the region.[23] The British also wanted control over the South Shetland and South Orkney Islands, which flank the passage.

Both the U.S. and the USSR have been active in the region, particularly since the International Geophysical Year of 1956/57, but neither has claimed sovereignty over Antarctic territory, and both emphasize the North Polar region as by far the more strategic of the high-latitude areas.

Despite the lack of current interest in Antarctica's strategic value, the U.S. was once apparently considering Antarctic naval operations seriously.

22. Hanessian (n. 9 above), p. 42.
23. Lovering and Prescott (n. 7 above), p. 153.

Operation Highjump, conducted in 1946, included an aircraft carrier, destroyer, and fleet submarine.[24] Only the carrier seemed to have a legitimate nonmilitary mission, to ferry long-range land-based aircraft to the continent. The planes then operated from ice and snow runways.

The submarine, USS *Sennet,* was deemed to be too frail to operate in ice-infested waters and was eventually towed back to open water by an ice-breaker.[25] Admiral Byrd, Commander of the operation, concluded that "submarines are not fitted for ice maneuvers and risk constant danger of disaster."[26] Since then, however, nuclear powered submarines have demonstrated their ability to operate in and under polar ice on several occasions, and more than one has surfaced at the North Pole. The question of submarine operations in Antarctic waters is no longer a question of capability but one of utility. The lack of surface ship targets, both military and commercial, in Antarctic waters makes the value of attack submarines decidedly questionable. It now seems apparent that submarine operations by the U.S. in the South Polar region in 1946 were for the purpose of testing their capability in polar environments generally and that the Arctic was envisioned as the eventual site of operations.

The destroyer, USS *Henderson,* seemed to have even less function in Operation Highjump than the submarine. It was used in conjunction with the seaplane tender *Currituck* and the naval tanker *Cacapon* presumably to "protect" the two vulnerable ships in the event of an "enemy" attack by submarines or other surface ships.[27] Since this was a typical peacetime U.S. Navy exercise, the use of a destroyer seems reasonable, even though it is highly unlikely that one would ever be actually employed in the Antarctic in a hostile enemy environment.

Among the Northern Hemisphere nations, Norway and France also claim parts of the continent. There is no evidence that France's claim has strategic overtones, and despite the timing of Norway's claim, to forestall a possible German claim in the same sector, the Norwegian claim apparently was more for economic than strategic reasons.

But any assessment of the strategic value of Antarctica must consider the views of Southern Hemisphere countries, which are almost certain to have different perspectives on the military value of Antarctica than do Northern Hemisphere nations. Argentina, Australia, Chile, New Zealand, and South Africa, which has claims to some subantarctic islands only, all have some interest in the possible use of the South Polar continent as a base for military operations. In reviewing national interests in Antarctica, Hanessian wrote of

24. A description of U.S. naval activities is contained in Richard E. Byrd, "Our Navy Explores Antarctica," *National Geographic Magazine* 92: 429–522.
    25. Byrd, p. 438.
    26. Ibid., p. 458.
    27. Ibid., p. 491.

the "important geopolitical position of this huge continent. Argentina, Chile, New Zealand, and Australia all have valid concern regarding the necessity of protecting their southern flank."[28] The four countries do not discount the possibility that the <700-mile-wide Drake Passage might become an important sea route if wartime activities should result in the closing of the Panama Canal.[29] Bertram points out that Deception Island "affords the sole near continental site for any form of Antarctic naval base or other substantial military establishment. It is moderately free of sea ice for much of the year and there are ample areas of flattish ice-free land."[30] It is no wonder then that the U.K., Argentina, and Chile all claim Deception Island and all three countries maintain scientific stations there.

The U.K. has already demonstrated that it considers the Falkland Islands important enough to fight for, but whether or not the British government feels strongly enough about its Antarctic claims to consider military operations there is conjectural, in view of the hostile natural environment and great expense of supporting military forces on the continent. It is also possible, of course, that Argentina or Chile might consider military actions to support their claims. It seems unreasonable to argue that the presence of scientists on Deception Island presents a security threat to any of the three claimants, but when nationalistic feelings run high the possibility of armed conflict over seemingly inconsequential and illogical issues is increased.

In addition to nationalistic feelings over their Antarctic claims, both Argentina and Chile have even more reason to consider strategic matters in formulating their Antarctic policies. Less than 1,000 km separates Tierra del Fuego from the northern extremity of the Antarctic Peninsula; control of the sea route through the Drake Passage is thus an important concern.[31] In Chile there has long been awareness of the importance of Antarctica to her security and "to some extent her interest in the Antarctic is derived from strategic considerations due to her close proximity to the continent."[32]

Australia has based her Antarctic policies on considerations of security, in addition to her long-term interests in scientific work and exploration on the continent. "Because Antarctica could be used by a hostile power as a base threatening Australian security, she has opposed the entrance of other nations, particularly the Soviet Union (and to a lesser extent Japan) into her polar 'back door.' She has consistently refused to consider renouncing her Antarctic claims, partly because of strong nationalistic feeling but partly because of the fear that the continent (although nearly 3,000 km distant) might be used as a base for attack by a hostile force."[33]

New Zealand's Antarctic claim is sizable, and both New Zealand and U.S.

28. Hanessian, p. 5.
29. Bertram, p. 9.
30. Ibid.
31. Hanessian, p. 12.
32. Ibid., p. 15.
33. Ibid., p. 14.

expeditions use New Zealand ports as take-off points for their annual scientific and logistic efforts on the continent. New Zealand "inherited" its Antarctic territory, the Ross Sea Dependency, from the British; the initial U.K. claim was probably made more for historical, scientific, and political motives than for strategic considerations. Both Scott and Shackleton used bases on Ross Island for their expeditions, and the Ross Sea and Ross Ice Shelf were discovered by the Britisher James Clark Ross.

New Zealand, as a Southern Hemisphere country, might be expected to take the possibilities of the continent's being used for military activity more seriously. "Although most New Zealanders have long felt that Antarctica is not strategically or economically important . . . the necessity of securing the southern frontier has often been asserted, and vague arguments concerning security have been made to explain New Zealand's sensitivity to Japanese and Soviet whalers in the adjacent subantarctic waters."[34] Although the strategic nature of Antarctica is apparently not a factor in official New Zealand military policy, "The strategic importance of controlling possible bases and air lanes in the Southern Hemisphere great circle route across Antarctica has also received considerable journalistic attention. It has further been argued that Antarctic bases could prevent raiders from moving through southern waters to prey on shipping as they did during World War II,"[35] a reference to German combat activities in the region.

Two German raiders, *Pinguin* and *Komet*, destroyed 193,000 tons of Allied shipping. Part of the Norwegian whaling fleet was intercepted at 59°S, 2°30'W, and two factory ships, a supply ship, and seven whale catchers were captured.[36] The German raiders used Kerguelen Island as a base, a fact confirmed by searches conducted by HMS *Neptune*. Subsequently, Australian naval forces laid mines in the approaches to Kerguelen Island. Southern Hemisphere nations are naturally more sensitive to the possibility that similar naval activity might take place during a future war.

The demise of the whaling industry has reduced the number of available targets, but there is an important shipping route for large tankers south of the Cape of Good Hope, brought about partially by the earlier closing of the Suez Canal and the subsequent building of very large, deep-draft vessels. Closing the Panama Canal, always possible in wartime, would create a vital sea lane south of Cape Horn.

Neither New Zealand nor South Africa is in a position to do much about protecting Southern Hemisphere sea lanes. The New Zealand navy is far too small to protect the two main islands, patrol the very large Exclusive Economic Zone that it now claims, defend its Antarctic territory, and exercise sufficient control of the Southern Hemisphere seas to insure the safe passage

34. Ibid., p. 20.
35. Ibid.
36. A description of German naval activities in the Antarctic during World War II can be found in Hanessian (n. 9 above), p. 47, and *Polar Record* 4, no. 32 (1946): 402, 403.

of ships in and out of its ports.[37] South Africa "cannot afford to ignore the strategic aspects" of the Antarctic region,[38] particularly in view of the already existing important shipping route for tankers around the Cape of Good Hope. However, recent decisions about the size and composition of the South African navy make it unlikely that naval operations in subantarctic waters are contemplated. The fleet now consists primarily of small coastal defense craft, vessels not capable of operations in the stormy South Polar sea.[39]

The Australian navy is much larger and more capable than those of either New Zealand or South Africa,[40] and with a claim to 42% of the continent and expressed concerns over the Soviet base at Mirnyy, which is within 3,500 km of Australia, the Australians might be inclined to operate in Antarctic waters. Moreover, Australia has memories of World War II German combat activities in the region. For the present, however, there is no evidence that USSR activities at Mirnyy are anything but scientific, and IGY and post-IGY exchanges of visits to the Soviet station have confirmed that the research work conforms to the Antarctic Treaty.[41]

The Southern Hemisphere view of the strategic nature of the Antarctic continent has been summarized, "To these four states (and to a lesser extent South Africa) the Antarctic is not a frigid ice mass, but a nearby continent on which hostile military activity could easily threaten national security. Australia for example has for some years been uneasy regarding Soviet Antarctic intentions."[42]

Others have expressed similar views, particularly the U.S. Central Intelligence Agency, which stated, "The Antarctic has been described as a remote ice sheet of no major strategic value, but to Argentina, Chile, Australia, and New Zealand, it has at times been considered a nearby continent from which a hostile power could threaten their security. National feelings concerning Antarctica are particularly strong in Argentina and Chile, since less than 1,000 kilometers separate Tierra del Fuego from the Antarctic Peninsula across Drake Passage. These and other nations have also been aware of the potential use of Drake Passage as an alternative to the Panama Canal."[43]

While some Southern Hemisphere nations may consider the Antarctic of strategic importance, the superpowers, the U.S. and USSR, do not. Both make it clear that they are to be consulted as signers of the Antarctic Treaty, but military strategy does not figure greatly in their Antarctic policies. As nonclaimants both would be in favor of the maximum extent of high-seas

---

37. A description of New Zealand naval capabilities can be found in Joseph R. Morgan, "Small Navies," *Ocean Yearbook 6,* ed. Elisabeth Mann Borgese and Norton Ginsburg (Chicago: University of Chicago Press, 1986), pp. 381, 382.

38. Bertram (n. 21 above), p. 14.

39. See Morgan, pp. 385, 386.

40. Information on the navies of Australia, New Zealand, and South Africa is from *Jane's Fighting Ships, 1988–89* (London: Jane's, 1989).

41. Hanessian, p. 32.

42. Ibid., p. 5.

43. CIA (n. 8 above), p. 5.

freedoms for the region, and both would tend to resist the actions of claimants in expanding their territory offshore in the form of Exclusive Economic Zones and Continental Shelves.

Employment of modern warships in the Antarctic Ocean would pose problems in logistics and upkeep. Ports would be exceedingly laborious to build, and the presence of sea ice would make operations difficult and dangerous. There is still relatively little commercial seagoing traffic in the region, although a wartime closing of the Panama and Suez canals would necessarily divert many ships into southern waters. A world war fought with conventional weapons would undoubtedly include the Southern Hemisphere, as it did in World War II. Hence, Southern Hemisphere nations would be concerned with protecting sea lanes and their coasts from attack.

Whether or not the Antarctic is strategically important depends on which nation is making the judgment. Australia, Argentina, Chile, New Zealand, and South Africa have good reason to consider military strategy in formulating their Antarctic policies. The U.K. has also demonstrated its strategic interests in the region. But strategic considerations, while they might influence general policy, probably will not be strong enough to make the region worth fighting for.

POSSIBILITIES OF NAVAL CONFLICT IN THE ANTARCTIC

Although the Antarctic Treaty contains a number of admonitions to avoid naval conflict in the waters surrounding Antarctica, and the strategic value of the continent does not seem to be great, either in terms of its location or the possibility that valuable living and mineral resources can be economically obtained, naval activities are possible under two circumstances. First, the waters south of 60°S latitude, although governed by the Treaty, are high seas, and naval vessels as well as merchant ships are accorded the usual freedoms of navigation. Therefore, there is nothing in the Treaty to prevent naval operations by any nation in Antarctic waters.

There is a possibility that worldwide changes in naval capabilities, particularly in submarine and antisubmarine operations, might make operations in far southern latitudes strategically valuable. Fleet ballistic missile submarines, which both superpowers as well as France and the United Kingdom operate, are now capable of launching their weapons at greatly increased ranges from targets in the Northern Hemisphere. Hence, it may be possible in the foreseeable future to operate submarines near Antarctica with missiles targeted in the Soviet Union or the United States. As antisubmarine warfare capabilities improve, it is conceivable that someday Northern Hemisphere submarine operating areas will become unsafe to the superpowers and they will have to deploy their ballistic missile subs much farther from prospective targets.[44] As

44. Parsons (n. 2 above), p. 99.

a matter of fact, the current range of submarine-launched missiles makes it possible to station the submarines very close to home waters, perhaps actually in port areas, and still hit enemy targets. Therefore, it is not inconceivable that Antarctic waters might be used in order to make the antisubmarine efforts more difficult with greatly increased areas of ocean to be covered.

The disadvantage of using the Southern Ocean waters south of 60°S as submarine operating areas is the increased travel time required to reach an operating area and the consequent loss of time on station.

The argument that increasingly more effective antisubmarine warfare technology, coupled with greater ranges of submarine-launched ballistic missiles, might make the Antarctic region more strategically valuable is contrary to the views of other writers on the subject. For instance, according to Deborah Shapley, "Military strategy has evolved so as to make the Antarctic even less important strategically than in 1961 because the United States has become less dependent on overseas bases. . . . In the last 23 years satellites have become less dependent on networks of ground tracking stations, lessening the need to use Antarctica to track satellites. . . . The Antarctic Treaty has had an advantage over other arms control accords, in that evolving technology has created less rather than more pressure to violate or change it."[45]

The second circumstance under which Antarctic waters might become scenes of naval combat is the previously mentioned possible closing of the Suez and Panama canals. This is not merely conjectural, since the Suez Canal has been closed for long periods as a consequence of wars in the Middle East. The Suez closing resulted in the construction of larger oil tankers, which still transit from the Middle East oil terminals to Western Europe via a route south of the Cape of Good Hope. If the Panama Canal were to be closed due to wartime activities or the decision of the government of Panama to restrict traffic in the canal to countries with which it was allied, a route south of Cape Horn would become important. A world war might then spread deep into the Southern Hemisphere as enemy navies attacked sea lanes in far southerly latitudes.

However, under the envisioned circumstances the Drake Passage, between Cape Horn and the Antarctica Peninsula, can scarcely be considered a genuine choke point, since it is more than 600 miles wide. Waters south of the Cape of Good Hope are even wider. Hence, it probably would not be considered strategically cost-effective to establish naval bases in Antarctica, and *extensive* naval operations by ships based in southern South America and South Africa, though possible, are not likely.

However, there are some military strategists who disagree with this contention. They argue that the United Kingdom's successful efforts to retake the Falkland Islands and South Georgia from Argentina in 1982 were for strategic purposes. Keegan and Wheatcroft, for instance, state, "The impor-

45. Deborah Shapley, *Bulletin of the Atomic Scientists* 40, no. 6 (1986): 31–32.

tance of the Antarctic region only becomes clear from reversing the normal geographical perspective. The Falklands and the other British possessions occupy a strategically dominant position across the Drake Passage, which controls access from the Pacific to the Atlantic—a vital link. So even if the commercial value of Antarctica proves illusory, no power interested in the control of the line of access can afford to allow dominance to slip into unfriendly hands. This is the key to Fortress Falklands."[46]

CONCLUSION

Although the peaceful purposes resolution of the United Nations Convention on the Law of the Sea is unlikely to have much effect in restricting naval operations in either peacetime or wartime circumstances, naval operations in the waters near Antarctica are unlikely. The Antarctic Treaty has more specific prohibitions against military uses of the continent and the waters south of 60°S latitude, but more important, these waters are simply not now of a sufficiently strategic nature to make the use of navies probable.

But what of the future? Will foreseeable changes in either military technology, geostrategy, or the Antarctic Treaty itself change the picture? In his concluding chapter Parsons wrote, "Although the question of SSBN [nuclear-powered fleet ballistic missile submarine] safe havens in Antarctica may be remote, it should be recognized that the Treaty does not appear to ban the deployment of SSBNs in the region. It is for others to consider whether the issue should be raised or left fallow. Given the extreme difficulty of policing a rule which kept out SSBNs, and given the general conflict of such a rule with the Law of the Sea and Rights of Passage, it may be judged best not to raise it. We must recognize too that in war the superpowers will act in their own security interests regardless of the provisions of international agreements."[47]

Whether or not Antarctic waters will continue to be used solely for peaceful purposes will be determined by the combined effects of military and strategic concerns and international law. For the foreseeable future, the rule of law seems to be dominant, particularly in view of the improvement in relations between the superpowers and the restraint exercised by nations with competing and overlapping claims to the continent. It is, of course, possible that technological change and innovation in military equipment and strategy might change the situation in Southern Ocean waters, but at the present time scientific and economic considerations heavily favor the maintenance of the peaceful status quo in the Antarctic region.

46. John Keegan and Andrew Wheatcroft, *Zones of Conflict: An Atlas of Future Wars* (New York: Simon & Schuster, 1986), p. 143.
47. Parsons, p. 107.

The Environmental North Sea Regime: A Successful Regional Approach?

Steinar Andresen

Fridtjof Nansen Institute, Lysaker, Norway

INTRODUCTION

Generally, the negotiations over the provisions for the marine environment during the Law of the Sea negotiations did not attract much attention, and the making of these rules was not one of the major objectives of UNCLOS III.[1] It has also been maintained that the environmental provisions are superfluous; Part XII of the Convention contains nothing that cannot be found in already existing environmental agreements.[2] This is generally true for states that are already bound by other regional or global agreements, but Part XII does provide a general framework for marine environmental protection. One of the more important features is the comprehensive character of the 1982 Convention: "the problems of ocean space are closely interrelated and need to be considered as a whole."[3] This comprehensiveness is also reflected in Part XII of the Convention.

The 1982 Convention underlines the importance of both global and regional cooperation to reduce and control marine pollution. Although global standards and cooperation may be important, the protection of the marine environment has promoted most activity at the regional level. Even by 1972 the Stockholm Conference had emphasized the need for regional coordination in reducing marine pollution. The essential role of international cooperation, especially at the regional level, in the management of the seas is also

1. Peter Hayward, "Environmental Protection: Regional Approaches," *Marine Policy*, 8, no. 2 (April 1984): 107.
2. See *The Law of the Sea: Official Text of the United Nations Convention on the Law of the Sea with Annexes and Index* (New York: United Nations, 1983), Sales No. E.83.v.5 (and hereinafter cited as the 1982 Convention, or simply the Convention). This question is discussed by Jan Schneider, "Protection and Preservation of the Marine Environment: What Is New about the Law of the Sea Convention?" in *The 1982 Convention on the Law of the Sea,* ed. A. Koers and B. Oxman, Proceedings of the Law of the Sea Institute, seventeenth annual conference (Honolulu: University of Hawaii at Manoa, Law of the Sea Institute, 1984), pp. 567–74.
3. 1982 Convention, preamble.

expressed in the Report of the World Commission on Environment and Development.[4]

In Europe, which has several enclosed or semienclosed ocean areas, the regional approach has been dominant in considering the particular economic and ecological circumstances of specific areas. In some cases, the comprehensive approach, advocated in the Convention, has been chosen. There is thus one convention covering all sources of pollution for the Baltic Sea,[5] and an action plan has been developed for the Mediterranean, based on "integrated planning of the development and management of the Mediterranean Basin."[6] This was the first major activity of the United Nations Environment Programme's (UNEP) Regional Seas Programme in controlling pollution of the seas, adopted in 1975.

However, the environmental regime for the North Sea has developed in a more piecemeal and ad hoc fashion; as new environmental problems were perceived, the North Sea states tried to solve them by setting up different agreements. One reason for this ad hoc approach may be the fact that the first agreements were being negotiated around 1970, which was early, considering the short history of international pollution control. Thus almost 2 decades have passed since the countries bordering the North Sea perceived the necessity of protecting the North Sea against pollution.

Today, the North Sea is probably one of the ocean areas where the input of contaminants is most extensively regulated. How successful have the North Sea countries been in reaching their declared aim of reducing the level of pollution in the North Sea? In a comparative perspective, the prospects for controlling pollution in the North Sea should be good. This assumption is based on three facts: (*a*) compared to most other marine regions, the countries bordering the North Sea are all fairly wealthy industrialized countries; (*b*) the North Sea is one of the ocean areas in the world that has been the object of the most extensive research with a view to mapping the extent and

4. World Commission on Environment and Development, *Our Common Future* (New York: United Nations, 1987).

5. Convention on the Protection of the Marine Environment of the Baltic Sea Area (Helsinki Convention; signed March 24, 1974, and entered into force May 1980); and Gerhard Peet, "Sea Use Management for the North Sea," in *The UN Convention on the Law of the Sea: Impact and Implementation,* ed. E. Brown and R. Churchill, Proceedings of The Law of the Sea Institute, nineteenth annual conference (Honolulu: University of Hawaii at Manoa, Law of the Sea Institute, 1987), p. 430.

6. Editors' note.—See Adalberto Vallega, "A Human Approach to Semienclosed Seas: The Mediterranean Case," *Ocean Yearbook 7,* ed. Elisabeth Mann Borgese, Norton Ginsburg, and Joseph R. Morgan (Chicago: University of Chicago Press, 1988), pp. 372–93, and see nn. 8, 13, and 15 for references to the UNEP and other programs that have been instigated to better integrate development and management of the Mediterranean Basin. See also "Symposium on Marine Co-operation in the Mediterranean Sea, Third Tunis Declaration, 28 November 1986," *Ocean Yearbook 7,* pp. 534–36.

effect of marine contamination;[7] and (c) the extensive legal and political cooperation between the North Sea countries to protect the North Sea environment is an indication of these countries' intention to pool resources at the
regional level to control marine pollution.

The North Sea states have chosen two separate strategies to control
marine pollution; a traditional legal (convention) approach and a new political
(conference) approach. What has been the separate and combined effect of
these strategies to reduce the level of contaminants entering into the North
Sea, and what can be learned from the North Sea experience concerning the
effectiveness of this specific regional approach and applied to other regional
seas? First, consider some basic facts about the North Sea.

THE NORTH SEA

The North Sea is a shallow basin, relatively enclosed, with a volume of about
47,000 cubic km.[8] The average depth is only 70 m, making it rich in a variety
of natural resources, although the volume of water is relatively small. The
North Sea has one of the world's most productive fisheries; it covers only 0.2%
of the world's ocean, though more than 4% of the world's catch of fish is taken
here. The North Sea is one of the most heavily trafficked shipping areas in the
world and has a considerable production of offshore oil and gas.[9] It is surrounded by a densely populated area with many large towns and long-
established industries around its coast. The combined effect of these factors is
that large amounts of wastes and contaminants reach the North Sea every day
from rivers, from atmospheric inputs, and from direct discharges. What has
been the response of the North Sea countries to this problem?

THE NORTH SEA ENVIRONMENTAL REGIME:
A SHORT DESCRIPTION

The Traditional Approach: The Conventions

The following will focus only on the regional arrangements.

The Bonn Agreement
The Agreement for Co-operation in Dealing with Pollution of the North Sea
by Oil (the Bonn Agreement), dates back to 1969 and was the response of the

7. Sunneva Saetevik, *Environmental Cooperation between the North Sea States* (London: Belhaven Press, 1988), p. 9.

8. Second International Conference on the Protection of the North Sea, Scientific
and Technical Working Group, *Quality Status of the North Sea, Summary* (London: Department of the Environment, September 1987), p. 5.

9. Brit Fløistad, *Conflicting Usage of the North Sea* (Lysaker, Norway: Fridtjof Nansen Institute, 1988), pp. 2–5.

North Sea states to the *Torrey Canyon* disaster in 1967. The eight states bordering the North Sea are, according to the Bonn Agreement, responsible for surveillance, reporting, and combating oil spills in their respective economic zones and for providing mutual assistance if necessary. The Bonn Agreement went into force only a couple of months after it opened for signature in 1969. "For several years the Bonn Agreement lay dormant,"[10] but disasters sparked life into the organization, specifically, the Ekofisk blowout on the Norwegian Continental Shelf in 1977 and the *Amoco Cadiz* disaster a year later. Following these incidents, the contracting parties have met annually, and a separate technical working group has been set up. In 1983, the eight states that signed the Bonn Agreement, with the addition of the European Economic Community (EEC), signed a new Bonn Agreement extending their cooperation to harmful substances other than oil.

The Oslo Convention[11]
The Oslo Convention for the Prevention of Pollution by Dumping, signed in 1972 and entered into force 2 years later, was the earliest of the regional conventions established to prevent, rather than to cope with the resulting problems of, marine pollution. Again, a specific incident was instrumental in speeding up the creation of a convention. A Dutch ship, the *Stella Maris*, intended to dump 650 tons of chlorinated waste into the northern North Sea. Because of public opinion and governments' protests, the ship had to return without dumping the waste, and the incident demonstrated the need for international regulation of the practice of ocean dumping.

The Oslo Convention introduced a new principle that has since been copied by most international treaties on marine pollution: the differentiation between a *grey* list and a *black* list. As a point of departure, dumping black-listed materials is prohibited, although gray-listed materials may be dumped but require special care. According to the Oslo Convention, all dumping operations require approval by national authorities, and the Convention also specifies certain provisions that must be applied by the contracting parties when permits are issued. A separate protocol regulates incineration at sea.[12]

The Paris Convention[13]
The Paris Convention for the Prevention of Marine Pollution from Land-based Sources was set up because the contracting parties to the Oslo Conven-

10. Hayward (n. 1 above), p. 106.
11. Editors' note.—For more information about the Oslo Convention, see "Excerpts from the ACOPS Annual Report 1982," *Ocean Yearbook 5*, ed. Elisabeth Mann Borgese and Norton Ginsburg (Chicago: University of Chicago Press, 1985), pp. 294–304, esp. at pp. 296–97.
12. Hayward, pp. 110–11.
13. Editors' note.—For more information about the Paris Convention, see "Excerpts from the ACOPS Annual Report 1982," pp. 297–98.

tion realized that this problem also needed to be dealt with. This Convention was opened for signature in 1974 and went into force 4 years later.[14] The parties to this Convention undertake to implement programs and measures for the elimination of pollution by black-list substances and to reduce or eliminate pollution by gray-list substances. The emission of black-listed materials are, however, not prohibited. Rather, the Convention requires the elimination of pollution by black-listed substances. Gray-listed materials can be discharged only after approval by each contracting party.[15]

Organization and Participation

The Oslo Commission and the Paris Commission are set up to implement the two respective Conventions. They are the permanent decision-making bodies, and normally they meet once a year. Different working groups are set up to facilitate their work and provide more specialized advice and information.[16] The two Commissions carry out their work with the help of a small secretariat in London, which also handles the administrative responsibilities for the Bonn Agreement. Thus, although there is no unitary approach or a single Convention covering the North Sea, the joint secretariat facilitates coordination between the different Conventions.[17]

There is an "inner" and an "outer" circle of countries participating in the North Sea environmental regime. The nine core North Sea countries taking part in all these arrangements are Belgium, Denmark, the Federal Republic of Germany, France, Ireland, the Netherlands, Norway, Sweden, and the United Kingdom. The "outer circle" countries, not directly bordering the North Sea, are Finland, Iceland, Luxembourg, Portugal, Spain, and Turkey, which participate in one or two of the three organizations. The European Common Market is a different kind of actor that does not readily fit this pattern. The Common Market was not a party to the orginal Bonn Agreement, but it signed the new Bonn Agreement of 1983. The EEC is also a party to the Paris Convention but not to the Oslo Convention. The EEC

14. For a list of when the different countries ratified, see Saetevik, p. 32.

15. Annex I to the Convention list the substances on the "grey" list while the "black"-listed substances can be found in Annex II.

16. The Oslo Commission has a Standing Advisory Committee for Scientific Advice (SACSA), and the Paris Commission has a Technical Working Group (TWG). A Joint Monitoring Group (JMG) provides information to both Commissions.

17. The secretariat is small—one secretary, two deputy secretaries, and three clerks—and yet it is considered very efficient, especially taking into account its limited resources. See Saetevik (n. 7 above), p. 117.

also has its own Environmental Action Programs that, in part, overlap the Conventions.[18]

The New Political Approach

The initiative to organize an International Conference for the Protection of the North Sea came from the Federal Republic of Germany (FRG) as a response to a report of the environmental problems of the North Sea by the German Council of Experts for Environmental Affairs.[19] The idea was not to develop new legal approaches; the existing international treaties already provided the necessary legal framework. The aim of such a Conference was of "reaching decisions on a noticeable reduction of pollution in the North Sea through harmonized action."[20]

The First International Conference on the Protection of the North Sea was convened by the FRG in Bremen in 1984. Eight of the "inner-circle" countries participated; only Ireland did not. The member of the Commission of the European Communities responsible for environmental protection also was a full participant. Observer status was given to some other states and international organizations. In contrast to the Commissions, where civil servants represent the different countries, the delegations were headed by the ministries responsible for the marine environment. In connection with the Conference, a report on the quality status of the North Sea had been made, representing the scientific base for the political discussions. A detailed Declaration in 10 parts was produced, and three focal points of joint action were identified: (1) reduction of pollution from land-based sources, including rivers and atmospheric inputs; (2) reduction of pollution at sea, including operational discharges; and (3) further development of the Joint Monitoring Program. It was then decided that a new Conference to evaluate the progress made was needed.

The Second North Sea Conference was held in London in November 1987. This time, the planning process had been much more thorough, and the political discussions focused mainly on questions that had not already found their solutions during the preliminary meetings. Another quality status report was published, and another Ministerial Declaration was produced. It was decided to have a new Conference in the Netherlands in 1990, indicating

18. Sonia Boehmer-Christiansen, "Marine Pollution Control in Europe: Regional Approaches, 1972–80," *Marine Policy* 8, no. 1 (January 1984): 45–52.

19. The background for the First North Sea Conference is discussed by Gerhard Peet, "The North Sea Conference: A Preview," *Marine Policy* 8, no. 3 (July 1984): 259–71.

20. Peet, "Sea Use Management for the North Sea" (n. 5 above), p. 432.

that these conferences have become institutionalized and a permanent part of the environmental regime of the North Sea.

EVALUATION OF THE NORTH SEA ENVIRONMENTAL REGIME

General Problems of Measurement

Thus, an elaborate regional, legal, and political environmental regime exists for the North Sea. But what have been the results of these conventions and conferences? Have they actually contributed to the reduction of pollution in the North Sea? Few studies have been made on the actual functioning of the environmental regime in the North Sea.[21] The methodological problems are severe, and it has been maintained that "there are severe limits to what we can expect from efforts to evaluate regimes in terms of the outcomes they produce."[22] This observation seems most relevant in evaluating the outcome produced by international environmental agreements. Although we have gotten a more precise—but still far from complete—picture of the state and effect of contamination of the North Sea over the last few years, little was known about this before the different measures were agreed upon. This makes it difficult to learn what the trends are. An additional complicating factor relates to the problem of isolating the effects of different types of factors from each other. "The quantity of contaminants entering the North Sea from year to year [depends] on a number of factors including natural variations . . . , economic changes . . . , industrial restructuring . . . and changes in pollution control."[23] Thus, if there is an improvement in the quality of the marine environment, often it is not known what has caused this development. Such factors need to be kept in mind when the performance and effects of the North Sea environmental regime are discussed.

The Legal Regime

The Bonn Agreement is the least ambitious in scope but has been a useful vehicle mainly for communication and exchange of mutual information be-

---

21. An extensive study of the Paris Commission has been done by Saetevik. Boehmer-Christiansen has done a comparative analysis of marine pollution control in Europe in the 1970s. More studies have been made of the legal obligations following from the different conventions applying to the North Sea. See, e.g., P. Fotheringham and P. W. Birnie, "Regulation of North Sea Marine Pollution," in *The Effective Management of Resources*, ed. C. Mason (London, 1979), pp. 168–223.

22. Oran Young, *Resources Regimes: Natural Resources and Social Institutions* (Los Angeles and Berkeley: University of California Press, 1982), p. 136.

23. *Quality Status of the North Sea, Summary* (n. 8 above), p. 7.

tween the North Sea countries. When it was put to a test during the Ekofisk blowout in 1977, it seemed to work satisfactorily, and it was also said to have been a useful channel for communication on matters like the use of dispersants.[24] It is a mechanism for mutual response regarding different kinds of pollution incidents, and, as such, it may have contributed somewhat to the reduction of the extent of pollution when accidents have occurred. The speed with which the Bonn Agreement went into force—only 2 months—shows strong support for the treaty but also indicates that it was easy to ratify because the benefits were perceived to outweigh the costs for all parties.

It is much more difficult to assess the effect of the Oslo and Paris Commissions. In a "self-evaluation" in the book *The First Decade: International Co-operation in Protecting our Marine Environment,* published by the Oslo and Paris Commissions in 1984, no clearcut answer is given. Whether the structure and efforts of the two Commissions "fulfils the aims formulated by the Conventions, . . . cannot be answered by a simple 'yes' or 'no,' "[25] which is probably a valid statement, regarding both the work of these Commissions as well as that of other international organizations.[26]

Knowledge of the effects of different contaminants was virtually nonexistent when the Oslo Commission started its work in the first half of the 1970s. Consequently "a period of struggle between those who wanted to control now and those who preferred to learn more first became unavoidable."[27] In order for the Oslo Convention to have any real significance in reducing dumping, a first step on the part of the contracting parties would be to set up a national permit system licensing marine waste disposal based on criteria laid down by competent national authorities under the guidance of the Oslo Commission. Not until 1982—10 years after the signing of the treaty—had this been fulfilled by all member states, an indication either that this task was not given very high priority by many North Sea countries or that bureaucratic and practical difficulties were significant—or perhaps a combination of both. Thus, the Oslo Commission did not have much practical effect during the first few years of its existence. Its greatest significance was probably the development of procedures for consultation prior to dumping and incineration, and joint implementation and monitoring of pollution levels.

As information was gathered and cooperation gradually developed, more concrete measures were agreed upon. Organic chemicals were now being incinerated instead of dumped, cadmium and mercury were being restrained as waste components, and the dumping of industrial wastes and

24. Boehmer-Christiansen, p. 45.
25. Oslo and Paris Commissions, *The First Decade: International Co-operation in Protecting Our Marine Environment* (London: Oslo and Paris Commissions, 1984), p. 36.
26. The difficulties of using "failure" and "success" as measurement criteria during international negotiations is discussed by Arild Underdal, "Causes of Negotiating Failure," *Internasjonal politikk* (1984): 81–97.
27. Boehmer-Christiansen, p. 46.

sewage sludge was being reduced and even stopped by some countries, whereas dumping permits had earlier been routinely renewed. In contrast, there were many indefinite juridical terms in the Convention "which leave a very wide scope for personal assessments."[28] Consequently, quite different standards are applied from country to country, and although "quite a number of the Contracting Parties to the Oslo Convention are of the opinion that the disposal of waste at sea (dumping and incineration) should be terminated, this view is not shared by all contracting parties."[29] Still, dumping is gradually being subjected to stricter national and international control. However, the Oslo Commission alone can hardly be credited with this development.

The Paris Commission faced an even more difficult task owing to the very nature of land-based pollution. As far as the Oslo Convention is concerned, pollution originates from a deliberate operation, but land-based pollution "has . . . widespread geographical origin emanating from all sorts of human activity and is not . . . the result of any single deliberate disposal operation."[30] The types of contaminants covered by the Paris Commission also account for—by far—most of the pollution of the North Sea.[31] The fact that it took 4 years before the Convention went into force (1978) and that it was not ratified by all parties until 1984 indicates that many countries were reluctant to participate because of the complexity of the issue and the high costs of reducing land-based emissions.

A major problem that has influenced the activities of the Paris Commission from its very start has been the dispute over the very term "pollution," as it is not precisely defined in the Convention. The crux of the argument is whether pollution is the introduction of harmful substances as such or the effects caused by their introduction into the sea. These differing interpretations have given rise to one "Continental" (and Scandinavian) approach to pollution control and one "Anglo-Saxon" approach. The former group advocates the use of uniform emission standards (UES), using the best available technical methods and equal standards for all countries. Great Britain advocates the environmental quality objective (EQO), arguing that the essence is the quality of the environment. In 1978, it was decided that both strategies should be applied over a period of 5 years. When the two methods were evaluated in 1983, it was impossible to say which method was most appropriate.[32] It was, therefore, decided that the dual strategy should continue; it was up to the parties to decide which method they would choose.

28. K. Sperling, "Protection of the North Sea: Balance and Prospects," *Marine Pollution Bulletin* 17, no. 6 (June 1986): 242.

29. Oslo and the Paris Commissions, p. 372.

30. Ibid., p. 39.

31. The assumptions vary, but most pollutants entering the sea—as much as 80%—originate from land or from the atmosphere; both are covered by the Paris Commission.

32. Saetevik (n. 7 above), p. 43.

However important this question has been, it still was more of an "academic dispute," at least in the 1970s, as most countries had little or no information available when they were requested to provide information on use, discharge, effects, and controls of the 15 substances on the first selected black list. "The response must have been somewhat shattering for it revealed enormous gaps in the information readily available to government departments. No country was able to make full response and only one was able to supply any information on lead and zinc."[33] To the extent that information was provided, it was often limited to usage only.

The combination of the lack of basic data and the disagreement over the term "pollution" has made it difficult for the Paris Commission to play an active role in reducing land-based emissions. A significant part of its work has been to organize monitoring programs, jointly with the Oslo Commission. Few decisions have been taken to reduce land-based emissions. In the period 1978–85 four binding decisions were taken. The number of *recommendations* has been more plentiful, and there has also been discussion on taking joint measures on certain substances, but it has not been possible to reach agreement (nondecisions).[34]

In sum, the work carried out by the two Commissions may have resulted in a certain reduction of the level of pollution of the North Sea. At least, the level of contaminants entering into the North Sea has probably been reduced compared to what it might have been had these Commissions not been established, but the ambitious aims of the two Conventions are still far from fulfilled. However, when judging the performance of these organizations, one must keep in mind that collective action is often limited to those measures acceptable to the "least enthusiastic party."[35] The international North Sea Conferences have provided an opportunity for the responsible ministers to discuss their differences. What has been the outcome of these discussions?

Evaluation of the Political Approach

Judged by the concrete measures agreed upon, the First North Sea Conference could hardly be labeled a success. According to one evaluation, "they agreed to nothing more than a declaration of intent."[36] Others claimed that "the final declaration . . . is long on principle, but short on specific new

33. Boehmer-Christiansen (n. 18 above), p. 49.

34. Saetevik, pp. 42–52.

35. For an elaboration of this point, see Arild Underdal, *The Politics of International Fisheries Management: The Case of the North East Atlantic* (Oslo: Universitetsforlaget, 1980).

36. "Tough North Sea Deal, but Sewage Slips Through," *New Scientist* (December 3, 1987), p. 3.

policies."[37] If this standard of evaluation is used, these statements seem reasonable. According to the Declaration adopted at the Bremen Conference, the North Sea states accepted the need to "take timely preventive measures to maintain the quality of the North Sea" and "to make every effort to protect the marine environment of the North Sea effectively and permanently." Generalities like these may have their value as declarations of intent, but they have limited value as guidelines for action.

But what could realistically be expected from this Conference? It was the first of its kind, and, although there had been some preparations, differences of opinion could hardly be expected to evaporate simply because the responsible ministers met in an international Conference for 2 days. Besides, the quality status report of the North Sea, representing the scientific basis of the Conference, indicated that, in general, the environmental health of the North Sea seemed to be good, thus making difficult the environmental demands for "sweeping clean-up measures."[38]

In this perspective, it seems premature to label the result of the Bremen Conference a failure. "Irrespective of results, it is important that it took place; that the Ministers of the Environment of the entire region met to discuss the problems of pollution in the North Sea. Taken at face value, this must be seen as an indication of higher political priority given to this issue; which is a necessary (but not sufficient) precondition for further international action."[39] Another observer claims that "an important effect is the new political commitment they give to the work of international intergovernmental organizations such as the secretariats of the Oslo, London, and Paris Conventions, and possibly even to the European Commission as far as the North Sea is concerned."[40] The Conference also highlighted pollution control of the North Sea outside the narrow range of scientists and civil servants of the different Commissions for the North Sea. The Commissions dealing with these matters have done so with hardly any public attention; "they [the Commissions] have not exactly sought the limelight."[41]

37. Environmental Data Services, Ltd. (ENDS), *UK's Defensive Efforts Pay Off at North Sea Conference*, Report 118 (November 1984). Author's note.—Environmental Data Services, Ltd. (ENDS), Unit 24, Finsbury Business Centre, 40 Bowling Green Lane, London, England. Their publications are printed by Southwell Press, Camberly, Surrey, England.

38. Environmental Data Services Ltd, *Opening Skirmishes on Health of North Sea*, Report 141 (October 1986).

39. Steinar Andresen, "A Comprehensive North Sea Convention: New Approach—Old Policy?" in *The Status of the North Sea Environment: Reasons for Concern*, ed. Gerhard Peet, Proceedings of the 2d North Sea Seminar 1986 (Amsterdam: Werkgroep Nordzee, 1987).

40. Peet, "The North Sea Conference" (n. 19 above), p. 262.

41. Konrad von Moltke, "International Commissions and Implementation of International Environmental Law," *International Environmental Diplomacy*, ed. John E. Carroll (Cambridge: Cambridge University Press, 1988), pp. 87–95.

Thus an evaluation of the Bremen Conference to a large extent depends upon the criteria of measurement. Judged by standards of concrete obligations for joint action to reduce pollution, the results were meager. Judged as a first and essential step in a political process of accomplishing the same goal, the Bremen Conference offered some prospects for the future.

The outcome of the second Conference has generally been described as a "success," a breakthrough for environmental cooperation in the North Sea, or even as a "historic event."[42] Some of the following measures were agreed upon: as to land-based pollution (inputs via rivers and estuaries), the North Sea states agreed to "take measures to reduce urgently and drastically the total quantity of such substances . . . , with the aim of achieving a substantial reduction [on the order of 50%] . . . between 1985 and 1995."[43] The North Sea states agreed on the same formula for inputs of nutrients, aiming "to achieve a substantial reduction [on the order of 50%] in inputs of phosphorus and nitrogen . . . between 1985 and 1995."[44] As for dumping and incineration, they agreed that no material should be dumped in the North Sea after January 1, 1989, "unless there are no practical alternatives on land and it can be shown to the competent international organisations that the materials pose no risk to the marine environment."[45] They also agreed to "phase out dumping . . . of industrial wastes by December 31, 1989" (with certain exceptions allowed for).[46] As for marine incineration, it was decided to phase out dangerous liquid waste by the end of 1994, following a reduction of 65% by 1991.[47]

The North Sea states also decided to continue strengthening the joint effort to monitor and reduce inputs of pollutants into the atmosphere, to carry forward action to improve the quality of (dumped) dredged materials, to reduce the contamination of sewage sludge, and to ensure that the quantites of such contaminants in the immediate future do not increase above the 1987 level.[48]

The most significant differences from the first Conference (and from the legal approach) are the time limits and quantifications set forward in this document. The higher level of specificity should imply a higher degree of concrete obligations on part of the North Sea countries, a major accomplishment of this Conference. The practical effect of the Ministerial Declaration on the marine environment, however, is still not clear. The uncertainty relates

42. This question is discussed by Folkert de Jung, "The Second Ministerial Conference on the Protection of the North Sea: An Historic Event? *North Sea Monitor* (January 1988), pp. 2–4.
43. Ministerial Declaration (London, November 1987), Art. 16, No. 2.
44. Ibid., Art. 16, No. 11.
45. Ibid., Art. 16, No. 21.
46. Ibid., Art. 16, No. 22(a).
47. Ibid., Art. 16, No. 24(b) and (c).
48. Ibid., Art. 16, No. 22(b) and (c).

partly to possible difficulties of implementation and partly to the lack of data on the contamination of the North Sea.

Although this is a trivial point, it should be kept in mind that this is an international declaration of intent only. The Ministerial Declaration is fairly specific on many points, but the eight North Sea states have no legal obligation to abide by its statements. If they feel a political obligation to do so, this need not be a major problem; but the question remains whether it is possible to know if some of the goals will actually be accomplished or how they will be measured. Although there has been a steady improvement in the data base since the Commissions started their work more than a decade ago, there is still considerable uncertainty as to the actual amounts of contaminants entering the North Sea. According to the last quality status report: "A great deal of information has been collected on the inputs of contaminants. . . . Unfortunately, the data sets from the different countries are not always consistent or equally accurate. It is therefore difficult, except for a few contaminants, to make valid comparisons between countries or to achieve a reliable overall synthesis."[49]

In table 1, figures are presented for 14 North Sea contaminants, and 7 different sources of contamination are listed, adding up to 98 separate categories. Of these 98 categories, information is not available for 28. While the data base in general seems to be pretty good for *substances,* like cadmium (Cd), copper (Cu), and nickel (Ni), hardly any information is available on substances like DDT and PCBs. If one focuses on *sources* of pollution, the same type of variation is discerned. While the data base appears to be reasonably good for river inputs and direct discharges, uncertainty is the best characteristic for atmospheric inputs, with a huge difference between maximum- and minimum-range figures.[50] Frequently, information is not available for atmospheric inputs at all, as is the case for many of the substances contained in the dredgings, sewage sludge, and industrial waste being dumped.

Even where data seem to be quite complete, "it should be noted that all figures for input, with the possible exception of dumping, are subject to considerable uncertainty, the extent of which is variable and difficult to quantify."[51] Table 2, which breaks down contaminating substances for the eight North Sea countries (where the aggregate information in table 1 seems quite complete), illustrates this point. None of the countries are able to provide information for all substances. The United Kingdom produces the most extensive information; data are given for 12 of the 14 materials. Norway is at the other end of the spectrum, providing information on only 4 of the 14 substances.

The uncertainty created by the lack of data from many countries is some-

49. *Quality Status of the North Sea, Summary* (n. 8 above), p. 6.
50. Ibid.
51. Ibid., p. 7.

what modified by the fact that the countries with the largest emissions—the Netherlands, the United Kingdom, and the FRG—provide the most extensive information. Still, data are almost nonexistent for substances like DDT and PCBs. There are also differences among these countries regarding how up-to-date the information is. Again, the United Kingdom has the highest "information score" with data from 1985/86, while countries like Denmark and Belgium provide information dating back to 1983.

It is a well-known fact that there have been disputes between scientists and countries regarding the effect of the contamination of the North Sea. Less attention has been paid to the uncertainty surrounding the actual amount of some of the contaminants entering the North Sea and the consequences this may have for the possibility of an efficient joint policy for the reduction of pollution. For example, it seems difficult to know whether a 50% reduction will be accomplished since the present level of important contaminants is not known. This is not an argument to reduce the importance of the 50% goal agreed upon, but a query as to whether such seemingly precise quantifications can be measured.

The question is, Who is going to bear the *costs* of the aimed reductions? For example, it is explicitly stated that the aim is a 50% reduction of the total quantity of pollutants from river inputs. In other words, those who pollute most must make the greatest—that is, one assumes, the costliest—reductions. This does not seem unreasonable. However, it raises the difficult question of the "free-rider" problem.[52] If, for example, an economically strong nation like the FRG or the presumably "environmentally conscious" Netherlands makes very strong reductions, other North Sea countries can be free riders; they do not have to reduce inputs at all. The goal, to the extent that it is possible to measure it, may still be reached. In short, the questions of distribution of costs (and benefits) are not dealt with. It will not be surprising if they surface on a later occasion and make implementation of the "neat" 50% reductions difficult.

The wide variety in information levels provided by different countries may also pose implementation problems. If table 2 is used as an illustration, it would be possible to trace the performance of countries like Great Britain and the Netherlands with regard to river input, but it would not be possible for countries like Norway and Sweden since they have so few figures available. In a somewhat cynical fashion, there may be certain advantages to providing scant information; then little will be known as to whether performance is

---

52. The "free-rider" problem has been clearly illustrated in connection with the role of Organization of Petroleum Exporting Countries (OPEC) and non-OPEC members concerning the prices of oil. The strong increase of oil prices in the 1970s was, at least in part, a result of the dominant role played by OPEC. However, countries like the United Kingdom and Norway have also benefited from the price hike without being OPEC members; they were "free-riders."

TABLE 1.—SUMMARY OF CONTAMINANT INPUTS TO THE NORTH SEA IN TONS PER YEAR[a]

Source	Nitrogen		Phosphorus		Cadmium		Mercury	
	Max	Min	Max	Min	Max	Min	Max	Min
River inputs	1,000,000	...	76,000	...	52	46	21	20
Direct discharges	95,000	...	25,000	...	20	20	5	5
Atmospheric dumpings:	400,000[b]	...	N/A	...	240	45	30	10
Dredgings	N/A	...	N/A	...	20	...	17	...
Sewage sludge	11,700	10,000	2,800	2,200	3	...	.6	...
Industrial waste[c]	N/A	...	N/A	...	.3	...	.2	...
Incineration at sea[d]	N/A	...	N/A	...	.1	...	trace	...
Total (rounded)	1,500,000	...	100,000	...	335	135	75	50

Source	Copper		Lead		Zinc		Chromium		Nickel	
	Max	Min	Max	Min	Max	Min	Max	Min	Max	Min
River inputs	1,330	1,290	980	920	7,370	7,360	630	590	270	240
Direct discharges	315	...	170	...	1,170	...	490	...	115	...
Atmospheric dumpings:	1,600	400	7,400	2,600	11,000	4,900	900	300	950	300
Dredgings	1,000	...	2,000	...	8,000	...	2,500	...	700	...
Sewage sludge	100	...	100	...	220	...	40	...	15	...
Industrial waste	160	...	200	...	450	...	350	...	70	...
Incineration at sea	3	...	2	...	12	...	1.7	...	3	...
Total (rounded)	4,500	3,000	11,000	6,000	28,000	22,000	5,000	4,200	2,100	1,450

	Arsenic		HCH[e]		Drins[f]		DDT		PCB	
	Max	Min	Max	Min	Max	Min	Max	Min	Max	Min
River inputs	360	320	3	3	.1	...	.1	...	3	3
Direct discharges	220	...	0	...	N/A	...	0	...	0	...
Atmospheric dumpings:	120	40	N/A	N/A	N/A	...	N/A	...	N/A	...
Dredgings	200	...	N/A	...	N/A	...	N/A	...	N/A	...
Sewage sludge	trace	...	N/A	...	N/A	...	N/A	...	N/A	...
Industrial waste	40	...	N/A	...	N/A	...	N/A	...	N/A	...
Incineration at sea	.1	...	N/A	...	N/A	...	N/A	...	N/A	...
Total (rounded)	950	820	3	3	.1	...	.1	...	0	3

SOURCE.—Second International Conference on the Protection of the North Sea, Scientific and Technical Working Group, *Quality Status of the North Sea, Summary* (London: Department of the Environment, September 1987). Reproduced with permission.

NOTE.—It should be noted that all figures for inputs, with the possible exception of dumping, are subject to considerable uncertainty, the extent of which is variable and difficult to quantify. N/A = not available.

a The figures used in this table are the "rounded" totals taken from tables 2B, 2D, and 2E of the quality status report. These totals do not include amounts entering the North Sea from the North Atlantic, the English Channel, and the Baltic Sea. The data are taken from different years, but mainly 1983–86.

b Estimates are as reported by the fourth meeting of the Paris Commission Working Group on Inputs.

c Includes liquid and solid.

d Refers to material burned and assumed to be discharged from the stack; this therefore represents the maximum amount that would enter the sea.

e HCH: Hexachlorocyclohexane—a chlorinated compound of which eight forms are possible, though in practice only three (α, β, and γ) are found in the environment.

f Drins are a group of nonsystemic and persistent insecticides containing chlorine. Bioaccumulated by marine organisms, they include aldrin, dieldrin, endrin, and telodrin.

TABLE 2.—RIVER INPUTS TO NORTH SEA IN TONS PER YEAR

Country	Year of Estimates	River Flow (Mm³/D)ᵃ	Nitrogen		Phosphorus		Cadmium		Mercury	
			Max	Min	Max	Min	Max	Min	Max	Min
Sweden	1984	47.1	17,600	...	281	...	.2	...	0	...
Federal Republic of Germany	1985	87.0	259,000	...	16,490	...	12.1	...	8.8	8.8
Belgium	1983	1.8	4,700	...	1,570	...	.4	.3	.1	.1
Netherlands	1984	259.0	599,000	...	52,000	...	25.4	...	6.3	...
Norway	1983/84	67.9	N/A	...	N/A	...	N/A	...	N/A	...
Denmark	1983	10.9	22,000	...	2,400	...	.1	...	0	...
United Kingdom	1985/86	89.7	110,740	...	3,439	...	14.1	7.7	5.3	4.3
Total (rounded)		565	1,000,000	...	76,000	...	52	46	21	20

Country	Copper		Lead		Zinc		Chromium		Nickel	
	Max	Min	Max	Min	Max	Min	Max	Min	Max	Min
Sweden	25.0	...	N/A	...	125.0	...	N/A	...	N/A	...
Federal Republic of Germany	290.0	...	259.1	...	2,098.7	...	N/A	...	N/A	...
Belgium	4.9	4.3	2.9	...	28.0	...	1.4	...	6.0	...
Netherlands	661.0	...	395.0	...	3,408.0	...	475.0	...	N/A	...
Norway	65.0	...	16.0	...	248.0	...	15.0	...	N/A	...
Denmark	2.4	...	.8	...	8.0	...	.4	...	N/A	...
United Kingdom	282.9	239.5	303.4	245.5	1,450.6	1,440.4	141.4	97.6	265.3	233.2
Total (rounded)	1,330	1,290	980	920	7,370	7,360	630	590	270	240

	Arsenic		HCH[b]		Drins[c]		DDT		PCB	
	Max	Min	Max	Min	Max	Min	Max	Min	Max	Min
Sweden	N/A	...	N/A	...	N/A	...	N/A	...	N/A	...
Federal Republic of Germany	N/A	...	.5	.5	N/A	...	N/A	...	2.9	2.9
Belgium	N/A	...	0	0	N/A	...	N/A	...	N/A	...
Netherlands	292.0	...	2.3	...	.1	...	N/A	...	.4	...
Norway	N/A	...	N/A	...	N/A	...	N/A	...	N/A	...
Denmark	N/A	...	N/A	...	N/A	...	N/A	...	N/A	...
United Kingdom	64.9	28.1	.4	.4	N/A	...	.1	...	N/A	...
Total (rounded)	360	320	3	3	.1	...	.1	...	3	3

Source.—Second International Conference on the Protection of the North Sea Scientific and Technical Working Group, *Quality Status of the North Sea, Summary* (London: Department of the Environment, September 1987). Reproduced with permission.

Note.—No French figures are included in the totals, as no French rivers discharge directly into the North Sea. It should be noted that these figures are subject to considerable uncertainty owing to, among others, analytical and hydrological variations. N/A = not available.

[a]Mm³/D = million cubic meters per day.

[b]HCH: Hexachlorocyclohexane—a chlorinated compound of which eight forms are possible, though in practice only three (α, β, and γ) are found in the environment.

[c]Drins are a group of nonsystemic and persistent insecticides containing chlorine. Bioaccumulated by marine organisms, they include Aldrin, Dieldrin, Endrin, and Telodrin.

improving or not. In this perspective, an information obligation—to the extent possible—may be of importance equal to a 50% rule.

Uncertainties as to the practical effect of the Declaration have also been pointed out by others.[53] One point relates to the fact that the Oslo and the Paris Commissions will be responsible for implementing essential parts of the agreement. There has been disagreement over key questions in these Commissions. Have these differences been solved through the 1987 Conference in London? According to one critical observer regarding dumping in the Oslo Commission: "It would be naive to expect a major change with respect to this issue."[54] This observer also points out that a possible implementation problem arises from the fact that there are member countries in the Oslo Commission that did not take part in the London Conference. Thus countries like Spain, Portugal, and Ireland may block the agreement.

As to land-based pollution, no list of substances that are considered persistent, toxic, and liable to bioaccumulate has been drawn up. Great Britain has announced its intention to draw up a "red list" of the most dangerous substances. The list is expected to contain some 25 or 30 hazardous substances. The European Commission, however, has "identified nearly 130 potentially hazardous substances which it believes should be tightly controlled."[55] These differing perceptions certainly pose a challenge, but, based upon the data that the North Sea countries are currently able to submit, this problem seems rather academic, at least in the short run. As noted above, most countries have limited data for the 14 substances listed in table 2. Before discussing whether a list of 30 or 130 substances should be made, more information must be provided on these substances. It is also uncertain what practical effect the Ministerial Declaration will have for important questions related to radioactive waste and atmospheric pollution.

Output, Interests, and Organization

These questions do not alter the main conclusion: from an environmental point of view, the London Conference was an essential step forward. Agreement was made on important points, and this could have significant impact on the work of the Oslo and Paris Commissions. Most observers claim that Great Britain, being somewhat isolated from the other North Sea countries on key questions, gave way on important points. Generally, Great Britain has been skeptical about supporting stricter control measures unless there is scientific

53. The analysis by de Jung (n. 42 above) points to several "loopholes" in the Declaration.
54. Ibid., p. 2.
55. Britain Calls for Task Force to Model North Sea Pollution," *New Scientist* (November 26, 1987), p. 26.

proof that contamination causes harm to the marine environment. Now it seems ready to accept the *precautionary approach* "which may require action to control inputs . . . even before a causal link has been established by absolutely clear scientific evidence."[56] Although the Ministerial Declaration reaffirms the use of both the EQO and the UES approaches, "Britain has . . . edged closer to her North Sea neighbours in adopting fixed limits for factory emissions to complement its reliance on more general environmental quality objectives."[57] Great Britain also made important concessions concerning dumping and incineration.

This is not to say that the other countries got all that they wanted. Britain, alone, dumps sewage sludge and may continue to do so, as "under this formula, Britain will be able to continue to dump colliery spoil off the coast of northeast England."[58] Some Scandinavian countries worked actively to include radioactive pollution in the Declaration, and they succeeded in doing so. The Declaration states: "To this end . . . the best available technology [should be applied] to minimize any pollution caused by radioactive discharges."[59] This wording is more strict than the provisions in the Paris Convention, but it is uncertain whether it will have any impact on the continued discharges from the reactor in Sellafield in Great Britain.[60]

Although the Declaration was a compromise, Great Britain made the most important concessions, thus paving the way for this historic event. But why has Great Britain been more reluctant to support stricter control than most of the other countries? This question can be answered at least partly by looking at the level of pollution and the water circulation in the North Sea. According to the latest quality status report: "Studies of water discharges . . . have shown . . . a long term anti-clockwise circulation and in particular the water discharges from the UK East coast . . . spreads throughout the central North Sea, while water discharging from continental estuaries tends to be more confined to the eastern coastal region."[61] This means that the level of contamination is highest in the eastern coastal zone of the southern North Sea, the Wadden Sea, and the German Bight. The continental countries in a sense have to live with their own pollution because it tends to remain in their waters, while the level of pollution is not very high in British waters. Traditionally, Great Britain has tended to argue that it was the responsibility of the continental countries to handle the high level of pollution, especially in the Wadden Sea. The contamination of Swedish and Norwegian waters are gen-

56. Ministerial Declaration, Art. 7.
57. *New Scientist* (December 3, 1987), p. 3.
58. Ibid.
59. Ministerial Declaration, Art. 16, no. 38 (n. 44 above).
60. De Jung, in his analysis of the London Conference, claims that it will have no effect. Norwegian officials, however, appear to believe that the "sharpening" in the text may have an impact.
61. *Quality Status of the North Sea, Summary* (n. 8 above), p. 5.

erally not high either, but, while the United Kingdom is mainly an exporter of contaminants, the Scandinavian countries probably import more pollution than they export, making these countries more favorable to strict international control than Britain.[62]

Thus, the United Kingdom, at least up until the London Conference, has been a "stumbling block" in the sense that it has been more negative towards a comprehensive joint action plan to reduce the level of contamination than most other North Sea countries. To put it in more neutral terms, because Great Britain has been less affected by pollution than many other North Sea countries, it has not had the same interest in contributing to the reduction of marine pollution.

While the United Kingdom has been the target of the other North Sea countries and the environmental groups, less attention has been paid to the role of the EEC on this matter. Studies indicate that the European Community—for different reasons—may have been an equally important "stumbling block." Besides, although the United Kingdom has tended to be the most vocal opponent of strict international rules, countries like Belgium and France have not necessarily been more progressive—but they have been less outspoken.[63]

What is the scientific basis for the drive to strongly reduce the level of pollution in the North Sea being advocated by most North Sea countries, the environmental organizations, and now by the Ministerial Declaration? Does the quality status report indicate that the North Sea marine environment is in danger, and how does this report compare to the status report prepared for the First North Sea Conference? According to the latest report: "In general, deleterious effects, at present, can only be seen in certain regions, in the coastal margins, or nearby identifiable pollution sources. There is as yet no evidence of pollution away from these areas."[64] According to the same report, however, "ongoing assessment is essential [and] . . . improved monitoring and scientific programmes need to be developed to provide more consistent data and to permit links between inputs, concentrations, and effects to be established with greater confidence."[65]

How does this quality status report compare to the one made in connection with the Bremen Conference? One observer has maintained that "compared to the Bremen Conference, the environmental 'diagnosis' was not much

62. Saetevik (n. 7 above), pp. 52–67, discusses the perceptions of the North Sea countries as to which countries are exporters and which are importers of land-based pollution.

63. This view on the role of the EEC, Belgium, and France has been expressed by Norwegian and Dutch officials involved in the North Sea environmental regime. To a large extent it is supported by the studies made by Boehmer-Christiansen and Saetevik. See, e.g., Saetevik, pp. 68–74.

64. *Quality Status of the North Sea, Summary,* p. 1.

65. Ibid.

altered, but more generally accepted."[66] To the extent that it was altered, however, the dangers of the effects from the various contaminants were more strongly underlined. Thus, at least to some extent, it is a scientific basis for the more strict approach advocated by the latest Ministerial Declaration. The slight difference between the two reports does not, however, explain the strong difference between the two declarations.

The new approach has been linked to the changing positions of the United Kingdom, which does not appear to be based primarily on a differing perception of the North Sea environment. "The political pressure activity of national and non-governmental actors, made especially effective by the 'stumbling block' of UK's role as a host of the conference, seems to be chief among these."[67] Thus, it may be that politics have mattered more than science in bringing about this change. Organizational factors—more specifically, the preparatory process—also seem to have played an important part.[68]

In connection with the Bremen Conference, the first preparatory meeting took place 8 months before the Conference, while the first corresponding meeting in the Scientific and Technical Working Group preparing for the London Conference took place close to 2 years in advance. The scientific preparatory work for the London Conference was more explicit in being politically "usable."[69] A Policy Working Group was also established prior to the London Conference. During a period of more than half a year, an agenda gradually emerged, "and a number of misconceptions and real disagreements were either done away with or, at least, reduced."[70] This is not to say that conflicts of interest disappear because of a thorough preparatory process. Combined with other factors, however, the structured form of the process leading to the London Conference seems to have made a difference.

LESSONS TO BE LEARNED FROM THE NORTH SEA

Almost 2 decades have passed since the North Sea countries started their work to protect the North Sea environment. How well have they succeeded, and what lessons can be derived from their experience? A first observation concerns time. Although the North Sea countries in a comparative perspective started the buildup of their environmental regime quite early, few direct measures to reduce the level of pollution were agreed on during the first

66. Jørgen Wettestad, "The Outcome of the 1987 North Sea Conference. Science Counts, but Politics Decides?" *International Challenges* 8, no. 2 (1988): 33.

67. Ibid., p. 34.

68. Jørgen Wettestad, "Science, Politics and Ocean Pollution: Explaining the Outcome of the 1987 North Sea Conference," *International Challenges* 8, no. 3 (1988): 26–32.

69. Ibid.

70. Ibid.

decade or so. Due to lack of knowledge about the level and effect of contamination, monitoring and information gathering constituted the most important work in the 1970s and the beginning of the 1980s. Disagreement over key questions, like the definition of the term "pollution," also contributed to the relative passivity of the Commissions. In short, the buildup of an action-oriented environmental regime takes a long time.

A second observation concerns the challenge and difficulty of measuring the amount and effects of contaminants on the marine environment. Although the level of knowledge has increased significantly since the early 1970s, the latest quality status report for the North Sea strongly stressed that uncertainty is still prominent. This applies to important contaminant inputs as well as to their deleterious effects. The former may pose a problem in the sense that it is difficult to measure trends and to learn whether goals are achieved. The latter has made it difficult to reach consensus on what measures to adopt, as there is disagreement on the health of the "patient." However, it now appears that a consensus is approaching. It is not presently feasible to reach perfect levels of scientific information; thus, the North Sea states have agreed to base their environmental North Sea policy on the "anticipatory" principle, implying that measures have to be taken when damage is suspected. It remains to be seen whether this principle will be implemented in practice, but the likelihood that this will happen has probably been strengthened by incidents after the London Conference, such as the invasion of poisonous algae and the death of seals in the North Sea. From an environmental point of view, this is good, but too little still is known about the balance between control of costs and benefits of reduced pollution.

A third observation applies to the organizational approach of combating marine pollution. After approximately 1 decade, the traditional legal approach was supplemented with a new political approach, at least partly due to dissatisfaction with the efficiency of the conventional approach. After the London Conference in 1987, it appears that the political approach has revitalized the North Sea environmental regime. Regardless of potential implementation difficulties, the North Sea states have committed themselves in an unprecedented manner to a reduction of pollution in the North Sea. There are many reasons why this change has come about, but the organizing of the planning process prior to the London Conference seems to have been quite decisive. The legal and political processes cannot, however, be viewed separately. During the next 2 years, the Commissions will be responsible for the implementation of the ambitious Declaration program. It remains to be seen whether they will succeed.

Does this mean that the combination of a thorough planning process and a top-level political conference in most cases stands the best chances of success? Should such a model be transferred to other regions as well? Neither of these questions can be answered with a simple "yes" or "no." While a thorough planning process is always important, the political relations in the region will represent one crucial intervening variable. As indicated earlier, the North Sea states, in a comparative perspective, are fairly homogeneous. They are all wealthy pluralistic states in the Western world with basically the same political ideologies. In a global perspective, this may be more the exception than the rule. In many regions of the world—for example, the Sea of Japan area—a top-level political conference would be unthinkable; but where political relations are strained, much could still be accomplished if scientists from riparian countries could meet free of political constraints.

Although the North Sea countries are fairly similar in important respects, the long and laborious process leading up to the London Conference has demonstrated beyond doubt that there have been strong differences in interests and opinions. After almost 2 decades of concentrating much of their energy on disagreements, the North Sea states now seem to be ready to focus on ways to reduce the level of pollution in the North Sea. However, it is still far too early to know whether or not they will be successful in reaching their declared aim. The dual approach and the high prominence that this question has been given on the political agenda of North Sea states offers encouraging prospects for continued improvement in the regulation of marine pollution in the North Sea.

Participation in Marine Regionalism: An Appraisal in a Mediterranean Context

Aldo E. Chircop
International Ocean Institute, Malta

INTRODUCTION

A student of marine affairs may be optimistic, pessimistic, perhaps even skeptical about the future of the United Nations Convention on the Law of the Sea 1982 (LOS Convention)[1] but he cannot ignore the reality of the growing interest in marine regionalism. Marine regionalism is not a creation of the Third United Nations Conference on the Law of the Sea (UNCLOS III), as it dates back at least to the first decade of this century. The concept found wide support among the fisheries organizations that emerged during the post–Second World War era, and it has received new vigor since the 1970s with the establishment of the United Nations Environment Programme (UNEP) and UNCLOS III.

The Mediterranean was one of the pioneer areas of marine regionalism. Marine regionalism made a first appearance there in the establishment of the International Commission for the Scientific Exploration of the Mediterranean Sea (ICSEM) by the International Geographical Union (IGU) in 1908 and its consolidation in 1919.[2] Among the aims of ICSEM was the promotion of cooperation among Mediterranean countries in marine scientific research. In the context of maritime security, the Nyon Agreement of 1937 may perhaps have been the first instance of regional cooperation in the maintenance of freedom of navigation in the Mediterranean. During the Spanish Civil War, 10 Mediterranean and Black Sea states formulated a short-term plan to clear the Mediterranean of pirate submarines. One might be tempted to argue that

---

1. *The Law of the Sea: Official Text of the United Nations Convention on the Law of the Sea with Annexes and Index* (New York: United Nations, 1983), Sales No. E.83.V.5 and hereafter cited as the LOS Convention.
2. International Commission for the Scientific Exploration of the Mediterranean Sea (ICSEM), Secretariat General, 16 Blvd. de la Suisse, MC-98030, Monaco. The publications of the ICSEM are: *Rapports et procès-verbaux des réunions de la CIESM*, and *Bulletin de liaison des laboratoires* (annuals). For a report on the IGU marine activities see "The International Geographic Union Commission on Marine Geography," Appendix A, in this volume.

the Treaty of Montreux (1936) on navigation through the Turkish Straits is also a form of marine regionalism, but with both regional and extraregional participation.[3] At its fourth session in 1948, the Food and Agriculture Organization (FAO) decided to establish a fisheries organization for the Mediterranean. The General Fisheries Council for the Mediterranean (GFCM) came into existence formally in 1952 with a mandate for living-resource development that eventually extended to include management. The International Ocean Institute (IOI), established in 1972, is situated in Malta and seeks to promote research on peaceful uses of ocean space and resources. The IOI holds annual *Pacem in Maribus* convocations and organizes training programs.[4] Last, but not least, is UNEP's Regional Seas Programme for the Mediterranean, which started with an action plan in 1975 and now consists of a complex network of policies, agreements, programs, institutions, and activities involving practically all the coastal states in the protection and preservation of the marine environment.[5] Some Mediterranean states are actively

3. The Montreux Convention resulted from a conference held in Montreux, Switzerland, June 22–July 20, 1936, by delegates from Bulgaria, France, Greece, Italy, Japan, Romania, USSR, Turkey, United Kingdom, and Yugoslavia. It was agreed at that conference to replace the 1923 Straits Convention (part of the peace treaty signed with Turkey at Lausanne, 1923), with the new Montreux Convention. Editors' note.— See Jon Van Dyke, "The Role of Islands in Delimiting Maritime Zones: The Case of the Aegean Sea," in this volume.

4. Editors' note.—See the report of the IOI, "Report of the International Ocean Institute: 1989," Appendix A, in this volume. Also, *Pacem in Maribus III, The Mediterranean Marine Environment and the Development of the Region*, ed. Norton Ginsburg, Sidney Holt, and William Murdoch, Proceedings of a Conference, Split, Yugoslavia, April 28–39, 1972 (Malta: Royal University of Malta Press on behalf of the International Ocean Institute, 1974).

5. Editors' note.—For further information about UNEP's involvement in the Mediterranean see UNEP, *The Health of the Oceans*, UNEP Regional Seas Reports and Studies no. 16 (Geneva: UNEP, 1982), and for a list of the publications of the UNEP Mediterranean Action Plan refer to the United Nations Environment Programme, *Catalogue of Publications for Ocean and Coastal Areas* (Nairobi: Oceans and Coastal Areas Programme Activity Centre [OCA/PAC], 1989). For information about the Mediterranean Action Plan, Priority Actions Programme see Arsen Pavasovic, "Coastal Zone Management in the Mediterranean Region—Practices of UNEP," *Coastal Zone '89* (New York: American Society of Civil Engineers, 1989) or write to Arsen Pavasovic, Director, Regional Activity Centre for the Priority Actions Programme, Mediterranean Action Plan, UNEP, Kraj Sv. Ivana 11, 58000 Split, Yugoslavia. See also "UN, Environmental Programme: Activities for the Protection and Development of the Mediterranean Region," *Ocean Yearbook 1*, ed. Elisabeth Mann Borgese and Norton Ginsburg (Chicago: University of Chicago Press, 1978), pp. 584–97; "Recommendations for the Future Development of the Mediterranean Action Plan," *Ocean Yearbook 2*, ed. Elisabeth Mann Borgese and Norton Ginsburg (Chicago: University of Chicago Press, 1980), App. B, pp. 547–54; Peter S. Thacher and Nikki Meith, "Approaches to Regional Marine Problems: A Progress Report on UNEP's Regional Seas Programme," *Ocean Yearbook 2*, ed. Elisabeth Mann Borgese and Norton Ginsburg (Chicago: University of Chicago Press, 1980), pp. 153–82 (the Mediterranean Action

lobbying for maritime security to be considered on a regional basis through both regional and global conference diplomacy. Although Mediterranean and other interested states are far from concluding a regional maritime security arrangement, the potential for an understanding exists.

The question that concerns the long-term success of regionalism in the Mediterranean and elsewhere is that of participation. Depending on the values at stake, some regional arrangements are inconclusive in allowing wide membership, whereas others are exclusive by restricting membership to identified participants. How wide membership is depends on the main actors in the process and practical considerations at a particular moment in time, but how extensive it should be relates to long-term perceptions of the problem in question. The two do not always coincide.

REGION AND REGIONALISM

"Region" and its derivatives (regional, subregional) constitute a set of concepts that serve to contain or define the physical (whether real or artificial) and intellectual limits of the subject matter in a given process. "Region" has two functional components: (i) subject matter (for example, a problem as a focal point); and (ii) context (that is, limits). Its use is justified on the basis of the relationship between the subject matter and the context. The subject matter may not always be explicit, but in defining a context one is implicitly characterizing the subject matter.

There may be as many uses of "region" as there are users. In a general sense, "region" may be used to identify a geographic area, or to denote economic or political groupings. In marine usage, Alexander identifies at least three uses of "region": (1) a physical area, such as a semienclosed sea; (2) a management region, which may coincide with a physical area but which is aimed at a problem; and (3) an operations region, in the sense of the site of a regional agreement.[6] These uses of "region" may appear independently, but

---

Plan is discussed on pp. 158–68); and a further update, "UNEP, An Update on the Regional Seas Programme," *Ocean Yearbook 4,* ed. Elisabeth Mann Borgese and Norton Ginsburg (Chicago: University of Chicago Press, 1982), pp. 450–61; and Adalberto Vallega, "A Human Geographical Approach to Semienclosed Seas: The Mediterranean Case," *Ocean Yearbook 7,* ed. Elisabeth Mann Borgese, Norton Ginsburg, and Joseph R. Morgan (Chicago: University of Chicago Press, 1988), pp. 372–93. "WHO Coastal Water Quality Program and Its Relation to Other International Efforts on Marine Pollution Control" (*Ocean Yearbook 2,* ed. Elisabeth Mann Borgese and Norton Ginsburg [Chicago: University of Chicago Press, 1980], pp. 435–44) also discusses the Mediterranean Action Plan and water pollution standards and controls.

6. Lewis Alexander, "Regionalism at Sea: Concept and Reality," in *Regionalization of the Law of the Sea,* ed. Douglas M. Johnston, Proceedings of the Law of the Sea Institute, Eleventh Annual Conference, November 14–17, 1977 (Cambridge, Mass.: Ballinger, 1978), pp. 3–16, at 5.

they may also form parts of a whole. The region may be broken down into subregions in localities that present divergences in subject matter.

Regionalism consists of decision makers' perceptions and actions in a process directed at a particular region, specifically at the subject matter in context. Regionalism may be broadly oriented toward a problem or, more narrowly, it may be directed at an issue. Depending on the nature of the subject matter, regionalism may result in settlements or arrangements.

In an era of marine affairs characterized by both cooperative and competitive behavior, a general recognition of marine interdependence gives rise to a cooperative ethic and norm that promotes regional arrangements not only for good neighborliness but also for the reconciliation of near and distant interests. Regionalism presupposes cooperative participation. Participation should not be construed narrowly, however, in a process of problem solving or management exclusive to neighboring coastal states. Should other interests be involved in the subject matter in question or should the subject matter extend beyond the intellectual limits identified, equity and efficiency suggest an inclusive approach.

REGION AND REGIONALISM IN THE LOS CONVENTION

The LOS Convention is a treaty with both global and regional implications.[7] The concept "region" is used frequently, particularly in relation to living resources, throughout the convention but its use is not symmetrical or consistent.

"Region" and its derivatives specifically appear 53 times in 25 different articles, relating to the Exclusive Economic Zone (EEZ), the high seas, enclosed and semienclosed seas, land-locked states, the area, the protection and preservation of the marine environment, and transfer of technology, as illustrated in table 1.[8]

The concept "region" seems to be used in five different senses. First, it is used to refer to participation in processes foreseen in the LOS convention. Participation in the exploitation of surplus fisheries in neighboring EEZs by land-locked and geographically disadvantaged states is conditional on presence in the region or subregion. The composition of the council of the International Sea-Bed Authority is based on equitable distribution among identified geographical (and geopolitical) regions, that is, Africa, Asia, Eastern Europe (Socialist), Latin America, Western Europe, and others.

Second, it designates the competent international organizations for a

7. UN document A/Conf.61/122, October 7, 1982.
8. The articles of the LOS Convention (n. 1 above) referred to are the following: 61–64, 66, 69–70, 98, 118–19, 123, 125, 156, 161, 163, 197, 200, 207–8, 210, 212, 270, 272, 276–77.

TABLE 1.—"REGION" AND "REGIONALISM" REFERENCES IN THE 1982
UNITED NATIONS LAW OF THE SEA CONVENTION

	EEZ	H.S.	E.S.	LLS	Area	Env.	T.T.	Total
No. of articles	7	3	1	1	3	6	4	25
No. of references	27	4	1	1	5	9	6	53

SOURCE.—*The Law of the Sea: Official Text of the United Nations Convention on the Law of the Sea with Annexes and Index* (New York: United Nations, 1983).
EEZ = Exclusive Economic Zones; H.S. = High Seas; E.S. = Enclosed and Semi-enclosed Seas; LLS = Land-locked States; Area = The Deep Sea-Bed Area; Env. = Protection and Preservation of Marine Environment; T.T. = Transfer of Technology.

given area. Thus, regional and subregional organizations are assigned roles in the conservation and management of living resources. Where these do not exist with reference to areas in High Seas, subregional and regional fisheries organizations are to be established.

Third, it refers to the level of action in a specific area. States are required to take action on bilateral, subregional, regional, and global levels in the setting of international minimum standards and exchange of information relating to the conservation of living resources. Coordination of measures relating to straddling stocks is to take place through subregional or regional organizations. Policies relating to land-based and sea-bed sources of pollution are to be harmonized at the regional level.

Fourth, it designates the physical dimension of certain rules. Thus land-locked and geographically disadvantaged states have a right of access to surplus fisheries in their subregion or region. Global and regional cooperation in the protection and preservation of the marine environment should take into consideration characteristic regional features.

Fifth, it refers to the type of action (modalities) required in a given situation. This action may take the form of regional agreements, arrangements, programs, or the establishment of centers. The modalities of access to surplus fisheries and rights of transit of land-locked states are to be set out in bilateral, subregional, or regional agreements. Through competent international organizations, states are to establish global and regional rules, standards, practices, and procedures to curb dumping.[9] Search and rescue on the High Seas should be established through regional arrangements. Transfer of technology is to be carried out inter alia through regional programs. Regional marine scientific and technology centers are to be established.[10] The International Sea-Bed Authority may also establish regional centers and offices.

9. Editors' note.—For information about the London Dumping Convention see "Report of the Eleventh Consultative Meeting of Contracting Parties to the Convention on the Prevention of Marine Pollution by Dumping of Wastes and Other Matter, October 3–7, 1988," Appendix A, in this volume.

10. Editors' note.—This is discussed in the report of *Pacem in Maribus XVII*, Appendix B, in this volume.

The concept discussed above centered on explicit use. There remain many instances where region and subregion are implicit, as in the case of enclosed and semienclosed seas[11], ice-covered areas, and especially sensitive areas. The definition of enclosed and semienclosed seas in the LOS Convention serves the function of identifying a physical area for the application of a set of rules relating to regional cooperation.

DEFINING "MEDITERRANEAN" AND PARTICIPATION IN REGIONALISM

It is reasonable to state that there is more than one "Mediterranean." The perspective that one chooses to adopt in relation to subject matter necessarily defines the boundaries of the so-called Mediterranean, Mediterranean Sea, and Mediterranean region.

Writing from a historical and geographical perspective, the great French historian Fernand Braudel faced the difficult task of setting the parameters for his monumental work on the Mediterranean. His warning to the historian of the immensity of the task is clear:

> Woe betide the historian who thinks that this preliminary interrogation is unnecessary, that the Mediterranean as an entity needs no definition because it has long been clearly defined, is instantly recognizable and can be described by dividing general history along the lines of its geographical contours. What possible value could these contours have for our studies?

> But how could one write any history of the sea, even over a period of only fifty years, if one stopped at one end with the Pillars of Hercules and the other with the straits at whose entrance ancient Ilium once stood guard? The question of boundaries is the first to be encountered: from it all others follow. To draw a boundary around anything is to define, analyze, and reconstruct it, in this case select, indeed adopt, a philosophy of history.[12]

Consequently, Braudel spends some 350 pages in discussing the role of the environment in terms of physical, economic, and human geography. His "greater Mediterranean" encompasses, among others, the Sahara, trade routes, Islam, the Russian, Polish, and German isthmuses, and the Atlantic.

11. Editors' note.—For further information about the Mediterranean semi-enclosed sea, see Vallega, pp. 372–93.

12. Fernand Braudel, *The Mediterranean and the Mediterranean World in the Age of Philip II*, 2 vols. (London: Collins, 1972), vol. 1, "Preface to the First Edition," pp. 17–18.

Even Braudel may have underestimated the difficulty that disciplines other than history also face in defining the Mediterranean, particularly with the advancement of scientific knowledge. "Nothing could be clearer than the Mediterranean defined by oceanographer, geologist, or even geographer." [13] And yet different disciplines and scholars within the same discipline adopt divergent definitions of the Mediterranean. To the marine scientist, who is conscious of the unity of, and consequent interdependencies within, the global marine environment, the Mediterranean Sea is but part of a whole. While its geographical marine contours can be easily defined, certain marine species, in effect, render the Mediterranean Sea *sans frontières.* Highly migratory species are not region bound, and tuna and swordfish do not respect academically defined boundaries. [14] New species, alien to the Mediterranean Sea, migrate to the Mediterranean. [15] The consequence is that, while it is possible to identify a generic Mediterranean marine management unit, a different management concept is required for certain Mediterranean species that would require the scholar to ignore artificial boundaries. [16] In other instances, "it is more productive from a management point of view to regard the Mediterranean as a collection of small ecosystems which are rather loosely linked in an ecological sense, though the larger ecosystem imposes constraints on the smaller systems." [17]

Alternatively, a broader conception of the Mediterranean is necessary to comprehend the wider ecological problem of aridity and desertization. [18] Deserts are both natural and man-made; the latter are dramatically increasing by several hundred thousand hectares each year. The consequence of this phenomenon is that climatologically the concept of Mediterranean is increasingly elastic.

Geology's concern with plate tectonics also forces this discipline to look at

13. Braudel, p. 17.

14. Tuna, for instance, migrate from the North Atlantic and North Sea into the Mediterranean to spawn and eventually head back to those northern regions. Editors' note.—Biological resources are referred to by Vallega at pp. 381–82.

15. For instance, since the opening of the Suez Canal in 1869, several Indo-Pacific species have migrated into eastern Mediterranean waters.

16. See, e.g., W. W. Murdoch and C. P. Onuf, "The Mediterranean as a System, Part I—Large Ecosystems," *International Journal of Environmental Studies* 5 (1974):275–84.

17. C. P. Onuf and W. W. Murdoch, "The Mediterranean as a System, Part II—Small Ecosystems," *International Journal of Environment Studies* 6 (1974): 29–34, quoted p. 29.

18. H. N. Le Houerou notes that "The Mediterranean arid zone comprises parts of the following countries (from west to east): Spain (the southeastern part), Morocco, Algeria, Tunisia, Libya, Egypt, Israel, Syria, Jordan, Iraq, Saudi Arabia, Kuwait, Iran and Afghanistan. According to this definition the Mediterranean arid zone thus includes some four million km² " ("Man and Desertization in the Mediterranean Region," *Ambio* 6 [1977]: 363–65, quote at p. 363).

greater and lesser boundaries. The Mediterranean Sea overlaps two great plates of the earth's crust, the African and Eurasian, which include large portions of both the African and European continents. Certain regions of the Mediterranean can also be defined in terms of the microplates that they cover. Moreover, crustal plates are continuously in motion, so that the concept of regional boundaries in geological time is also elastic.

The social sciences, too, have elaborated functional definitions. In a recent collection of studies on the Mediterranean by social scientists, "the Mediterranean world consists of those nations and peoples who physcially are linked to the Mediterranean basin and those nations especially who are tied to that geographic area through a high level of interdependent policy interaction."[19] Consequently, the European Economic Community (EEC), Jordan, the United States, and the Union of Soviet Socialist Republics (USSR) feature prominently in this conception of the Mediterranean.

Beyond academic definitions lie operational designations for regional action. Based on the facts of the situation, these should have realistic possibilities for practical application. The question of participation is raised at this stage.

For the purposes of the General Fisheries Council for the Mediterranean (GFCM), the Food and Agriculture Organization (FAO) designated the area of operations to include the Black Sea and the Seas of Marmara and Azov, in addition to the Mediterranean Sea.[20] For statistical purposes, the GFCM region has four large geographical sectors with eight divisions.[21] These divisions, however, do not apply to tuna, which is the responsibility of the International Commission for the Conservation of Atlantic Tunas (ICCAT). Participation is not restricted to the riparian states of these seas. Any FAO member nation or associate member and non-FAO members who are members of the United Nations, its specialized agencies, and the International Atomic Energy Agency may participate in GFCM.[22]

---

19. C. F. Pinkele and A. Pollis, eds, *The Contemporary Mediterranean World* (New York: Praeger, 1983). Editors' note.—See also Vallega.

20. The GFCM region is identified as "the Mediterranean and the Black Sea and connecting waters," Preamble and Article IV, "General Fisheries Council for the Mediterranean: Agreement and Rules of Procedure," *Basic Texts*, Vol. III (Rome: Food and Agriculture Organization, 1982).

21. The geographical areas and divisions are as follows: the WESTERN subarea, which includes three divisions (Balearic, Gulf of Lion and Sardinia); The CENTRAL subarea, which includes two divisions (Adriatic and Ionian); the EASTERN subarea, which includes two divisions (Aegean and Levant); The BLACK SEA subarea which coincides with the division Black Sea. See General Fisheries Council for the Mediterranean, GFCM Statistical Bulletin no. 5, *Nominal Catches, 1972–82* (Rome: FAO, 1984), p. iv.

22. The GFCM members to date are Algeria, Bulgaria, Cyprus, Egypt, France, Greece, Israel, Italy, Lebanon, Libya, Malta, Monaco, Morocco, Romania, Spain, Syria, Tunisia, Turkey, and Yugoslavia.

Conversely, the United Nations Environment Programme's (UNEP) Regional Sea Programme for the Mediterranean resulted in a convention that defines geographical coverage for the purposes of protecting this sea from pollution in terms of the Mediterranean basin.[23] Unlike GFCM, participation is effectively restricted to the Mediterranean coastal states that are also parties to the Barcelona Convention.[24] While it is forseen that other states may even-

23. This geographical coverage is defined as follows:
1. For the purposes of this Convention, the Mediterranean Sea Area shall mean the maritime waters of the Mediterranean Sea proper, including its gulfs and seas bounded to the west by the meridian passing through Cape Spartel lighthouse, at the entrance of the Straits of Gibraltar, and to the east by the southern limits of the straits of the Dardanelles between Mehmetcik and Kumkale lighthouses.
2. Except as may be otherwise provided in any protocol to this Convention the Mediterranean Sea Area shall not include internal waters of the Contracting Parties.
See Article 1, Convention for the Protection of the Mediterranean Sea against Pollution, Done at Barcelona on February 16, 1976, in *International Legal Materials* 15 (1976): 290–300. In two subsequent protocols the geographical coverage has been extended for specific purposes to include waters on the landwand side of territorial sea baselines up to the freshwater limit, saltwater marshes communicating with the sea, and, at the discretion of the coastal state, also wetlands and coastal areas. See Article 3, Protocol for the Protection of the Mediterranean Sea against Pollution from Land-based Sources, Done at Athens on May 17, 1980, in *International Legal Materials* 19 (1980): 869–78; Article 2, Protocol concerning Mediterranean Specially Protected Areas, Done at Geneva on April 3, 1982; Kenneth R. Simmonds, ed., *New Directions in the Law of the Sea* (New York: Oceania, 1984).
24. Article 24, Barcelona Convention, 1976. The Action Plan for the protection of the Mediterranean Sea was agreed on at a conference of representatives of 16 Mediterranean coastal states and other interested parties held in Barcelona, January 28–February 4, 1974, under the auspices of UNEP. A further conference was held in Barcelona, February 2–16, 1976, also under the auspices of UNEP, which resulted in the signing of a convention and two protocols by 12 of these states. The Barcelona Convention entered into force February 12, 1978, after being ratified by 16 states (but not Albania or Turkey—who later signed, in 1980). For further information about the conference and text of the Barcelona Convention, see "Conference of Plenipotentiaries of the Coastal States of the Mediterranean Region for the Protection of the Mediterranean Sea" (Barcelona, February 2–16, 1976), *Ocean Yearbook 1*, ed. Elisabeth Mann Borgese and Norton Ginsburg (Chicago: University of Chicago Press, 1978), pp. 702–33; the text of the "Protocol for the Protection of the Mediterranean Sea against Pollution from Land-based Sources," reprinted in *Ocean Yearbook 3*, ed. Elisabeth Mann Borgese and Norton Ginsburg (Chicago: University of Chicago Press, 1982), pp. 489–96; "The International Convention concerning Pollution and the Mediterranean," *Ocean Yearbook 6*, ed. Elisabeth Mann Borgese and Norton Ginsburg (Chicago: University of Chicago Press, 1986) pp. 572–73; and "Symposium on Marine Cooperation in the Mediterranean, Third Tunis Declaration, November 28, 1986, " *Ocean Yearbook 7*, ed. Elisabeth Mann Borgese, Norton Ginsburg, and Joseph R. Morgan (Chicago: University of Chicago Press, 1988), pp. 534–36. With the exception of Albania, the Mediterranean coastal states are full participants in this regional arrangement. Albania has recently started to attend meetings as an observer. The question of

tually participate,[25] the latest protocols have been open for signature and ratification or accession only to invited participants of regional conferences, that is, the Mediterranean coastal states.[26] The results of scientific research on land-based sources of pollution, however, indicate that in addition to riparians of rivers with outlets on the Mediterranean, increasing amounts of long-range atmospheric pollution is reaching this sea from extraregional sources. The regional effort may indeed be seeing its limitations.

According to the LOS Convention, the Mediterranean may be defined as a semienclosed sea, meaning a "sea surrounded by two or more States and connected to another sea or the ocean by a narrow outlet or consisting entirely or primarily of the territorial seas and exclusive economic zones of two or more coastal States." [27] This definition entails important juridical consequences that impinge on participation in regional processes. States bordering semienclosed seas are governed by a cooperative ethic and norms relating to living resources, the marine environment, and scientific research. In their pursuit of these responsibilities, however, Mediterranean coastal states are required "to invite, as appropriate," other interested states and international organizations.[28]

GUIDELINES FOR PARTICIPATION IN REGIONALISM

How is participation to be determined? And, once determined, what is its measure? Cost apart, the starting point is the subject matter, that is, the problem or issue at stake. It is possible to identify three major types of marine problems according to whether they relate to development, management, or

---

broader participation has been raised by the USSR. In 1977, Anatoly Kolodkin remarked,

> It seems to me that the problem of regionalization is very important and we should not underestimate or diminish its significance. No doubt it has positive features, and positive consequences for the developing and developed countries, including the USSR itself. For example, there is a serious regional problem in the Mediterranean in the field of the environment, and we could not understand why the Soviet Union, Bulgaria, Romania, and Black Sea Congress were not invited to the Barcelona Conference as participants, but only as observers. It seems to me that the Black Sea is a subregional zone of the Mediterranean, and I cannot imagine how there can be full compliance with the provisions of that conference without the Black Sea Congress.

(Johnston, ed. [n. 6 above], quoted at p. 45.)

25. Article 24, Barcelona Convention, includes "*any* State entitled to sign any protocol in accordance with the provisions of such protocol" (emphasis added).

26. Article 16, Athens Protocol; Article 18, Geneva Protocol.

27. Article 122, LOS Convention.

28. Article 123(*d*), LOS Convention.

security. A contextual consideration serves to clarify the extent of the subject matter and the interdependencies involved.

Once the problem or issue has been characterized, the interests and competencies affected require identification. It is suggested that national interests and competent international organizations affected may be considered according to a set of criteria flowing from the concepts of user and nonuser participation.

User Participants

Decision making as a criterion apart, user participation in the Mediterranean Sea may be classified according to three criteria: geographical proximity, entitlement to both maritime zones and uses, and actual marine use.

Geographical proximity denotes the physical relationship of a state to the Mediterranean Sea. A state may be central coastal, adjacent coastal, adjacent landlocked, or distant. The Mediterranean riparians are obviously central coastal. Adjacent coastal are those states that border on neighboring seas and have a direct physical relationship to the Mediterranean Sea, namely the western Atlantic approaches, the Black Sea, and the Red Sea. Adjacent landlocked are those landlocked states that border the central coastal states. All other states are considered distant.

Entitlement refers to a state's right under modern international law to claim maritime zones, its rights and freedoms with reference to certain ocean uses, and its prerogative in certain situations to living-resource surplus and transit rights. Three categories of states are identifiable: zone claimants, resource and nonresource users, and nonresource users. Central coastal states are zone claimants. Adjacent landlocked and distant fishing nations qualify as potential resource and nonresource users. All others may qualify at a minimum as nonresource users.

Not all states in the region are actual marine users, but those that are may be subdivided according to whether they are multiple or single users. Central coasters are presumed to be multiple users because of their geographical relationship to the Mediterranean. Other states may also qualify as multiple users. It is assumed that most single users are concerned with marine transportation.

Having stated the above criteria, it is possible to identify user states according to whether they are core or peripheral participants, with further distinctions as to inner or outer core. Utilizing the criteria established above, tables 2 and 3 propose a point range to classify state users.

Table 2 assigns range in each criterion from maximum to minimum. Table 3 provides the point ranges that identify which user participants are inner or outer core and inner or outer periphery.

Only the central coastal states, that is, the 18 Mediterranean coastal states

TABLE 2.—GUIDELINES FOR PARTICIPATION IN REGIONALISM

Criterion for Participation	Range (within each criterion)[a]
Geographical proximity:	
Central coastal	4
Adjacent coastal	3
Adjacent landlocked	2
Distant	1
Entitlement:	
Maritime zones	3
Resource and nonresource uses	2
Nonresource uses	1
Actual uses:	
Multiple	2
Single	1

SOURCE:—Aldo Chircop, "Cooperation Regimes in Ocean Management: A Study in Mediterranean Regionalism" (J.S.D. thesis, Dalhousie University, 1988), p. 172.

[a] Maximum down to minimum range.

and the EEC, qualify as inner core participants, may claim maritime zones, and are multiple marine users all at the same time. Generally, this physical relationship, which is the basis of the maritime claim, is immutable. This may not be the case with the probable nineteenth central coastal state, the United Kingdom, whose presence in Gibraltar and in the sovereign military bases on Cyprus may not be indefinite. It is possible that there may be new candidates for inner-core group membership (e.g., the Turkish Federated State of Cyprus and a new Palestinian state), or that the political geography of the current inner-core participants may need to be slightly modified.[29] Two other important but exceptional user participants are the USSR and the United States. Neither power is central coastal, but their continuous naval presence in the Mediterranean Sea over a long period of time has, in effect, given them a factual physical presence tantamount to floating states. They are in a position of physically making maritime claims (although not resource oriented) and counterclaims. Accordingly, these two exceptional participants are notionally treated as central coastal in geographical proximity, but they still appertain to the outer-core user group.

Depending on whether and how they utilize their resource rights, land-locked states may qualify for either outer core or inner periphery. Adjacent coastal states are in a similar category as land-locked states although, generally, they are expected to be more intensive users than the latter. All other states generally belong to the inner or outer periphery.

Admittedly, this model may seem to oversimplify reality. A factor to be borne in mind is time. The model is dynamic. Marine uses and users vary in

29. For example, the Spanish enclaves of Ceuta and Melilla.

TABLE 3.—IDENTIFICATION OF USER PARTICIPANTS BY POINT RANGE

User Participants	Range
Inner core	9
Outer core	6–8
Inner periphery	4–5
Outer periphery	1–3

SOURCE.—Aldo Chircop, "Cooperation Regimes in Ocean Management: A Study in Mediterranean Regionalism" (J.S.D. thesis, Dalhousie University, 1988), p. 172.

time. The eventual EEZ mosaic in this sea will have the effect of pushing noncentral coastal users farther toward the edge of the outer periphery unless access and other rights are acquired from the central coastal states.

Another factor to be borne in mind constantly is that marine use is relative to the user in intensity and extensiveness. The same use may be minor to one actor but crucial to another, even though the use is more extensive and intensive for the former than for the latter. From a different perspective, the model may seem to treat all marine uses equally but, in reality, one specific use may have a much greater effect than a cluster of other uses would. For instance, waste discharge, marine transportation, and military uses are certainly the most significant in affecting the well-being of other actors today. Consequently, the enquirer embarking on sectoral analysis would have to adjust the criteria of entitlement and actual use.

The classification of user participants into core and periphery is important for several reasons. It is hypothesized that cooperation or competition in Mediterranean marine affairs among core participants will either improve or worsen the health, wealth, and security of the Mediterranean Sea and its dependent peoples. Inner-core participants are essentially the managers of the Mediterranean. Their geographical proximity and entitlements give them a special responsibility for its well-being. They are also the actors that are most directly affected by value pursuits in the Mediterranean Sea. In the near future they will be in a position to regulate community access to that sea for value pursuits and may have to answer the demands of peripheral participants for value sharing or for noninterference in value pursuits. In a way, the inner-core group may, in most cases, be in a position to determine a particular extraregional actor's status as user participant. Such an attempt is, in fact, being made by certain central coastal states to terminate the USSR and U.S. naval presence in the Mediterranean and thus to relegate these two states to the periphery.[30]

The new law of the sea emphasizes regional cooperation in the develop-

30. See *Strengthening of Security and Co-Operation in the Mediterranean Region,"* Report of the Secretary General to the United Nations General Assembly in its Forty-First Session, Agenda Item 67, UN document A/41/486, 20 October, 1986.

ment and management of enclosed or semienclosed seas. This facilitates the identification of the Mediterranean riparians as an inner-core group with certain responsibilities and prerogatives that necessitate concerted intragroup decision making in marine affairs. It is conceivable that, beyond national sovereignty, a new kind of territoriality will emerge as a result of the mobilization of the regional conscience. Whether this territoriality will be exclusive with reference to specific functions remains to be seen.

Nonuser Participation

International organizations are participants in marine affairs in the Mediterranean, although they are not marine users. They may be classified in terms of core and periphery but, for obvious reasons, on a different basis than user participants. As nonuser participants, international marine organizations may be classified according to the spatial reference of their mandate and their activities. Unlike nation states, international organizations are institutions with limited functions that determine the type, extent, and place of their participation in processes.

Core participants have mandated or assumed geographical references for their activities. They may be either regional organizations, set up to deal specifically with problems of a transnational nature, or global organizations that have developed active programs for the Mediterranean. The former includes GFCM and ICSEM. The latter includes UNEP, through the Regional Seas Programme for the Mediterranean, and the International Maritime Organization (IMO) by way of its presence in the Regional Oil Combating Centre (ROCC) in Malta.[31] Core nonuser participants are actively involved as direct participants in, or facilitators of, functional decision-making processes.

There are innumerable other international governmental and nongovernmental organizations that have a global mandate and have been peripherally involved in marine affairs in the Mediterranean by way of information input, communication, expertise, and funding. For example, the World Health Organization (WHO), the United Nations Development Programme (UNDP), and the United Nations Educational, Scientific and Cultural Organization (Unesco) directly, and through the Intergovernmental Oceanographic Commission (IOC).[32] Through their functional inputs, these organizations may be instrumental in setting up regional programs.

31. The second protocol of the Barcelona Convention provided for coordinated action in the event of oil spills and for the establishment of regional headquarters (ROCC) which opened on Manoel Island, Malta on December 11, 1976, to direct such operations. The ROCC publication, *ROCC-INFO* is a thrice-yearly newsletter.

32. Editors' note.—See report of the IOC, "Intergovernmental Oceanographic Commission (of Unesco): 1989," Appendix A, in this volume.

CONCLUSION

The model outlined above suggests guidelines for expanded participation in current regional arrangements and initiatives dealing with development, management, and security. Beyond participation, either formal or de facto, lies the subsequent stage—the commitment to act, which this model does not address.

Equity and efficiency in a new law of the sea imbued with a cooperative ethic suggest inclusion of affected interests. Not all instances of exclusion are attributable to the nature of the arrangement or the values shared by the actors. Lack of participation may be due to disinterest or competing value pursuits. Growing knowledge about the nature and processes of marine problem increases the opportunity for effective management of the Mediterranean Sea, with a participation that reflects accurately the extent of the interdependence. Facilitation of the management for a regional problem may mean that the geographical limits of the region need to be stepped over.

Reports from Organizations

The reports and abridged reports contained in this appendix represent reviews of only some of the activities by organizations which deal with ocean-related matters. They are intended to provide the reader with basic coverage of important programs and directions being pursued by selected major international organizations.

THE EDITORS

Asian Wetland Bureau: 1989*

There are 120 million hectares of wetlands in East and South Asia. However, as much as 85% of wetlands surveyed are threatened by degradation or destruction. Destruction of wetlands has caused severe declines in fishery and forestry resources and biodiversity, increased flooding, seasonal droughts, coastal erosion and other socioeconomic problems.

The Asian Wetland Bureau (AWB) is an independent, international, nonprofit organization which aims to promote protection and sustainable utilization of wetland resources in Asia.

AWB is based at the Institute of Advanced Studies in the University of Malaya, has national offices in Indonesia and the Philippines, and has project workers in five other Asian countries.

AWB's work is split into four major themes: biodiversity; water resources; institution building and awareness; and management and policy.

To maximize effectiveness of regional work, AWB's approach is to involve and work with local agencies, both government and nongovernment. To date, AWB has worked for or with agencies in Indonesia, Malaysia, Philippines, Brunei, Thailand, Vietnam, Hong Kong, China, Taiwan, South Korea, Bangladesh, and Pakistan.

The Asian Wetland Bureau is keen to cooperate with or provide technical assistance to other institutions or individuals who have an interest in wetlands.

AWB currently channels US$500,000 a year into projects in the region. Additional funds are constantly needed for new projects.

PERSONNEL RESOURCES

AWB currently comprises 27 full-time staff, 8 consultants, and 12 part-time or volunteer staff, and has 13 counterparts from government agencies. Staff expertise ranges from ecology, aquaculture, fisheries, limnology, water chemistry, ornithology, foresty, and remote sensing to training, conservation planning, and resource management.

MAIN FUNDING BODIES (CONTRIBUTING MORE THAN US$10,000)

Asian Development Bank; Australian National Parks and Wildlife Service; the British government (ODA); British Petroleum; Conservation Treaty Support Fund; Danish International Development Agency; the Ford Foundation; ICBP-International/Australia/Japan; IWRB; Lady Y. P. McNeice; the government of The Netherlands;

*Report supplied by Mr. Roger Jaensch, Project Manager, Asian Wetland Bureau Headquarters, Institute of Advanced Studies, University of Malaya, Lembah Pantai, 59100 Kuala Lumpur, Malaysia.

the government of New Zealand; Selangor state government; UNEP; USAID; U.S. Fish and Wildlife Service; WWF-International/Denmark/Malaysia/Netherlands/U.K./U.S.

MAIN COLLABORATING AGENCIES

Department of Environment and Natural Resources, Philippines; Department of Wildlife and National Parks, Peninsular Malaysia; Directorate General of Forest Protection and Nature Conservation, Indonesia; East China Normal University; Forest Department of Bangladesh; Forest Research Institute, Malaysia; Institute of Advanced Studies; University of Malaya; Jiangxi Forest Department, China; Kyung Hee University, Seoul, South Korea; National Parks and Wildlife Office, Sarawak Forest Department; Prince of Songkia University, Thailand; Royal Thai Forest Department; State Planning Unit, Selangor, Malaysia; University of Hasanuddin, Indonesia; University of Hanoi, Vietnam; University of Sriwijaya, Indonesia; Wild Bird Society of Japan; Wildlife Fund, Thailand.

SOME AWB ACHIEVEMENTS

1. The completion of over 60 projects and publication of over 40 reports on many aspects of wetlands in Asia
2. Assisting the preparation of National Wetland Inventories in Malaysia, Indonesia, and the Philippines
3. Completing detailed studies on peat swamp and mangrove forests in Malaysia and Indonesia, and preparing recommendations for their management
4. Identification of key sites for migratory water birds in Southeast Asia and providing assistance for their protection
5. Cooperating in a major programme with the Indonesian Ministry of Forestry to develop management recommendations for 35 million ha of wetlands throughout Indonesia
6. Completion of 18 training courses in 7 countries, involving wetland scientists from 12 Asian countries
7. Monitoring of development projects funded by international agencies which affect wetlands in Asia
8. Preparation of guidelines to minimize the impact of coastal aquaculture on mangrove forests

SOME FUTURE PROJECTS

—Development of manuals describing techniques for wetland evaluation
—Identification of priority management areas on the Bangladesh coastline, with the Bangladesh Forest Department
—Evaluation of the status of fish species in Asia
—Production of a mangrove ecology film in Indonesia
—Development of sustainable forestry management techniques for peat swamp forests

—Indus delta survey and training course, Pakistan
—Long-term ecological study of the Yangtze estuary, China, by East China Normal University
—Integrated management and sustainable development of coastal wetlands in the Philippines
—Evaluation of coastal wetlands in South Korea
—Development of a wetland training program in Thailand
—Development of local language training and educational material
—Collaboration with development assistance agencies on large-scale implementation of sustainable wetland management

Ocean Governance Program, Environment and Policy Institute, East-West Center: 1989

The Ocean Governance Program of the Environment and Policy Institute, East-West Center, is becoming known among those in the Asia/Pacific region concerned with marine policy. This has been a consequence of its previous work in Southeast Asia and its current associations with researchers in the East Asian countries. Two recent conferences on the Yellow Sea and the Sea of Japan have been instrumental in Ocean Governance Program project research, and colleagues in Japan, South Korea, and China, some of whom have been active participants in its research, have expressed appreciation for the Program's efforts.

The Pacific and Indian Oceans cover a large part of the Asia/Pacific realm with which the Center is concerned. These waters contain living and mineral resources and are used for such activities as shipping, waste disposal, naval defense, fishing, and exploitation of offshore hydrocarbon resources. Moreover, the evolving Law of the Sea, which gives coastal nations the right to exercise control over resources within 200 nautical miles of their coastal baselines, has added additional impetus to concerns of countries in the East-West Center region to manage their marine resources wisely and in accordance with the 1982 UN Convention on the Law of the Sea.

In addition to conveying rights to both living and non-living resources in a large offshore area, the Law of the Sea Convention requires that nations cooperate with each other to protect the seas from pollution and that the rights to navigate the oceans, both within and outside coastal state jurisdictions, be protected.

The principal functions of the Ocean Governance Program are to conduct research on the policy implications of the newly acquired entitlement to large offshore resources jurisdictions and to offer guidance, in the form of policy options, to countries in the Asia/Pacific realm. The Program is coordinated by a working group composed of Joseph R. Morgan, Chairman, Norton Ginsburg, John Bardach, Toufiq Siddiqi, and Daniel Dzurek, of the Environment and Policy Institute, and Mark Valencia of Resource Systems Institute, East-West Center.

ACTIVITIES OF THE OCEAN GOVERNANCE PROGRAM

The Program focuses on three separate but related activities: production of policy-relevant publications focused on the principal marine regions (Southeast Asian Seas, East Asian Seas, South Asian Seas, and the Southwest Pacific Ocean); production of *Ocean Yearbook;* and the Pacific Maritime Collegium.

The studies of marine regions, which are the heart of the Ocean Governance Program, are on an approximately 12-year cycle; each substantive study and atlas production takes about three years. While it would be highly desirable to speed up this cycle, personnel and financial resources are limiting factors. To enrich and enlarge the Program, scholars in the fields of Law of the Sea, marine policy, marine resources, and environmental concerns such as ocean pollution are brought to the Center.

Production of Marine Policy Atlases

Focus is given to each region of Asia/Pacific in turn with the objective of producing a substantial publication, in the form of an atlas, that describes and analyzes the relevant marine policy concerns of regional countries. While each marine region within the East-West Center area of interest has unique features, based on the geography of the countries within the region, they all have similar concerns with making the best use of increasingly valuable marine resources while managing or governing the areas under their jurisdiction in accordance with applicable international law. Atlas contents focus on regional topics such as the natural environmental setting, scientific research, maritime jurisdictions, maritime defense, fisheries, hydrocarbons, vulnerable resources, pollution, shipping, and a final section that attempts to integrate some of the above topics to analyze specific policy concerns.

Thus far, an *Atlas for Marine Policy in Southeast Asian Seas* has been published, and work on a similar type of publication on East Asian Seas is in progress.

East Asian Seas Project
The three semi-enclosed seas in East Asia (Yellow Sea, East China Sea, and Sea of Japan) are areas in which conflicts over resources are likely if the surrounding coastal states do not cooperate in establishing regional arrangements for such activities as fishing, shipping, pollution control, protection of threatened and endangered species, and extraction of marine mineral resources.

The objective of this project is to assist in achieving the necessary regional cooperation by bringing together at conferences representatives from China, the Koreas, Japan, Taiwan, and the USSR, and by publishing research results in the form of an atlas for marine policy, with accompanying text of policy analysis. The atlas will cover the following topics:

(1) The natural setting, including oceanographic features, natural hazards, and oceanic climate.
(2) Scientific research in the form of a summary of results now available.
(3) Valuable and vulnerable resources, including endangered and threatened species, pristine environments worthy of special protection, and research stations.
(4) Maritime jurisdictions and boundaries, with particular emphasis on overlapping and conflicting claims.
(5) Fisheries, including catch statistics, optimum yields, and mariculture.
(6) Shipping, including trade routes and flows, and ports.
(7) Distribution of potential hydrocarbon resources in the marine region, exploration and exploitation activities, and lease arrangements.
(8) Pollution sources, both land-based and from transiting ships.

The publication will utilize the work of scholars from the region as reported in the conferences and workshops and will serve an audience of East Asian managers and scholars concerned with marine resources and governance. Research is directed toward a comprehensive assessment of marine resource endowments, their value and significance, and the claims to them by riparian states, as well as to the formulation of policy parameters for regional international cooperation leading to conflict resolution and more efficient resource management practices.

The *Atlas for Marine Policy in East Asian Seas* will be completed and prepared for publication during 1990. Funding is provided by both the University of Hawaii/East-West Center Collaborative Research Grant and the East-West Center. The Soviet-hosted conference in September 1989 focused on the Sea of Japan and the Sea of Okhotsk. It was a follow-on to the successful conference on the Sea of Japan held in Niigata, Japan, in October 1988. A conference on the East China Sea may be held if Chinese sponsors are agreeable.

Production of *Ocean Yearbook*

The editorial offices of *Ocean Yearbook* were relocated in the Ocean Governance Program of the Environment and Policy Institute, East-West Center, in 1987 soon after the appointment of Norton Ginsburg, a founding editor of the *Yearbook,* to the position of Director, Environment and Policy Institute. Professor Elisabeth Mann Borgese, Dalhousie University, Canada, continues as co-editor and Professor Joseph R. Morgan, Chairman of the Ocean Governance Program, has become the third co-editor.

As readers of this publication are aware, the *Yearbook* is a journal published by the University of Chicago Press. It has considerable value for academic institutions that have graduate programs in ocean-related subjects, as well as for decision-makers in international agencies and governments. It is sponsored by the International Ocean Institute in Malta, a non-governmental policy-oriented body, with an international reputation for originality and objectivity in the field of marine policy. The East-West Center provides editorial and financial support for the publication of the *Ocean Yearbooks.*

Each *Yearbook* contains invited articles on various policy-relevant topics, reports from international governmental and non-governmental organizations, selected documents and proceedings, and updated tables on living and non-living resources, transportation and communication, military information, and general information tables such as status of ratifications of the UN Convention on the Law of the Sea and the Preparatory Commission activities.

Pacific Maritime Collegium

The Collegium has been operating since 1986. It organizes and sponsors periodic seminars and colloquia and publishes an informal newsletter which is usually devoted to a summary report of the most recent seminar presentations.

Intergovernmental Oceanographic Commission (of Unesco): 1989,[1] Excerpts from the Fifteenth Session of the IOC Assembly, Unesco, Paris, 4–19 July 1989, "Report of the Secretary on Intersessional Activities" (1 January–31 December 1988)*

STRENGTHENING OF NATIONAL AND REGIONAL RESEARCH CAPABILITIES

Training, Education and Mutual Assistance in the Marine Sciences (TEMA)

Some of the highlights of the activities under TEMA include progressive involvement of training activities in both developed and developing countries fostering marine scientific and technological transfer relative to the ongoing programmes of the Commission; the increased collaboration with other international organizations in TEMA activities; and steady progress in the implementation of the Unesco-IOC Comprehensive Plan for a Major Assistance to Enhance Marine Capabilities in the Developing Countries.

During the intersessional period, about 122 scientists received training of which 83 were in marine pollution research and monitoring, 25 in international oceanographic data exchange, and the remaining in other fields. As part of the Unesco-IOC Comprehensive Plan, Marine Science Country Profiles (MSCPs) for 8 countries have been developed and 6 project proposals for technical assistance for extra-budgetary funding have been, or are in the process of being formulated, at the request of Member States.

Regional Training Courses
In support of WESTPAC Programme activities, the Government of Japan, through its Funds-in-Trust contribution, offered the seventh intensive training course on Oceanographic Data Management at the Japanese Oceanographic Data Centre (JODC) in Tokyo, from 26 September to 8 October 1988. The course was announced to Member States in the WESTPAC region by means of IOC Circular Letter No. 1185 (May 1988). Three scientists (from Republic of Korea, Thailand and Viet Nam) attended the course which involved lectures by the JODC staff on IODE System and data exchange in the WESTPAC region, data flow in JODC data processing methods for various data items such as geological, geophysical and wave data, GF3 and MDG77, as well as practical exercises for the use of microcomputer software for oceanographic data management.

*This report has been excerpted from the Intergovernmental Oceanographic Commission (of Unesco), Fifteenth Session of the IOC Assembly, July 4–19, 1989, Report of the Secretary of Intersessional Activities (January 1–December 31, 1988), and was supplied by Mr. Gunnar Kullenberg, Secretary, IOC, Unesco, 7, Place de Fontenoy, 75700 Paris, France.

The IOC-Unesco Training Course on the Use of Microcomputers for Oceanographic Data Management, which is taking place at the Asian Institute of Technology, Bangkok, 16 January–3 February 1989 was announced through the IOC Circular Letter NO 1195 (July 1988). Fifteen scientists from the WESTPAC region are participating in the training course.

Individual Study Grants and Internship
The following study grants were provided by the IOC to scientists to facilitate their participation in activities relevant to IOC programmes:
A Peruvian scientist received a study grant to conduct research on a sub-superficial undercurrent along North Peru at Nova University, Florida (USA), from 1 to 30 May 1988.
A scientist from Thailand participated in the Symposium on Biogeochemical Studies of Chanjing Estuary (Hangzhou, China, 20–25 March 1988) and visited Bedford Institute of Oceanography (Dartmouth, Canada, 10–14 October 1988).
An Egyptian student was assisted in his study on a "simulation model for mercury pollution in fish" at the Royal Danish School of Pharmacy (Copenhagen) (3 weeks in June 1988).
Support was provided to one trainee from Turkey to participate in the course on Egg Production Method, from 5 to 16 December 1988, Bremerhaven (Federal Republic of Germany).

Shipboard Training/Participation in Oceanographic Cruises
In support of WESTPAC activities the Government of Japan, through its Funds-in-Trust contribution, offered opportunities for the participation of scientists from the region in the following research cruises on board Japanese research/survey vessels:
Two scientists from China and the Philippines participated in the cruise of S.V. TAKUYO organized by the Maritime Safety Agency in the Western Pacific, from 12 February to 15 March 1988, for oceanographic observations.
One scientist from China participated in the cruise of R.V. HAKUHO-MARU organized by the Ocean Research Institute of the University of Tokyo in the Northwestern Pacific, from 22 September to 31 October 1988, for biological studies, particularly in the trench, marginal seas and subtropical waters.
Two scientists from the Philippines and the Republic of Korea participated in the cruise of R.V. HAKUHO-MARU in the waters of the southern part of Japan and the East China Sea, from 6 to 26 December 1988, for the studies on the mechanism of the transportation and dispersal of eggs and larvae of important fish.
The following group training courses in co-operation with other international organization (e.g., UNEP) and Member States were supported:

Marine Pollution

Training Course in Marine Pollution at Bermuda Biological Station, January 1988	3
Training Course on bioassays and toxicity tests, Cartagena, Colombia, 11–22 April 1988 (CPPS, UNEP)	20 Southeast Pacific/CPPS region

Training Course on Analysis of Benthic data (Mediterranean) Piran, Yugoslavia, 15–24 June 1988 (FAO, UNEP-MAP)	25 (Mediterranean countries)
Second Biological Effects Workshop Bermuda, 10 September–2 October 1988 (IOC)	10 (for several regions)
Training Course on Ecological Impacts of Pollutants on the Coastal Marine environment, Guayaquil, Ecuador, 18–28 October 1988 (CPPS, UNEP)	25 S.E. Pacific/CPPS region

Ocean Services

Training Course on the use of IOC GF3 Format and GF3-Proc software package, Bidston, UK, May 1988	2
Training Course on Ocean Data Management Course for WESTPAC region, Tokyo, Japan, 26 September–8 October 1988	3
Training Course on Oceanographic Data Management, Ottawa, Canada, 14 November–9 December 1988	5

In addition, the IOC course module on scientific basis for global conventions as the London Dumping Convention and MARPOL 73/78 was presented at the World Maritime University, May 1988, for 102 students.

IOC Research Fellowship Scheme (IOC-RFS)

Three fellowships contributed by the Government of the Federal Republic of Germany for 1988 under the IOC-RFS were announced through IOC Circular Letter No. 1173 (February 1988). From among a total of 26 applicants, one scientist from Ecuador was awarded a fellowship for a period of 12 months starting 23 September 1988 for polar marine biology and benthos research, one Indian scientist for a period of 12 months starting January 1989 for polar and subpolar paleo-oceanography, and one Chilean scientist for a period of 8 months starting February 1989 for polar marine microbiology.

In response to a request from IOC, the Rector of the Vrije Universiteit, Brussels (VUB), in consultation with the authorities of the General Administration for Co-operation and Development (GACD), has agreed to contribute four of its allocated fellowships to the IOC under the IOC-RFS. In addition to this, it is suggested that VUB in co-operation with the IOC will establish training programmes in Marine Geology under the International Training Course on Fundamental and Applied Quaternary Geology (IFAQ) to scientists from South-east Asia and Africa. Professor Dr. R. Paepe, Director of IFAQ, visited the Secretariat on 27 January 1989 for discussion on this subject. The Secretariat Staff will soon be visiting the authorities in Brussels (Belgium) to finalize the programme of co-operation in those fields.

COMPREHENSIVE PLAN FOR A MAJOR ASSISTANCE PROGRAMME TO
ENHANCE MARINE SCIENCE CAPABILITIES IN THE DEVELOPING
COUNTRIES

Marine Science Country Profiles (MSCPs)

During the intersessional period the IOC Secretariat, in collaboration with Unesco
OPS/SC, using UNDP funds, and with the assistance of national experts, prepared
drafts of the Marine Science Country Profiles (MSCPs) for the following countries:
Barbados, Curaçao, Guyana, Jamaica, St. Lucia, Trinidad and Tobago. In addition,
MSCPs for Malta and Portugal have been completed. MSCPs for Pakistan and the
Philippines are in process.

Formulation of Technical Assistance Projects for Extrabudgetary Support

The IOC initiated the following project proposals at the request of Member States
which are either submitted for extrabudgetary support or in the process of being
prepared:

(i) Project on "Monitoring and Prediction of the El Niño Phenomenon in the
Southeast Pacific." The application to Development (RLA/88/0-10) was, after
receiving the consultation from UNDP and the Government authorities concerned, revised and submitted officially to UNDP Headquarters. The document has been well accepted and is now in process of obtaining the final
approval. However, UNDP recently informed IOC that although an official
endorsement of at least three participating governments was expected, only
one government to date has transmitted its endorsement to UNDP Headquarters.

(ii) Project on "Strengthening of National and Regional Capabilities in Ocean
Science and Ocean Services in Support of Ocean Development in South-East
Asia." The project which was submitted by Unesco to UNDP has now been
short-listed by the UNDP for consideration and final decision at the Aid Co-
ordinator Meeting (MAC-IV) Session to be held in Jakarta, March 1989.

(iii) Project on "Tsunami Warning System in the Southwest Pacific." This project
which was submitted by Unesco to UNDP has now been short-listed by UNDP
for consideration and decision at MAC-IV Session, scheduled in Jakarta,
March 1989.

(iv) The Government of Pakistan requested IOC to field a mission to develop a
technical assistance project to be funded by UNDP. The UNDP Resident
Representative, Islamabad, formally requested Unesco/IOC to submit the
proposal for field experts mission to be funded by UNDP.

(v) The Federal Republic of Germany has agreed to provide contribution to the
IOC Trust Fund to organize a regional training course in marine geology in
the South-east Asian region.

(vi) At the request of the Democratic People's Republic of Korea's Government, a
staff member visited the country to investigate the possibility of an UNDP-
assisted project. Following the advice of the UNDP Office, a project document on the "Establishment of Marine Pollution Monitoring System and

Improvement of the Oceanographic Data Centre in the DPRK" is in preparation.

At the request of Member States, and following the Advisory Mission to the Eastern Caribbean in 1987, a regional technical assistance project proposal for strengthening ocean science and ocean services is in the process of being developed for possible extrabudgetary funding. Arrangements for a workshop of experts, to be designated by the governments concerned, will soon be negotiated with the CARICOM Secretariat and UNDP for funding in order to complete the project proposal.

Progress on the Preparation of a Guide on Operation and Management of
Research Vessels

The IOC was represented at the Third Meeting of the ICOD-IOC-FAO Steering Committee on the above-mentioned subject (Rome, 2–4 March 1988), and hosted the Fourth Meeting (Paris, 26–30 September 1988) in order to review the draft prepared. A thorough discussion on the draft led to concrete suggestions for revision. The text is being accordingly improved by ICOD.

Internship

Dr. Victor Georgievich Ussov, Head of the Division of International Problems of the Marine Sciences Development in the Pacific, at the Pacific Oceanological Institute, USSR Academy of Sciences, worked as an intern at the IOC for a period of 2 months, starting 23 February 1988, in order to acquaint himself with the programme and activities of the Commission.

Mr. N'Diawar Assane Cisse of Senegal, after completing post-graduate studies in France in the field of marine law, worked at the Secretariat as an intern, from 2 November to 16 December 1988, in order to acquaint himself with the programmes of the Commission as well as to study the provisions of the UN Convention on the Law of the Sea relevant to the marine scientific research under the legal regimes.

Other Activities

The IOC co-sponsored jointly with Unesco the International Conference on Computer Modelling in Ocean Engineering (Venice, 19–23 September 1988). One staff member representing the Secretary of IOC delivered the opening address. About 140 Delegates from 20 countries from different parts of the world attended the Conference. Altogether, 96 original papers on different topics were presented at the Conference.

Taking advantage of the presence of eminent experts at the Conference, a short consultation of ECOR officials was held with a view to obtaining views on training and education opportunities for ocean engineers and to identify areas of interest and possible collaboration with ECOR in its future activities. Participants were issued a questionnaire in order to obtain information about their interest in ECOR.

REGIONAL ACTIVITIES

Caribbean and Adjacent Regions (IOCARIBE)

Major activities in the region were carried out in accordance with recommendations of the Second Session of the Sub-Commission for IOCARIBE (Havana, December 1986), progress on which will be reviewed at IOCARIBE-III, forecast for early December 1989, in Venezuela, together with a regional scientific seminar. The formal host arrangements regarding the IOCARIBE Secretariat have now been agreed with the Colombian government, and the agreement on the Secretariat for IOCARIBE has been formally signed.

Within marine pollution research and monitoring, an IOC-UNEP regional task team for preparation of a regional review on the state of the marine environment has prepared a draft report now being edited. The UNEP-IOC regional task team to evaluate possible effects, on coastal zones and areas, of sea level and temperature rises, induced by possible climatic changes, has completed its task. Follow-up to these proposals for further studies are being initiated in collaboration with the relevant parts of the IOCARIBE programmes. Considerable activities have been undertaken in the regional component of GIPME-MARPOLMON (CARIPOL) and the joint IOC-UNEP project including intercalibration exercises, training courses, data collection, and an *ad hoc* expert group meeting to prepare a regional workshop (Kingston, Jamaica, 11–13 May 1988).

The IOC Workshop on IBCCA Data Sources and Map Compilation was held in Boulder (Colorado, USA, 18–19 July 1988) in conjunction with the Second Session of the Editorial Board for IBCCA. Experts from the IOCARIBE region, actively participating in the development of the bathymetric charts, reviewed the following subjects: data available for IBCCA; synthesis of data of different scales, projections and with different accuracies; use of modern techniques and other subjects relating to the scientific and technical problems of IBCCA compilation.

Several activities have also been accomplished within the regional components of OSLR and OSNLR (see Sections 1.2 and 1.3, respectively).

Western Pacific (WESTPAC)

During the intersessional period there has been a steady progress in the implementation of the programme approved by the Fourth Session of WESTPAC (Bangkok, Thailand, 22–26 June 1987), calling for co-operative research amongst Member States in target-oriented projects, adopted by WESTPAC-IV.

The Chairman of WESTPAC, in consultation with institutions, designated selected experts as programme co-ordinators and project leaders, respectively, for various programmes and co-operative research projects approved by WESTPAC-IV.

In support of these co-operative research projects, a number of workshops were organized in 1988 as follows:

The IOC-FAO Workshop on Recruitment of Penaeid Prawns in the Indo-western Pacific Region (CSIRO Marine Laboratories, Cleveland, Australia, 24–30 July 1988) was sponsored by the Australian International Development Assistance Bureau (AIDAB), the Australian Fisheries Service (AFS), IOC, FAO and the South East Asian Fisheries Development Centre (SEAFDEC). Attended by 20 participants from Australia, Indonesia, Malaysia, Papua New Guinea, Philippines and Thailand, the Workshop adopted a number of recommendations, including an operational plan for the participation of designated institutions. The Report of the Workshop is in the process of being published by the IOC. The Workshop also reviewed a project document for assistance which is in the process of being negotiated by IOC and FAO for consideration by UNDP.

The IOC Workshop on Banding of Porites Corals as a Component of Ocean Dynamics was held at the Australian Institute of Marine Science (AIMS), Townsville (Australia, 15–17 August 1988). The objective of the Workshop was to examine all facets of coral boring programmes and projected national research to establish a network of research institutions in the WESTPAC region, as well as to develop a strategy for the execution of the project. A full report on the Workshop is being prepared to be published in the IOC Workshop Report Series.

The IOC Secretariat prepared a preliminary draft on the project on Continental Shelf Circulation in the Western Pacific which was circulated to Member States through IOC Circular Letter No. 1197, dated 12 August 1988, requesting comment. Replies were received from Australia, China, Indonesia, Philippines and Viet Nam. An informal consultation on the draft project was held on 30 November 1988 between the Programme Co-ordinator for Ocean Dynamics, Professor K. Taira (Japan), the Project Leader, Dr. M. Bungapong (Thailand) and the Secretariat staff responsible for the project. It was recommended that the project should start with a planning and presentation Workshop to discuss and to develop the observational and operational plans for the implementation of the project. Action is being taken to implement this proposal later in 1989.

The Workshop on Marine Geology/Geophysics in the WESTPAC Region was held from 6 to 7 September 1988, in conjunction with the First International Conference on Asian Marine Geology (Shanghai, China, 7–10 September 1988). Twenty-two participants from eight WESTPAC countries attended the Workshop. The Workshop identified six research projects under the framework of the project on Palaeogeographical Mapping in the WESTPAC region and five more under the project in Margin of Active Plates. These will be further improved and submitted with workplans to the Fifth Session of WESTPAC.

The IOC jointly with Unesco and in close collaboration with the Asian Institute of Technology, Bangkok, has organized the Training Course on the Use of Microcomputers for Oceanographic Data Management, 16 January to 3 February 1989. The Training Course is being attended by 15 participants from the WESTPAC region.

At the invitation of the Chairman of the Committee on Marine Sciences of the Pacific Science Association, Academician V. I. Ilyichev, the IOC Senior Assistant Secretary, also acting as Technical Secretary for WESTPAC, attended the Second Pacific Symposium on Marine Sciences, held in Nakhodka, USSR, 12–19 August 1988. The Symposium adopted recommendations for future action, which *inter alia* called on the Chairman of the Symposium to negotiate with the IOC to treat the future Symposium of the Marine Science Committee of the Pacific Science Association "as serving the objectives of WESTPAC" and in organizing the future Sessions of the Symposium, if possible, in conjunction with the IOC scientific seminars preceding each WESTPAC Session.

Project Proposals for Extra-budgetary Funding
At the request of the Member States of the WESTPAC region, the IOC Secretariat formulated two project proposals which were submitted by Unesco to UNDP for funding.

Missions
An expert mission (23 January–10 February 1988) was organized under the leadership of the Project Leader for the IOC-FAO Penaeid Recruitment Project (PREP), Dr. Derek Staples (Australia), accompanied by the Chairman of SCORRAD, Mr. Purwito Martosubroto, and Mr. R. N. Harriss, who visited Australia, Indonesia, Malaysia, the Philippines and Thailand. They discussed with relevant authorities and institutions their participation in the project and explored the possibility of establishing a network of institutions in the region. All the countries expressed interest in participating and most of them designated Co-ordinating Agencies and National Co-ordinators for the project, who are expected to assist in developing the proposed network. The mission also provided the required background information for organizing a training workshop to follow later at CSIRO Laboratories (Cleveland, July 1988).

The IOC Technical Secretary for WESTPAC visited Indonesia, Malaysia, the Philippines and Thailand (14–28 December 1988) to consult with senior officials and the institutions in each country dealing with WESTPAC activities. The purpose of the mission was to determine priority areas of technical assistance required by each country, so as to provide input to the IOC-Unesco project on "Ocean Science and Ocean Services in Support of Ocean Development in South-east Asia," submitted by Unesco to UNDP. The mission helped identify areas of common interest to serve as the basis for the formulation of technical assistance projects. During the mission the authorities in Thailand were consulted on the progress made with regard to the establishment of a Secretariat for the Sub-commission for WESTPAC in Bangkok, offered by Thailand at WESTPAC-IV. The details of the findings will be submitted before the Assembly under the relevant item.

North and Central Western Indian Ocean (IOCINCWIO)

Following recommendations of the Second Session of the Regional Committee for the Co-operative Investigation in the North and Central Western Indian Ocean (IOCINCWIO), (Arusha, Tanzania, 7–11 December 1987), follow-up action has been taken in the fields of ocean observing systems, ocean dynamics and climate, ocean mapping, marine pollution research and monitoring, and marine information management.

The CCCO Indian Ocean Climate Studies panel meeting (Islamabad, Pakistan, 27 June–1 July 1988) considered relevant actions in the region and the First Session of the Regional Committee for the Central Indian Ocean (Islamabad, Pakistan, 3–7 July 1988) suggested appropriate interfacing of its proposed climate research and monitoring activities with those of IOCINCWIO.

An expert mission in the region in the field of marine pollution research and monitoring, (31 August–16 September 1988), identified competent institutions to participate in the regional project oil pollution monitoring, which re-confirmed their interest in participating in the project, and identified needs to initiate the projects. These results are being followed up.

In June–July 1987, under the aegis of IOC and with support of the Federal Republic of Germany, a Training Course on Ocean Mapping was held on board the R. V. METEOR; it was attended by 15 participants from countries of the region. This course included lectures and coastal field work around Nosy-Bé (Madagascar), lasting ten days, and onboard training in between Nosy-Bé and Mombasa (Kenya) for another ten days. These activities have helped lay the basis for the preparation of the International Bathymetric Chart of the Western Indian Ocean (IBCWIO), the editorial board for which will be held in Antananarivo (Madagascar, 4–7 April 1989).

Member States were invited by IOC Circular Letter No. 1200 to nominate experts for a regional Expert Steering Group on OSLR, the first meeting of which is scheduled for the first half of 1989.

Progress is being made in development of RECOSCIX-WIO as an information network within the region which will be strengthened by the appointment in early 1989 of an associate expert offered by the Belgian Government.

Central and Eastern Atlantic (IOCEA)

At the initiative of the Ghana Commission for IOC, a regional seminar was organized on the Conservation and Utilization of Marine Resources, in Accra, 4–9 September 1988. IOC supported the participation of two experts.

A preliminary meeting on the implementation of the Joint IOC-UN Programme on Ocean Science in Relation to Non-Living Resources (OSNLR) in the Central Eastern Atlantic was held in Abidjan, Côte d'Ivoire, from 2 to 5 November 1988, attended by 10 participants. The following areas of research were considered: (i) Coastal erosion, particularly in Benin Bay; (ii) Effect of damming rivers on the transfer of sediments within the coastal zones; (iii) Relationship between upwellings and phosphorite deposits.

The meeting was focussed on coastal erosion as a priority project. The need to provide a scientific basis for long-term and large-scale coastal management and protection within the region resulted in the adoption for implementation of the project "Sediment Budget along the West African Coastline." For effective execution, the project was divided into two sub-projects: (a) the effect of dams on the sediment flux of rivers reaching the coastline, and (b) the hydrography and dynamics of the coastal zone including the inner continental shelf.

In respect of (b), two communal cruises have been planned. The first, to cover the Gulf of Guinea sector, is slated in the last quarter of 1989. In this regard, the offer of Nigeria to commit one of her oceanographic research vessels, R. V. SARKIM BAKA, was accepted. The second cruise, with the study of the upwelling phenomenon in this sector as its primary objective, will cover the northwestern sector, possibly onboard a research vessel made available by Morocco.

Regarding the development of the regional component of the IOC Global Sea-Level Observing System (GLOSS), a preparatory mission of a consultant from the

Federal Republic of Germany was arranged (November–December 1988) to prepare the installation of tide gauges provided by Sweden (countries visited: Ghana,
Mauritania, Nigeria, Senegal, Sierra Leone). One expert from Ghana participated in
the Sea-Level Training Course in June 1988 at the Bidston Observatory, UK.

Evaluation of the first phase of the project on the Monitoring of Pollution in the
Marine Environment of the West and Central African Region (WACAF/2) was made
at the Second Workshop of participants organized in Accra (Ghana, 13–17 June
1988), in co-operation with FAO, IAEA, WHO and UNEP, and a recommendation for
the continuation of the project was adopted.

A Unesco-IOC Training Course on Physical Oceanography for African French
speaking countries was held in Conakry, at CERESCOR (Centre de Recherche
Scientifique de Conakry, Rogbane) (Guinea, 15–22 December 1988), attended by 37
participants, with the IOC support to 1 participant coming from outside the country
and also to 1 lecturer.

Central Indian Ocean (IOCINDIO)

The IOC Regional Committee for the Central Indian Ocean at its First Session (Islamabad, Pakistan, 3–7 July 1988), reviewed and endorsed in principle several regional co-operative project proposals and agreed on an implementation plan. The
eight project proposals had been considered during *ad hoc* Preparatory Expert Consultations prior to the Session (Islamabad, 28–30 June 1988). Follow-up has been initiated in several fields: marine pollution research and monitoring, planning a regional
training workshop in 1989 or 1990; coastal water dynamics through a meeting of the
Steering Group (Karachi, Pakistan 15–20 January 1989); and ocean science in relation
to living and non-living resources.

Prior to the IOCINDIO Session, the CCCO Indian Ocean Climate Studies Panel
(Islamabad, Pakistan, 27 June–1 July 1988) and the IOC-UNEP Task team met to
prepare a study on the potential impacts on coastal zones and areas from possible sea
level and temperature rises induced by expected climate changes in the South Asian
Seas region.

The Regional Committee also considered data and information exchange and
management, regional ocean observing networks, and the TEMA requirements for
enhancing the marine science capabilities in the Region, together with the regional
implementation of the Comprehensive Plan for a Major Assistance Programme to
Enhance the Marine Science Capabilities of Developing Countries.

Southern Ocean (IOCSOC)

Upon the instructions of the IOC Executive Council at its Twenty-first Session, particular attention in 1988 was paid to the development of ocean observing and data
management systems in this region, through the existing programmes (IGOSS,
GLOSS, IODE and DRIBU). Development of sea-level observing network in the region was considered at the Workshop on Sea-Level Measurements in Hostile Conditions (Bidston, UK, 28–31 March 1988). The IOC-WMO Working Committee on
IGOSS at its Fifth Session (Paris, 14–23 November 1988), approved the application of

Australia for the accreditation of an IGOSS Specialized Oceanographic Centre for the Indian and South Pacific Oceans including relevant sectors of the Southern Ocean. Drifting buoy activities in the region were discussed at the Fourth Session of the WMO-IOC Drifting-buoy Co-operation Panel (New Orleans, USA, 18–21 October 1988). Member States of IOC and Members of WMO that operate any kind of ship in the Southern Ocean have been requested through the Joint IOC-WMO Circular Letter IGOSS No. 88-100 of 1 April 1988 to consider ways and means of providing the required data within the framework of IGOSS and to increase the volume and quality of IGOSS data from this region and adjacent areas.

The Scientific Committee on Antarctic Research (SCAR) at its Twentieth Session (Hobart, Australia, 12–16 September 1988), made a proposal to establish a small joint *ad hoc* committee consisting of SCAR, WMO, and IOC representatives to work out proposals for improving marine meteorological and ice information services for the Antarctic Treaty area of the Southern Ocean, and to organize a scientific meeting jointly sponsored by SCAR, WMO and IOC to discuss the objective and types of such information services. That meeting will be held in Leningrad in February 1989. The Chairman of SOC was requested to represent IOC at the joint *ad hoc* Committee at which the IGOSS International Coordinator will also participate.

Some other proposals made by SCAR at its Twentieth Session are of direct international relevance to IOC activities in this region. It is proposed, in particular, to convene in 1991 a special conference on Antarctic Science to demonstrate the role of Antarctica in global processes in atmosphere, ocean and land.

The CCAMLR meeting was held in Hobart, Australia, from 23 October to 4 November, 1988. The IOC Secretary requested the Executive Secretary of CCAMLR to bring to the attention of the meeting the recommendation 2 of the Regional Committee for SOC regarding possible future joint efforts on the proposed research projects.

Core project 2 of the World Ocean Circulation Experiment is a Study of the Southern Ocean. It will constitute, in fact, within the period 1990–1995, one of the major cooperative efforts of oceanographers and include an extensive observational programme and modelling.

South-eastern Pacific

The already considerable cooperation with the Permanent Commission for the South East Pacific is increasing and broadening in the programmes on marine pollution research and monitoring, ocean science in relation to non-living resources, and to living resources with WMO in studies on the "El-Niño" phenomenon, through the joint IOC-WMO-CPPS Working Group of the Investigations of "El Niño" which held its Sixth Session in Vina del Mar (Chile, 24–26 November 1988).

The CPPS-IOC-UNEP regional task team on preparation of a regional review on the state of the marine environment has completed its task, the report has been edited and is under publication. The CPPS-IOC-UNEP task team on a regional review of the possible effects of sea level and temperature rises, induced by expected climate changes, on coastal zone areas, has also completed its task and the report is under publication.

A proposal for a SARP project within the OSLR programme was formulated by Chilean scientists at a meeting in Santiago (Chile, 18 November 1988).

During 1988, the principal activities of IGOSS in the South-eastern Pacific were concerned with finding methods to improve data availability from that region, especially through the Ship-of-Opportunity Programme. The South-eastern Pacific represents a data-sparse area which is, at the same time, a region of great scientific interest. The IGOSS Operations Coordinator travelled to Chile (7–19 December 1988) on a mission to assess the status of marine technology in that country and to determine ways in which Chilean scientists might contribute more oceanographic data to national and international user groups.

South-west Atlantic

The regional IOC-UNEP Task Team preparing a review on the state of the marine environment in the South-west Atlantic has completed the draft report presently being edited. Follow-up to proposals made at an *ad hoc* consultation (Rio de Janeiro) regarding regional cooperation in marine chemistry studies have been initiated, particularly with respect to training and intercalibration exercises.

The IOC Expert Consultation and the Scientific Workshop on Ocean Dynamics and Climate in South America (Buenos Aires, Argentina, 18–22 July 1988) recommended, *inter alia,* an increasing effort on training in matters related to ocean dynamics and climate; support to participation in workshops; and certain national actions to strengthen the participation in the ocean climate research. These are being followed up to the extent possible within existing resources. During an expert mission to Argentina, Brazil, and Uruguay in March 1988, the feasibility for developing SARP in this region was discussed with the contact points and local scientists. The main result of this mission was a SARP project proposal to investigate cooperatively the recruitment of the SW-Atlantic anchovy. As a first step for the joint SARP investigations, a course on the "Egg Production Method" was carried out in the Alfred-Wegener-Institut, attended by six scientists of the region.

Mediterranean Sea

In pursuance of the decision of the Twenty-first Session of the IOC Executive Council (Paris, 7–15 March 1988) that a group of experts should work out proposals for cooperative studies of the western Mediterranean Sea, a provisional document was prepared by a drafting group. Initial discussions of the document on a Programme de Recherche Internationale en Mediterranee Occidentale (PRIMO) took place at an *ad hoc* meeting of experts, hosted by Centro Richerche Energia Ambiente (CREA) (La Spezia, Italy, 29–31 March 1988). IOC supported the participation of five experts.

This document was presented to and discussed at a special Session within the framework of the Thirty-first Congress and Plenary Assembly of the International Commission for Scientific Exploration of the Mediterranean Sea (ICSEM) (Athens, Greece, 17–22 October 1988). The Secretary of IOC sent a message to the Congress and IOC was represented by two members of the Secretariat. IOC supported the participation of two experts.

Contacts have been established with ICSEM for the joint sponsorship of this project within the framework of the Memorandum of Understanding between the ICSEM and the IOC.

The IOC programme for the ocean circulation in the Mediterranean Sea has been presented at the Plenary Session of the European Association of Remote Sensing Laboratories (EARSEL) (Capri, Italy, 19–20 May 1988).

Concerning the Programme on Physical Oceanography of the Eastern Mediterranean (POEM), a meeting of the Steering Committee was held in Paris, at Unesco Headquarters, 12–13 January 1988, and a Workshop was organized and hosted at Osservatorio Geofisico Sperimentale (OGS) (Trieste, Italy, 31 May–4 June 1988) supported jointly by IOC and the Division of Marine Sciences. This second POEM Scientific Workshop had three major objectives: (i) pooling of data; (ii) analysis with synthesis of the physical processes as focus; and (iii) future planning with a focus on the implications of POEM physical oceanography for biology, chemistry, etc.

CO-OPERATION WITH OTHER INTERNATIONAL ORGANIZATIONS

In practicality all the IOC global programmes and their regional components cooperation with other international organizations is extensive and generally broadening. The specifics of the cooperation are presented under the different programmatic sections in the report.

General aspects of cooperation in a broad context have been discussed with the United Nations Environment Programme (UNEP), and proposals for restructuring some of the collaborative activities are presented in IOC-XV/8 Annex 1. These proposals are essentially based on the Memorandum of Understanding signed in August 1987, and the experiences made from the on-going programmatic cooperation.

A consultation was held at ICSU, 24 November 1988, between officers of SCOR, the Chairmen of IOC, OPC and CCCO, members of the IOC Secretariat and the Executive Secretaries of ICSU and SCOR, considering in particular the interaction between SCOR and the IOC with respect to the work and future role of CCCO. It was agreed that an important coming role for CCCO was to provide a forum for connection and liaison at the scientific and non-governmental levels between programmes involved in studies of the role of the ocean in climate change and other climate programmes, and in helping to provide an international scientific link between the oceanographic community and other global change programmes, as well as maintaining a strong oversight role in its own major activities (TOGA and WOCE).

The IOC was represented at the First Meeting of the Scientific Advisory Council for the IGBP (Stockholm, Sweden, 24–28 October 1988), and has also held discussions on various occasions concerning the development of the Global Change Programme. Further information on the possible role of IOC in this context is given in Document IOC-XV/8 Annex 3.

IOC and Unesco worked in close co-operation with SCOR and other non-governmental bodies in making arrangements for the Joint Oceanographic Assembly (Acapulco, 23–31 August 1988). Through the JOA Logistics Committee and with funds provided by IOC, Unesco and WMO, travel support was offered to 34 participants, both invited speakers and persons presenting contributed papers, most from developing countries. Selection of the grantees, drawn from over 130 applications, was carried out in consultation with, and complementary to support offered SCOR, CMG,

IABO and IAPSO. The local organizing committee assisted in setting up a display on ASFIS, in conjunction with exhibits on information (CICH) and one on the Unesco marine science programme.

Upon the invitation of the UN Under-Secretary-General, Special Representative of the Secretary-General for the Law of the Sea, an informal inter-agency meeting was held in Geneva, 11–13 July 1988, at which IOC was represented by the Secretary IOC. A wide exchange of information and views on current and planned activities took place during the meeting and participants agreed that it would be helpful to have periodic consultations of the same nature in order to discuss matters in the field of ocean affairs that fall outside the mandate and composition of the Inter-Secretariat Committee on Scientific Programmes Relating to Oceanography (ICSPRO). A second meeting may be arranged in conjunction with the summer session of ECOSOC in 1989.

ENHANCING THE ROLE OF THE COMMISSION

Under the Chairmanship of Dr. Manuel Murillo, First Vice-Chairman IOC, the First Session of the *ad hoc* Study Group on Measures to Ensure Adequate and Dependable Resources for the Commission's Programme of Work was held at Headquarters from 21–25 November 1988. The *ad hoc* Study Group made substantial progress in responding to its mandate, in particular regarding suggestions on modalities for implementation of Article 10 of the IOC Statutes; improvement of documentation and conference services for IOC; guidelines for summary reports and draft resolutions; formulation of programme requirements; and implications of functional autonomy.

Among other matters considered by the *ad hoc* Study Group as requiring further consideration are: responsibilities of Member States towards the Commission; binding agreements and other legal instruments, including pledging commitments; continuing review of the Rules of Procedure; and other subjects referred to the Group by the IOC Executive Council and Assembly.

Future work of the *ad hoc* Study Group will be discussed at the Fifteenth Session of the Assembly in conjunction with presentation of its first Report (Document IOC/FURES-I/3).

PUBLICATIONS

The number of titles issued within the standing IOC series increased considerably in 1988, as can be seen in the following list. A full list of available titles is found in Document IOC/INF-700 rev. 3.

IOC Technical Series

33	Time Series of Ocean Measurements. Volume 4, 1988. 75pp. (English).
34	Bruun Memorial Lectures 1987: Recent Advances in Selected Areas in the Regions of the Caribbean, Indian Ocean and the Western Pacific. 1988 (Composite English/French/Spanish). (In press)

IOC Manuals and Guides

3 rev.	Guide to Operational Procedures for the Collection and Exchange of IGOSS Data. Second revised edition, 1988. 68 pp. (English, French, Spanish, Russian).
6 rev.	Wave Reporting Procedures for Tide Observers in the Tsunami Warning System. 1988. 30 pp. (English).
19	Guide to IGOSS Specialized Oceanographic Centres (SOCs). 1988. 17 pp. (English, French, Spanish, Russian).
20	Guide to Drifting Data Buoys. 1988. 71 pp. (English, French, Spanish, Russian).

IOC Workshop Reports

37 *Suppl.*	Submitted Papers to the IOC-Unesco Workshop on Regional Co-operation in Marine Science in the Central Indian Ocean and Adjacent Seas and Gulfs. Colombo, Sri Lanka. 1985. 369 pp. (English). (Published in 1988)
40	IOC Workshop on the Technical Aspects of Tsunami Analysis, Prediction and Communication. Sidney, B.C., Canada. 1985. 35 pp. (English). (Published in 1988)
40 *Suppl.*	Submitted Papers to the First International Workshop on Tsunami Analyses Prediction and Communications. Sidney, B.C., Canada. 1985 (English). (Published in 1988)
47	IOC Symposium on Marine Science in the Western Pacific: The Indo-Pacific Convergence. Townsville, Australia. 1986. 170 pp. (English). (Published in 1988)
52	SCPR-IOC-Unesco Symposium on Vertical Motion in the Equatorial Upper Ocean and its Effects Upon Living Resources and the Atmosphere. Paris, 1985. 39 pp. (English). (Published in 1988)
53	IOC Workshop on Biological Effects of Pollutants, Oslo, Norway, 1986. 74 pp. (English). (Published in 1988)
54	Workshop on Sea-Level Measurements in Hostile Conditions. Bidston, U.K., 1988. 97 pp. (English).
55	Second Workshop of Participants in the Joint FAO-IOC-WHO-IAEA-UNEP Project on Monitoring of Pollution in the Marine Environment of the West and Central African Region. Lagos, Nigeria, 1988. 46 pp. (English). (In press)

Brochures	IODE: Ocean Data for Science, Industry, Education. 1988 (English, French, Spanish, Russian).

SECRETARIAT

Staff

A number of staff changes took place in 1988, both in regard to Unesco posts and seconded personnel.

Agreement has been reached for an Associate Expert provided by Belgium to be posted to Mombasa, Kenya, to work on marine information management for the IOCINCWIO region and an appointment is expected early in 1989.

John Withrow returned to the United States in July, and was replaced in June by Carl Berman (USA) in the position of IGOSS Operations Co-ordinator, with an overlap in appointment to ensure transition of functions. One of the two Associate Experts posted at Headquarters for Marine Pollution Research and Monitoring, Peter Bjornsen returned to Denmark in September and is expected to be replaced during early 1989 by another Associate Expert provided by Denmark. Yihang Jiang (China) joined the Secretariat in March 1988 as a seconded staff member and is assisting in the Ocean Science programme. Jurgen Alheit (Federal Republic of Germany) was appointed to a post funded by his government through an IOC Trust Fund contribution and has assumed responsibility for the Programme on Ocean Science in Relation to Living Resources since October 1988.

Youri Oliounine returned to the USSR in October, leaving vacant post SC-255 which is now in the final stages of recruitment. Ray Griffiths (P-5) was recalled to FAO as of 31 December 1988. Mario Ruivo (D-2) retired on 31 December 1988 and Gunnar Kullenberg assumed the functions of Secretary IOC on 1 January 1989. Announcement for recruitment of his former position, SC-251 (P-5), is pending.

As has been noted in previous Reports of the Secretary, there remains a critical shortage of personnel in the Secretariat and, with the presently three vacant Unesco professional posts and the departure of the FAO seconded post, the situation can be seen as deteriorating. The Director General has been alerted to the probable negative impact on IOC programme implementation and the severe burden being placed on the Secretariat staff at a time when there is increasing expansion of Member State participation.

Missions of the Secretary

During the year, the Secretary undertook several missions both specifically associated with the development of the policy and programmes of the Commission and with matters concerning Unesco as a whole in accompanying the Director-General on Country visits.

In addition to the missions referred to in the relevant subject areas of the text, following missions can be mentioned:

Geneva, 9–11 February 1988.—To attend in his capacity of Secretary of ISCPRO the Twenty-sixth Session of the Inter-Secretariat Committee on Scientific Programmes Relating to Oceanography.

Geneva, 20–21 April 1988.—In conjunction with the UN Administrative Committee on Co-ordination (ACC), to attend, with the Director-General of Unesco, a high level informal consultation of the Organization Members of ICSPRO.

Lisbon, 27 April–5 May 1988.—To attend a Junta Nacional de Investigaçao Cientifica e Tecnologica (JNICT) meeting on final preparation of the Marine Science Country Profiles for Portugal.

Rome, 13–18 May 1988.—To have consultations with the FAO Secretariat and Italian authorities.

Nouakchott, Nouadhibou, Mauritania, 11–17 June 1988.—Following the official invitation of the authorities of Mauritania to follow-up on ocean related matters discussed during the visit of the Director-General to Mauritania, and to have preliminary consultations with local authorities and UNDP Resident Representative on a sub-regional project aimed at strengthening the capacities of Member States concerned in marine sciences and related aspects.

Islamabad, Pakistan, 29 June–8 July 1988.—To attend the First Session of IOC-INDIO and to have consultations with authorities of Pakistan so as to mobilize support to the further strengthening of ocean research and development in Pakistan as well as the participation of Pakistan in IOCINDIO programme.

Stockholm, Sweden, 23–28 October 1988.—To represent IOC at, and to act as Head of the Delegation of Unesco to the First Meeting of the IGBP Scientific Advisory Council. He also had consultations with senior officials of the Ministry of Foreign Affairs of Sweden on mobilizing support for assistance to developing countries in the field of TEMA, particularly through the implementation of the Unesco/IOC Comprehensive Plan for a Major Assistance Programme to Enhance the Marine Science Capabilities of Developing Countries.

Geneva, 9–11 November 1988.—To attend the First Session of the WMO-UNEP Intergovernmental Panel on Climate Change.

Porto, 11–14 November 1988, and Lisbon, 15–16 November 1988, Portugal.—To participate with the Director-General in the Conference on Twenty Years of Scientific Policy (Porto, 11–14 November 1988) and to have discussions with Portuguese authorities on measures to be adopted in the framework of TEMA to assist Portuguese-speaking African countries (Lisbon, 15–16 November 1988).

Reports from Organizations

International Council of Scientific Unions, Proceedings of the Scientific Committee on Ocean Research (SCOR): Report of the Nineteenth General Meeting of SCOR*

ACAPULCO, MEXICO

AUGUST 23 AND SEPTEMBER 1, 1988

The Nineteenth General Meeting of SCOR was held in two sessions during the Joint Oceanographic Assembly, which took place from August 23 to September 1, 1988, at the International Center of Acapulco, Mexico. The President of SCOR, Professor Gerold Siedler, presided over the meeting. The Joint Oceanographic Assembly was organized under the auspices of SCOR, by the Mexican Committee for SCOR and its chairman, Dr. A. Ayala-Castanares. In accordance with its usual practice, SCOR held the General Meeting in conjunction with the JOA.

REPORT OF THE PRESIDENT OF SCOR

The President of SCOR, Professor Gerold Siedler, made the following statement in which he reviewed the major activities of his term of office which concluded at the General Meeting:

"In the course of individual agenda items, most of the activities of SCOR since the last General Meeting in 1986 in Hobart will be discussed, and I do not wish to lengthen our discussion by reporting on these items separately now. Since, after the end of the Acapulco General Meeting on 1 September, I will have reached the end of my term as President, I thought I would try, instead, to single out some activities and events of these five years of my presidency and indicate the directions I think SCOR may take in the future. This will necessarily be a subjective view, and I will try not to take up much time of the meeting for this report.

When I stepped in suddenly in 1983, I found that the matter which had long occupied much of Eric Simpson's time and energy again required immediate attention. This was the question of finding an appropriate mode of operation

* EDITORS' NOTE.—This report is excerpted from International Council of Scientific Unions, *Proceedings of the Scientific Committee on Oceanic Research*, vol. 24 (1989) that was supplied by Elizabeth Tidmarsh, Executive Secretary of the Scientific Committee on Oceanic Research, Department of Oceanography, Dalhousie University, Halifax Nova Scotia, B3H 4J1, Canada. An earlier report of SCOR's activities may be found in "Report of the Twenty-fifth Meeting of the SCOR Executive Committee," *Ocean Yearbook 5*, ed. Elisabeth Mann Borgese and Norton Ginsburg (Chicago: University of Chicago Press, 1985), pp. 330–42.

between IOC and SCOR regarding the work of CCCO. It turned out to be a complex task, and Dr. Stewart will remember the numerous discussions that were necessary to solve the problems with which we were faced. The matter was resolved by the signature of Principles of Agreement between JCSU and Unesco and a Memorandum of Understanding between SCOR and IOC on CCCO. These documents have since successfully provided the basis for our cooperation with the IOC in the climate program. The work related to TOGA and WOCE has been one of the major activities of SCOR together with other organizations in recent years, and will be so in the future.

CCCO, however, despite its great importance, is not a typical example of SCOR's mode of operation, which had previously taken place through Working Groups. In 1983, we had a peculiar situation with respect to these WG's: there were physically and chemically oriented groups, a few biological groups, but no geological groups at all. We tried hard to obtain a better balance in disciplines in the following years, and when you look at the list of our present groups, you will find a noticeable addition of WG's in the fields of geoscience, especially marine geology, and of biological oceanography.

At the last General Meeting in Hobart, the proposal was made to convene a meeting of experts interested in global flux studies. It led to the establishment of the JGOFS Committee at the Executive Committee meeting in Zurich last year. This can be expected to be the start of a major new international program in oceanography, and in spite of uncertainties with respect to the science plan and funding, many exciting questions are being posed. I am pleased SCOR was able to respond so quickly in setting up the international framework for JGOFS.

SCOR's work can only be successful if sufficient funding is available to provide the necessary seed money for projects and programs. When you look at the budget later in this meeting you will find that SCOR has a good financial base, despite some of the recent problems other organizations have had to face in the wake of the United Nations financial situation. It is to the credit of the Executive Secretary that arrears in membership fees are no longer a problem in the SCOR budget.

I was particularly pleased to be involved with two events concerning SCOR membership. It took a considerable effort and the understanding and help of the participants in the Roscoff General Meeting to change the SCOR Constitution to make the membership of China possible without losing our colleagues in Taiwan. The Chinese SCOR Committee has been amazingly active and quick in its response to SCOR initiatives, and I consider the involvement of the Chinese scientists to be a major asset to SCOR. The second event is the renewed membership of Spain. Following the formal establishment of a Spanish SCOR Committee, and with the authorization of the Executive Committee meeting, I was recently able to welcome our Spanish colleagues again in SCOR after a long absence.

What directions might SCOR take in the future? What might have to change in SCOR's approach in the future? I will only make a few points.

Changes may develop in the operation of SCOR. Getting people together may no longer always be necessary with the new means of communication available today. Telemail and Fax have already changed completely the day-to-day work of the SCOR Offices and Secretariat, and in a time when scientists use electronic communication to do science together, it is foreseeable that the operations of SCOR WG's and Committees will be considerably influenced by computer links. Unfortunately, the availability of such links today is restricted mostly to western industrialized countries, but we can hope for advances on a larger scale during the forthcoming years.

I believe it will be important to preserve a good balance of short-lived, specific task-oriented groups and longer-term committees directly related to the major international programs. I think we should not let all SCOR activities become part of these big programs.

But the most important requirement is the commitment of first-rate scientists to SCOR activities. This is our main strength, and only if this commitment can be preserved in the future, will SCOR be able to continue to contribute to international marine science."

ELECTION OF OFFICERS

The SCOR Constitution states that the President of SCOR shall serve for one four-year term only and that the three Vice-Presidents and the Secretary are elected for two-year terms with the possibility of serving three such terms consecutively. The Nominations Committee proposed the following slate of nominees for election as SCOR Officers:

President	Prof. J. O. Stromberg (Sweden)
Secretary	Prof. R. O. Fournier (Canada)
Vice-Presidents	Prof. G. R. Heath (USA)
	Prof. T. Asai (Japan)
	Dr. A. P. Kuznetsov (USSR)

The General Meeting accepted this proposal.

SUBSIDIARY BODIES

1. Arising from Former Working Groups

The Executive Secretary reported that the final reports of the following working groups had been published during the 1987–88 period.

WG 51—Evaluation of CTD Data—Final Report, "The Acquisition, Calibration and Analysis of CTD Data." *Unesco Technical Papers in Marine Science* No. 54, 1988.

WG 56—Equatorial Upwelling Processes—"Vertical Motion in the Equatorial Upper Ocean: Proceedings of the International Symposium on Vertical Motion in the Equatorial Upper Ocean and its Effects upon Living Resources and the Atmosphere," Unesco, Paris, France, 6–10 May, 1985. *Oceanologica Acta Special Volume No. 6*, 1987.

WG 65—Coastal-Offshore Ecosystems Relationships—"Coastal Offshore Ecosystem Interactions: Proceedings of a Symposium Sponsored by SCOR, UNESCO, San Francisco Society, California Sea Grant Program and U.S. Dept. of Interior, Mineral Management Service, held at San Francisco State University, Tiburon, California, April 7–22, 1986." *Springer-Verlag Lecture Notes on Coastal and Estuarine Studies* No. 22, 1988.

In addition to these publications, the following are in the final stages of preparation:

WG 46—River Inputs to Ocean Systems—the final report of this group is in press at Unesco and will appear as "River Inputs to Ocean Systems: Status and Recommendations for Research" in the series *Unesco Technical Papers in Marine Science.*

WG 73—Ecological Theory in Relation to Biological Oceanography—the proceedings of a workshop will be published in early 1989 in the *Springer-Verlag Lecture Notes on Coastal and Estuarine Studies* with the title "Flows of Energy and Materials in Marine Ecosystems: Network Analysis."

2. Current Working Groups

WG 69   Small-scale Turbulence and Mixing in the Ocean (with IOC)
The Chairman of WG 69, Dr. K. N. Fedorov, noted that the tasks of his Working Group had been fulfilled with the publication of the Proceedings of the 1987 Liege Colloquium and the preparation of the final report which will be published by SCOR in the near future. Dr. Fedorov noted that many divergent points of view had been aired among the members of the WG during the discussion of certain terms related to the turbulent diffusivities of temperature, salinity and momentum, but that these had eventually been resolved during the preparation of the glossary of terms and definitions used in turbulence studies. He urged the adoption of these terms by scientists in this field.

The General Meeting agreed to disband WG 69 after congratulating its Chairman on the successful completion of a challenging assignment.

WG 71   Particulate Biogeochemical Processes (with UNESCO)
The report of WG 71 was presented by its Chairman, Dr. S. Krishnaswami, who reported on the group's activities, especially the final meeting which had taken place in France shortly before the General Meeting. The final report of WG 71 will be published in the near future.

The Working Group proposed that SCOR consider the establishment of two new groups, one on the role of ocean margins in the marine geochemical cycles of carbon and other elements, and another on the use of satellite ocean color imagery to derive estimates of standing crops and rates of primary/new production. It was acknowledged that such groups would be very closely related to JGOFS' interests, however, the General Meeting was also of the opinion that such topics could also be addressed by independent SCOR subsidiary bodies. Dr. Krishnaswami accepted the request of the General Meeting that he take the lead in developing these two proposals for further consideration by SCOR.

The General Meeting thanked Dr. Krishnaswami for his report, and noting that a comprehensive final report from WG 71 was ready for publication, agreed to disband the group.

WG 72   The Ocean as a Source and Sink for Atmospheric Constituents (with IOC)
Professor Heath, as Executive Committee Reporter for WG 72, informed the General Meeting that the group's final meeting had been held in November 1987 at Schloss Ringberg, Bavaria.

The group conducted a discussion of the micrometeorological methods suitable for measuring air-sea exchange fluxes. It was emphasized that the complexity of the wet deposition process to the oceans is usually underestimated, and combined tower-aircraft-ship experiments to investigate gas exchange processes were recommended. For the determination of particle fluxes, size-dependent eddy correlation techniques together with studies of the chemical composition of the aerosol appear appropriate. The physiology of sulphur compounds in phytoplankton was also reviewed, and potentially underestimated sources of sulphur gases, such as, picoplankton, macrophytes and microbial mats were identified.

The Chairman of WG 72, Professor Andreae, presented an overview of the changes in our view of the air/sea exchange of sulphur since the last meeting. Of special interest here are recent determinations of H_2S and CS_2 in seawater and the marine atmosphere and the develpment of a hypothesis linking marine sulphur emissions and global climate.

Other topics discussed by WG 72 included:

1. The role of carbon and nutrient fluxes to the ocean. For example, it was pointed out that the inputs of some nutrients, especially Fe, may be growth-limiting under some circumstances;

2. The variability in time and space of particle fluxes to the ocean and the resulting problems of scaling and extrapolation;

3. Recent advances in our views of the role of the microlayer and the microbial communities in this layer for gas exchange; and

4. The processes responsible for the production of hydrocarbons and halogenated hydrocarbons in the sea. In view of the potential importance of some of these compounds for atmospheric chemistry, they deserve more intensive study.

The Working Group agreed to produce a final report containing a state-of-the-art review and a set of recommendations. Publication as a UNESCO Technical Report was found to be an appropriate medium.

The group therefore proposed the formation of a new SCOR Working Group: "Biogeochemical Air/Sea Exchange Studies" (BASES) with the following terms of reference:

1. To identify research needs and define scientific planning approaches related to:
   a. Emissions of biologically and photochemically produced trace species from the surface ocean.
   b. Deposition of atmospheric materials to the ocean surface and its influence on marine chemistry and biology, including their use as tracers.
   c. Measurement technologies for air/sea exchange fluxes.

2. To provide a means of interaction and guidance for the planning activities of the International Global Chemistry and the Joint Global Ocean Flux Studies programs related to air/sea exchange.

It was proposed that the membership of such a group be comprised equally of atmospheric and marine scientists nominated by the JGOFS and IGAC planning committees.

In discussing this recommendation of WG 72, the General Meeting agreed that the proposal should go to the JGOFS Committee for detailed consideration at its meeting in September at which Dr. Andreae would be present.

The General Meeting agreed to disband WG 72 pending publication of its final report.

WG 75   Methodologies for Oceanic CO_2 Measurements (with UNESCO)
The Chairman of WG 75, Dr. C. S. Wong, informed the General Meeting that WG 75 has been working on the technical aspects of the world-wide implementation of a CO_2 measurement program as part of the WOCE and JGOFS programs. He noted that the group held three meetings and that its final meeting would take place at Woods Hole in mid-October. He made special reference to the collaboration between WG 75, the JPOTS Sub-Panel on CO_2 Standards and the ICES Marine Chemistry WG on a CO_2 intercalibration exercise. The results of this will be discussed at the Woods Hole meeting; aliquots of homogenized seawater were distributed to a number of laboratories for intercalibration of the methodology for DIC and alkalinity.

Dr. Wong reviewed other items on the agenda for the final meeting of his Working Group and noted that a final report would be prepared which would incorporate a revision of previous WG 75 reports. Given the fact that WG 75 has already been in existence for the normal six years, the Executive Committee wished to recommend that the group be disbanded following submission of its final report.

WG 76   Deep-Sea Ecology (with IOC)
The Executive Committee Reporter for WG 76, Professor Stromberg, introduced the report and noted that the Working Group met for the second and last time in late June 1988 immediately before the Fifth Deep Sea Biology Symposium in Brest, France. In accordance with the report submitted to SCOR in 1987, a discussion document had been prepared by Dr. Rice, based on submission from the WG members and this had been circulated several weeks prior to the meeting of WG 76. The earlier decision to divide the document into structural and functional aspects had proven to be impractical and discussions in Brest centered on a revised scheme. The final report of WG 76 will address the following topics:
1. Introduction: This section will define the problem, that is, the need to be able to recognize and detect the effects on deep-sea communities of past perturbations and to forecast the likely consequences of future ones;
2. Natural variability of deep-sea communities: This section will outline the background "noise" resulting from the known spatial and temporal variability of the deep sea and will review some of the factors known, or believed, to be responsible for this;
3. Sampling strategy: Taking into account the known or suspected variability, this section will emphasize the need to adopt an appropriate sampling strategy, the need for intercalibration of techniques and adequate descriptions of methods to validate the comparison of results;
4. Sample treatments:
    a. Taxonomic composition: An outline of the importance of the specific identification of deep-sea biological material and the difficulties associated with this. This section will also distinguish between what might be achieved in the short-term (months) and the long-term (years).
    b. Measurements: This section will review the attributes of a community which might be measured (abundance, biomass, chemical constituents (or organisms), metabolism, etc.), and the "taxonomic" level at which these might be partitioned (ranging from the species to the total community).
5. Specific objectives: All of the above could apply to the study of natural, unperturbed communities. This final section would focus on objectives specifically associated with expected or actual anthropogenic disturbances distinguishing between predictive and retrospective activities:

a. Predictive: Controlled perturbation experiments to elucidate immediate and chronic effects on population structure and on the functioning of the ecosystem; and

b. Retrospective: Assessment of the effects of mining or waste-disposal activities on natural populations ranging from (*a*) short-term, localized, high-impact situations, to (*b*) long-term, large-scale, low-impact scenarios. Specific examples of passed or planned impact studies will be reviewed, with an assessment of the results obtained and the difficulties anticipated.

The General Meeting approved of the intention of the group to complete the compilation of its directory of deep-sea biologists with the information received in response to a widely distributed questionnaire. Lastly, the meeting recommended that WG 76 be disbanded once these final tasks have been completed.

WG 77   *Laboratory Tests Related to Basic Physical Measurements at Sea (with IOC)*

Dr. Fedorov, the Executive Committee Reporter for WG 77, informed the General Meeting that it had, regrettably, been necessary to postpone the CTD intercalibration experiment from Kiel in June 1988 and it will now take place late in 1988 and involves the use of the high pressure laboratory of the Institute of Applied Physics at the University of Kiel to compare a number of CTD's supplied by several institutes around the world, both with reference instruments and with each other, under a wide range of temperature, pressure and salinity conditions.

In his report, the Chairman of WG 77, Dr. Striggow, had requested approval for two meetings of the group in 1989; one for working up the results of the intercomparison experiment, and another to begin planning of a "dynamic CTD-sensor intercomparison experiment" to take place in 1990 or 1991. While the 1988 experiment will test sensors under conditions of mechanical, thermal and electrical equilibrium, the most recent developments of these instruments have led to a dramatic increase in the speed of oceanographic measurements, and the response of the sensors to sudden changes in pressure, temperature, and salinity requires more attention.

The General Meeting urged WG 77 to finalize the results of the first intercomparison experiment before proceeding to plan a second. It therefore gave approval for a single meeting of WG 77 in 1989 and urged the group to submit a slightly more detailed statement of its plans for other activities for consideration by the Executive Committee at its meeting in late 1989.

WG 78   *Determination of Photosynthetic Pigments in Seawater (with UNESCO)*

At its meeting in 1987, the Executive Committee gave approval for a three-week pigment methodology workshop to be held in Hobart, Australia in late 1988. The purpose of this workshop was to: *a*) quantitatively compare existing spectrophotometric and fluorometric procedures with newer HPLC and thin layer chromatographic techniques for the determination of a variety of pigments; and *b*) to conduct intercalibrations of HPLC's and to standardize the current procedures for pigment analysis so as to permit the production of a reasonable pigment data set and the improved identification of detrital and pigment breakdown products in seawater samples.

Discussion between members of the group, however, had led to the conclusion that it would not be possible to achieve all of these goals in a single, large three-week workshop involving all of the WG 78 members, equipment, and technical support. The Chairman of WG 78 sought the approval of the General Meeting for a series of three more restricted "mini-workshops" on chlorophylls, carotenoids and field applications in quick succession. After consideration of the financial implications of this change in plans, the General Meeting accepted this revised proposal, while urging the group to ensure that the results of workshops are disseminated as quickly as possible since they are of great interest to JGOFS which is attempting to develop sampling protocols for the Pilot Study in 1989 and to establish sea truths for a satellite ocean color instrument to be launched in 1990 or 1991.

WG 79   Geological Variations in Carbon Dioxide and the Carbon Cycle
This working group was discussed together with WG 81. See below.

WG 80   Role of Phase Transfer Processes in the Cycling of Trace Metals in Estuaries (with UNESCO)
The report from the Chairman of WG 80, Dr. M. Whitfield, informed the General Meeting that the group has been working in correspondence for about eighteen months, conducting an evaluation of the state of knowledge of particle/water reactions for key metals and metalloids under a range of estuarine conditions, and defining the most useful properties for the characterization of estuarine particulates. These two topics correspond to two of the terms of reference of WG 80. Good progress has been made in a review of the knowledge of particle/water interactions driven by biological and geochemical processes. However, the reviews are not quite complete, especially with respect to the controlling processes in major Asian estuaries. The key papers related to the measurements of important properties of estuarine particles are still in preparation and will be required in order to provide a sound basis for discussion by WG 80 at its planned first meeting. In view of this, the Chairman proposed that the meeting planned for October 1988 be postponed until May 1989 in order to ensure the availability of a complete discussion document. This will permit completion of the last term of reference, which is to recommend experimental and theoretical research needed to develop models of particle/water interactions and to predict the behavior of trace metals in various estuarine regimes. The General Meeting approved this request and suggested that WG 80 attempt to complete its work at the 1989 meeting.

WG 81   Deep Water Palaeoceanography and WG 79 Geological Variations in Carbon Dioxide and the Carbon Cycle (both with IOC)
At its meeting in 1987, the Executive Committee noted that the interests of WG's 79 and 81 appeared to be converging, particularly in the areas of the use of ice core records of variations in atmospheric CO_2 and in modelling. The Chairmen of these two groups were asked to consider whether the scientific objectives of their groups could be better achieved by constituting a new group which would combine the efforts of WG's 79 and 81. In response to this request, Dr. Shackleton (Chairman of WG 81) and Dr. Sundquist (Chairman of WG 79) met in May 1988 in Louvain-la-Neuve, Belgium, to consider ways in which the overlapping interests of the two groups might be combined. The two chairmen agreed that the Executive Committee's suggestions could best be answered by proposing a new Working Group on "Ocean/Atmosphere Palaeochemistry." The proposed meeting will be a unique opportunity to enhance the interdisciplinary communication that is essential to understanding ocean/atmosphere palaeochemistry.

The revised activities for the new working group are:
1. To study past chemical interactions—particularly carbon-cycle interactions— between the oceans and atmosphere over time scales of 10^3 to 10^6 years;

2. To organize and publish the results of a meeting on the palaeoceanographic implications of ice core records; and

3. To assess and recommend improved methods for the stratigraphic correlation of marine and ice core records, and for the development of appropriate time-dependent models of the carbon cycle on 10^3 to 10^6 year time scales.

The General Meeting agreed to disband both WG 79 and WG 81 and to establish the proposed new group as WG 92, under the Chairmanship of Dr. Sundquist. Drs. Sundquist and Shackleton were urged to continue with their organization of the 1988 AGU special session on "Palaeoceanographic Implications of Ice Core Records" as planned. Since the session will involve several individuals suggested for membership in WG 92, the occasion may also provide an opportunity for an organizational meeting of the new group.

WG 82 Polar Deep-Sea Palaeoenvironments (with IOC)

The Chairman of WG 82, Dr. J. Thiede, presented information on the status of the group's activities. He noted that progress in polar palaeoceanographic research had been reported in detail during the session on the "Polar Seas Geological Record" organized by WG 82 during the Second International Conference on Palaeoceanography in Woods Hole in September 1986. The hostile environments of the high latitude areas present many technical challenges to deep-sea drilling, however, the contributions of this session were seen as a major step forward in this field. The papers presented will be published in *Palaeoceanography*.

Dr. Thiede also reported that a conference on Polar Palaeoceanography and a workshop on Northern High Latitude Drilling, both being held in Bremen, FRG, in October 1988, would provide a forum for detailed discussion of the feasibility of Arctic Ocean deep-sea drilling under the aegis of an international cooperative project from 1993 to 1996 known as the Nansen Arctic Drilling Program. The final meeting of WG 82 will take place in conjunction with these events and it was expected that the final report of the group would take the form of a publication addressing many of these questions. It was agreed that the group should remain in existence until the next General Meeting of SCOR in order to bring this publication to completion.

WG 83 Wave Modelling (with IOC)

In response to the request of the Executive Committee, a report had been submitted to SCOR on the present status of various activities in surface wave research in order to clarify the relationships between the groups active in the field. The report was to focus on the topic of medium-range wave forecasting, which had been identified as an area in which WG 83 has special expertise, and it was to establish a plan of work for the group, leading to some final product.

The Executive Committee Reporter for WG 83, Dr. Fedorov, advised the General Meeting that, in his view, the report was an appropriate response to the request of the Executive Committee at this stage of the group's existence. He noted that the group expects to achieve its goals in 1992, two years after the launch of the ERS-1 satellite. It was hoped that the report of the 1989 meeting would contain somewhat more scientific information and that WG 83 would prepare a more detailed summary report on the state-of-the-art of wave modelling towards the end of its term.

WG 84 Hydrothermal Emanations at Plate Boundaries (with IOC)

The Chairman of WG 84, Professor Suess, made a brief presentation to the General Meeting about the scientific issues of concern to his working group. Professor Suess sought approval for the first meeting of WG 84 to be rescheduled to meet in Kiel in May 1989.

Professor Suess noted that fluid venting processes at plate boundaries is a current research frontier and that it was one of the most vigorously pursued fields of oceanic and marine geologic research in 1987 and 1988. Numerous project planning activities and exchange of scientific results are beginning to shape several national and international research efforts for the coming decade. All of these efforts touch on the mandate of SCOR Working Group 84.

As a first step, WG 84 has identified these project planning and research activities, documented institutions and participants, and has formally or informally established collaboration and joint programs.

The second step for WG 84, according to Professor Suess, is to have this report published by SCOR as a "white paper."

As a third and final step, WG 84 intends to seek funding from the Dahlem Foundation Berlin, Germany and to organize a conference on Fluid Venting at Plate Boundaries under the auspices of and utilizing the successful model of the Dahlem Konferenzen. Professor Suess expected that if this effort were successful, the conference could be held as soon as the spring of 1991 or later during the fall of that year.

WG 85   Experimental Ecosystems (with UNESCO)
Professor Fournier informed the General Meeting that WG 85 had held its first meeting in Hamburg in June 1988. Each member of the group had prepared a review paper addressing its first term of reference, "to examine previous studies involving experimental ecosystems; critically evaluate the results and the application of such techniques to estuarine, coastal and open sea problems." The final versions of these papers will be submitted to SCOR in a format suitable for publication in the primary scientific literature.

The second term of reference of WG 85 calls upon it to make recommendations regarding the application of experimental ecosystems to suitable research problems in biological oceanography.

Lastly, WG 85 began a consideration of its third term of reference which is to specify design criteria for experimental ecosystems to be used in a variety of conditions. The group agreed to prepare an "Experimental Ecosystem Manual" to serve future users of these systems. In addition, the group wishes to organize a colloquium on the most recent results from experimental ecosystems, possibly in conjunction with the Fifth International Congress on Ecology in Japan in August 1990.

WG 86   Sea Ice Ecology (with SCAR and AOSB)
Professor R. O. Fournier introduced the discussion of WG 86, one of the most recently established, and commended the group's Chairman, Professor C. Sullivan, for the promising start which his group has made. It is expected that the first formal meeting of WG 86 will take place in January 1990. Dr. I. Melnikov will be invited to join the working group.

WG 87   Fine-scale Distribution of Gelatinous Planktonic Animals (with UNESCO)
Dr. Harbison, the Chairman of WG 87, reported that several members of the group had not actively participated in the preparation of written contributions to a document which would form the basis for the group's work. As a result, Dr. Harbison had sought contributions from a number of experts who were not members of his working group.

The General Meeting suggested that inactive members be urged to fulfill their commitments in order to participate in the meeting of the group and suggested that other individuals could be invited to become Corresponding Members of WG 87 with participation in the meeting to be determined on the basis of contributions to the document to be submitted to SCOR.

WG 88 *Intercalibration of Drifting Buoys (with IOC)*
On behalf of WG 88, one of its members, Dr. G. Cresswell presented a brief report on the following topics of interest to the group:
1. Examples of the use of satellite-tracked drifters in the tropical Pacific Ocean and in the waters around Australia;
2. The standards to be met for drogue slippage and sensor accuracy;
3. Drifter designs that are presently being used and intercalibrated by the WG; and
4. An exciting study of the tropical Pacific Ocean that is about to commence.

The last item refers to the Pan Pacific Surface Current Study under the direction of WG 88 member, Professor P. Niiler of Scripps, with the participation of WG members, D.Hansen, G. Reverdin and G. Cresswell, as well as by Dr. K. Takeuchi. The study aims at understanding the changing SST pattern in the tropical Pacific, for which the competing hypotheses are horizontal advection and vertical mixing. Some 150 drifters will be deployed between October and December 1988. The study will run for two years and the desired grid spacing is 2° latitude by 10° longitude.

Dr. Cresswell concluded by pointing out that there will be benefits for many countries in the region through a better understanding of ocean currents. The study will contribute to TOGA and will assist in the planning of satellite drifter studies for WOCE.

WG 89 *Sea Level and Erosion of the World's Coastlines*
Professor Heath reminded the participants in the General Meeting that WG 89 had just been established at the last meeting of the Executive Committee. The representative of the New Zealand SCOR Committee reiterated the importance of a significant involvement of scientists from developing countries in WG 89 because of the critical importance of sea level changes in many regions with low-lying coastal areas. A number of other participants, including an observer from Nigeria, also referred to this point.

The Chairman of the UK SCOR Committee, Sir Anthony Laughton, noted that the terms of reference for WG 89 (see SCOR Proceedings, Vol. 23) have two distinct thrusts; sea level rise in general, and the effects of sea level rise on coastlines in particular. Following a brief discussion, the General Meeting agreed that the terms of reference should be revised as follows:
1. To examine the applicability of the existing models for prediction of coastal erosion dependent upon sea level rise, and formulate a program of investigations for their verification or rejection;
2. To evaluate differences between short and long-term sea level rises on beach erosion;
3. To recommend strategies for monitoring programs on coastal erosion for coastlines which lack a data base; and
4. To produce a report for SCOR which addresses these questions.

The following individuals will be invited to join WG 89: P. Komar (USA) (Chairman), N. Lanfredi (Argentina) (Vice-Chairman), R. Dean (USA), T. Sunamura (Japan), B. Thom (Australia), K. Dyer (UK), M. Baba (India), T. Healy (New Zealand), M.A. Ortez Perez (Mexico), A. Ibe (Nigeria).

3. Committees and Panels

Joint SCOR/IOC Committee on Climatic Changes and the Ocean
The Past-Chairman of CCCO, Professor R.W. Stewart, made a verbal presentation on recent CCCO activities and, in particular, highlighted the items arising from the ninth session of CCCO which took place in Paris in May 1988. He noted that the TOGA is progressing well and that WOCE is rapidly approaching the implementation phase. Both experiments are being planned and conducted under the aegis of strong, scientifically expert groups, the JSC/CCCO Scientific Steering Groups, or SSG's, for TOGA and WOCE. This has caused the CCCO itself to consider its future role in relation to these major experiments which it initiated. The results of this review will be the topic of discussion between the Committee's parent organizations, SCOR and IOC before the end of the year.

Professor Stewart reported that a great deal of progress is being made in TOGA and that the date obtained can actually be used in predictive, rather than hindcast, models, even if these are not yet perfected. He made special mention of developments in the Indian Ocean, an area of earlier concern to CCCO due to the poor availability of data, and for which groups from Florida State and Oxford Universities have recently made substantial progress in the modelling field. The TOGA observational program in the Atlantic Ocean has expanded since 1987 with the incorporation of USSR "Sections" data into near-real-time models by French oceanographers. The growing ability to use data immediately in such models to provide operational climate prediction, to test hypotheses and to feed back to the actual experimental design is an exciting development which can change the face of oceanographic science, according to Professor Stewart.

With regard to WOCE, he noted that the detailed WOCE Implementation Plan is now available and will be the subject of a detailed review at the International WOCE Scientific Conference which will take place at Unesco Headquarters in Paris from November 28 to December 2, 1988. The sponsoring organizations of the CCCO and the WOCE SSG recognize that the organizational arrangements for the experiment must change as it moves from the intensive planning to the implementation phase. A primary concern is to develop a structure which allows the scientific community to continue to direct WOCE while at the same time providing appropriate mechanisms for the contribution of resources to the experiment by governments and by non-governmental organizations.

The General Meeting gave its endorsements to the program of CCCO's activities.

Committee for the Joint Global Ocean Flux Study
The President of SCOR reviewed the background to the decision taken at the 28th Executive Meeting of SCOR (Zurich, October 1987) to establish an international scientific planning Committee for the Joint Global Ocean Flux Study.

The Chairman of JGOFS, Professor B. Zeitzschel, made a presentation to the General Meeting in which he briefly described the scientific rationale for the newly established Joint Global Ocean Flux Study. He gave the main goals of JGOFS, in particular, the scientific aspects of the so-called "biological pump" which is considered to be responsible for the "biological" removal of CO_2 from the atmosphere, and stressed the importance of the use of sophisticated instrumentation such as airborne platforms or satellite ocean color scanners to monitor, e.g., chlorophyll on a global scale. A great deal of effort has to be put into the development of suitable algorithms to convert the remotely sensed signal to the standing stock of phytoplankton and primary production. A valuable tool to monitor the dynamics of the euphotic zone/ mixed layer might be satellite-tracked drifting sediment traps with additional chemical and optical sensors. The collected material can be used to give information on the coupling between phyto- and zooplankton in this zone. Another field of intensive work will be the development of models of the global carbon budget.

The first meeting of the JGOFS Committee, which took place in Miami in January 1988, concentrated on two major items: the development of a Science Plan for JGOFS and the establishment of planning and management strategies for the Pilot Study.

The Science Plan, to be produced during 1988, will provide the scientific background and rationale for JGOFS, will identify and elaborate upon its detailed scientific objectives, discuss the technical problems to be solved in order to meet these objectives, and introduce preliminary proposals for strategies to meet them.

The Committee approved the recommendation of the North Atlantic Planning Workshop that there should be a coordinated program on the 20°W transect (see above). The JGOFS North Atlantic Pilot Study, which will begin in early 1989, is being coordinated by a group of national chief scientists from the participating countries (Canada, FRG, The Netherlands, UK, and USA).

At its first meeting the Committee established two working groups on topics which it considered needed to be addressed on an urgent basis: Data Management and Modelling. The Data Management group will give first priority to issues which can be resolved before the Pilot Study while the Modelling group will begin by reviewing existing models to determine which biogeochemical and physical parameters and fluxes appear to be most critical in controlling the global distribution of CO_2, nutrients, and other biogeochemical tracers. This is expected to yield results which will influence the design of sampling strategies for JGOFS.

Professor Siedler informed the General Meeting that close relationships are being formed between JGOFS and ICSU's International Geosphere-Biosphere Programme (IGBP—also known as A Study of Global Change). Discussions had taken place in Acapulco between the Officers of SCOR and the Chairman of ICSU's Special Committee for the IGBP, Professor McCarthy, who is also a Corresponding Member of the JGOFS Committee. Clearly, JGOFS will be an essential contribution to the marine component of IGBP and it was agreed that close liaison must be maintained between these two programs, but that the details of the formal relationship between JGOFS and IGBP should be discussed by the JGOFS Committee at its second meeting in September 1988 and specific recommendations forwarded to SCOR for action.

The General Meeting ratified the decision of the 28th Executive Committee Meeting to establish the JGOFS Committee and congratulated the Committee on the pace of its activities to date.

UNESCO/SCOR/ICES/IAPSO Joint Panel on Oceanographic Tables and Standards
Dr. Selim Morcos of the Unesco Division of Marine Science informed the General Meeting that recent JPOTS activities have involved two of its sub-panels. The Editorial

Panel on the Oceanographic Manual held its second meeting at the same time as the last meeting of SCOR in October 1987. A report of that session appears in SCOR Proceedings, Volume 23. In addition to its meeting, the panel has continued its work in the intersessional period, and several scientists were requested to contribute or comment on specific aspects of the manual. In preparation for a third meeting scheduled to take place in Tokyo, from 13 to 19 September, 1988, a second draft had been circulated to the Panel members by Dr. H. Dooley (ICES), who will produce a third draft for editorial work during the forthcoming meeting. Unesco is envisaging the publication of the manual in 1989 in its series "UNESCO Monographs in Oceanographic Methodology" provided the publication is in conformity with this series.

The JPOTS sub-panel on standards for CO_2 measurements met first between August 13–17, 1987, i.e., during the IUGG General Assembly in Vancouver, Canada. The sub-panel discussed the current state of carbon dioxide measurements in seawater.

The second meeting of the CO_2 sub-panel, August 29–31, 1988, was held during the Joint Oceanographic Assembly in Acapulco, Mexico. At this meeting, Drs. Poisson and Culkin reported the preliminary results of the intercomparison exercise. The Panel agreed that the results of the exercise emphasized the need for certified reference materials (CRMs) for oceanic carbon dioxide measurements. The remainder of the 1988 meeting was devoted to formulating specific recommendations for the certification and proposing a timetable for the preparation and distribution of an interim reference material based on natural seawater. This material needs to be available in limited quantities by July 1990, i.e., at the start of the proposed decade-long global carbon dioxide program under WOCE and JGOFS.

The draft recommendations are:

1. Standard reference materials should be used by the workers involved in the global carbon dioxide programs so as to ensure the quality control and coherence of the various data sets;
2. These reference materials should be available for total alkalinity and total dissolved inorganic carbon in July 1990;
3. To achieve the proposed target date, we recommend that SCOR and other international committees endorse the concept of using standard reference materials for the global carbon dioxide programs; and
4. Funds need to be secured by the appropriate global programs to set up a mechanism to bottle, certify, and distribute these standard reference materials. The certification should be carried out by two or more selected laboratories.

Editorial Panel for the Ocean Modelling Newsletter
The General Meeting was pleased to note the continuing success of the Ocean Modelling Newsletter with the seventy-eighth issue having been published in July and distributed to more than seven hundred readers. Financial support from the U.S. Office of Naval Research has been renewed for a three-year period.

4. Proposals for New Working Groups

Chemical and Biological Oceanographic Sensor Technology
This proposal, originally submitted to SCOR by the Australian Committee with the title "Methodologies Available for the Development of Biological Oceanographic

Probes," was widely circulated for comment and was discussed by the SCOR Executive Committee at its 1987 meeting. It was deferred, however, due to the death of its author, until his replacement could be found. The Executive Committee also recommended that the original proposal should be revised. Accordingly, the General Meeting had before it a revised proposal submitted by Dr. Denis Mackey on behalf of the Australian SCOR Committee, with more narrowly defined terms of reference and a list of membership suggestions which reflected the emphasis on optics and microelectrodes. It received broad support, especially from those participants who spoke of the need to improve the efficiency of acquisition of biological oceanographic data. The General Meeting accepted revisions to the proposed terms of reference in order to reflect this approach, as follows:

1. To review current technologies that may be suitable for measuring chemical and biological properties with high resolution in time and space;
2. To assess which of these technologies holds most promise for widespread deployment as oceanographic probes in the next five years; and
3. To evaluate ways in which the data from such probes can be calibrated, standardized, and integrated into the data base from standard hydrographic instruments.

In order to avoid the impression that the working group would be involved in the actual construction of instruments, the title proposed was changed to "Chemical and Biological Oceanographic Sensor Technology." It was agreed to establish this group as WG 90 and to invite Dr. D. J. Mackey (Australia) to serve as Chairman. The following tentative membership list was drawn up using some suggestions contained in the original proposal and others received from SCOR Committees and participants in the General Meeting: A. Zirino (USA), M. Atkinson (USA), D. Turner (UK), H. P. Hansen (FRG), A. Herman (Canada), P. Williams (UK), and Y. A. Sorokin (USSR).

Chemical Evolution and Origin of Life in Marine Hydrothermal Systems
This proposal was also first considered by the Executive Committee at its meeting in 1987, having been submitted by the Swedish SCOR Committee with the title "Neo-Abiogenesis and the Origin of Life in Marine Hydrothermal Systems." The Executive Committee requested that the proposal be revised to make it more oceanographic in content, rather than purely chemical.

The General Meeting agreed to establish this group as WG 91 with the following terms of reference:

1. To determine likely constituents necessary for neoabiogenesis according to the state of the art of the origin of life sciences and thermodynamic calculations;
2. To review available data concerning primordial organic monomers and polymers already observed in hydrothermal system (for example, carboxylic acids, amino acids, cyano- and heterocrylic compounds); compile a list of potential substances that have to be searched for; and differentiate compounds formed abiogenically and biogenically;
3. To evaluate the role of different classes of possible inorganic catalysts which may be required for the synthesis of organic compounds in hydrothermal systems; and
4. To sponsor a symposium and publish a set of papers in 1992 summarizing the state of knowledge and identifying research opportunities in this field.

It was agreed that Professor Holm should be invited to chair WG 91 and that the membership list would include some names submitted with the original proposal and additional nominations received at the General Meeting, as follows: R. M. Daniels (New Zealand), J. P. Ferris or B. Simoneit (USA), H. Yanagawa (Japan), G. Cairns-Smith (UK), and R. Hennet (Switzerland).

Dr. Kuznetsov agreed to identify an appropriate Soviet member for WG 91.

Pelagic Biogeography
The Committee for SCOR in The Netherlands submitted a proposal for this working group as a result of a recommendation of the International Conference on Pelagic Biogeography which took place in 1985. The proposal was presented to the General Meeting by Dr. A. Pierrot-Bults who noted that a knowledge of distribution patterns on a much larger scale than is available at present is required to assist with the development of a global model for the ocean and climate. After discussion in the General Meeting and revision of Dr. Pierrot-Bults and others, the proposal is as follows:

1. To review recent developments in biogeographic theory and their application to oceanic pelagic biogeography;
2. To recommend new approaches to the future studies on pelagic biogeography emphasizing the mechanisms that result in observed distribution patterns and the interactions of organisms and their physical-chemical-biological environment and their impact on the observed patterns;
3. To examine the possibilities of more adequate sampling techniques and the interpretation of available data and the use of existing plankton and nekton collections for biogeographical studies;
4. To prepare a manual in 1982 of existing collections as a guide to all interested scientists. These collections are of extreme importance as they reflect conditions of the past, evidence of which could never be collected again; and
5. To hold appropriate workshops, followed by a second international conference on pelagic biogeography in cooperation with other interested organizations.

The General Meeting had before it another, related proposal (see next item). There was concern expressed, not only about the possible overlap between these two proposals, but also with the OSLR program of IOC and certain activities within ICES. Accordingly, it was agreed that they should be referred to these organizations, and to JGOFS for advice before proceeding to the decisions on their establishment.

Physical Processes Affecting Biological Variability
This proposal was sent to SCOR by the US Committee for SCOR just before the General Meeting and had not been circulated for review by SCOR Committees. It requested SCOR to consider the establishment of a working group to examine the interactions between population dynamics and ocean variability.

The proposal was considered in conjunction with the one on Pelagic Biogeography (see above item). It noted that a central problem in biological oceanography is the cause of variability in populations, especially the effects of changes in the physical environment. Questions about population dynamics and productivity are central to development of community and ecosystem concepts. The study of the relationship between physical processes in the ocean and biological variability is not a new topic, but it has greatly increased in significance in the context of the climate change/ocean flux programs. The global emphasis of WOCE and JGOFS raise questions about the impact of physical and chemical change on the marine biosphere. The most sensitive indicators are species abundance and community structure but we require causal relations between the physics and biology if changes in these indicators are to be seen in the correct perspective. Further, changes in open ocean populations can in turn affect the biochemical fluxes and feedback to the problems of global changes. The emphasis proposed for the Working Group is broader than "living resources" but more focused than general "marine ecology." It is an important topic in its own right but would complement other SCOR WG's.

The terms of reference proposed for the WG were:

1. To review existing case studies of physical/ecological interactions;
2. To examine the existing suite of population dynamics and ecological models and propose new opportunities for generalizations between environments;
3. To determine sampling and experimental technologies required to link theory and observations;
4. To assist in planning cooperative programs at the international level; and
5. To hold appropriate workshops and produce a report on strategies for international cooperation by 1991.

The General Meeting wished to know whether this proposal overlapped with the interests of the IOC's program on Ocean Science in Relation to Living Resources. The representative of IOC, Dr. Kullenberg stated that while the OSLR program could generate data of interest to the proposed group, it was primarily concerned with the coupling between physical processes and recruitment to marine populations. While some participants felt that there was considerable duplication between this proposal and the one on Pelagic Biogeography, since both appeared to be related to biological communities and physical processes in a continuum of scales, others were of the opinion that they were quite distinct, with one concentrating on large-scale patterns of population distributions and the other addressing the effects of physical processes on population variability.

A debate in the General Meeting as to whether this proposal would overlap with the interests of the IOC's program on Ocean Science in Relation to Living Resources resulted in a request that each proposal be revised to define more clearly the distinct nature of the questions to be addressed. As noted above, this proposal will also be sent to IOC, ICES and JGOFS for advice.

The International Geographical Union Commission on Marine Geography*

INTRODUCTION

The establishment by the International Geographical Union (IGU) of a Commission on Marine Geography (CMG) at the twenty-sixth IGU Congress in Sydney, Australia, in August 1988 marked the "end of the beginning" of a process begun at the twenty-fifth congress in Paris four years previously, at which the idea was first informally mooted. It marked the international institutionalization of geography as a discipline in the study of the sea, rather late in the day compared to many other disciplines, yet foreshadowed by the prior establishment of national study groups of professional geographers—both physical and human geographers—concerned with marine or maritime geography in, for example, the United States, Canada, France, the Federal Republic of Germany, Italy, and the U.S.S.R. These national developments in turn signify a coming together of an increasing number of academic geographers, with interests in the marine environment, engaged in the academic development of the discipline, the inauguration of formal educational programs usually in a multidisciplinary context, and the application of the subject to the development and management of the world's oceans and seas. This last concern has particularly inspired the foundation of the CMG during its early stages.

This report outlines the establishment of the CMG and development of its program, considers the objectives, meetings, and research projects, and discusses the role of the CMG in geography and marine studies. It concludes with a brief look at future developments.

ESTABLISHMENT AND DEVELOPMENT OF THE WORK PROGRAM:
THE STUDY GROUP ON MARINE GEOGRAPHY

The CMG originated as the Study Group on Marine Geography (SGMG) established at the IGU Regional Congress in Barcelona in September 1986. This move was preceded by meetings in Genoa (1985) on the theme of coastal zone management[1] and at Nice in 1986 on the theme of man in the marine environment.[2] The prime movers were

*Editors' Note.—This report was supplied by Dr. Hance D. Smith, Secretary, IGU Commission on Marine Geography, University of Wales College at Cardiff, Department of Maritime Studies, P.O. Box 907, Cardiff CF1 3YP, Wales, United Kingdom.

1. C. Da Pozzo, P. Fabbri, and A. Vallega, eds., *Planificazione marittimo litoranea: Realta e prospettive/Coastal Planning: Realities and Perspectives* (Genoa: Universita di Genova, Commune de Genova/International Geographical Union, 1985).

2. Colloque international sur l'integration des activités de l'homme dans le milieu océanique, *Analyse Spatiale, Quantitative et Appliquée*, No. 22 (1987): 3-107, published by the Laboratoire de Géographie Raoul Blanchard, Nice, France; and Colloque international sur l'integration des activités de l'homme dans le milieu océanique, *Cahiers Nantais*, No. 29 (1987): 3-83, published by the Centre Nantais de Recherche pour l'Aménagement Régional, Université de Nantes, France.

Professor Adalberto Vallega, Director of the Institute of Geographical Sciences at the University of Genoa (and now Chairman of the CMG) and Professor André Vigarié, then Head of the Institute of Human Sciences of the Sea in the University of Nantes (now Vice-Chairman of the CMG). The next move was to convene a Round Table meeting at the IGU Regional Conference in Barcelona entitled, "New Frontiers in Marine Geography,"[3] chaired by Professor Jesse Walker of Louisiana State University at Baton Rouge. It was at this juncture that the Executive Committee of the IGU approved the establishment of the SGMG, chaired by Professor Vallega, for a period of two years leading up to the twenty-sixth Congress in 1988. The establishment of the SGMG was underwritten by a program of work planned beyond 1988 as far as 1992.

The first meeting of the SGMG took place at the University of Wales Institute of Science and Technology (now the University of Wales College of Cardiff) in July 1987 and had as its theme, "The Integration of Human Activities in the Management of the Sea."[4] The second full meeting was held in Sydney during August 1988 with the theme, "The Role of Geographers in the Interpretation and Application of the Law of the Sea: The Use and Management of the Sea and Development of Economic, Political, and Strategic Regions." Meanwhile a policy of participation in the work of related bodies and meetings began. In January 1988 a special session held in honor of Mr. Leopold Senghor on developing countries and maritime management was held in Genoa and was devoted to African cultures and North-South cooperation.

Both main meetings, in Cardiff and Sydney, respectively, were very successful. This success, coupled with a large initial membership (by January 1988 there were 11 full members [11 countries], 61 corresponding members [33 countries plus the United Nations], 24 institutions [15 countries], and 179 scholars [33 countries])[5] a regular Newsletter[6] and the planned program of work until 1992,[7] together with the growing general interest in the geography of the sea, ensured a firm decision by the IGU General Assembly to establish a Commission for the period 1988–92.

The program of work is based on a series of meetings and research projects, discussed below. The main theme is entitled, "Christopher Columbus 2000: The Future of the Oceans." Annual meetings have been scheduled in the Federal Republic of Germany (1989), the People's Republic of China (1990), Spain (1991), and the United States (1992), supplemented where necessary with additional meetings. The first of the CMG annual meetings was held at Wilhelmshaven in May 1989 in association with meetings of the IGU Commission on the Coastal Environment, the Working Group on Seas and Coasts of the Association of German Geographers, and MareTech 89, a meeting on Maritime Technology and International Cooperation. The joint meetings were sponsored by the City of Wilhelmshaven and the Government of Lower Saxony.

3. H. D. Smith and A. Vigarie, eds., *The New Frontiers in Marine Geography/Les nouvelles frontieres de la geographie de la mer* (Rome: Consiglio Nazionale della Ricerche, Gruppo di coordinamento Geografia Umana, 1988).

4. University of Wales, Institute of Science and Technology (UWIST), *The Integration of Human Activities in the Management of the Sea,* conference preprints (Cardiff: UWIST, 1987); and H. D. Smith, ed., *The Development of Integrated Sea Use Management* (London: Routledge, 1989).

5. *SGMG Directory.*

6. *Newsletters 1–5* (No. 6 is in press).

7. *Newsletter 3.*

OBJECTIVES, MEETINGS, AND RESEARCH PROJECTS

The challenge of developing a body such as the CMG is to establish worthwhile practical objectives and research plans. This is done by drawing together a diversity of growing expertise through the meetings, which in turn generates further research themes and links both within and beyond the Commission. It is thus instructive in the first instance to analyze the pattern of work emerging in the meetings before considering the nature of the research projects, which are still in very early developmental stages.

The preliminary meetings primarily established reference points that provided, in part, starting points for the establishment of objectives at the Barcelona meeting. The Genoa and Nice meetings of 1985 and 1986, respectively, focused on coastal and nearshore themes, where some of the most pressing and land-related problems exist. At the Barcelona meeting the focus was upon the possible research frontiers, including the major sea-use groups and the theory and practice of sea-use management as a whole.

The basic objectives and working contexts settled at the Barcelona meeting were relatively clear-cut. These concerned establishment of work in the geopolitical implications of the Law of the Sea; the geoeconomy of the ocean; sea-land connections; nearshore planning; and deep-sea regionalization. These objectives are set within a broader framework that involves developing basic academic thinking on marine geography (the thematic approach); the application of the field to sea-use management (which involves a regional approach); and the encouragement of educational programs (the educational approach). It is too early to assess the extent to which these aims and objectives are being achieved. Suffice it to say at the present juncture that, as already noted, initial interest has focused upon sea-use management.

The first real directions of development can be gauged from the Cardiff, Sydney, and Wilhelmshaven meetings, as revealed by the nature of the papers offered. The central theme at Cardiff inolved the development of integrated management of sea uses. Major subthemes emerging at the meeting included the development of research in sea-use management; technical management issues including maritime boundary delimitation; key issues in the spatial framework of management; the management of individual uses including navigation, fisheries, oil and gas, and waste disposal; and the regional bases of integrated management. At the Sydney meeting important themes that emerged included multiple-use management developments; development and management of specific ocean uses at subcontinental/oceanic scale; and the regional development of sea-use management in the Australian region, where of course advanced examples exist of international boundary delimitation, multiple-use management (the Great Barrier Reef Marine Park), and oceanic management (Commission for the Conservation of Antarctic Marine Living Resources [CCAMLR]), as well as measures taken to conserve the traditional maritime cultures of the aborigines and Torres Strait islanders.

At Wilhelmshaven a very strong focus on the management of the European seas understandably emerged, with a clear division in the papers presented between the metropolitan core coastal and nearshore waters, where every major use was considered in some way, and the outer coasts and seas, including the North Sea, Baltic, and Mediterranean. In the core seas major themes included maritime boundary issues and settlement; navigation, river basin and related environmental management; and conservation; with overall emphasis on multiple-use management issues. By contrast, in the outer coasts and seas the focus was upon major-use groups: strategic, fisheries, recreation; maritime boundaries; and rural coastal management. Another major theme of further papers concerned specific-use groups at the global scale—navigation, strategic uses, and mineral resources, particularly.

The development of the research projects forms the second aspect of the work. There is a certain logic involved in the projects. Most basic is the project on information technology concerned with compilation of data bases on marine geography; a second grouping of three projects dealing with shipping, strategic use, and fisheries—the "old" uses which most clearly reflect regional development patterns and pressures on the marine environment both at global and regional scales; and a final group concerned essentially integrated management approaches and dealing with sea-use planning and semienclosed seas where the most immediate multiple-use management problems and solutions have to be worked out.

The information technology project is concerned essentially with two tasks: the compilation of sources related to major sea-use groups in both development and management contexts, and the preliminary investigation of data-base management implications of this for marine geography as a whole.

The second group of projects is development oriented. The first deals with geo-strategy and maritime transport and is thus concerned with the global economy itself. Two major aspects concern, respectively, relationships between shipping and trade flows and the organizational shift from the shipping focus of maritime trade to the physical distribution management system approach in which shipping interests no longer necessarily dominate the system, which, in turn, has geostrategic implications of both global and regional dimensions.

The second project is concerned with the relationship between the structural changes in the global shipping industry, particularly with regard to the international division of labor. There is a strong focus on the developing countries' role in maintaining shipping development and reactions in shipping management by developed countries including the flagging out or open registry systems.

The sea-use planning and semienclosed seas projects focus upon the beginnings of integrated management approaches. The first is concerned with a comparative analysis of local and regional planning and development concepts and practice of European coastal states. This involves consideration of the levels of authorities involved, from local authority types upward, and relationships with current land planning systems.

The semienclosed sea project involves a survey of the management systems applied to these areas worldwide, conceived in terms of objectives, technical and general management approaches, and relationships with regional development that produces distinctive patterns of sea-use conflicts and environmental impacts. A publication dealing with the overall regional patterns of semienclosed seas, and a special case study of the Ligurian Sea, sponsored by the Ligurian Regional Authority (Regione Liguria) are the initial products of the project.[8]

Taking the objectives, meetings, and research projects together, a considerable amount of work has already been accomplished, and major thrusts can be identified.

8. H. D. Smith and A. Vallega, eds., *The Management of Semi-enclosed Seas: The Global Pattern and the Ligurian Sea* (Genoa: Regione Liguria, 1989).

The first is the emergence of a development and management geography for the major sea use groups, where shipping and navigation, strategic use, and fisheries have been advanced furthest. Second is the development of integrated management approaches, both conceptually and applied to specific regional case studies. Third is the building up of maritime geographies of specific major regions: so far European and Australian seas have been considered in considerable detail, with clear patterns beginning to emerge.

THE ROLE OF THE CMG IN GEOGRAPHY AND MARINE STUDIES

Having established the emerging patterns of work of the CMG to date, it is useful to consider the wider disciplinary and institutional contexts in which this work is evolving before making a closer assessment of the present and future contribution being made. There are three immediate aspects which draw attention at this early stage. These are, respectively, the development of marine geography in relation to geography as a whole; second is the relationship between marine geography and other disciplines with a strong presence in the marine field; third is the institutionalization of other, essentially nonacademic interests in the study of the sea which can have a profound influence on the development of the CMG and marine geography as a whole.

The development of marine geography in a geography context can be usefully considered from the history of geographical thought perspective, and it is planned to develop cooperation with the IGU Commission on the History of Geographical Thought for this purpose. Pending the outcome of this initiative, it is worth analyzing the recent background of events within the IGU which gives some food for thought in this field (see the Appendix).

Central to the reorganization of the IGU's structure and activities in 1988[9] has been the identification and promotion of the major thrusts in international effort in geography. There appears to have been a move away from traditional disciplinary subdivisions as a basis for organization of work in international geography towards a broader-based approach in which concepts and linkages among these subdisciplines have become more central. Thus it is possible to identify a specific environment group with a more modern physical geography structure; a human group which highlights contemporary social geography developments; a development group which has moved outwards from traditional economic geography and in which man-environment links are central, including settlement, modern sectors, and economic development; and a development of geography group concentrating on recent more sophisticated approaches to the discipline itself.

Marine geography belongs to the development group and is extending both development and management themes in man-sea interactions albeit, as already noted, with emphasis on the management aspects at present. Concern with data-base development and human organization dimensions of management also places it within mainstream development, as does preliminary work on global environmental change,[10] which will link the Commission's work with the International Geosphere-Biosphere

---

9. *International Geographical Union, 1988–1992*, leaflet (1988), 6 pp.

10. A Vallega, *Ocean Change in Global Change: Introductory Geographical Analysis* (Genoa: Universita degli studi di Genova, Instituto di Scienze Geografiche, 1989).

Program and Human Dimensions of Global Change Program inaugurated by the International Council of Scientific Unions (ICSU) and the International Social Science Council (ISSC), respectively.

Beyond geography and the IGU, which is, of course, grounded in the collective efforts of individual geographers, including marine geographers, marine geography finds itself part of a palimpsest of disciplines which have developed marine interests to varying degrees. The strongest are naturally the marine sciences—especially oceanography, marine meterology, marine geology, and marine biology; maritime and the public international law of the sea, which have developed considerably in the scale of effort in response to the expansion of international maritime trade and the development of the Law of the Sea Conventions, respectively, since World War II. Also important is shipping and marine resource economics dealing with fisheries and offshore petroleum. History and anthropology and archaeology also have strongly developed maritime studies.

The study of the sea is also being expanded by nonacademic interests in response to current changes, including increasing awareness of the economic, social, and political possibilities concerning the sea. These include technology, urban administration, as well as already established user groups such as shipping interests; and environmental groups and international and national governmental agencies. The CMG already has within its membership some of these bodies, which will form a valuable link, especially in the development of applied and educational work.

TOWARDS A GEOGRAPHY OF THE SEA

The CMG provides considerable opportunities for the development of marine geography in both industrial and developing countries and for strengthening links with other disciplines both for research and education. Links with organizations also provide the opportunity for development of practical applications of the subject.

The development of marine or maritime geography, while in many respects rather limited, has a relatively long history, the early part of which belongs to the period before the early twentieth century when geography and related disciplines were less fragmented. Thus physical geographical/oceanographical papers sometimes appeared in the *Geographical Journal* and *Scottish Geographical Magazine,* for example. The second aspect was the development of systematic marine studies in coastal geomorphology, ports and shipping, fisheries, and, more recently, oil and gas.[11] Further, general geographic approaches date from the educational origins of the subject as specialist courses in the tertiary sector[12] and the setting up of the specialist study groups already referred to. The SGMG and CMG have thus come at a time when the subject is at a relatively early stage of development. Major undeveloped systematic gaps exist, for example, in strategic, recreation, and conservation sea-use groups, and the field is wide open for the development of integrated approaches, be these ecological, spatial, regional, or behavioral. The SGMG and CMG have already made contributions to sea-use development studies, especially in shipping, strategic use, and fisheries.

11. H. D. Smith, "The Geography of the Sea," *Geography* 71, No. 4 (1986): 320–24.

12. H. D. Smith, "Maritime Geography: Applications in Coastal and Sea Use Management," *Cambria* 14, No. 2 (1987): 105–14.

Applied studies can be both development and management oriented. Here the principal contributions of the SGMG and CMG have so far been systematic, concerned with the development of approaches to applied studies, including technical and general management concepts which have the potential to expand sea-use management horizons based on traditional sectoral economic organization. Major contributions have also been made to the study of maritime regions—the European and Australian seas. In the regional studies, the ocean regions tend to be treated in broad economic, strategic, and political contexts, whereas the European and Australian regional studies focus more on technical and general management issues, including maritime boundary delimitation and regional development of specific sea uses such as fisheries, environmental management, and integrated sea-use management.

It may be through the IGBP and HDGC programs that the most fruitful avenues for developing links with other disciplines and organizations will be found, in both academic and applied contexts. The concept of environmental and human historical change is important both for academic frameworks and practical management studies. It is the context within which the relationship between man and the sea becomes meaningful, both within and beyond academic circles, to the general public. It is also a good educational vehicle, as well as a research focus for dealing with the problems of marine mineral resource development, fisheries management, spatial organization of the marine economy, and environmental management issues in which the seas and oceans are central, including marine pollution, the greenhouse effect, sea-level rise, and the scientific study and conservation of the marine environments of the world. The Christopher Columbus 2000 Program of the CMG is firmly focused upon these challenging futures.

APPENDIX

ANALYSIS OF IGU COMMISSIONS AND STUDY GROUPS STRUCTURE

Development of Geography:
  Geographical education
  Geographical monitoring and forecasting
  Mathematical models
  History of geographical thought
  Geographic information systems
Environment:
  Measurement, theory, and applications in geomorphology
  Mountain geoecology and resource management
  Coastal environment
  Climatology
  Frost-action environments
  Rapid geomorphological hazards
  Historical geography of global environmental change
  Environmental change in karst areas

Development:
   Urban systems and urban development
   Changing rural systems
   Industrial changes
   Geography of leisure and recreation
   Geography of commercial activities
   Geography of telecommunication and communication
   Marine geography
   Third World development
   Early industrial regions
   Famine research and food production systems
Human:
   Population geography
   Health and development
   World political map
   Geography and public administration
   Gender and geography

The International Maritime Satellite Organization (INMARSAT)*

INTRODUCTION

INMARSAT is an international, intergovernmental organization which was established in order to provide the space segment (satellites and ground control facilities) for global maritime satellite telecommunications. The Organization, which has its headquarters in London, has at present fifty-six member states.

ORIGINS

By a Resolution of November 23, 1973, the Assembly of the Inter-Governmental Maritime Consultative Organization (now the International Maritime Organization) decided to convene an international conference to decide on the principle of setting up an international maritime system, and if it accepted this principle, to conclude agreements to give effect to this decision. The International Conference on the Establishment of an International Maritime Satellite System was held in London. Three sessions were convened, between April 23, 1975, and September 3, 1976. The Conference adopted the text of two international instruments: the Convention on the International Maritime Satellite Organization (INMARSAT), and the Operating Agreement on the International Maritime Satellite Organization (INMARSAT). Both instruments entered into force on July 16, 1979. The Organization commenced operations in 1982.

ACTIVITIES

The purpose of INMARSAT, as stated in Article 3 of the Convention, is "to make provision for the space segment necessary for improving maritime communications, thereby assisting in improving distress and safety of life at sea communications, efficiency and management of ships, maritime public correspondence services and radiodetermination capabilities." The Convention was amended in 1985 so as to give the Organization the additional competence to provide aeronautical satellite telecommunications services. It is expected that these amendments will come into effect in 1989. The Convention was further amended in January 1989 to give the Organization the further competence to provide land mobile-satellite services. These amendments are also not yet in effect. The Organization is required to act for exclusively peaceful purposes.

*EDITORS' NOTE.—This report was supplied by Phillip Dann, Assistant General Counsel, International Maritime Satellite Organization, 40 Melton Street, Euston Square, London NW1 2EQ, England.

A wide variety of telecommunications services are offered to shipping and the off-shore industry using the INMARSAT system. These include telephone, telex, fac-simile, data, compressed video and slow-scan television. A recent addition is the En-hanced Group Call (EGC) service. EGC enables weather and navigation safety information to be transmitted to a large number of ships simultaneously. It is possible to address all ships within a given geographical area, all ships in a particular fleet, etc. The service can also be used to distribute fleet management information or commer-cial broadcast services such as subscription news broadcasts.

INMARSAT is now developing navigation, radiodetermination, and position re-porting services, and the INMARSAT system has a role of particular importance in relation to maritime distress and safety communications. These matters are referred to in more detail below.

INMARSAT also provides secretariat services for the COSPAS-SARSAT satellite system. This is a system of polar-orbiting satellites which is able to detect and locate radio distress signals transmitted by emergency position-indicating radio beacons (EPIRBs). The system is established and operated by the governments of the United States, the USSR, Canada, and France. The carriage of satellite EPIRBs is also an element of the GMDSS.

Regular aeronautical services will be introduced in the near future. Certain in-terim services have already commenced, including satellite telephone services for air-line passengers and a voice service for the Ontario air ambulance service. It is antici-pated that satellite telecommunications will be used for air traffic services and airline operational communications, as well as public correspondence services.

The INMARSAT system has already been used for certain land-based communi-cations. These include fixed communications in remote regions where no reasonable alternative telecommunications facilities are available. There has also been frequent use of light, transportable earth stations for communications in the aftermath of natu-ral disasters, such as earthquakes and hurricanes, where the existing telecommunica-tions infrastructure has been disrupted.

MEMBERSHIP AND STRUCTURE

Membership of INMARSAT is open to all States. In order to join INMARSAT it is necessary for a State to become a party to the INMARSAT Convention. The Party must in addition either sign the Operating Agreement itself or designate a competent entity, public or private, subject to the jurisdiction of that Party, to sign the Operating Agreement. A Party or entity which has signed the Operating Agreement is referred to as a Signatory. At present the INMARSAT Signatories consist of state enterprises, national PTT organizations, private commercial organizations, and a small number of Parties.

The INMARSAT space segment is open for use by ships and other mobiles of all nations on a nondiscriminatory basis. It is therefore not necessary for a country to join INMARSAT in order to make use of the system.

INMARSAT has three organs: the Assembly, the Council and the Directorate. The Assembly is composed of all the Parties. Regular Sessions are held every two years, normally in London. Each Party has one vote in the Assembly. The Assembly considers and reviews the activities, purposes, general policy, and long-term objectives of the Organization, and expresses views and makes recommendations thereon to the Council. It ensures that the activities of the Organization are consistent with the Convention and with the purposes and principles of the United Nations Charter. It also decides upon questions concerning formal relationships between the Organization, States, and other international organizations.

The Council is the main policy-making body of the Organization. It determines mobile satellite telecommunications requirements and adopts policies and programs for the design, development, and acquisition of the INMARSAT space segment. It adopts criteria and procedures for the approval of earth stations and also adopts procurement procedures, financial policies, and the annual budget of the Organization. The Council consists of twenty-two representatives of Signatories. Of these, eighteen are representatives of those Signatories which have the largest investment shares in the Organization. Two or more Signatories may group together for the purpose of representation on this basis. The four remaining places on the Council are taken by representatives of Signatories elected by the Assembly, irrespective of their investment shares, in order to ensure that the principle of just geographical representation is taken into account, with due regard to the interests of developing countries. The Council normally takes decisions by consensus. Where this is not possible, voting is weighted according to the investment share of each Signatory represented on the Council.

The Directorate has an international staff of over 200. It is headed by the Director General and is located at the Organization's Headquarters in London.

FINANCIAL PRINCIPLES

INMARSAT is financed by the contributions of its Signatories. Each Signatory contributes to the capital requirements of the Organization in accordance with its investment share. This is based on the amount of utilization of the INMARSAT system attributable to the country of the Signatory, calculated as a proportion of the total utilization of the system.

The total utilization of the INMARSAT system is calculated and apportioned in the following way. Utilization is measured in terms of the charges made by INMARSAT for the use of the space segment: these are generally expressed in terms of so much per minute for telephone and so much per kilobit for data. For the purpose of attribution, the charge levied for each communication is divided into two equal parts: a ship part and a land part. The ship part is attributed to the country of the flag of the ship where the communication originates or terminates, and the land part is attributable to the country in whose land territory the communication originates or terminates.

The capital contributions of Signatories attract compensation for use of capital at a commercial rate which is fixed by the Council and at present stands at fourteen percent.

The Council also establishes space segment utilization charges, which have the objective of earning sufficient revenues for the Organization to cover its operating, maintenance, and administrative costs, the provision of operating funds as necessary, the amortization of investment made by Signatories (return of capital contributions) and compensation for use of capital. The Convention requires that the rates of utilization charge for each type of utilization shall be the same for all Signatories for that type of utilization. It should be noted that there is no provision for the Organization to make any surplus profit or to provide any return on investment to Signatories other than the fixed compensation for use of capital.

The Organization is required to operate on a sound economic and financial basis having regard to accepted commercial principles. Certain detailed provisions of the Convention supplement this: for example, the procurement of goods and services required by the Organization is to be effected by the award of contracts, based on responses to open international invitations to tender. There is no requirement that contracts be distributed among Member States in accordance with investment shares or on some other basis. There are also provisions in the Convention specifying the intellectual and industrial property rights which the Organization is required to take in inventions and technical information resulting from research or development contracts financed by the Organization.

THE INMARSAT SYSTEM

INMARSAT may either own or lease its space segment. At present the Organization leases capacity on INTELSAT V satellites in the Atlantic, Pacific, and Indian Ocean regions; MARECS satellites operated by the European Space Agency in the Atlantic and Pacific Ocean regions; and MARISAT satellites operated by the Communications Satellite Corporation in the Atlantic, Indian, and Pacific Ocean regions.

The Organization has placed orders for four Second Generation satellites with an international consortium led by British Aerospace PLC. It is expected that the first of these satellites will be delivered in 1989 and will be launched in the first half of 1990. INMARSAT has agreements for the launch of its satellites on the Delta, Ariane, and Space Shuttle launch systems.

INMARSAT has also placed contracts for ground control facilities. There will be a Satellite Control Centre located at the headquarters in London, and TT&C stations at Fucino in Italy, Beijing in China, and possibly other locations.

It is expected that, during the second half of 1989, INMARSAT will issue a Request for Proposals for its Third Generation space segment. These satellites will make use of the entire radio spectrum allocated for mobile-satellite services at the 1987 World Administrative Radio Conference of the International Telecommunication Union. The satellites will each have spot beams in addition to a global beam.

Communications to and from ships are routed via coast earth stations which provide the interface between the satellite link and the terrestrial communications network. The coast earth stations are not owned and operated by INMARSAT but are required to comply with technical and operational standards established by the Council. There are at present twenty coast earth stations operating in the INMARSAT system. All are owned by INMARSAT Signatories, although this is not a requirement of the Organization.

The end-user charges for use of the INMARSAT system are established by each coast earth station operator. This charge will typically consist of three elements, namely, the space segment utilization charge for which the coast earth station must account to INMARSAT, the charge for use of the terrestrial network, and the charge added by the coast earth station operator for the use of its facility. Because coast earth station operators have the freedom to establish their own end-user charges and charging structures, there exists in practice considerable intrasystem competition between the operators to the benefit of end-users.

UTILIZATION OF THE INMARSAT SYSTEM

At present over 8,500 ship earth stations have been commissioned for access to the INMARSAT space segment, of which more than 7,500 are on ships, off-shore drilling rigs, and platforms. This represents a growth in the number of users of the system by some 58 percent during the past two years. It is expected that this growth will be even more rapid in the future with the introduction from 1989 of Standard-C services. The Standard-C ship earth station provides data and messaging services, although not the voice and telex services provided by the existing Standard-A ship earth station. However, Standard-C ship earth stations will be significantly smaller, lighter, and cheaper than Standard-A, thereby enabling many smaller vessels to use satellite telecommunications.

In the calendar year 1988 the Organization's total revenues were almost US$100 million, an increase of 34 percent over the previous year. It is expected that, although revenues from maritime services will continue to increase rapidly, aeronautical and land-mobile services will in due course provide a significant portion of total revenues.

RADIODETERMINATION AND NAVIGATION SERVICES

INMARSAT has established a program to develop a radiodetermination and navigation capability. This program is intended to be integrated with or complementary to existing navigation satellite systems, and it comprises the following elements:

a) near-term establishment of a full range of position reporting and position surveillance services for maritime, aeronautical, and land mobile applications;

b) provisions of GPS/GLONASS differential correction transmissions via INMARSAT; and

c) an international civil satellite navigation capability comprising an independent GPS/GLONASS augmentation overlay consisting of GPS/GLONASS health status and/or "integrity" information and a complementary "navigation reference" signal from INMARSAT's geostationary satellites. These will enable international civil users to use the national satellite navigation systems (GPS, GLONASS) with greater confidence and reliability.

DISTRESS AND SAFETY SERVICES

INMARSAT was represented at the international conference convened by the International Maritime Organization in December 1988, which adopted amendments to the Safety of Life at Sea (SOLAS) Convention 1974. These amendments are designed to implement the new Global Maritime Distress and Safety System (GMDSS). The SOLAS Convention applies to all passenger ships and cargo vessels of 300 GRT and above making international voyages. The amendments provide that the INMARSAT system may be used to satisfy the mandatory radio carriage requirements for ships, as follows:

a) all SOLAS vessels in Areas A1 and A2 may use INMARSAT ship earth stations optionally for accomplishing GMDSS communications functions;
b) all SOLAS vessels sailing in Area A3, that is to say, outside the range of VHF or MF coastal radio, may carry INMARSAT Standard-A or Standard-C ship earth stations as an alternative to HF radio with narrow-band direct printing and digital selective calling;
c) EPIRBs operating at 1.6 GHz (L-band) may be fitted as an alternative to EPIRBs operating at 406 MHz in the COSPAS/SARSAT system, subject to availability of appropriate receiving and processing ground facilities for each ocean region; and
d) vessels sailing beyond the coverage of 518 KHz Navtex systems should fit INMARSAT Enhanced Group Call equipment to permit reception of maritime safety information through the INMARSAT SafetyNET™ service.

INMARSAT's operational plans place a high priority on the closure of the partial "gap" in coverage between the Atlantic and Pacific Ocean Regions. This is in order to satisfy the requirements of both the GMDSS and maritime users generally.

INMARSAT has convened two meetings of experts from INMARSAT Parties to discuss the funding of international satellite distress and safety communications services. This question is still under review by the Organization and other interested bodies.

USE OF SHIP EARTH STATIONS WITHIN THE TERRITORIAL SEA AND PORTS

Many States prohibit or restrict the use of radio transmitters, including ship earth stations, on ships within ports and the territorial sea. Historically, these restrictions have been based on considerations of national security, the desire to maintain revenues from the use of local telecommunications facilities, and the fire risk created by the use of spark transmitters in ports. This last consideration has little relevance to the use of satellite telecommunications equipment.

These restrictions are inconvenient to ships' crews and create barriers to the efficient management of ships, which typically spend substantial periods of time in ports, moored off-shore or passing through the territorial sea. Local, terrestrial telecommunications facilities can only be used when the ship is in port. Even then, such facilities may be inadequate because of language problems or delays due to manual operation. The International Conference which established INMARSAT recommended in its Final Act that all countries should be invited to permit ship earth stations to operate within harbor limits and other waters under national jurisdiction. There was support for this approach at the 1983 World Administrative Radio Conference for the Mobile Services and also at the June 1983 meeting of the International Maritime Organization's Maritime Safety Committee.

In 1985 the INMARSAT Assembly adopted the text of an International Agreement on the Use of INMARSAT Ship Earth Stations within the territorial sea and ports. The Agreement is open for signature or accession by any State, whether or not an INMARSAT Party. The Agreement provides that parties shall permit in their territorial sea and ports the operation of approved ship earth stations appertaining to the maritime space communications system provided by INMARSAT and properly installed aboard ships flying the flag of any other party. It follows that the Agreement imposes obligations on the basis of reciprocity. Ship earth stations are required to use the maritime mobile-satellite frequencies and to comply with the applicable Radio Regulations. It is also provided that the operation of ship earth stations shall not be prejudicial to the peace, good order, and security of the Coastal State. There is provision for a party to restrict, suspend, or prohibit the operation of ship earth stations in ports and areas of territorial sea specified by them. The Agreement does not apply to warships and other government ships operated for noncommercial purposes. Neither does it apply to any internal waters, including navigable rivers.

Twenty States have now signed, ratified, or accepted the International Agreement, although in some cases the signatures remain subject to ratification. The Agreement will enter into force thirty days after the date on which twenty five States have become parties.

RELATIONS WITH OTHER INTERNATIONAL ORGANIZATIONS

INMARSAT continues to work in cooperation with numerous other international Organizations, including the International Maritime Organization, the International Civil Aviation Organization, the International Telecommunication Union, the United Nations Committee on the Peaceful Uses of Outer Space, the World Meteorological Organization, the International Hydrographic Organization, INTELSAT, the European Space Agency, and INTERSPUTNIK.

COOPERATION WITH DEVELOPING COUNTRIES

INMARSAT is implementing a program of cooperation with developing countries, with the aim of promoting self-help to meet particular needs rather than merely providing "external" assistance. The Program comprises the following elements:

    a) a Professional Assignee program to enable nominees to receive training or gain experience at INMARSAT;

    b) a Demonstration Program to provide free use of Standard-C for innovative applications in developing countries; and

    c) technical assistance.

Applications for participation in the Program have been invited and will be considered in consultation with a number of Signatories and intergovernmental bodies.

Reports from Organizations

Report of the International Ocean Institute: 1989*

INTRODUCTION

Since the last Report on the activities of the International Ocean Institute, a great deal has taken place within the various components of its work.[1] The following is a summary of the projects and programs which have been carried out or are on-going in the 1988–1989 period.

From its beginnings in 1972, the work of the IOI has aimed at keeping several years ahead of current developments in order to serve as a stimulus—but not too far ahead which would be utopian. To do this it has been necessary to anticipate trends in the implementation and further development of the Law of the Sea and its impact on the building of a new international order, including a New International Economic Order. In trying to achieve this end, the work of the IOI has been organized in four major sections: Training, Conferences and Seminars, Research, and Publications.

All four components are strictly interconnected. The result of the research projects are the basis for discussion at IOI conferences and seminars. The conferences and seminars serve to refine and correct the results of research. This work gives direction and a distinct character to the Training Program, as well as to the Institute's publications. The publications provide access to the latest data and insight to new interdisciplinary approaches to ocean management for use in the Training Program. Publications, training, conferences, and seminars serve to widen the circle of dialogue on the uses and conservation of the oceans in the context of building a New International Economic Order.

TRAINING PROGRAM

The adoption of the UN Convention on the Law of the Sea, and the concurrent ongoing social, economic and technological developments have set into motion new trends of economic growth and created needs, especially in the Third World. The new Law of the Sea has brought extensive rights, benefits and obligations. This in turn has added a different dimension to the administration and management of the vast areas of ocean space which now fall under the control of coastal States, as well as for international maritime activity beyond the limits of national jurisdiction. It also extends special rights to the landlocked and geographically disadvantaged States.

In order to meet these responsibilities, the developing countries around the world

*EDITOR'S NOTE.—This report has been supplied by Ms. C. F. Vanderbilt, Executive Director, International Ocean Institute, Malta.

1. Editor's note.—For further information about the International Ocean Institute, see "Report of the International Ocean Institute," *Ocean Yearbook 7*, ed. Elisabeth Mann Borgese, Norton Ginsburg, and Joseph R. Morgan (Chicago: University of Chicago Press, 1988), pp. 433–37.

saw the need for the formation of a new policy which would maximize the benefits to be derived from the integration of ocean management into their national and international development strategy. To be able to follow an integrated approach to the problems of ocean management, States needed to augment, strengthen, and alter their institutional structures. They also needed to formulate an ocean policy, enact and enforce appropriate legislation, formulate, implement and evaluate programs, collect, store and disseminate information, license and regulate resource exploitation, monitor, control and prevent pollution, and conserve and preserve resources whether living or nonliving.

As Third World nations became increasingly aware of the enormous wealth of the marine environment and the potential this could hold for economic development, they also felt the lack of necessary scientific equipment, technological skills, information systems, and relevant expertise—at the national as well as the international level.

It was in response to this need that the IOI developed its Training Program on the Management and Conservation of Marine Resources. The purpose of the Training Program is to deepen the understanding of the ever-increasing importance of the oceans and their resources in world politics and world economics; to assist developing countries in the formation of a core of decision-makers fully aware of the complex issues of ocean management; and to maximize the benefits to be derived from the proper integration of ocean management into national and international development strategy.

This Program, which began in 1980, originally consisted of three types of courses: Class A on Ocean Mining; Class B on EEZ Management; and Class C which focuses on issues of regional cooperation in marine affairs. During 1985 and 1986, after sixteen courses for over 250 participants had taken place, an in-depth evaluation and recasting of the entire program was undertaken. Surveys, seminars, and numerous discussions were conducted over an 18-month period. The result was the complete revision and updating of the syllabuses of Classes A and B with similar appropriate changes being made in the Class C—Regional approach. The frequency of Class A was reduced and a new intensive 5-week course was introduced: Class A2 on Development and Management of Marine Technology.

By the end of 1988, 25 courses had taken place. In this nine-year period, 421 participants from 86 developing countries attended. Of these, 17, or 4%, have come from landlocked or geographically disadvantaged States, and 63, or 15%, have been female.

In the period since the IOI Report for *Ocean Yearbook 3,* 5 courses have taken place. Chronologically these were: Class B2 in the P.R. China (5 October–11 December 1987); Class A2 in Colombia (22 February–25 March 1988); Class B in Halifax, Canada (13 June–19 August 1988); Class C in Malaysia (10 October–16 December 1988); Class A-2 in Madras, India (16 January–17 February 1989); and Class B in Canada (12 June–August 1989).

Class A2-88: Columbia

For the second year, the Class A2 course was held. Again, the Course Director was Dr. Krishan Saigal but the venue this time was Cartagena, Colombia. The collaborating institute was the Centro de Investigaciones Oceanograficas e Hidrograficas (CIOH) of the Direccion General Maritim y Portuaria of the Armada Nacional of Colombia. The syllabus of the course concentrates on the high-tech aspects of marine technology which are an integral part of the new industrial revolution. During the first four weeks, the Participants develop skills in futurecasting, project design, management and evaluation of high-tech ventures. At the same time they develop alternative structures, processing systems, and work programs for the Enterprise along with the associated requirements of human, material and financial resources. The final week of the course was held in Jamaica where the group presented their work to the delegates at the Preparatory Commission meeting.

Class A2-89: India

This A2 course was held in cooperation with the Ocean Engineering Centre of the Indian Institute of Technology (IIT) in Madras and was sponsored by the Department of Ocean Development (DOD) of the Government of India. The syllabus is very similar to that of the A2 conducted in Colombia with the obvious exception of the final week. In this case, the Participants joined the scientific crew of an Indian deep seabed mining vessel and spent several days observing the scientific research techniques of ocean mining. The Course Director was Dr. Krishan Saigal.

Class B-88: Canada

As in the past, Class B on EEZ Management took place at Dalhousie University in Halifax, Nova Scotia, Canada under the direction of Dr. S. P. Jagota. For the second year the course was carried out in cooperation with the Marine Affairs Program and was open to regular students of Dalhousie University.

Class B2-87: P.R. China

This special Class B course was conducted in cooperation with the State Oceanic Administration (SOA) of the Ministry of Foreign Affairs. The Course Co-Directors were Dr. John Vandermeulen of the Bedford Institute of Oceanography, Canada, and Prof. Xhou Qiulin of the Third Institute of Oceanography, SOA, in Xiamen, PRC. The course took place in Beijing with field trips to Qingdao and Tianjin.

Class C-88: Malaysia

The second in the Indian Ocean series, this course was under the direction of Dr. Krishan Saigal of India. The course was held in cooperation with the Indian Ocean Marine Affairs Cooperation (IOMAC) Secretariat with funding from CIDA, ICOD, and UNDP. It was held in Kuala Lumpur and field trips were made to Port Dickson, Ipoh, and Penang.

WMU Workshop

In addition to the Training Program, a workshop on the UN Convention on the Law of the Sea is held annually at the World Maritime University in Malmö, Sweden. This one-week course is designed for the first-year students of that University. The workshop is attended by about 100 participants each year.[2] The course coordinator and chief lecturer was Ambassador M. C. W. Pinto.

CONFERENCES AND SEMINARS

Pacem in Maribus[3]

The IOI's annual convocation, *Pacem in Maribus*, began in 1970. Since that time, six such convocations have taken place in Malta (1971–74, and 1987), in addition to Japan (1975), Mexico (1976), Algeria (1977), Cameroon (1978), Austria (1980), Mexico (1982), Sri Lanka (1983), Sweden (1984), USSR (1985), and Canada (1988).

Each year the conference concentrates on a special area of discussion of international ocean affairs, based on the concept of ocean space as the common heritage of mankind. The area of consideration is looked at from a strictly interdisciplinary approach, on a national, regional, international, transnational, and global level. *PIM XVI* took place this year in Canada. In addition, a brief reference is made to a few *PIMs* planned for the future.

PIM XVI—OCEAN TECHNOLOGY, DEVELOPMENT, TRAINING AND TRANSFER

The sixteenth *Pacem in Maribus* conference of the International Ocean Institute was devoted to a theme that was central to former Canadian Prime Minister Lester B. Pearson's concerns—that without technology transfer and the parallel development of human resources there can be no technology development, and without technology development, no economic growth.

The world has become increasingly aware of and concerned with technology transfer—indeed, this expression is now common jargon in the development field, and considerable global funding is directed to the "transfer" of technology from north to south.

This sixteenth *Pacem in Maribus*, however, was convened to examine the notion that "transfer of technology" is a misnomer—and that two decades of confrontational debate on the issue have led nowhere. *Pacem in Maribus XVI* tested the concept of "joint technology development"—that developing countries must be given access to the public/private international R&D consortia which increasingly dominate the R&D

2. Editor's note.—For further information about the World Maritime University, see "Report of the World Maritime University: 1989," in this volume.

3. For further information about *Pacem in Maribus*, see "*Pacem in Maribus XV*," *Ocean Yearbook* 7, pp. 501–5.

scene in the "north." This new, and more dynamic concept of "technology co-development," responding to the technological challenges of "High Technology" and to the economic challenges of the "Service Economy," was tested in the context of two concrete proposals for action: the establishment of a Mediterranean Centre for R&D in Marine Industrial Technology, first proposed by the IOI in 1986 and already passed onto the official intergovernmental agenda; and the project for joint deep-sea exploration and exploitation technology development through an "International Enterprise" to be established under the auspices of the Jamaica Preparatory Commission, in accordance with proposals by the Delegations of Colombia and Austria.

Another important aspect of *Pacem in Maribus XVI* was its emphasis on the development of human resources as an essential part of the technology development process. On this point, *PIM XVI* linked the theoretical concerns and the practical activities of the IOI, by presenting a working paper elaborated by a group of 22 young civil servants from 19 countries who had attended the ten-week IOI training programme at Dalhousie just preceding the conference. These young experts from the developing countries added a particular note of verve and activism to the proceedings.

Through *PIM XVI*, some 75 marine affairs experts from 16 countries, North, South, East and West, and representing all facets of ocean affairs, offered and discussed various views on "Technology Transfer." As in previous *PIM* conferences, these experts were senior government bureaucrats, cabinet ministers, diplomats, international civil servants, and ocean scientists. The week's programme was organized in three sections:

— Marine Technology;
— Transfer of Technology;
— The Development of Human Resources: an Essential Part of Technology Development and Transfer.

These were further broken down into eleven panels.

During the three months prior to *PIM XVI*, special study projects had been carried out on each of the subjects. These were presented as special lead-off papers, with further presentations rounding out each panel.

Another important aspect of this particular conference was that part of it was held at Canada's world-class Bedford Institute of Oceanography. This afforded the participants the occasion to consider the various ocean issues in the very real environment of active ocean scientific research. Also, this *PIM* further benefitted from a two-day Ocean Technology Exhibit that was organised as part of the five-day agenda. This, plus side visits to a private geophysics research vessel and "spur-of-the-moment" tours of individual research laboratories at B.I.O., further enhanced the mood of *PIM XVI*.

In retrospect, the conference more than achieved its original goal. The participants examined the various current ocean issues of fisheries, aquaculture, nonrenewable and renewable resources. More importantly, the conference carved out new ground in its discussion on future needs and approaches.

Pacem in Maribus XVI was attended by 106 participants from 29 countries.

Pacem in Maribus XVI was made possible thanks to the generous contributions of the International Development Research Centre, Canada; the Canadian International Development Agency; the International Centre for Ocean Development, Canada; the Atlantic Canada Opportunity Agency; the Dept. of Fisheries and Oceans, Canada; the Dept. of Industry, Trade and Technology, Province of Nova Scotia; the United Nations Environment Programme; and UNESCO.

Eighteen Chief Recommendations from *Pacem in Maribus XVI:*

On the Training of Fishermen
"Development of human resources capability should be the major thrust of fisheries technology assistance to developing countries."

On Assistance to Artisanal Fishermen
"Governments of developing countries should make available short- to mid-term loans to small and medium-size enterprises in coastal fishing communities in order to (a) help them modernize artisanal methods and technology; and (b) encourage craft fishermen not to abandon either their vocation or their home community."

On the Future of Artisanal Fisheries
"Artisanal fishermen should be assisted, wherever possible, in retraining for aquaculture which may be the key to their future."

On the Internationalisation of OTEC Research
"Action should be taken to create an international testing facility where industry and research teams could be supplied with deep cold water from the depths for R&D work relating to Ocean Thermal Energy Conversion applications."

On the Importance of Information Technology
"The formulation of national policies and plans on information technologies designed to facilitate technology transfer through the development of appropriate information technologies relevant at the national or regional level should be a Government priority in every country. Donor Agencies and competent international organisations should assist this process as appropriate."

On the Need for New Economic/Ecological Indicators
"We will need to develop and use new indicators of value added: calculations that include, for example, the value deducted from societal benefits by the production process itself."

On the National Basis of International Cooperation
"Technology 'transfer'—more appropriately described as 'joint technology development' must have a strong national basis to make international cooperation meaningful and effective."

On the Building of National Infrastructure
"At the national level, developing countries should be encouraged to:
    (a) set up a policy making and implementing agency;
    (b) build up, under the auspices of such an agency, a strong industrial information system;
    (c) set up engineering design and consultancy organisations; and
    (d) establish R&D laboratories to provide specialised advanced training, do applied research, assist the policy-making agency and industrial enterprises in identifying, selecting, and negotiating with foreign technology suppliers."

On Financing Indigenous Technology Development
"Every developing country should earmark a certain percentage of its educational budget for the advancement of science and technology, including marine technology. The Third World Academy of Science recommends that 4 percent of the educational budget should thus be earmarked for fundamental science; another 4 percent for applied research, and 10 percent to Research and Development (R&D)."

On Regional Centres for R&D
"Developing countries should initiate steps for the timely implementation of Articles 276 and 277 of the United Nations Convention on the Law of the Sea, calling for the establishment of regional centres for the advancement of marine sciences and technology."

On Global Cooperation
"All countries should promote
    (a) the adoption of an international code of behaviour for technology transfers;
    (b) the creation of an International Fund for Technology Development;
    (c) the establishment of an International Register of Technological Data; and
    (d) the identification of mechanisms providing mediation services between patent holders and users."

On the Enhancement of Mutual Interests
"In creating new forms of industrial/scientific cooperation between industrialised and developing countries, the interests of both parties must be taken into due consideration. To be successful, a joint enterprise must benefit all parties involved."

On the Mediterranean Centre for R&D in Marine Industrial Technology
"The Governments of Mediterranean countries and the U.N. Specialised Agencies should respond promptly and positively to the Secretary General's inquiry with regard to the next steps towards the establishment of the Mediterranean Centre for R&D in Marine Industrial Technology."

On Regional Centres in Other Oceanic Regions
"Experts in other oceanic regions should be encouraged to examine the feasibility study to see how it could be adapted to their regional circumstances, with a view towards implementing Articles 276 and 277 of the United Nations Convention on the Law of the Sea, strengthening the institutional infrastructure of Regional Seas Programme, and enhancing South-South cooperation in technology development."

On the Broader Context of Technology Training
"Training on technology needs to be supplemented by the creation of an awareness of the social, political, and cultural implications which the increased use of technology may generate."

On Special Clauses to Be Included in Technology Transfer Agreements
"All new technology supplied to the Third World must carry a 'joint research and development' clause, ensuring that the next generation of the technology is jointly

developed, as well as an 'update clause' to make new versions of the original technology available to the less developed countries."

On Special Clauses to Be Included in Technology Transfer Agreements
"Training of personnel in the use, repair, and further development of high-tech marine industrial equipment should be a part of the development and transfer of marine technology, and provisions to this effect should be included in the relevant agreements or contracts."

Conclusion
"Human resources development should be at the core of the development of marine science and technology and of the programmes of cooperation between the developed and the developing countries."

PIM XVII—PEACE IN THE OCEANS: THE NEW ERA

This meeting, which is scheduled for June 1989 in Moscow, USSR, will be co-sponsored by the Soviet Maritime Law Association (SMLA). The conference will concentrate on the promotion of peace, legal order, and environmental protection in the seas and oceans. Along with traditional sources of pollution, considerable damage is incurred to the marine environment and living resources through States' naval activities and tests of new types of arms, especially nuclear ones. *PIM XVII* will examine ways of reducing military activities in specific sea areas, as well as the establishment of confidence-building measures and "zones of peace" and "nuclear free zones" which can greatly contribute to the protection of the marine environment.

PIM XVIII—PORTS AS NODAL POINTS

This conference is scheduled for 27–31 August 1990 and focuses on ports as Nodal Points in an emerging global shipping system. This will look at ports in their newly emerging physical, economic, and organizational terms, as well as in the context of the global transport system. The venue will be Rotterdam, The Netherlands.

PIM XIX—REGIONAL SEAS AND INTERREGIONAL COOPERATION

The subject for *PIM XIX* will include a comparative study of the eleven Regional Seas Programmes, and the linkages between regional and global systems. The venue will be Nairobi, Kenya.

PIM XX—OCEAN DEVELOPMENT 2000

Since 1992 marks the 20th *PIM* as well as the 20th Anniversary of the IOI, Malta will be the venue. The theme will be more fully developed in 1989.

RESEARCH

The following are the principal on-going research projects underway at the IOI at present. Further information can be obtained directly from the Institute. The progress of the research projects will be covered in the next issue of *Ocean Yearbook*.

Mediterranean Centre for R&D in Marine Industrial Technology

There has been growing concern about the impact of the ongoing new industrial revolution on the marine environment, industrial structures and societal systems, especially of the developing countries. To meet this challenge, consensus has been emerging for the need for restructuring institutions (global, regional and national) in order to reduce the technological "gap" between the industrialized and the industrializing nations; and for having technological trajectories which are both socially and environmentally harmonious. On the basis of these considerations, UNIDO called for the setting up of regional centers in the spheres of high-tech, including marine technology. The UN Convention on the Law of the Sea 1982 mandated the establishment of regional centers of marine science technology. And the Brundtland Commission felt that organized efforts to diffuse the new emerging technologies are necessary if the future developmental paths are to be structurally harmonious and environmentally acceptable.

Drawing upon these ideas, the Government of Malta took the view that the Mediterranean, being a semi-enclosed sea bordered by nations both from the "North" and the "South," was a good place for undertaking a feasibility study aiming at the establishment of a Regional Center for marine technology. A proposal was put forth by the Government of Malta at the Fifth Ordinary Meeting of the Contracting Parties to the Barcelona Convention (Athens, 7–11 September 1987).

This feasibility study has now been completed and published. The Secretary General of the United Nations distributed it to the governments of all Mediterranean states and all competent United Nations agencies and institutions. IOI, which has been instrumental in moving the project forward to this point, continued to provide all necessary advisory and research support to ensure its successful progress.

In April 1989 UNIDO, in cooperation with the Office of the UN Secretariat for the Law of the Sea and Ocean Affairs and the UN Environment Programme, convened a Meeting of Experts on the Subject. IOI prepared the annotated agenda and background documentation. In this Meeting, representatives of eight Mediterranean countries found consensus on the need for such a center and in the ensuing report, UNIDO was given the mandate to take the necessary steps towards its implementation.

Distance Learning Project in Marine Resources Management

The IOI, in collaboration with the Commonwealth of Learning in Canada and the Centre for Distance Learning at the University of Malta, has developed a proposal for a distance learning project in marine resources management. The proposal concerns a core course on marine resource management through a distance education method.

The course is global in perspective and aims at fostering awareness of the need for long-term policies of rational management of the oceans and their resources.

The course would comprise twelve Distance Learning Units that examine different aspects of ocean use, development and management. Each unit would consist of a 35–40 minute film or video and a 120–130 page reader or textbook, including still photos, transparencies, graphs, audio-tapes, etc., making up an integrated package of instructional materials. The units would begin with a general introduction to the oceans and its human uses, and then go on in following units to cover the physical character of the oceans, law of the sea, utilization and management of living resources, aquaculture, utilization of nonliving resources, deep seabed mining, ocean energy, navigation, shipping and ports, tourism and recreational uses, protection and preservation of the coastal and marine environment and finally conclude with a unit on the management challenge. It is hoped that the course would be accompanied by additional peripheral units addressing specific concerns of different marine regions.

The timescale for this project anticipates the completion of three units per year and the budget is US$2,000,000. A meeting on the next preparatory phase is scheduled to be held at the University of Malta in June 1989.

Resource Person Data Base for Training Purposes

This project is a collaborative effort by the IOI, the International Centre for Ocean Development (ICOD) in Canada, and the World Maritime University (WMU) in Sweden. It involves the building of a computerized and multi-indexed global data base of resource persons with expertise and experience in marine affairs. The data base will be useful for training program planners to identify lecturers, consultants etc., both for general and specialized training purposes and covering all marine regions. With the experience gained through the holding of its Training Program over the past ten years, the IOI has developed an extensive teaching faculty. ICOD and WMU also have considerable experience in the field of training in marine affairs. Together their expertise could combine to form an extensive data base.

Alternative Modes of Funding the Mediterranean Trust Fund

This project aims at identifying new ways of funding the Mediterranean Action Plan (MAP), particularly through tourism sources. MAP suffers chronic cash shortages through delayed payment of national contributions and this often disrupts on-going activities. The 100 million annual visitors to the Mediterranean coastal areas could be a source of funding. Mediterranean tourism is heavily concentrated on the coast and in some countries coastal tourism represents up to 90% of all tourism. No Mediterranean country, however, has had any success in their attempts to improve tourist distribution throughout its national territory. This project is looking at current charges on tourist activities in the region in view of proposing an environmental tax.

Marine Affairs Directory for Malta

As a maritime nation, Malta has historically developed several maritime interests and activities which influence its very livelihood. These include shipping, ports and harbors, navigation, fishing, aquaculture, marine scientific research, offshore oil and gas exploration, desalination, tourism, marine archaeology, maritime boundaries, treaty obligation and regional cooperation. Past and present national development plans and private enterprise have had to contend with this reality. As a consequence, there is now a proliferation of institutional and individual interests, often overlapping and subject to cooperative, competitive or autonomous initiatives. While a mixture of cooperation and competition is always healthy, there is a lack of organized dissemination of information on the interests and activities concerned among government, industry, academia and the public at large.

This project, undertaken in collaboration with the Mediterranean Institute of the Foundation for International Studies, aims at the development of a multiple indexed inventory of human and institutional resources in marine affairs in Malta. The purpose of the Directory is to disseminate vital information for the development and rational utilization of human resources of a small island State.

PUBLICATIONS

The IOI produces several different publications: *Ocean Yearbook,* Proceedings of Pacem in Maribus Convocations, Newsletters, and Occasional Papers.

Ocean Yearbook

Since its beginnings in 1978, eight volumes of *Ocean Yearbook* have been published. Publication will be continued, roughly in 15-month intervals, with *Ocean Yearbook 9* expected in Spring 1991 and *Ocean Yearbook 10* in Summer 1992. Volume 10 will be planned in relation to the 20th Anniversary of the IOI and Ocean Development 2000.

PIM Proceedings

The Proceedings of *PIM XIV:* "Maritime Transport and the Law of the Sea" have been published by the Government of Malta Press. They will be distributed by the Soviet Maritime Law Association in Moscow on the occasion of *PIM XVII.* This 600 page volume contains all the papers which were presented at the conference in 1985 as well as the Summary and Conclusions.

The first volume of the Proceedings of *PIM XV* on "The Law of the Sea: The Past Two and the Next Two Decades" has been published by the University of Malta Press.

The first volume of the Proceedings of *PIM XVI* on "Ocean Technology, Development, Training, and Transfer" has been issued. This consists of the Summary and Conclusions. The second volume which contains the papers of the conference will be available in the coming months.

IOI Newsletters

Across the Oceans, the first IOI newsletter which began publication in 1982, comes out semi-annually and is currently into its seventh volume. Its main purpose is to maintain contact with Training Program alumni by publishing their communications and keeping them up to date on IOI activities.

It also gives readers an idea of what new issues are being discussed in the current training programs. In addition, it lists the participants who have attended the IOI Training Program during that given year.

IOI News, the latest newsletter, began in the Summer of 1988 and is a quarterly publication. It is distributed free of charge to anyone who wishes to be placed on the mailing list. Its aim is to make the IOI and its work better known and to forge new friendships and cooperative arrangements to work together to protect and wisely manage the common heritage of mankind—the oceans. It gives details of projects which IOI is undertaking and provides information regarding the Institute's current activities.

IOI Occasional Papers

The IOI Occasional Paper series appears irregularly and on an ad hoc basis. Ten have been published to date. Although no new Occasional Papers were published in 1988, a proposal has been made that a prestigious IOI lecture could be developed during the coming years which should be delivered by a world-renowned personality at the University of Malta, or at other venues in connection with *Pacem in Maribus*. Such IOI lectures could be published as Occasional Papers.

CONCLUSION

The above Report on the activities of the International Ocean Institute has endeavored to summarize the work which has taken place over the past eighteen months or is on-going in the four principal sectors of IOI activities—training, conferences and seminars, research, and publications. The IOI Report in *Ocean Yearbook 9* will follow up on these projects and programs.

Reports from Organizations

The International Whaling Commission—Forty-first Annual Meeting, June 12–16, 1989, San Diego, California

I. PARTICIPATION

The forty-first annual meeting of the International Whaling Commission was attended by delegates from 28 member states, 58 international nongovernmental organizations, and several government and intergovernmental organizations, and several government and intergovernmental observers.[1] The main subjects of the meeting were as follows: Indian Ocean Sanctuary, comprehensive assessment of whale stocks, whale stocks, aboriginal subsistence whaling, national research programs involving whaling under special (scientific) permit, socioeconomic implications of the moratorium, so-called small cetaceans, revision of the Convention, finances, and coming meetings.

II. THE INDIAN OCEAN SANCTUARY

The Indian Ocean Sanctuary was to cease as of October 24, 1989, unless the forty-first meeting of the Commission decided otherwise. While Japan and Iceland held the opinion that the sanctuary should be discontinued, the Technical Committee approved and amended schedule paragraph 7 to read, "The prohibition will apply until 24 October 1992 until the Commission decides otherwise," and the Commission approved this amendment by consensus.[2] Thus, the Indian Ocean Sanctuary duration has been extended another 3 years.

 The Commission was informed of the Scientific Committee's observation that there was inadequate coordination between scientists in the Indian Ocean, and research vessels encountered access problems into waters under national jurisdiction. The Commission finally approved by consensus a recommendation on scientific coordination in the Indian Ocean as follows:[3]

> The initiation, development, and fulfillment of research, especially nonlethal research, on whale populations in the Indian Ocean is a long-term process calling for cooperation among states, research groups, and organizations at several levels:
> 1. for assisting countries with little previous experience in cetacean research to acquire skills and develop national research groups;

---

 1. This report is an abridged version of the Observer's Report of the forty-first annual meeting of the International Whaling Commission prepared by Ole Lindquist, Akureyi, Iceland, for the International Ocean Institute, Valletta, Malta, June 1989.
 2. IWC/41/23.
 3. IWC/41/5:2 and Appendix 2; and IWC/41/5:21.

2. for coordinating methods, exchanging materials, pooling data and cooperatively evaluating results; and

3. facilitating access by scientists, including scientists from other countries.

The Commission recommends that the Commission specifically authorize the Secretariat to provide good offices:

a. to assist with arrangements for such cooperation, and to consult with the secretariats of other organizations such as UNEP, Unesco/IOC, and appropriate regional bodies, when it is thought this would be helpful;

b. to encourage Indian Ocean coastal states, and other interested states, to take appropriate steps to strengthen and better coordinate their efforts in this matter.

III. COMPREHENSIVE ASSESSEMENT OF WHALE STOCKS

A. Stock Identity and Stock Boundaries: North Atlantic Fin and Minke Whales—New Management Procedures

The Scientific Committee stated that "if asked to consider stock boundaries today it would not make the same decision as in 1976. However, there is still insufficient evidence to determine where, if anywhere, stock boundaries should be drawn."[4] The Scientific Committee was presented with an analysis of morphometric data from minke whales caught throughout the North Atlantic and once again agreed that "resolving the question of stock identity was the highest priority for future research" of the West Greenland/North Atlantic minke whales.[5]

The West Greenland fin whale stock was unable to be classified by biological stock identity boundaries.[6] There is a continuous distribution of fin whales observed around southern Greenland, and one discovery mark from a fin whale marked in Newfoundland waters in 1979 was recovered from the whale in Icelandic waters.[7] "Achievement of management objectives will require a more flexible and sophisticated approach to the questions of 'stocks' than in the past. Stock delineations should be evaluated regularly, as new techniques (for example in the field of genetics, individual identification, and satellite tagging) provide new information on the uniqueness of populations and rates of interchange between them."[8]

B. Demographic Parameters of Stocks Depleted to Low Levels: The Southern Hemisphere Right Whales; Blue and Humpback Whales off Western Iceland

"It was agreed that useful information could be obtained from the estimation and monitoring of demographic parameters of stocks depleted to low levels, (such as the

4. IWC/41/4:9; IWC/41/15:4.
5. IWC/41/4:12.
6. IWC/41/4:13.
7. IWC/41/4:50.
8. IWC/41/4:19.

right whales) and of the gray whale, for which it is anticipated that a reduction in population growth rate will soon occur. Identified parameters were: calving interval, age at first parturition, and adult and juvenile survival rates. Because of the time involved for such studies, they will not be of assistance for the 1990 Comprehensive Assessment, but will be of use in the near future."[9]

"New rates of population increase were reported from several stocks. South African right whales were estimated to have increased at 7.1% p.a. from 1971 to 1987 (SC/41/PS3), Argentinian right whales at 7.6% p.a. from 1971 to 1986 (SC/41/PS17), and blue and humpback whales off western Iceland at 4.9% and 11.6% respectively, from 1970 to 1988 (SC/41/0:22). In the last case, the authors considered the rate of increase of the humpback whales not to represent that of the population."[10]

C. Benign Research

The trend toward increasing importance and diversification of benign research techniques so clearly established during the past few years continues. Radio telemetry, both conventional and satellite based, can provide information of animals' distribution; acoustic methods can be successfully applied to population estimation; chemical contaminants and fatty acid profiles can be determined from biopsy samplings; and genetic information from morphometric studies are all useful and complementary methods to increase our understanding about whales. Regarding the recovery of discovery marks from whales (two feet long steel rods shot into the whales), the Scientific Committee agreed "that whale mark data can provide information on current stock boundary validity, but the small sample sizes and the fact that marking and recovery usually took place in the same area limits the value of the results . . . [however,] as in the case of other techniques, marking results should not be seen in isolation, but combined with information obtained from available benign techniques, particularly molecular genetics and photoidentification."[11]

D. Management Procedures

The Scientific Committee has made significant progress in the development of a revised management procedure for whale stocks in spite of the difficulty of the task, and the Committee expects to be able to present a recommended management procedure to the Commission at the 1991 meeting.[12] The Committee noted in its report to the Commission that, in the event of the Commission deciding to allow some commercial whaling from 1990, "any interim [management] procedure so developed might be considerably inferior to the one envisaged at the end of the development process outlined in [IWC/41/4] Table 4, in part since efforts would have to be diverted to allow time for the interim procedure to be validated before the 1990 meeting."[13]

9. IWC/41/4:29.
10. IWC/41/4:30.
11. IWC/41/4:21.
12. IWC/41/15:1.
13. IWC/41/4:27.

In the plenary debate, Iceland, Japan, and Norway expressed sympathy for considering such an interim management procedure.

IV. WHALE STOCKS

A. Southern Hemisphere Minke Whales

Results from the International Whaling Commission and the Interntional Decade of Cetacean Research (IWC/IDCR) assessment cruise were presented. An analysis of these results and an assessment of this minke stock will be given priority at the Scientific Committee meeting in June 1990.[14]

B. Antarctic Blue, Fin, Sei, Humpback, and Sperm Whales

Estimates were provided by the Scientific Committee based on sightings made during the 1978/79–1985/86 IWC/IDCR Southern Hemisphere minke whale assessment cruises:

blue	453	(CV 0.84)
humpback	4,047	(CV 0.28)
sperm	3,059	(CV 0.56)
fin	2,096	(CV 0.47)
sei	1,498	(CV 0.46)

The Committee noted that a large proportion of the catchings and sightings of fin and sei whales were made north of 60° S, and these values probably underestimate the abundance of the species. However, for the blue and fin whales at least, "stocks are only a very small fraction of their unexploited levels."[15]

In the Joint Technical and Scientific Committee Working Group on the Comprehensive Assessment of Whale Stocks, the Seychelles concluded that these analyses "show that blue, fin, sperm, humpback, and sei whales are much more depleted that previously believed. This result highlights the over-optimistic nature of earlier [catch per unit effort] CPUE-based assessments."[16]

C. Northeast Atlantic Minke Whales

The Scientific Committee accepted the provisional estimate of 19,112 (CV 0.163) (or 17,014 [CV 0.179]). This stock will be part of the assessment of North Atlantic minke whales, which the Scientific Committee will give priority at the June 1990 meeting.[17]

14. IWC/41/4:46, 81.
15. IWC/41/4:17–18. Note that CV = the estimated statistical variance.
16. IWC/41/15:6.
17. IWC/41/4:9.

D. Central North Atlantic Minke Whales

Central North Atlantic minke whales will be part of the priority assessment for the Scientific Committee's June 1990 meeting.

E. West Greenland Minke Whales

The Scientific Committee agreed that the best estimate of the number of minke whales off West Greenland was 3,266 (CV 0.31).[18] The Committee has given highest priority of future research to stock identity and stressed the importance of continuing to monitor the number of whales off West Greenland. The Committee welcomed the aerial survey conducted as part of the 1989 North Atlantic Sightings Survey (NASS-89) and noted that similar surveys were likely to form the bases for future management of West Greenland minke whales.[19] This stock is also part of the priority assessment of North Atlantic minke whales for the Scientific Committee meeting in June 1990.

F. West Greenland Fin Whales

The best estimate the Scientific Committee could provide for this stock was between 763 and 2,950 whales. "There were insufficient data available to attempt an assessment of this stock."[20] The small number is cause for concern and the Committee was unable to decide whether or not the stock is above the minimum level below which aboriginal catches should not be taken.[21]

G. East Greenland–Iceland Fin Whales

The Scientific Committee accepted an estimate of 11,563 (CV 0.261) as the best available estimate. The Committee noted with pleasure that the second survey of North Atlantic whale stocks (NASS—89) was planned for the summer of 1989.

H. Iceland-Denmark Strait Sei Whales

This stock was a priority for the NASS-89 survey.[22]

18. IWC/41/4:10.
19. IWC/41/4:12.
20. IWC/41/4:13–14.
21. Ibid.
22. IWC/41/4:15.

I. Eastern North Pacific Gray Whales

It was reported that in 1988 150 gray whales were taken from this stock off Chutkotka by the USSR and that the Alaskan Eskimos had also struck and lost one gray whale during the 1988 hunt which in total is below the catch limit of 179 set by the Commission for this stock.[23]

A special Scientific Committee meeting on the comprehensive assessment of gray whales will be held prior to the 1990 annual meeting of the Commission.[24] "The Scientific Committee noted that entanglement in fishing gear and subsequent loss of gray whales may be a significant source of mortality to consider in the assessment. It therefore recommends that the governments of Canada, Mexico, and the U.S.A. be requested to collect and provide past and current information on gray whale entanglements in fishing gear in time for the Special Meeting."[25]

Because of the possibility that there may be some exchange between the western and eastern Pacific stocks of gray whales, the Scientific Committee recommended that historical and current sightings data on the western Pacific stock be requested from the authorities of the Republic of Korea, the People's Republic of China, and the USSR for use at the special meeting.[26]

J. Bering-Chukchi-Beaufort Seas Bowhead Whales

A total of 29 bowhead whales were struck from the allowed quota of 44 strikes in 1988—23 were landed, and 6 lost, which gave a struck and lost rate for 1988 of 21 percent (slightly lower than for 1987 and 1986).

The estimate for this stock has not changed from the previous estimate of 7,800 (with a confidence level of 5,700–10,600 for this stock in 1988).[27]

K. North Atlantic Humpback Whales

No humpback whales were taken off St. Vincent and the Grenadines during the 1988/89 season. The Scientific Committee noted with appreciation that no females or females accompanied by calves have been taken in this operation. "The Commissioner [for St. Vincent and the Grenedines] stated that no attempt is being made to improve the methods used in the hunt because the Government does not wish to encourage its continuation."[28] When challenged about the specific reservation St. Vincent and the Grenadines lodged in the Convention on the International Trade in Endangered Species of Wild Fauna and Flora (CITES), the Commissioner stated that the country would stand by its IWC commitments.[29]

23. IWC/41/4:16.
24. IWC/41/4:15; IWC/41/8:13; and IWC/41/15:8.
25. IWC/41/4:15.
26. IWC/41/4:15.
27. IWC/41/4:16.
28. IWC/41/5:18.
29. IWC/41/13:1–2; IWC/41/5:10–11.

L. North Atlantic and North Pacific Pilot Whales

A reported total catch of 2,319 so-called small cetaceans took place in the Faeroe Islands in 1988: 1,690 long-finned pilot whales, 3 northern bottlenose whales, 1 orca, 11 harbor porpoises, and the rest apparently being Atlantic white-sided dolphins.[30]

The Scientific Committee commended the great efforts embodied in the international program of research on pilot whales taken in the drive hunt at the Faeroe Islands. Evidence of the existence of two or more stocks or breeding populations is not available, although there are indications that individual pilot whales do not routinely move between major regions in the North Atlantic. The Committee recommended that genetic comparisons be carried out between pilot whales of the Faeroes and from other regions in the North Atlantic using both analyses of isoenzyme allelic frequencies and appropriate analyses of DNA. The area will be surveyed during the NASS-89 survey.

In the pilot whales analyzed, levels of the contaminants DDT, PCBs, and mercury were found to be comparable to those in similarly sized whales from other parts of the North Atlantic, while cadmium levels were very high in some pods.[31]

In the western North Atlantic, the population of pilot whales off Newfoundland was estimated in 1982 at 6,700–19,600. It has not been possible to evaluate the surveys off the northeastern USA and estimates range seasonally from 4,000 to 10,000 and 10,000 to 12,000 due to lack of estimates of variances and details concerning survey design and data analysis.

Noting that catches of pilot whales in Japanese waters are changing in stock composition with the Japanese small-type whaling industry expanding to include the southern form stock, "the [Scientific] Committee considers it urgent that the stock-structure question be addressed before the catches are increased."[32]

V. ABORIGINAL SUBSISTENCE WHALING

Block quotas are currently in force under the aboriginal subsistence proviso for the Alaskan Eskimo bowhead hunt (1989–91), the USSR/Siberian Eskimo gray whale hunt (1989–91), and the St. Vincent and the Grenadines/Bequia humpback hunt (1987/88–1989/90). Therefore, this year the commission had only to set aboriginal subsistence whaling catch limits for the Greenlandic take from the central Atlantic stock of minke whales and the fin and minke whales from the West Greenland stocks.

The Commission approved by consensus that for the years 1990–92 an annual catch limit of 12 whales may be taken from the central Atlantic management stock of minke whales, and for the two seasons 1990–91, a total of 190 minke and 42 fin whales may be taken from the West Greenland stocks with a maximum of 100 and 23, respectively, in each year.[33]

---

30. IWC/41/4/Annex H. Table—Reported Catches of Small Cetaceans in 1988; J. J. Gibson-Lonsdale, *Pilot Whaling in the Faeroe Islands*, a fourth report by the Environmental Investigation Agency (EIA)(London: EIA, 1988), p. 5.

31. IWC/41/4:74.

32. IWC/41/4:75.

33. IWC/41/13:1–3.

Denmark proposed to the Technical Meeting (seconded by Iceland and Japan) that the schedule classification of West Greenland stock of minke whales should be changed from "Protected Stock" to "Unclassified" as the Scientific Committee agreed that it was not possible to recommend a classification for this stock.[34] Consensus could not be reached, and the matter has been deferred to next year's meeting.

VI. NORWEGIAN RESEARCH PROGRAM INVOLVING WHALING UNDER SPECIAL (SCIENTIFIC) PERMITS: RESOLUTION ON THE NORWEGIAN PROPOSAL

The Norwegian representative presented to the Scientific Committee a research program involving the take of 20 minke whales in 1989 and a possible future annual catch of some 300 whales. The Scientific Committee, however, limited its discussion only to the 1989 proposal.[35] The proposal attracted strong criticism from most member representatives for being imprecise in terms of research objectives and methodologies, but it received support from Icelandic and Japanese scientists.[36]

The Scientific Committee also discussed at length the experiments carried out off Norway last year and the proposed work for the coming year concerning the marking and chemical immobilization by darting of minke whales together with subsequent observation from helicopter with a view to the surgical implantation of satellite transmitters.[37]

The Commission adopted a "Resolution on Norwegian Proposal for Special Permits" (see table 1, Vote 1; IWC/41/27/Rev. 1) that concluded

> [The COMMISSION] CONSIDERS; that the proposed take of minke whales in the North Atlantic under the research programme described in SC/41/NHMi12 does not satisfy all the criteria specified in both the 1986 Resolution on Scientific Research programmes, particularly in that the proposed research is not adequately structured so as to contribute to or materially facilitate the Comprehensive Assessment; neither has it been established that the proposed research addresses critically important research needs;
>
> and accordingly INVITES the Government of Norway to reconsider the proposed take on minke whales in 1989 under Special Permit.[38]

VII. ICELANDIC RESEARCH PROGRAM INVOLVING WHALING UNDER SPECIAL (SCIENTIFIC) PERMITS: RESOLUTION ON THE ICELANDIC PROPOSAL

The Scientific Committee has discussed essentially the same research program over the last 4 years.

In 1988, Iceland took 68 fin whales and 10 sei whales under a scientific permit.

34. IWC/41/5:10; IWC/41/4:11.
35. IWC/41/4:63.
36. IWC/41/4:63–66.
37. IWC/41/4:7–9.
38. IWC/41/27/Rev. 1.

TABLE 1.—VOTING RECORD OF THE TECHNICAL/PLENARY SESSION,
FORTY-FIRST INTERNATIONAL WHALING COMMISSION ANNUAL
MEETING, JUNE 25, 1989, SAN DIEGO

Country	Vote 1[a]	Vote 2[b]
Antigua (present but no voting rights)	. . .	. . .
Argentina	Yes	Abstain
Australia	Yes	Yes
Brazil	Yes	Yes
Chile	Abstain	Abstain
People's Republic of China	Abstain	Abstain
Costa Rica (absent, voting rights suspended)	. . .	. . .
Denmark	Yes	Yes
Egypt (absent)	. . .	. . .
Finland (absent)	. . .	. . .
France	Abstain	Abstain
Federal Republic of Germany	Yes	Yes
Iceland	No	No
India	Yes	Yes
Ireland (absent)	. . .	. . .
Japan	No	No
Kenya (absent/suspended)	. . .	. . .
Republic of Korea	Abstain	Abstain
Mexico	Abstain	Abstain
Monaco (absent)	. . .	. . .
The Netherlands	Yes	Yes
New Zealand	Yes	Yes
Norway	No	No
Oman	Yes	Yes
Peru (absent/suspended)	. . .	. . .
St. Lucia	No	No
St. Vincent	No	No
Senegal (absent/suspended)	. . .	. . .
Seychelles	Yes	Abstain
Solomon Islands (absent/suspended)	. . .	. . .
South Africa	Yes	Abstain
Spain	Abstain	Yes
Sweden	Yes	Yes
Switzerland	Yes	Yes
USSR	No	No
United Kingdom	Yes	Yes
U.S.A.	Yes	Yes
Uruguay (absent/suspended)	. . .	. . .

Source.—"Votes, Day 4," *ECO Whales and Whaling* 46, no. 5 (June 16, 1989): 1.

Note.—Vote 1 Total: 15 Yes, 6 No, 6 Abstain; motion passed. Vote 2 total: 13 Yes, 6 No, 8 Abstain; motion passed.

[a]Resolution on Norwegian Proposal for Scientific Whaling Permits (IWC/41/27/Rev. 1). Proposed by Australia, France, Germany, Netherlands, New Zealand, Oman, Seychelles, South Africa, Sweden, Switzerland, U.K., and U.S.A.

[b]Resolution on the Proposed Take by Japan of Whales in the Southern Hemisphere Under Special Permit (IWC/41/28/Rev. 2). Proposed by Australia, Federal Republic of Germany, Netherlands, New Zealand, Sweden, Switzerland, U.K., and U.S.A.

Iceland reported to the Scientific Committee on its current cetacean research, inter alia: upward trends in age at sexual maturity in fin whales; body condition of fin and sei whales; abundance of blue and humpback whales; ecological role of whales in Icelandic waters; food-web model; studies on genetic variation in fin and sei whales; and alpha-globin HVR DNA fingerprinting.[39]

The Icelandic representatives stressed the importance of the biological parameters obtained from the catch for management purposes. However, while studies such as the work on genetic markers would provide a useful contribution to the Comprehensive Assessment, this study requires the killing of the whale. It was noted by the Scientific Committee that nonlethal methods of examining genetic markers should be considered.[40]

It was concluded that the most valuable information from the Icelandic program was coming from the nonlethal sections, particularly the sighting surveys.[41]

The Commission adopted a "Resolution on the Icelandic Proposal for Scientific Catches" by consensus, and concluded

> Now THEREFORE the Commission:
>
> INVITES the Government of Iceland to reconsider the proposed take of fin whales in 1989 under Special Permit, in the light of the critiera specified in the 1986 Resolution on Special Permits for Scientific Research and the 1987 Resolution on Scientific Research Programmes and the comments of the Scientific Committee.[42]

The Resolution IWC/41/26 was adopted just around noon on Wednesday, June 14, 1989; some 5 hours later the Icelandic commissioner announced that the Government of Iceland in the meantime had reconsidered its plans as requested by the Commission and had decided to reduce the catch to 68 fin whales for which permits had been issued. While the United States delegation instantly applauded the Icelandic response (which many delegates and observers felt had been long prepared), the United Kingdom and the Netherlands voiced their disagreement with the statement by the United States.

VIII. JAPANESE RESEARCH PROGRAM INVOLVING WHALING UNDER SPECIAL (SCIENTIFIC) PERMITS: RESOLUTIONS ON THE JAPANESE PROPOSAL

The Japanese 1988/89 studies in the Antarctic took place between January 12 and March 31, 1989. A paired-vessel sampling strategy was employed this season as opposed to the single vessel strategy of the previous season. The Japanese delegates presented to the Scientific Committee analyses of biological parameters, statistical values, and sightings data.

Members of the Scientific Committee commented that the data presented did not address the key issue that had been brought up in previous Scientific Committee

39. IWC/41/4:50–51.
40. IWC/41/4:62.
41. IWC/41/4:51.
42. IWC/41/26.

discussions on estimation of age-specific natural mortality and trends of age and abundance data, sought under the Japanese Research Program.[43] That was, not whether catch-at-age data can provide estimates of net recruitment rates when used in conjunction with abundance data, but whether such estimates are superior to those obtained from abundance data alone.

The Japanese research permit proposal for 1989/90 noted that the proposal should be viewed in conjunction with the original proposal submitted by Japan in 1987[44] and that it contained modifications and plans for the taking of 400 minke whales in the Antarctic Management Area IV.[45]

The Scientific Committee commented that

the program implied that estimates of biological parameters could be used to estimate parameters of direct management interest, such as MSY [maximum sustainable yield], MSYL, and RY [replacement yield]. The Committee stated that the proposal needed to show how the research will answer questions which need to be answered, and how other components of the problem can be solved. Accordingly, regardless of whether age-dependent mortality could be reliably estimated or not, the proposal did not show that the whole problem of estimating the relevant management parameters constituted a feasible objective.[46]

The Committee concluded that the proposed research catch would neither contribute information, which is essential for the future rational management of these stocks, nor contribute to progress on the Comprehensive Assessment.[47] They noted that, while improvements had been made in the methodology in order to secure random sampling, the important estimation of the absolute levels of mortality could only be achieved with a large variance at best, and any estimated replacement yields would have enormous variances and be of impractical value.[48]

In September 1988 Japan submitted the proposed Special Permits 1988/89 for review by the Scientific Committee under its intersessional mail procedure. Permits were issued in December 1988. In January 1989, a resolution was introduced by the United Kingdom, seconded by Australia and New Zealand on the matter, and a postal vote was requested. The resolution recommended that the Government of Japan withdraw its permits until the program had been revised and reassessed by the Commission. While the vote received support from a large number of voting states, it failed to be adopted on a technical matter.[49]

A draft resolution on the proposed take by Japan of whales in the Southern Hemisphere under special permit was introduced to the Plenary and was adopted (see table 1, Vote 2; IWC/41/28/Rev. 2) by 13 votes in favor to 6 against, with 8 abstaining.[50] The resolution concluded

43. SC/41/0/1; IWC/41/4:54–55.

44. SC/39/0/4.

45. IWC/41/4:56–57.

46. IWC/41/4:57.

47. IWC/41/4:Annex, p. 2; IWC/41/4:58.

48. IWC/41/4:60.

49. IWC/41/2:7–8; IWC/RG/VJH/17487/3, April 1989; and IWC Rules of Procedures and Financial Regulations, December 1988:4.

50. IWC/41/28/Rev.2.

Now, THEREFORE, the Commission

ACCEPTING that the Scientific committee was not unanimous in its view of the research programme described in SC/39/04, including the improvements described in SC/41/SHMi13(IWC/41/4);

CONSIDERS that the programme does not fully satisfy the criteria specified in both the 1986 Resolution on Scientific Research Programmes, more particularly in that the proposed research is not structured to provide or demonstrate that any existing methodology can solve the problems or satisfy the objectives which have been set, and therefore the proposed research does not contribute information essential for rational management of the stock, neither will the proposed take of minke whales in the Southern Hemisphere in 1989/90 under Special Permit materially facilitate the Comprehensive Assessment, nor has it been established that the proposed research addresses critically important research needs:

INVITES the Government of Japan to reconsider its research programme in the light of the criticisms based on the above-mentioned criteria.

IX. SOCIOECONOMIC IMPLICATIONS OF THE MORATORIUM: SMALL-TYPE WHALING; JAPANESE INTERIM RELIEF ALLOCATION REQUEST

At this year's annual meeting, all work and discussions regarding socioeconomic implications of a zero-catch limit and various kinds of small-type whaling fused.[51]

Iceland introduced a provisional description of the potential long-term socioeconomic impact of a zero-catch limit. This varies from a reduction in large-scale whaling, with the whalers absorbed into fishing and other employment, to small-type whaling, at present temporarily suspended, in which some whalers have shifted into fishing but unemployment in some small coastal communities has increased. The Icelandic delegation pointed out the potential implications of a zero-catch limit for Iceland's overall marine research and management in the context of ecological and multispecies approach.[52]

Norway stated that it had a very long tradition in small-type coastal whaling.[53]

Japan introduced several documents to the working groups, pointing out inter alia "that the moratorium affected the spiritual, psychological, and cultural well being of people who depended upon whaling."[54] Japan has repeatedly called for "respect of the human rights" of the people of the communities in Japan with "coastal whaling tradition." Japan recognizes the "justified exemption of aboriginal subsistence whaling from the moratorium due to the human rights and cultural needs of the dependent communities which it believes are in many respects similar to its own coastal communities."[55]

Many delegations in the Working Group, Technical Committee, and Plenary meetings expressed sympathy for individual cases of hardship but stated that changes in circumstances that lead to socioeconomic difficulties were not confined solely to

51. IWC/41/5/Appendix 3 Revised.
52. IWC/41/19:2; IWC/41/5:12.
53. IWC/41/16:5–6; IWC/41/5:12.
54. IWC/41/19:1.
55. IWC/41/5:12, 10.

whaling but were also found in other activities and occupations and that governments had various means by which they could successfully mitigate such difficulties even though it might mean some permanent changes had to be made.[56]

Discussion similar to last year's debates attempted to define a category of small-type whaling that was neither commercial nor aboriginal subsistence whaling. However, it was not possible for the group to reach consensus. Iceland and Norway stated that they believed that there was a lack of understanding of the issues involved and that the Working Group should continue its efforts to define the meaning of small-type whaling for small coastal communities.[57]

Japan then tabled an Interim Relief Allocation Request:[58]

> For this reason, in order to respect the human rights of those peoples, to ease human suffering, and to reduce the pressure on other cetacean species, Japan requests the following interim relief allocation.
> An interim relief allocation for one year from July 1, 1989, of 320 minke whales from the Okhotsk-West Pacific stock, to be taken off the coast of Japan and entirely within Japan's 200-mile zone.

Some governments reiterated their position from the debate over Japan's similar request at last year's meeting. The Commission did not want to either prejudge the Comprehensive Assessement or compromise the present moratorium, and the Japanese request has been deferred until the next annual meeting of the Commission.

X. SO-CALLED SMALL CETACEANS

A. IWC Competence Over So-called Small Cetaceans: Baird's Beaked Whales; Faeroese Hunt of Pilot Whales; and Atlantic White-sided Dolphins

The Japanese delegation in the Scientific Committee reiterated its position concerning small cetaceans: "That is, the activities of the Commission with the population management of whales should be limited to the matters concerning whales listed on the Nomenclature of the International Whaling Conference Final Act (1946). However, we can support the Resolution adopted by the IWC at its 32nd annual meeting in 1980 that the Scientific Committee of the IWC is recommended to consider the status of small cetaceans and provide such scientific advice as may be warranted to Contracting Governments."[59]

Japan "believes that the Convention was intended to regulate large whale species which therefore does not include Baird's beaked whale since this was not included in the Annex of Nomenclature to the Convention. The 200-n.mile Zone jurisdiction gives coastal states responsibility and it has its own domestic measures."[60]

There was much debate and disagreement with the Japanese assertion and the

56. IWC/41/19:2–3.
57. IWC/41/5:12–13.
58. IWC/41/29.
59. IWC/41/4/Annex U.
60. IWC/41/5:3.

majority of states took the view that the Convention applied to all whales, that the Schedule was not restrictive, and that the Annexed Table of Nomenclature was for information only and has no legal status of its own. Article 65 of the United Nations Law of the Sea Convention was also discussed, and the majority believed that the IWC is the competent international organization and can regulate all whales.[61]

The Human Killing Working Group discussed the Faeroese pilot whale hunt. Denmark refused to give current information about the hunt to the group on the grounds that it had provided member governments with that information over a prior 3-year period but agreed to provide information on a bilateral basis if requested.[62] Denmark reiterated its opinion that the IWC had no regulatory competence with regard to pilot whales. It was noted in the meeting that Denmark refused to give information to the Commission and that the Faeroese authorities did not encourage outsiders to observe the pilot whale hunt. Implicit in this controversy is the concern about the Faeroese take of great numbers of other so-called small cetaceans, for instance, the Atlantic white-sided dolphins. The three bottlenose whales taken by the Faeroese in 1988 were from a Protected stock.[63] The result of the current police investigation is to be forwarded to the Secretariat of the IWC as soon as possible.

B. Increased Japanese Take of Dall's Porpoises, Baird's Beaked Whales, and Pilot Whales

Japan stated to the Technical Committee:

> Although it [Japan] had asked the IWC for an interim allocation for its small-type coastal whaling operation last year, no assistance had been given and it had interpreted the results of postal responses to its request as leaving the matter to the discretion of Japan. It had therefore increased the national catch limit for Baird's beaked whale from 40 to 60 whales as an emergency measure to alleviate the severe local cultural and socio-economic impacts of the loss of the minke whale catch. It would reduce the catch if minke whales were permitted to be taken again.
>
> The Japanese Government had taken measures to alleviate local hardship by increasing national quotas for Baird's beaked whales and carrying over an unused pilot whale quota as an emergency measure.[64]

The Scientific Committee also expressed grave concern about the unsustainable catch of Dall's porpoise in the Japanese hand-harpoon fishery. In 1987, 13,000 Dall's were harpooned; in 1988, 39,000.[65] The porpoise meat is being substituted for large-whale meat. Two stocks of Dall's, of unknown population size, are being targeted, and, while there may be some overreporting in the 1988 catch estimate, no account has been taken of deaths resulting from the international gill-net fishery which traps great

61. IWC/41/5:3–4.
62. IWC/41/18:4; IWC/41/5:18.
63. SC/41/Prog. Rep. Denmark; IWC/41/4:75; IWC/41/6/Appendix 1; and IWC/41/5:3.
64. IWC/41/5:3; IWC/41/5:11.
65. IWC/41/Prog. Rep. Japan.

numbers of Dall's each year. The Committee felt urgent measures must be taken to reduce the take of Dall's and recommended that catch statistics be collected and reported on a stock-by-stock basis by Japan and that the Republic of Korea be requested to report to the IWC by-catches of Dall's and other cetaceans in its squid gillnet fishery.[66]

Japan commented that the 1982 commercial whaling moratorium had caused whale meat prices to increase and that it was difficult to change traditional eating habits. Japan stated that one possible consequence of such a shortage might be an increase in porpoise catches.[67]

C. By-catches of Small Cetaceans: Drift Gill-Net Fisheries; Effects of Oil Spills

Increased cooperation between states and industries, and with the assistance of the international observer program under the Inter-American Tropical Tuna Commission (IATTC) in the eastern North Pacific, has produced an estimated kill of 78,927–84,881 dolphins in 1988, which is 20%–25% down on the 1987 estimates and 40% down on the 1986 estimates.

The Scientific Committee noted the continued high kill of eastern spinner dolphin (about 20,000 in both 1987 and 1988) and expressed concern about the possible impacts of the kills of this stock and the stocks of offshore spotted and common dolphins and strongly urged that all agencies and states involved continue their efforts to reduce the kill.[68]

The Committee discussed the potential problems for cetaceans in the drift gill-net fisheries in the Tasman Sea and the southwestern Pacific Ocean to the east of New Zealand. These waters are the migratory routes for sperm, humpback, right, Bryde's, and other whales. Cetacean entanglement in drift nets off the east coast of New Zealand was first reported to biologists in 1987. The Committee urged the Commission to request involved countries to place scientific observers onboard the fishing vessels to monitor cetacean mortality (there was a reservation to this by the Japanese government).

Japan commented to the committee that cetacean kill in drift gill nets is not just a Japanese problem since there are many fleets operating on the High Seas. Japan stated that it intended to participate positively in appropriate international conferences on the problem.

Australia supported international action and has stopped all gill-net fisheries in waters under its own jurisdiction.[69]

The Technical Committee was informed by the Chairman of the Scientific Committee, at the request of Argentina, that 10–12 states were involved in the tuna fishery in the eastern Pacific. The Seychelles suggested that a listing of these states would help the secretariat to direct matters, but this was strongly opposed by Mexico, France, St. Vincent, Spain, Japan, and Korea, who all believed that dolphins were not in the competence of the IWC.

66. IWC/41/4:77; IWC/41/5:15.
67. IWC/41/16:5.
68. IWC/41/4:76.
69. IWC/41/5:14.

An Ad Hoc Working Group on the Effect of Oil Spills on Cetaceans reported to the Scientific Committee.[70] The Working Group discussed short-term effects of recent well-documented oil spills (*Bahia Paraise*, Western Antarctic Peninsula, January 1989; and *Exxon Valdez*, Prince William Sound, Alaska, March 1989) and the significant long-term degradation problem reported in Panamanian coastal waters as a result of an oil spill.[71] The report concluded: "Present emergency plans for oil spill containment and clean up have not proved to be effective. In general, experience shows that responses to oil spills are slow and inadequate. In remote areas with severe weather conditions (e.g., polar regions), even well-designed contingency plans are likely to be difficult to implement."[72]

The Scientific Committee "recognized the need for appropriate and practical measures to prevent and mitigate the impacts of petroleum exploration, development, and transportation" and "called for the formulation of appropriate and realistic contingency plans for containment of oil spills and mitigation of their effects on the marine environment in general and cetaceans in particular." The Commission approved its recommendation that "detailed data on oil spills and their effects be acquired in a timely manner and be made readily available to provide documentation of the effects of oil spills on wildlife and to allow for appropriate rescue and rehabilitation programs for cetaceans."[73]

XI. REVISION OF THE CONVENTION

A Working Group had been established to examine questions related to the operation of the International Convention on the Regulation of Whaling. Several governments had stated at last year's meeting that the Convention needs to be revised because of changed circumstances. Some member governments cautioned that to open the Convention to revision might lead to even less satisfactory results. The group prepared a list of questions at the last meeting to guide the group's deliberations, and it was decided that member governments should be invited to comment on these questions. The subject is retained on the agenda of the 1990 annual meeting.

Norway expressed to Plenary that the Comprehensive Assessment is a highly significant constitutional development of the Convention and that it will be an important test of its flexibility.

The USSR tabled a draft resolution on the Improvement and Updating of the International Convention for the Regulation of Whaling which was referred for consideration to next year's annual meeting.[74]

---

70. IWC/41/4/Annex T; IWC/41/4:83; and IWC/51/4:16–17.
71. J. B. C. Jackson et al., "Ecological Effects of a Major Oil Spill on Panamanian Coastal Marine Communities," *Science* 243 (January 1989): 37–44; and IWC/41/4/Annex T:2.
72. IWC/41/4/Annex T:1.
73. IWC/41/4:83; IWC/41/5:16–17; and the Scientific Committee detailed aspects which it urged member governments to pay special attention to in this respect.
74. IWC/41/25, based on IWC/40/25.

XII. FINANCES AND ADMINISTRATION

Seven contracting governments (Antigua and Barbuda, Costa Rica, Kenya, Peru, Senegal, Solomon Islands, and Uruguay) had their voting rights withheld because of arrears in contributions to the Commission. Four other governments (Argentina, Chile, Egypt, and France) are late in paying their 1989 dues. This amounts to a total shortfall of 25% (UK Sterling £105,000) of the projected 1988/89 income and expenditure based on 100% payment of dues. Accumulated debts total more than £341,000 with some £85,000 interest to be added. Three further members (Belize, Dominica, and the Philippines) owe the Commission a further £84,000 plus approximately £11,000 in interest. Mauritius and Jamaica have paid their contribution arrears in full up to their withdrawal from the Commission.

At the recommendation of the Finance and Administration Committee, the Commission approved substantial increases in contributions and fees to the Commission.[75]

XIII. COMING MEETINGS

These are the upcoming meetings:

Workshop on the Genetic Analysis of Cetacean Populations, La Jolla, California, September 27–29, 1989.

A Special Scientific Committee Meeting on the Comprehensive Assessment of Gray Whales is scheduled to take place before next year's ordinary Scientific Committee meeting.

The forty-second annual meeting of the Commission will be held in Noordwijk, June 25–July 7, 1990, at the invitation of the Government of the Netherlands with the preceding Scientific Committee meeting held at the same venue.

The Government of Iceland will host a special Scientific Committee meeting to assess the East Greenland–Iceland stock of fin whales in October 1990.

The United Nations Environment Programme (UNEP) plans to hold the Meeting to Review the Status and Problems of Small Cetaceans Worldwide in the Soviet Union in early 1990.

A Meeting on Mortality of Cetaceans in Fishing Nets and Traps is proposed to be held about mid-January 1991 at the invitation of the U.S. Southwest Fisheries Center, La Jolla, California.

75. IWC/41/9:2,10,13, and IWC/40/9:3.

Reports from Organizations

Report of the Law of the Sea Institute, 1988 and 1989*

The highlights of the year for the Institute were, in order of occurrence, the Washington, D.C., workshop in January on the tanker war in the Persian/Arabian Gulf; the twenty-second annual conference of the Institute at the Narragansett Bay Campus of the University of Rhode Island, the Institute's birthplace; and finally, the Moscow workshop, held in November and December.

The 1989 conference, as well, can be encompassed in this report—it is to be held at Noordwyke aan Zee, the Netherlands, in June, and will concentrate on the roles of international organizations in the ocean regime.

WORKSHOP ON THE PERSIAN/ARABIAN GULF TANKER WAR

In late January 1988, just under 100 persons from the legislative and executive branches, the diplomatic corps, and certain nongovernmental public organizations met at the Carnegie Endowment for International Peace Building to analyze the fate of the world's merchant fleet in the Persian Gulf. The meeting was under the combined sponsorship of the Law of the Sea Institute and the Council on Ocean Law.

The President of the Law of the Sea Institute, Professor Thomas Clingan, moderated the discussion, which featured, in alphabetical order, Burdick Brittin of the Council on Ocean Law, speaking on "The Need for Modernization of the Law of Naval Warfare"; Richard J. Grunawalt of the Naval War College, speaking on "The Rights of Neutrals and Belligerents"; Louis Henkin, of the faculty of law at Columbia University, offering commentary; Philip J. Loree, Chairman of the Federation of American Controlled Shipping, speaking on "Compensation for Attacks against Non-belligerent Vessels: The Hercules Case"; Thomas McNaugher, of the Brookings Institution, speaking on "The Evolution of U.S. Policy in the Persian Gulf"; and Elliot Richardson, Chairman of the Council on Ocean Law, speaking on "U.N. Protection for Non-belligerent Vessels."

Disregard by belligerent nations of the rights of neutral merchant vessels in international trade is a problem as old as the law of the sea itself, intimately intertwined with the fate of nations in war and in peace. That it persists today is perhaps not too surprising, but it is complicated by the density of international relations as compared with former years, the legal separation of many ships from naval protection through the use of flags of convenience, and the heavy international economic dependence upon oil from that strategic waterway.

*EDITORS' NOTE.—This report was supplied by Dr. Scott Allen, Associate Director, Law of the Sea Institute, William S. Richardson School of Law, University of Hawaii at Manoa, 2515 Dole Street, Honolulu, Hawaii 96822.

When the rights of merchant vessels are transgressed by belligerents, the best protection is probably naval power; but such protection is limited, under normal circumstances, to ships of the same flag. How to provide effective security for peaceful trade was the essential question before the panel. The discussants sought solutions for these old problems, reset in the modern political and economic context, through new principles of state practice, the moral authority of the United Nations, and traditional principles of neutrality.

VISITATION TO THE INSTITUTE BY SOVIET SCHOLARS

In January/February of 1988, the Institute was visited by four members of the Soviet Academy of Sciences: Artemy Saguirian, Nikolai Scherbina, Victor Vrevsky, and Alexander Vorontsov. The Institute Director, John Craven, and the Associate Director, Scott Allen, accompanied them in Washington, D.C., for contacts with the Hawaii Congressional delegation, the staff of the Law of the Sea Office of the Department of State, and the National Academy of Sciences prior to their arrival in Hawaii.

While in Hawaii, these Soviet scholars joined with Institute director, John Craven, former board member Jon Van Dyke, Thomas A. Clingan, Institute President and formerly Deputy Chair of the U.S. Delegation to the Third U.N. Conference on the Law of the Sea, and Edward Miles, Institute Board Member and Director of the University of Washington's Institute for Marine Studies, in a discussion of disagreements relating to freedom of navigation. Though this panel had been scheduled long before, it occurred in fact the day after the *USS Caron* and the *USS Yorktown* were physically "shouldered" by a Soviet vessel while in innocent passage in the Soviet territorial sea 9 miles south of Sevastopol.

This was no doubt the first nongovernmental discussion between U.S. and Soviet scholars of the issues raised by this event. That same evening, these U.S. and Soviet scholars appeared on the educational television program, *Dialogue,* giving public exposure to the same issues.

THE TWENTY-SECOND ANNUAL CONFERENCE: *NEW DEVELOPMENTS IN MARINE SCIENCE AND TECHNOLOGY: ECONOMIC, LEGAL, AND POLITICAL ASPECTS OF CHANGE*

Edward Miles had said that ocean policy is driven by advance in ocean science and technology and perceptions about the effects of such advance on prospects for obtaining wealth from the sea.

The theme of the twenty-second annual conference of the Institute focused on this central truth in its many ramifications. Meeting at its birthplace for the first time since its departure in 1977, the Institute could with justice review the changes that scientific advance had wrought since its early meetings, the guidance that science might yield to policy today, and the predictive skills of science.

After welcoming remarks from University of Rhode Island President Edward D. Eddy, the conference heard a keynote address from His Excellency Satya Nandan, Undersecretary General of the United Nations, who, in reviewing the suitability of the 1982 Convention in view of scientific changes, recalled that the 1958 Convention on

·the Continental Shelf contains a "consent regime," affirmed the character of the treaty as a "package deal," and asserted that the existence of differences regarding the seabed regime should not be taken to mean that the entirety of Part XI should be revised at a new conference called for the purpose. Rather, he said, there are indications that changed conditions since December 1982 would make revisions possible without modification of the fundamental structure of the seabed regime as contained in the treaty.

Panel I: New Directions in Marine Science and Technology

Chaired by Robert Duce, Thomas Malone, Robert Corell, and John Craven examined global environmenal change, its policy implications, and applications of new science and technologies for exploiting resources and carrying on activities in and under the sea. Craven compared his current view of the state of ocean science and technology with his April 1966 article in the *Proceedings of the U.S. Naval Institute,* in which he predicted, among other things, that extensions of national sovereignty at sea would make neighbors of nations formerly separated by the sea, and new technologies would enable occupation and exploitation of the seabed.

Today, Craven has predictions concerning incipient arms control and inspection at sea, the true (as compared with perceived) major ocean resources, and the elimination of the need to patrol large areas of the EEZ.

Panel II: Nonliving Marine Resources

In this panel, chaired by Tullio Treves, Conrad Welling reviewed new trends in the exploitation of deep seabed minerals; Rick Hoos and Roger Herrera looked at oil exploitation in the North American Arctic; and Dana Yoerger looked at the technical problems and possibilities in exploring historical and archaeological treasures, using the experience of the Woods Hole Oceanographic Institution with the *Titanic* as an example.

Panel III: Living Marine Resources

Edward Wolfe, chair, reported in his introductory remarks that U.S. Secretary of State Schultz and U.S.S.R. Foreign Minister Shevardnadze had recently signed a comprehensive new fisheries agreement, and that the United States would, on the following day, deposit its instrument of ratification of the South Pacific tuna treaty at Port Moresby.

Kenneth Sherman and Martin Belsky discussed, respectively, the scientific and legal aspects of large marine ecosystems, with commentary by Joseph Morgan.

Tucker Scully discussed the operation of the conservation of Antarctic fisheries in respect of the 1982 Convention and other treaties.

Three scholars followed a different substantive trail, that of law enforcement in the EEZs of developing nations. Peter Varghese covered the situation in Australia and the South Pacific island states; Don Aldous investigated the use of new technologies in the island states; and Jahara Yahaya discussed law enforcement in the Malaysian EEZ.

Panel IV: Non-resource Uses of the Ocean

Alastair Couper moderated this panel, in which Edgar Gold examined new developments regarding vessel source pollution and the continuing push for safer ships and cleaner seas. Awni Benham, by way of commentary, discussed open registries, the aspirations of developing nations to participate in world shipping, policy recommendations for developing and developed nations, and maritime fraud. Richard Grunawalt spoke on the legal issues surrounding the Persian Gulf Tanker War. Both he and Edgar Gold admonished the conference that the seamen who sailed through these troubled waters are at the cutting edge of the law of neutrality and the law of the sea, and that they are at great personal risk.

James Baker discussed the impacts of new remote sensor technology, making the point that satellite sensors, while not necessarily as accurate as point sensors, are nevertheless superior in collating regional and global data, providing a summary picture of great value in sensing physical and biological relationships. Further, cost improvements over time will make both satellite data and imagery processing capability cheaper and, therefore, more widely available.

Ralph Chipman commented that remote sensors have heretofore been used primarily for research, rather than operations. Thus, he said, political support for these sensors may dwindle unless concrete and useful results can be demonstrated.

Finally, Kenneth Hinga and Clifton Curtis discussed the technical and legal aspects, respectively, of the emplacement of high-level radioactive wastes on the seabed.

James Griffin spoke at dinner, while cruising in Narragansett Bay, of the relationship of science and technology (the latter tends to drive the former) and the consequent implications for the making of policy.

Panel V: Scientific Issues

John A. Knauss led this panel, which covered issues of access to ocean areas of scientific interest which are now restricted by the EEZ and the extension of the consent regime; and in the second phase, covered issues of maximization of the benefits of new ocean technologies.

Professor Alfred Soons reviewed the specific problems of marine science research as viewed from the perspective of the Netherlands, including the unique experience of the "Snelling II expedition." The perspective of scientists in the United States was presented by David Ross and Judith Fenwick of the Woods Hole Oceanographic Institution.

Opportunities presented by advances in oceanographic data management were presented by Gregory Withee and Douglas Hamilton of NOAA's Oceanographic Data Center. S. I. Rasool of NASDA Headquarters spoke of the efforts of the International Council of Scientific Unions (ICSU) to track global changes. Raymond Godin of the International Oceanographic Commission spoke about the programs of that institution in the same area. Dai Raguang of the State Oceanic Administration in Beijing spoke of the management of oceanographic data in China. Ferris Webster presented a preview of a forthcoming report by the National Research Council's Committee on Geophysical Data, which makes recommendation for improvements in the availability and distribution of such data.

Panel VIA: Ocean Management Concerns of Special Interest

Alasdair McIntyre of the University of Aberdeen considered anthropogenic impacts upon the oceans: dumping of wastes, the injuries caused by shipping, the hazards of exploitation of oil and other nonliving resources, and the changes wrought by fishing in their effects upon the public health and damage to natural habitats.

Stella Vallejo of the Ocean Affairs Office of the United Nations reviewed the history of the first two decades of U.N. involvement in ocean management. Eric Carlson spoke on the role of geographic data bases in coastal and ocean planning. Stephen Olsen of the University of Rhode Island reported on a URI/USAID project aimed at applying U.S. coastal zone planning experience to developing nations.

Panel VIB: Trends in the Law of Maritime Boundaries

Lewis Alexander chaired this panel, in which Philipe Cahier, of the Graduate Institute of International Studies, Geneva; Jean-Pierre Quenedec, University of Paris; and Francisco Orrego Vicuna, Institute of International Studies, Santiago, Chile, examined in some detail the reasonings of courts and arbitration panels in deciding boundary disputes and, from this data, made their analyses of the trends of law in this regard.

Panel VII: Today's Changing LOS Issues

Thomas Clingan chaired this panel, which examined the changing technological, scientific, economic, legal, and political international environment, and evaluated the effectiveness of the 1982 Convention in addressing these constantly changing conditions.

Captain Geoffrey Grieveldinger of the Office of the Secretary of Defense presented the military perspective. His conclusion held that in some areas the Convention is adequate, while in others, it presents a needed framework for further development of international law.

Thomas Mensah of the International Maritime Organization presented the perspective of shipping. He, too, found certain areas in which the rules of the Convention can be used to build new and more adequate rules and international practice.

The fisheries perspective was presented by Lee G. Anderson of the University of Delaware. Professor Anderson found problems in fisheries but believed them solvable with the tools placed in the hands of nations, whether they could be said to be wholly solved at the present or not.

Richard Greenwald of Deepsea Ventures, Inc., presented new information on the progress of seabed mining regulation within the United States.

Bryan Hoyle, of the U.S. Department of State, noting that the Convention is not yet in force, nevertheless found that it is threatened by "creative" interpretation and application of its provisions by nations wishing to expand their powers thereunder.

Finally, in his speech to the assembled conference at the closing banquet, Edward L. Miles of the University of Washington's Institute for Marine Studies discussed the reasons why opening a so-called UNCLOS IV would be a poor means to settle the remaining issues which prevent industrialized nations from signing or ratifying the 1982 Convention. His alternative strategies will entrance readers of the proceedings.

WORKSHOP IN MOSCOW

In the closing days of November and the first few days of December 1988, the Law of the Sea Institute and the Soviet Maritime Law Association sponsored a workshop in Moscow which covered U.S. and Soviet positions relating to navigation, freedom of marine scientific research, and straddling fisheries stocks.

TWENTY-THIRD ANNUAL CONFERENCE, NOORDWIJK AAN ZEE, THE NETHERLANDS

The timing for this submission is such that a brief report can be included of the upcoming twenty-third annual conference of the Institute, which will be held in June in the Netherlands. Its theme, "Implementation of the Law of the Sea Convention through International Institutions," is designed to examine the work of many international institutions since the 1982 Convention's signature six years ago, taking stock of progress made so far, and surveying what remains to be done.

OTHER MATTERS

The Institute's program of publication of the proceedings of each of its meetings continues, comprising a continuous record of the development of international ocean policy and law since its inception in 1965. This past year, the Institute has added *Los Lieder* to its professional correspondence circulars, relating the activities of the international ocean policy community.

The Institute was pleased to add to its governing body, the Executive Board, the following members of 1988:

Professor R. P. Anand, of Jawaharlal Nehru University, Delhi, India;

Dean Jeremy T. Harrison, William S. Richardson School of Law, University of Hawaii (University of Hawaii appointee);

Mr. Philip T. Major, Ministry of Agriculture and Fisheries, New Zealand; and

Professor Louis Sohn, Faculty of Law, University of Georgia.

Report of the Eleventh Consultative Meeting of Contracting Parties to the Convention on the Prevention of Marine Pollution by Dumping of Wastes and Other Matter, October 3–7, 1988

INTRODUCTION

The Eleventh Consultative Meeting of Contracting Parties to the Convention on the Prevention of Marine Pollution by Dumping of Wastes and Other Matter, 1972,[1] convened in accordance with Article XIV(3)(a) of the Convention, was held at IMO Headquarters, London, from 3 to 7 October 1988, under the chairmanship of Mr. G. L. Holland (Canada). Ms. S. Nurmi (Finland), and Vice-Admiral H. A. da Silva Horta (Portugal) were Vice-Chairmen.

CURRENT STATUS OF THE LONDON DUMPING CONVENTION

The Meeting took note of the report of the Secretary-General, prepared on 26 July 1988 (LDC 11/2), concerning the current status of the London Dumping Convention and of the 1978 and 1980 amendments thereto, noting that as of that date sixty-two Governments had ratified or acceded to the Convention. The Meeting further noted information provided by the Secretariat concerning steps currently being taken to ascertain the status of Costa Rica and San Marino in respect of which no formal notification as to their deposit of instruments of acceptance of the Convention had been received by the Secretary-General from the Governments of Depositary States. These two countries had been frequently listed as Contracting Parties to the London Dumping Convention in tables prepared by other United Nations organizations (LDC 11/2/2).

CONSIDERATION OF THE REPORT OF THE SCIENTIFIC GROUP ON DUMPING

The Secretariat briefly summarized the reports of the tenth and eleventh meetings of the Scientific Group on Dumping, drawing attention to those parts of the reports

---

1. This report is excerpted from "The Report of the Eleventh Consultative Meeting" November 8, 1988 (LDC 11/14). The Convention on the Prevention of Marine Pollution by Dumping Wastes and Other Matter is also known as the London Dumping Convention and is referred to in this report as the London Dumping Convention (LDC) or simply, the Convention. For further information about the London Dumping Convention, contact The Secretariat, London Dumping Convention, International Maritime Organization, 4 Albert Embankment, London SE1 7SR, England.

(LDC/SG 10/11, LDC/SG 11/13) which require particular action by the Consultative Meeting (LDC 11/3). The outgoing Chairman of the Scientific Group, Mr. R. Boelens (Ireland), provided a comprehensive review of activities carried out since the Tenth Consultative Meeting, highlighting the main developments and recommendations emanating from the Scientific Group. These are reflected in the following paragraphs, together with the actions taken thereon by the Consultative Meeting.

Review of the Position of Substances in Annexes I and II of the Convention

The Meeting recalled that it had previously adopted in principle the recommendations of the Scientific Group with respect to the position of lead and organosilicon compounds in the Annexes to the Convention. The Meeting noted that no new evidence had since been submitted which would change these recommendations.

Review of the Guidelines for the Allocation of Substances to the Annexes

The Consultative Meeting recalled that, when evaluating the hazard potential of substances and wastes in accordance with the Guidelines for the Allocation of Substances to the Annexes (resolution LDC.19(9)), six key characteristics of a substance would need to be considered, one of which was the bioaccumulative potential, and another the toxicity of a substance. Preliminary advice on the measurement of these parameters referred to the significance of "bioavailability"—a term which had itself not been well defined. The Meeting noted that the Scientific Group had refined its advice on the meaning and measurement of bioavailability and had proposed a new wording for inclusion in the Allocation Guidelines.

Similarly, the Scientific Group had refined its earlier advice on the matter of "environmental exposure," clarifying the fact that exposure becomes significant where concentrations and time elements facilitate harmful effects by substances with potentially harmful properties. A new text on environmental exposure had been prepared by the Scientific Group for inclusion in the Allocation Guidelines.

The Consultative Meeting nevertheless agreed that the revised texts provided sufficient clarification of the terms "bioavailability" and "significant exposure" and, following minor editorial changes, adopted resolution LDC.31(11).[2]

Review of the Guidance for Annex III of the Convention

In accordance with the emphasis given to "bioavailability" in the revised Allocation Guidelines, the Scientific Group had recommended that this term should also be added to the Annex III Guidelines, Sections A4–A6.

In this context it was also recalled that the Tenth Consultative Meeting had agreed to adopt an additional consideration under Annex III, Part A, concerning the ade-

---

2. Resolution LDC.31(11). The resolution is attached to this report as Annex 3, without the supporting guidelines for the allocation of substances to the convention Annexes, owing to page limitations.

quacy of data used to characterize wastes proposed for sea disposal. A draft guideline for the interpretation of the amendment to Annex III (Sec. A9) had been prepared by the Scientific Group. The Consultative meeting adopted the changes to the Annex III guidelines proposed by its Scientific Group.[3]

MATTERS RELATING TO THE INCINERATION OF WASTES AND OTHER MATTER AT SEA

As a result of the work undertaken by the joint LDC/Oslo Commission Meeting, and in the light of additional debate by the Scientific Group, it had been agreed at the eleventh meeting of the Scientific Group (LDC SG 11/13) that there was a need to encourage further research on certain aspects of incineration at sea, including

1. concepts for evaluating destruction efficiency of marine incinerators;
2. the effects on marine ecosystems due to possible impacts with the sea-surface microlayer; and
3. the collection of more data on the composition, persistence, toxicity, and levels of organic emissions.

Amendments to Guidelines of the Annex III of the Convention

In order to reflect within the operational procedures of the Convention, the conditions and considerations that are relevant to decisions on the use of incineration at sea, the Scientific Group recommended that a new guideline should be added to the Annex III Guideline C4—that is the guideline which relates to the practical availability of alternatives to sea disposal of wastes (LDC 11/4/3, paragraph 2 and Appendix). The proposed C4 Guideline on incineration at sea has, as its prime objective, a progressive reduction in the amounts of wastes that require destruction by incineration on land or at sea. The Guideline clearly indicates that incineration at sea should only be considered in the context of an active national waste management programme. In such a context, its use may only be justified on an interim basis pending the availability of other environmentally more acceptable land-based alternatives. The C4 Guideline also emphasizes that incineration at sea must always conform with the Regulations and the Interim Technical Guidelines established under the Convention to control the practice. The Consultative Meeting adopted the proposed changes to Section C4 of the Annex III Guidelines.[4]

Proposed Amendments to the Interim Technical Guidelines

The outgoing Chairman of the Scientific Group noted that his Group, in its efforts to ensure that the Interim Technical Guidelines on the Control of Incineration of Wastes

---

3. Resolution LDC.32(11). The resolution is attached to this report as Annex 4, without the supporting guidelines for the implementation and uniform interpretation of Annex III of the Convention, owing to page limitations.
    4. Ibid.

and Other Matter at Sea are continuously updated to reflect recent knowledge of incineration technology, has proposed a number of amendments to the Guidelines (LDC 11/4/3, LDC 11/4/3/Corr.1).

The Meeting discussed these proposed amendments at length and finally adopted the proposed amendments to the Interim Technical Guidelines on the Control of Incineration of Wastes and Other Matter at Sea.[5]

The Secretariat informed the Meeting of the request of the Scientific Group on Dumping that a composite document on guidance regarding incineration at sea (LDC 11/4/4) be prepared. The Secretariat suggested, and the Meeting agreed, that this task should await the outcome of the Eleventh Consultative Meeting so as to provide the most up-to-date information possible.

Guidelines for the Surveillance of Cleaning Operations

The Secretariat drew attention to the proposed resolution on new Guidelines for the Surveillance of Cleaning Operations Carried out at Sea on Board Incineration Vessels (LDC 11/4/2/Rev.1). That resolution had been considered at the Tenth Consultative Meeting and deferred for final adoption until the Eleventh Consultative Meeting to allow for further consideration of the text. The meeting adopted the referenced guidelines as set out in Annex 6 to this report.[6] It was further noted that Contracting Parties having ratified MARPOL 73/78 would apply the MARPOL requirements for the surveillance of cleaning operations carried out at sea on board incineration vessels, and Contracting Parties not having ratified MARPOL 73/78 would apply the newly adopted LDC guidelines.

Plans for Terminating Incineration at Sea

The observer from the Oslo Commission emphasized that the Commission's decision to terminate incinceration at sea by Contracting Parties to the Oslo Convention and within the Oslo Convention area by 31 December 1994, contained additional controls which were integral parts of that decision. In this connection it was noted that parties to the Oslo Convention should not export wastes intended for incineration in marine waters outside the Convention area, nor allow their disposal in other ways harmful to the environment.

The delegation from Denmark introduced its proposal to phase out incineration at sea (LDC 11/4 and LDC 11/4/Corr.1). That delegation outlined its rationale for phasing out incineration at sea as soon as possible.

A number of delegations supported the Danish proposal to terminate incineration at sea. In order to achieve more support for its proposal, the Danish delegation modified its proposal by changing the termination date from 1989 to the end of 1994.

---

5. Resolution LDC.33(11). The resolution is attached to this report as Annex 5, without the supporting interim technical guidelines on the control of incineration of wastes and other matter at sea, owing to page limitations.

6. Resolution LDC.34(11). The resolution is attached to this report as Annex 6, without the supporting guidelines on the conditions for surveillance of cleaning operations, owing to page limitations.

The Chairman of the Meeting, prior to requesting comments on the Danish proposal, noted that there appeared to be general agreement that incineration at sea was indeed considered as being an interim disposal method which might eventually be phased out and replaced by safer and more environmentally acceptable waste treatment and disposal options. In this connection he also drew the attention of the Meeting to its endorsement of a waste management hierarchy within the Annex III Guidelines and expressed his hope that a solution to this issue could be reached by consensus.

The Secretariat noted that there was an important distinction to draw between incineration at sea of all wastes (including garbage and oil residues as well as noxious liquid wastes) and incineration at sea of noxious liquid wastes only. The Meeting agreed to focus on noxious liquid wastes and requested the Secretariat to provide an assessment to the next Consultative Meeting on the possible implications on the incineration at sea of other wastes or matter, including the current MEPC examination of air emissions from the shipping industry.

The Working Group developed a draft resolution on the status of incineration at sea which the Meeting adopted with some minor changes.[7]

The delegation from the United States welcomed this development; it looked forward to the re-evaluation of incineration at sea as envisaged by the above resolution to be carried out by 1992. Such an evaluation should provide a sound basis for future decisions on this issue. That delegation also stated that this re-evaluation will be extremely important in assessing the scientific and technical aspects of incineration at sea and practicable land-based altenatives. The United States delegation, whilst supporting in principle the provisions of the resolution concerning the export of wastes for incineration at sea, expressed the need to review its domestic law on this matter to see how it could implement this provision. It would report its final views on this particular matter to the Secretariat.

The delegation of Argentina felt that the operative paragraphs 3 and 4 of resolution LDC.35(11) should be interpreted in such a way that it covers not only the export of wastes to a State not Party to the Convention but also the transportation of wastes to an overseas territory of a Contracting Party for the purpose of incineration at sea. Overseas territories are often located in relatively pollution-free environments.

ANNEX 3

RESOLUTION LDC.31(11)

Amendments to the Guidelines for Allocation of Substances to the Annexes to the
London Dumping Convention

The Eleventh Consultative Meeting,
RECALLING Article XIV(4)(b) of the Convention on the Prevention of Marine Pollution by Dumping of Wastes and Other Matter which emphasizes the importance of scientific and technical advice for Consultative Meetings when considering the review of the Annexes to the Convention,
    RECALLING FURTHER that Criteria for the Allocation of Substances to the Annexes of the Convention had been adopted together with guidelines thereto by the Ninth

---

7. Resolution LDC.35(11). The resolution is attached to this report as Annex 7.

Consultative Meeting of Contracting Parties (resolution LDC.19(9)) and that these called for a continuing review for the purpose of ensuring their revision in the light of new scientific and technical developments,

RECOGNIZING the role of the Scientific Group on Dumping as the scientific body responsible for keeping under review the provisions of the Annexes to the Convention,

NOTING the proposals made by the Scientific Group on Dumping regarding clarification of the terms "bioavailability" and "significant exposures" used in the Guidelines for the Allocation of Substances to the Annexes to the London Dumping Convention:

1. AGREES to the proposals of the Scientific Group on Dumping that the text of the Guidelines relating to "bioavailability" and to "significant exposures" be amended.
2. AGREES FURTHER that the attention of all Contracting Parties should be drawn to the amended guidelines as shown in the Annex to this resolution,
3. INVITES its Scientific Group on Dumping to continue the review of the Guidelines for the purpose of ensuring their revision as and when appropriate.

ANNEX 4

RESOLUTION LDC.32(11)

Amendments to the Guidance for the Application of
Annex III (Resolution LDC.17(8))

The Eleventh Consultative Meeting,

RECALLING Article I of the Convention on the Prevention of Marine Pollution by Dumping of Wastes and Other Matter, which provides that Contracting Parties shall individually and collectively promote the effective control of all sources of pollution of the marine environment,

RECALLING FURTHER that amendments to Annex III had been adopted by resolution LDC.26(10) concerning problems which had been encountered with ill-defined wastes that had been proposed for disposal at sea, and the impact of such wastes to marine life and human health,

EMPHASIZING the need that, in accordance with Annex III to the Convention, Contracting Parties, before considering the dumping or incineration of wastes at sea, should ensure that every effort has been made to determine the practical availability of alternative land-based methods of treatment, disposal or elimination of the wastes concerned,.

NOTING the discussion which took place within the Scientific Group on Dumping on the need for Contracting Parties, when establishing criteria governing the issue of permits for the dumping of matter at sea, to be guided in their application of the provisions of Annex III to the Convention,

HAVING CONSIDERED the Guidelines for the Implementation and Uniform Interpretation of Annex III to the London Dumping Convention (Resolution LDC.17(8)) and the proposed amendments to these guidelines prepared by the Scientific Group on Dumping,

1. ADOPTS amendments to sections A4 to A6, A9 and C4 of the Guidelines for the Implementation and Uniform Interpretation of Annex III to the London Dumping Convention,
2. RESOLVES that Contracting Parties to the Convention shall take full account of the amended Guidelines for the Implementation and Uniform Interpretation of Annex III as shown in annex when considering the factors set forth in that Annex prior to the issue of any permit for disposal and incineration of matter at sea.

ANNEX 5

RESOLUTION LDC.33(11)

Amendments to the Interim Technical Guidelines on the Control of Incineration of
Wastes and Other Matter at Sea

The Eleventh Consultative Meeting,
RECOGNIZING that Contracting Parties to the Convention when issuing permits for
incineration at sea should take full account of the Interim Technical Guidelines on the
Control of Incineration of Wastes and Other Matter at Sea, which had been adopted
by the Fourth Consultative Meeting and were subsequently amended by the Fifth,
Seventh and Eighth Consultative Meeting,
    NOTING that the Scientific Group on Dumping after consideration of the report of
the Joint LDC/OSCOM Group of Experts on Incineration at Sea (LDC/OSCOM/IAS 2/
9, LDC/OSCOM/IAS 2/9/Corr.1) agreed that further amendments to the Interim
Technical Guidelines on the Control of Incinceration of Wastes and Other Matter at
Sea were warranted to better reflect the current incineration operational techniques
and practices,
1. ADOPTS amendments to the Interim Technical Guidelines on the Control of Incineration of Wastes and Other Matter at Sea
2. RESOLVES that Contracting Parties to the Convention should:
       take full account of the new Interim Technical Guidelines on the Control of
       Incineration of Wastes and Other Matter at Sea as shown in annex;
       give preference to "no waste" and "low waste" technologies when considering
       individual proposals on incineration at sea.

ANNEX 6

RESOLUTION LDC.34(11)

Guidelines for the Surveillance of Cleaning Operations
Carried Out at Sea on Board Incineration Vessels

The Eleventh Consultative Meeting,
RECALLING Article I of the Convention on the Prevention of Marine Pollution by
Dumping of Wastes and Other Matter, which provides that Contracting Parties shall
individually and collectively promote the effective control of all sources of pollution in
the marine environment,
    RECALLING FURTHER that Regulations for the Control of Incineration of Wastes
and Other Matter had been adopted at its Third Meeting as set forth in an Addendum
to Annex I to the Convention and that this constitutes an integral part of that Annex,
    RECOGNIZING that in issuing permits for incineration at sea Contracting Parties
shall take full account of Technical Guidelines on the Control of Incineration of
Wastes and Other Matter at Sea,
    BEING AWARE that cleaning operations of incineration systems and of tanks of
incineration vessels may have to take place at sea,
    RECOGNIZING that the Technical Guidelines on the Control of Incineration of
Wastes and Other Matter at Sea provide that:
       tanks washings and pump room bilges contaminated with wastes should be in-
       cinerated at sea in accordance with the Regulations for the Control of Incineration

of Wastes and Other Matter at Sea and with the Technical Guidelines, or discharged to port facilities; and that

residues remaining in the incinerator should not be dumped at sea except in accordance with the provisions of the Convention,

RECOGNIZING FURTHER that the Marine Environment Protection Committee of the International Maritime Organization concluded that Annex II of MARPOL 73/78 applies to tank cleaning operations conducted on board incinerator ships and that it adopted interpretations to clarify the requirements for the specialized operations of incinerator ships and to reduce duplication of requirements,

NOTING that there should be consistency on surveillance procedures developed under the London Dumping Convention and MARPOL 73/78,

NOTING FURTHER that, in accordance with Article VII, paragraph 1 of the London Dumping Convention, each Contracting Party shall apply the measures required to implement that Convention to all vessels registered in its territory or flying its flag, or loading in its territory or territorial seas matter which is to be dumped,

1. ADOPTS the guidelines on the surveillance of cleaning operations carried out at sea on board incineration vessels [as described in the annex to the present resolution]
2. RESOLVES that Contracting Parties should take full account of the guidelines on the surveillance of cleaning operations carried out at sea on board incineration vessels.

ANNEX 7

RESOLUTION LDC.35(11)

Status of Incineration of Noxious Liquid Wastes at Sea

The Eleventh Consultative Meeting,

RECALLING Article I of the Convention on the Prevention of Marine Pollution by Dumping of Wastes and other Matter, which states that Contracting Parties shall individually and collectively promote the effective control of all sources of pollution of the marine environment,

REAFFIRMING that incineration at sea is an interim method of waste disposal, and RECOGNIZING that Contracting Parties should give priority to no waste and low waste technology within the hierarchy of waste management,

ACKNOWLEDGING that the Scientific Group on Dumping has considered the report of the Joint LDC/OSCOM Group of Experts on Incineration at sea (LDC/OSCOM/IAS 2/9) and advised the Eleventh Consultative Meeting that the information available provides an adequate basis to assess the environmental acceptability and safety of incineration at sea, and recognizing the need to continue to improve the controls and environmental safeguards in the use of incineration at sea,

RECOGNIZING ALSO the concerns of several Contracting Parties that incineration at sea, as a means of disposal of noxious liquid wastes which may contain highly toxic substances, is considered to represent subsequent risks of marine and atmospheric pollution,

RECOGNIZING FURTHER the potential risk of interference with other legitimate uses of the sea which could arise from incineration operations at sea,

NOTING the need to urge States, which have not previously carried out incineration operations at sea, that instead of starting such operations alternatives to incineration at sea should be considered and that particular attention should be given to developing land-based alternatives, providing they are safer and environmentally more acceptable,

AGREES
1. to take all steps possible to minimize or substantially reduce the use of marine incineration of noxious liquid wastes by 1 January 1991;
2. that Contracting Parties shall re-evaluate incineration at sea of noxious liquid wastes as early in 1992 as possible with a view to proceeding towards the termination of this practice by 31 December 1994. The re-evaluation shall take into account the scientific and technical aspects of incineration at sea, and the practical availability of safer and environmentally more acceptable land-based alternatives. The re-evaluation shall also take into account any other related information that may be brought forward, with particular attention given to the Oslo Commission experience while phasing out incineration at sea;
3. that Contracting Parties shall not export noxious liquid wastes intended for incineration at sea to any State not Party to the Convention, nor allow their disposal in other ways harmful to the environment;
4. that it is preferable that noxious liquid wastes from coastal States which are to be incinerated at sea be loaded in a harbour of the country from which they originate, and under full control of such a country, instead of being exported to another country; and
5. to employ the revised interim technical guidelines on incineration at sea (Resolution LDC.33(11)), reflecting the most recent scientific advice in this field, and the new Guidelines to Annex III C4 (Resolution LDC.32(11)) setting out the necessary consideration relevant to the use of incineration at sea.

UNEP: East Asian Seas Action Plan, 1989[1]

INTRODUCTION

On the initiative of States of the region, the Governing Council of the United Nations Environment Programme (UNEP) in 1977 decided that "steps are urgently needed to formulate and establish a scientific programme involving research, prevention and control of marine pollution and monitoring" for a Regional Seas Programme in Asia [Decision 88 (V), Oceans]. This decision was followed up by a series of preparatory projects and various meetings leading, in April 1981, to the adoption of an action plan for the protection and development of the marine and coastal areas of the East Asian Seas region by the Governments of Indonesia, Malaysia, Philippines, Singapore, and Thailand.[2]

Subsequently, in December 1981, the representatives of the same Governments:
—adopted the programme priorities for the action plan;
—established the Co-ordinating Body on the Seas of East Asia (COBSEA) to co-ordinate the implementation of the action plan;
—endorsed the establishment of a Trust Fund to provide financial support to the action plan; and
—requested UNEP to continue providing secretariat services to the action plan and technical coordination for its implementation.

The first meeting of COBSEA (Bangkok, 3 April 1982) decided to initiate the implementation of the action plan through six priority projects, each implemented by a network of national institutions and co-ordinated by a member of the network.

Through annual meetings of COBSEA (Yogyakarta, 25–26 March 1983; Genting Highlands, 5–6 April 1984; Manila, 22–23 April 1985; Singapore, 25–26 April 1986; Bangkok, 27–29 April 1987; Yogyakarta, 17–19 July 1988) the progress of the action plan was reviewed, and its future course and budget were decided upon together with the level of contributions to the Trust Fund.

The annual meeting of COBSEA proved to be an effective mechanism for determining priorities in the programme, for reaching an agreement on the contributions to the Trust Fund and for determining financial allocations to various activities. How-

1. This is an abridged version of the report of the Third Meeting of Experts on the East Asian Seas Action Plan, United Nations Environment Programme (UNEP), "Report of the Meeting of the Experts on the East Asian Seas Action Plan" (Quezon City, the Philippines, February 7–10, 1989) [Doc. UNEP(OCA)/EAS.WG.3/8, February 24, 1989].
2. UNEP, "Action Plan for the Protection and Development of the Marine and Coastal Areas of the East Asian Seas Region," UNEP Regional Seas Reports and Studies No. 24 (Nairobi: Programme Activity Centre for Oceans and Coastal Areas, UNEP, 1983).

ever, the meetings of COBSEA were unable to fully evaluate the scientific results achieved through these activities and for the analysis of the scientific merits of newly proposed activities. Therefore, the Executive Director decided to organize meetings of experts from the region in order to obtain independent scientific advice on the environmental problems of the East Asian Seas region and on the scientific aspects of activities carried out in the framework of the action plan. In addition, the Executive Director encouraged the experts from the region to form an Association of Southeast Asian Marine Scientists (ASEAMS) which could in the future assume the responsibility of acting as an advisory body to UNEP on environmental problems relevant to the East Asian Seas action plan.

Two meetings of experts were held in the past (Bangkok, 8–12 December 1986; Bangkok, 30 November–4 December 1987).[3] The Executive Director considers that both meetings contributed positively towards the furthering of the goals of the action plan and, therefore, decided to convene this third meeting of experts in consultation and co-operation with ASEAMS.[4] The participants of the meeting, attending in their individual expert capacity as advisors to the Executive Director, were invited to:

(*a*) review the environmental problems of the region covered by the action plan and to advise the Executive Director on their possible solution or mitigation;

(*b*) review the scientific aspects of the activities supported, or planned to be supported, by UNEP through, or as contributions to, the action plan;

(*c*) discuss ways and means to strengthen the co-operation among the environmental experts of the region and to increase their contribution to the protection of the marine and coastal environment of the East Asian Seas region.

The United Nations, UNCTAD, UNDP, UNU, ESCAP, UNCHS, ILO, FAO, Unesco, WHO, WMO, IMO, UNIDO, IAEA, IOC, IUCN, Greenpeace, ASCOPE, ICLARM and SEAFDEC[5] were invited to send their observers to the meeting.

The meeting was held at the Sulo Hotel, Quezon City, Metro Manila, Philippines on 7–10 February 1989.

This document is an excerpt of some of the items discussed at the meeting and reported in doc.(OCA)/EAS WG.3/8.

3. UNEP/WG.154/6 and UNEP(OCA)/EAS WG.2/9.

4. The co-operation and assistance of the Chairman of ASEAMS in the preparation of the meeting is gratefully acknowledged.

5. United Nations Conference on Trade and Development (UNCTAD); United Nations Development Programme (UNDP); United Nations University (UNU); Economic and Social Commission for Asia and the Pacific (ESCAP); United Nations Centre for Human Settlements (UNCHS); International Labour Organization (ILO); Food and Agriculture Organization of the United Nations (FAO); United Nations Educational Scientific and Cultural Organization (Unesco); World Health Organization (WHO); World Meteorological Organization (WMO); International Maritime Organization (IMO); United Nations Industrial Development Organization (UNIDO); International Atomic Energy Agency (IAEA); Intergovernmental Oceanographic Commission (IOC); International Union for the Conservation of Nature and Natural Resources (IUCN); ASEAN Council on Petroleum (ASCOPE); International Centre for Living Aquatic Resources Management (ICLARM); Southeast Asian Fisheries Development Centre (SEAFDEC).

AGENDA ITEM 4: ENVIRONMENTAL PROBLEMS OF THE EAST ASIAN
SEAS REGION

(*a*) State of the Marine Environment

The secretariat recalled that, as a contribution to the global scientific review on the
state of the marine environment sponsored by UNEP through GESAMP,[6] preparation
of regional reviews was initiated by UNEP in all regions covered by the Regional Seas
Programme. The review relevant to the East Asian Seas region has been prepared by a
Task Team of scientists from the East Asian Seas region, working under the technical
co-ordination of FAO and with the technical assistance of IOC and UNEP. The drafts
of the Task Team's report were presented to, and reviewed by, the first and second
meetings of experts on the East Asian Seas action plan held in Bangkok, 8–12 December 1986 and 30 November–4 December 1987, and by the two previous meetings of
COBSEA in Bangkok (27–29 April 1987) and in Yogyakarta (17–19 July 1988), respectively. The comments received from these meetings were used by the Task Team
in preparing the final text of the report.

The Rapporteur of the Task Team, Mr. Edgardo Gomez, in charge of the preparation of the regional report on the state of the marine environment in the East Asian
Seas region, presented the final draft of the Task Team's report [UNEP(OCA)/EAS
WG.3/5]. The highlights were as follows:
   —the unregulated exploitation of marine resources is causing the destruction of
      mangroves, coral reefs, and coastal fisheries;
   —high silt/sediment flux resulting from deforestation, improper land use prac-
      tices and mine tailings disposal is adversely affecting coastal ecosystems;
   —domestic wastes including raw sewage pose a threat to ecological processes and
      human health;
   —contaminants from industrial, agricultural and other activities, including oil
      exploration, exploitation and transport, need to be investigated further in or-
      der to assess their impacts; and
   —the present institutional capabilities to assess marine pollution in the region are
      limited and need to be strengthened if a more accurate evaluation of environ-
      mental problems is to be achieved.

The secretariat informed the meeting that the report on the state of the marine
environment in the East Asian Seas region was planned to be finalized, printed and
distributed by UNEP, in co-operation with FAO and IOC, in the UNEP Regional Seas
Reports and Studies series during the second quarter of 1989, together with similar
reports from other regions.

(*b*) Implications of Expected Climatic Changes in the East Asian Seas Region

The secretariat informed the meeting that the preparation of the regional report on
the implications of expected climatic changes in the East Asian Seas region, as re-

6. IMO/FAO/UNESCO/WMO/IAEA/UN/UNEP Joint Group of Experts on the Scientific
Aspects of Marine Pollution.

ported at the second meeting of experts [UNEP(OCA)/EAS WG.2.9, paragraphs 24–27], was delayed due to the unexpected illness of the Co-ordinator of the Task Team (Mr. A. Kapauan) in charge of the preparation of the report. For this reason, at the seventh meeting of COBSEA, a request was made for the nomination of possible members of the Task Team and of a new co-ordinator. Pursuant to this request, Mr. Chou Loke-Ming was selected as the new Co-ordinator of the Task Team and three members were also nominated by the focal points of the East Asian Seas action plan in Malaysia, the Philippines and Thailand.

The Co-ordinator of the Task Team on the implications of climatic changes in the East Asian Seas region presented his report [UNEP(OCA)/EAS WG. 3/4]. He informed the meeting that the Task Team was comprised of eight members, three of whom were identified by the national focal points and the rest selected by the Co-ordinator in consultation with UNEP. He presented the working schedule of the Task Team which would see the finalization of the report by the end of 1989. Each member of the Task Team has been given a topic for the preparation and presentation of a sectoral report at the first meeting of the Task Team in May 1989.

The meeting, after examination of the topics presented in the document UNEP(OCA)/EAS WG.3/4, recommended the inclusion of the following additional topics and provided the Co-ordinator of the Task Team with the names of potential Task Team members who could provide relevant inputs:

—coastal erosion;
—ground water resources;
—salt-water intrusion;
—mean sea level rise; and
—legal implications of the sea level rise.

The Co-ordinator of the Task Team thanked the meeting for their recommendations and informed the participants that after discussions with and clearance by UNEP their recommendations would be included in his Task Team report.

(c) Immediate, Short-Term and Long-Term Environmental Problems of the East Asian Seas Region

The secretariat recalled the ranking of the immediate, short-term and long-term environmental problems of the region, together with the possible approaches towards their solution of mitigation, which was prepared on the basis of the analysis carried out by the second meeting of experts [UNEP(OCA)/EAS.2.9, paragraphs 32–42].

The meeting was invited to re-examine and review and re-evaluate the environmental problems and their possible solution or mitigation.

The meeting further introduced three topics and separation of the "rise in sea level" from "other natural hazards," then recalculated and revised the severity of the immediate (present), short-term (in 10 years) and long-term (in 50–100 years) environmental problems of the region and ranked them on the scale of 1 (most severe) to 13 (least severe). The results of this re-evaluation are presented in the table 1.

In conjunction with this table, discussion ensued with a view towards the expansion and revision of the possible solution or mitigation measures proposed at the second meeting of experts [UNEP(OCA)/EAS WG.2/9, paragraph 41].

TABLE 1

Problem	Immediate	Short-Term	Long-Term
Destruction of support ecosystems specifically coral reefs and mangroves	1	1	1
Sewage pollution	2	2	3
Industrial pollution	3	3	2
Fisheries over-exploitation	4	4	6
Siltation/sedimentation	5	5	4
Oil pollution	6	6	8
Agricultural pollution	7	7	5
Red tides	8	8	12
Coastal area development	9	9	9
Coastal erosion	10	11	11
Hazardous waste	11	10	7
Natural hazards	12	13	13
Rise in sea level	13	12	10

The following solutions or mitigation measures for the identified problems were recommended for consideration:

   (i) As concerns the destruction of support ecosystems, there is a need to clarify the area of jurisdiction over the natural resources and to establish and enhance management measures, including planning and environmental impact assessment. To relieve pressure on these threatened ecosystems, alternative sources of livelihood should be sought for the people. Finally, some efforts should be exerted to rehabilitate damaged resources such as mangrove reforestation and transplantation of corals and other organisms.

   (ii) Sewage treatment or disposal facilities are required for controlling sewage pollution. Where critical, sewage treatment plants should be established, operated and maintained. In some areas, the proper construction of sewage outfalls will minimize impacts. With these in place, enforcement of regulations on waste disposal must follow.

   (iii) Industrial pollution is fast becoming critical as many countries in the region are rapidly developing into industrialized economies. Both preventive and curative measures have to be instituted. Countries should be more selective in approving industrial projects, preferably those with clean technologies. The project planning instruments such as prior requirement of Environmental Impact Assessment studies should be introduced and effectively implemented. In order to promote certain types of industries, the relevant national authority should ensure the necessary provisions of common toxic and hazardous waste treatment and facilities. Industries should also be encouraged, perhaps be given incentives, to exchange their waste products as raw materials with other interested parties. For curative measures, comprehensive legislation including effluent standards and strict law enforcement including planning, follow-up and monitoring are necessary to address the problem of industrial pollution. Proper waste treatment facilities are needed to eliminate contaminants at their source. Finally, manufacturers and other relevant industrial concerns should assume a greater social responsibility towards communities.

(iv) Fisheries over-exploitation should be addressed by proper stock assessment and management, including fishing seasons, other limits and strict licensing schemes. To release pressure on coastal fisheries, a shift to deep sea fishing may be encouraged as well as a shift to alternate sources of livelihood including non-destructive aquaculture.

(v) Siltation/sedimentation resulting from soil erosion can be limited by strict enforcement of control measures. This would include the control of soil erosion from earthworks, land development, the proper management of lowland and upland resources such as forests and the proper disposal of mine tailings.

(vi) Oil pollution from land-based sources may be addressed in the same way as industrial pollution. In addition, specific mitigation measures include the provision of reception facilities for waste oil and its collection. Serious consideration must be given on what to do with the waste oil particularly from automobiles and how to handle residual toxic wastes. As regards oil pollution from ships, fishing boats and small craft, this can be minimized by strict compliance with safety measures and by imposing heavier penalties on intentional oil discharges. Finally, oil spill contingency plans should be finalized at the national and regional levels. Ratification of oil pollution control-related international conventions and the enactment of national legislation in support of the same should be seriously considered by ASEAN countries. In addition, the adoption of a regional convention for the protection and development of marine and coastal areas will provide a legal framework for the East Asian Seas Programme. In all of these, political will is necessary to ensure effective implementation and enforcement.

(vii) Pollution from agricultural activities may be tackled by the regulation of farm inputs. This basically involves restrictions on the use of persistent pesticides and inorganic fertilizers and the application of reasonable fertilizer levels. In addition, the adoption of integrated pest management and organic farming must be encouraged.

(viii) There is a need to understand fully the cause(s) and phenomenon of toxic red tides. Regular monitoring of toxic dinoflagellate populations, and paralytic shellfish poisoning (PSP) toxin levels could provide an early warning system to mitigate the adverse effects of toxic red tides on human populations. Depuration measures could help mitigate economic losses to the shellfish fishery. Measures must be sought to eliminate/minimize the red tide phenomenon.

(ix) Problems arising from coastal areas development could be addressed by land use policy and management including zoning (limiting the number and type of structures). Where there is a high population density, treatment facilities for wastes can mitigate contamination. Reduction of population density in coastal areas can be effected by resettlement or transmigration.

(x) Coastal erosion induced by man's activities can be prevented by proper control measures such as the strict implementation and enforcement of management schemes. These would include the use of buffer zones, proper land use and the use of environmental impact assessments for projects before they are undertaken in the coastal zone.

(xi) The transport, disposal and dumping of hazardous wastes must be strictly controlled. Industries producing these wastes must prepare and establish waste disposal and management systems duly approved by the government before they are allowed to operate. There must be monitoring of the quantity and mode of disposal of these wastes to prevent their entry into the natural environment.

(xii) Hydraulic works for flood control may be useful for minimizing the effects of natural calamities such as typhoons, tsunamis and storm surges. Mitigation measures, including early warning systems for other natural hazards, should also be applied.

(xiii) The negative effects of sea level rise might be mitigated by careful long-term coastal zone planning. In low-lying areas, the continued extraction of ground water has to be examined as this results in accelerated land subsidence. This problem would be greatly aggravated by a rise in sea level. Consideration has to be given also to contingency plans such as the construction of dikes and other flood control measures to minimize coastal land loss and sea water intrusion. Due consideration should be given to the reduction of emission of greenhouse gases.

In general, the meeting recognized the need to improve national and regional efforts to enhance public awareness of the various marine environmental problems of the region. It was observed that there should be more educational efforts to increase the peoples' appreciation of the value of the environment and of solutions or mitigation measures to conserve or protect it.

AGENDA ITEM 5: SCIENTIFIC ASPECTS OF THE ACTIVITIES SUPPORTED, OR PLANNED TO BE SUPPORTED, BY UNEP IN THE EAST ASIAN SEAS REGION

(*a*) Ongoing Activities Supported through the East Asian Seas Action Plan

The secretariat briefly reviewed the decisions of the seventh meeting of COBSEA in connection with the ongoing and new projects within the framework of the East Asian Seas action plan [UNEP(OCA)/EAS IG.1/7, Table 1] and then introduced the document UNEP (OCA)/EAS WG.3/5 on the status of the ongoing projects.

The Chairman invited the coordinator of each project, or the person nominated by the relevant national focal points to represent the coordinator, to inform the meeting of the present status of each project with particular emphasis on the scientific aspects and achievements. The status and the scientific aspects of these projects as presented together with the observations and comments of the meeting are summarized as follows:

EAS-14: Study of Maritime Meteorological Phenomena and Oceanographic Features of the East Asian Seas Region
In the absence of the coordinator or his representative to present the status of this project the meeting could not be provided with any further information other than the

relevant material presented by the secretariat in the document UNEP(OCA)/EAS WG.3/5.

EAS-16: Assessment of Concentration Levels and Trends of Non-oil Pollutants and Their Effects on the Marine Environment in the East Asian Seas Region
In addition to the information provided by the secretariat in connection with trace metal and organochlorine assessments in document UNEP(OCA)/EAS WG.3/5, the meeting was informed by the project coordinator that with the assistance of UNEP, IOC and IAEA (ILMR), the supporting organization was contacting various consultants for the next phase of activities with respect to the heavy metal and organochlorine analysis. It was mentioned that all the activities of the project were behind schedule by several months due to the long delay in the preparation and submission of the present project revision by UNEP.

For future trace metal and organochlorine analysis, the following recommendations were made:

(a) Reference standards should be established to determine the applicability of the methodology(ies) adopted;

(b) A written protocol should be formulated to include: the procedures to be used, the preservation techniques and handling of samples, the setting of sensitivity and detection limits, the number of determinations and the format of the report; and

(c) More laboratories should be encouraged to participate.

EAS-19: Development of Management Plans for Endangered Coastal and Marine Living Resources in East Asia: Training Phase I
The coordinator of the project presented the progress report including the project consultant's report on the terms of reference for the development of the training modules for the training of managers and field staff in the management of marine parks in East Asia. The following three distinct categories of marine park staff were identified as requiring training in the framework of this project:

(a) Directorate staff;

(b) Marine Park Officers;

(c) Rangers.

The meeting was also presented with detailed training course content, and felt that the terms of reference for the training course module developed was quite comprehensive and adequate. Furthermore, it was generally observed that project EAS-19 was an important project and could lead to the initiation of a mechanism providing support in safeguarding the region's present and future marine parks, and in particular, its coral reefs.

EAS-21: Assessment of Land-based Urban, Industrial and Agricultural Sources of Pollution, Their Impact and Development of Recommendations for Possible Control Measures
The representative of the co-ordinator of the project provided additional information in connection with the present status of EAS-21. The attention of the meeting was drawn to the content of the training manual on the assessment of the quantities and types of land-based pollutant discharges into the marine and coastal environment.

The meeting was informed that after the training course, the participants carried out assessment programmes in their respective countries and then returned to Singapore to present and discuss their results at a 3-day seminar held on 25–27 January

1989. The project's consultant was engaged to analyze the national reports and prepared a draft regional report. The draft regional report on the land-based sources of pollution in the EAS region was discussed and amendments to the report were suggested by the participants:

(*a*) Develop techniques to carry out a quantitative assessment of:
  (i) silt discharge from land as a result of soil erosion;
  (ii) oil and grease discharge;
  (iii) heavy metals;
(*b*) Set up a network for exchange of information knowledge or experience in environmental problems with the ASEAN countries; and
(*c*) Meet again to finalize the amended report.

The meeting noted that the approach taken would provide an overall estimate of the land-based pollution load entering the marine environment of the region. Based on the information obtained further refined analysis of various land-based pollutants could be undertaken by the respective countries as and when necessary. In view of this, the meeting agreed that the project was being competently managed and the project's consultant had developed a comprehensive and suitable training module.

(*b*) Evaluation of Completed Projects (EAS-12, EAS-13, EAS-15 and EAS-17)
Prepared by IOC and MARC

The secretariat presented the document on the evaluation of the projects carried out by IOC and MARC [UNEP(OCA)/EAS WG.3/6]. The major conclusions of the evaluation reports were summarized as follows:
  —Some projects failed to meet their objectives due to poor planning at their formulation;
  —In some cases, project management/coordination both by UNEP and the implementing organization was inadequate;
  —Considerable training was conducted in the projects which was useful to the region;
  —Cost effectiveness of the results of the projects was generally poor;
  —There was virtually a complete lack of scientific data in the half-yearly progress reports submitted to UNEP;
  —Not enough attention was paid to relevant activities that other organizations carried out in the region.

The main recommendations of the consultants were also summarized as:
  —Project proposals should be reviewed and evaluated more carefully before approval;
  —Progress reporting, as well as the analysis and evaluation of these reports, should be on regular basis envisaged in the project documents;
  —Management must take an active part in the work;
  —Due consideration should be given to the work done outside the region on topics of a similar nature.

The meeting was of the opinion that the evaluation report [UNEP(OCA)/EAS WG.3/6] was prepared quite comprehensively and was a very useful document, addressing both the achievements and high points of the projects evaluated as well as their shortcomings. The critical comments were seen to have good constructive values, which if adopted and used by UNEP and the implementing organizations could lead to improvements in project implementation in the region. The meeting concurred with the conclusions and recommendations of the evaluation report and additionally urged UNEP and COBSEA to consider the following recommendations with a view towards the improvement of the implementation of projects within the framework of the action plan:

(a) More care should be taken in the project formulation stages in order to avoid problems with over-ambitious workplans and timetables as well as paying due consideration to the scientific aspects of the project. A technical group or ASEAMS could play a major role in the proper formulation of project proposals;

(b) The present format of the standard half-yearly reports of UNEP does not render itself to scientific reporting. Therefore, either the format of the half-yearly report should be changed or a mandatory scientific report should also be included in the reporting procedure of the projects;

(c) Progress and scientific reports should also be copied to all other national focal points for comments and information. This would serve to promote a clearer understanding among the participating agencies;

(d) Any scientific papers which are directly related to or come as spin-offs of the projects should accompany the half-yearly or scientific reports;

(e) The future ASEAMS symposia and the experts meetings should be as far as possible organized back to back so that the former becomes the forum for the presentation of the scientific results of the projects, particularly by the coordinator of each project;

(f) For each project, particularly for those implemented by the national focal points, a scientific/technical project co-ordinator should be identified. The name of the scientific/technical co-ordinator should be clearly stated in the project document. Correspondence/communication on technical issues should be made directly with the co-ordinator with copies to the implementing organizations and/or the national focal points;

(g) Direct implementation of the projects by competent national institutions should be encouraged;

(h) Each project revision should not only concentrate on the budget and the timetable but consideration should also be given for refining or changing the objectives and activities with a view towards the strengthening of the technical and scientific aspects of a project;

(i) ASEAMS should be encouraged to take part in the future evaluation of the completed East Asian Seas projects; and

(j) UNEP should consider the establishment of a Regional Co-ordinating Unit or posting a programme officer responsible for the EAS action plan in the region.

(c) Proposals for Activities to be Supported through the East Asian Seas Action Plan

The meeting was presented with four project proposal outlines which took into consideration:

(i) the long-term strategy for the action plan adopted at the sixth meeting of COBSEA; and

(ii) the immediate, short-term and long-term environmental problems of the region as re-evaluated by this meeting (paragraphs 25–30).

The meeting, after discussing each project outline in detail, decided to recommend the implementation of these projects in the following order of priority and requested UNEP to bring this recommendation to the attention of the forthcoming eighth meeting of COBSEA:

(i) development of management plans for endangered coastal and marine living resources in East Asia: training phase (Extension of Phase I);

(ii) monitoring and research to mitigate the adverse impacts of red tide in the East Asian Seas (EAS) Region;

(iii) quality-assurance for non-oil pollution monitoring; and

(iv) dynamic process of eutrophication in coastal waters—impact on living resources and development of management models.

The meeting also discussed a proposal to establish a UNEP consultancy mission to examine the possibility of setting up a regional network of observation stations to facilitate the study of marine meteorological phenomena and oceanographic features in the East Asian Seas. It was decided that the implementation of this mission should be recommended to UNEP.

The secretariat informed the meeting that the development of the two COBSEA approved project proposals, (i) development of a data exchange system for the East Asian Seas region (EAS-20) and (ii) the "umbrella" project on oil pollution control in the East Asian Seas region (EAS-23), have met with long delays.

AGENDA ITEM 6: WAYS AND MEANS OF STRENGTHENING THE CO-OPERATION BETWEEN ENVIRONMENTAL EXPERTS OF THE EAST ASIAN SEAS REGION AND OF INCREASING THEIR CONTRIBUTION TO THE EAST ASIAN SEAS ACTION PLAN

Under the subject of ways and means of strengthening the cooperation of environmental experts of the region and of increasing their contribution to the action plan, the newly-elected Chairman of the Association of Southeast Asian Marine Scientists (ASEAMS), Mr. Edgardo D. Gomez, presented a report on the highlights pertaining to this organization [UNEP(OCA)/EAS WG.3/7]. He traced the beginnings of the Association from a decision made at the first meeting of experts on the East Asian Seas action plan held in Bangkok, 8–12 December 1986. The early development of ASEAMS consisted of the drafting of its Charter or terms of reference by Mr. Gomez, then Interim Chairman, and a call for members to which some eighty individuals responded within the first year.

A major undertaking on which ASEAMS embarked almost immediately upon its inception was the publication of a quarterly newsletter, "LAUI." Other activities consisted of the editing of the proceedings of the Bali Workshop on oil pollution which took place on 19–23 October 1987 and the distribution of selected publications. At present, the Association on behalf of UNEP is providing the back-up for the Task Team on the Impact of Climatic Change in the East Asian Seas region.

AGENDA ITEM 7: ACTIVITIES OF INTERGOVERNMENTAL,
INTERNATIONAL AND REGIONAL ORGANIZATIONS RELEVANT TO
THE PROTECTION OF THE EAST ASIAN SEAS ENVIRONMENT

The FAO representative stated that the organization was involved in the preparatory phase of the East Asian Seas action plan. Although it has not been directly involved in project activities since the adoption of the action plan, it is interested in the implementation of some of the proposed projects in the region concerned with marine environmental protection, whether within the COBSEA purview or not. One inceptive project is the ASEAN/UNDP project on the role of seagrass in fisheries productivity. UNEP by providing support to an expert from the region was actively involved in the formulation of this project proposal. The representative also mentioned the role that FAO played in the coordination work of the Task Team for the preparation of the report on the state of the marine environment in the East Asian Seas region.

The representative of IOC informed the meeting that within the framework of GIPME, through the IOC-UNEP Group of Experts on Methods, Standards and Inter-calibration (GEMSI), a practical workshop on the use of sediments in marine pollution research and monitoring is being organized at the Institute of Marine Environmental Protection, Dalian, China, in the second half of 1989. An output of the Workshop is expected to be draft guidelines (manuals), and the establishment of a network of laboratories involved in using sediments in monitoring and research. The workshop is primarily chemical but some sampling will also be carried out for biological studies on a preliminary basis, through the IOC-UNEP-IMO Group of Experts on Effects of Pollutants.

He also mentioned that within the framework of WESTPAC, regional components of GIPME and the Marine Pollution Monitoring System (MARPOLMON) are gradually being developed. One project addresses the question of river inputs of contaminants building on the GESAMP Report (WG No. 22) and on the IOC River Input Workshop (Bangkok, Thailand, April–May 1986). The aim is to build a network of laboratories studying and monitoring river inputs. The other project aims at being a regional component of the possible global sentinel organisms programme presently being developed through a joint IOC-UNEP project. The IOC Regional Committee for WESTPAC, at its Fourth Session (Bangkok, Thailand, 22–26 June 1987), endorsed the organization of a training course on the use of micro-computers for oceanographic data management. The course was organized jointly with the Unesco Marine Science Division at the Asian Institute of Technology (Bangkok, Thailand, 16 January–3 February 1989).

Finally, the IOC representative stated that as the meeting was earlier informed through a consultant an evaluation of the petroleum hydrocarbon projects of EAS has been completed (supported by UNEP and IOC). The IOC is prepared to give technical backing, advice, in kind and some cash contribution to the new "umbrella" project on oil pollution monitoring proposed by Indonesia, and IOC has contributed to the formulation of the project document. In addition, IOC is willing to provide technical and scientific advice and backing to the regional review of impacts on coastal zones and areas of expected sea-level and temperature changes induced through climatic changes. IOC is prepared also to cooperate in regional or sub-regional projects falling within its mandate and would also wish to bring information about EAS projects to WESTPAC so as to increase the overall exchange of information among scientists and create support for co-operation.

The ASCOPE coordinator for environment and safety discussed major activities within ASCOPE that are relevant to the East Asian Seas environment. An environment information/data bank manual has been prepared and is now computerized into diskettes and distributed to member-countries. Presently being collected are accident and offshore personnel data for 1980–1987 for incorporation into a safety information/data bank manual. In connection with an objective of ASCOPE to integrate its contingency plan for the control and mitigation of marine pollution with similar plans in the region, ASCOPE will co-sponsor a workshop on oil spill contingency planning and response with IMO/NECOR in August 1989 in Indonesia.

He mentioned that as a component of the "umbrella" project on oil pollution control in the East Asian Seas region, a joint oil spill modeling project with COBSEA is nearing completion and will be installed in a computer either in Thailand (NEB/AIT) or at the Norwegian Meteorological Institute (DNMI) in Oslo. Two scientists from Thailand and one from Malaysia will undergo orientation and training in Norway on the development and operation of the model. Consequently, a training session will be conducted for ASCOPE and COBSEA members to familiarize themselves with the model. This model is planned to be presented at the ASCOPE '89 Conference and Exhibition in Singapore from 14–16 November 1989.

Finally, citing the need for closer cooperation in the prevention and control of marine pollution arising from petroleum operations in the region and consistent with the equal need for a convention on the handling and management of hazardous wastes as specifically discussed in this Meeting, the ASCOPE representative referred to "the establishment of a regime of equal right of access and non-discrimination regarding transfrontier pollution from offshore operations within ASCOPE and other countries of the region" which was proposed at the seventh COBSEA meeting and observed as necessary. This need was reiterated in the present Meeting.

The ICLARM representative expressed thanks to UNEP for being invited to attend this important meeting. He informed the group that after 12 years of operation in the field of fisheries research and management in the tropical region, ICLARM has now expanded its activities to cover research and management of the coastal areas in response to the urgent need to improve and protect the environmental quality of the coastal area in order to avert the current threats of environmental degradation and resource depletion. The Coastal Area Management (CAM) Program will focus its activities in five areas:

(1) develop CAM plans for implementation,
(2) conduct research on management issues affecting coastal resource utilization/ management,
(3) develop coastal area research and management information systems,
(4) manpower development, and
(5) information dissemination.

Specifically, ICLARM is presently implementing an ASEAN/USAID project on Coastal Area Management. The project focuses on three components:

(1) development of coastal area management plans at specific pilot sites in the six ASEAN countries. The plans include specific action plans to address management issues pertaining to coastal area development.
(2) strengthening of regional CAM capability through training programs in which 102 scientists in the region have been trained. Upcoming training courses for 1989 include:

(*a*) monitoring and training workshop on *Pyrodinium* red tides, (*b*) integrated methodology on CAM, and

    (*c*) economic valuation of coastal resources.

(3) promoting information dissemination and public awareness related to coastal area management.

A number of publications have been released including environmental profiles of specific pilot sites of the six countries, educational materials on marine parks, coral reefs, seagrasses; other forthcoming publications include a directory of institutions and individuals involved in research and management of the coastal areas and a bibliography on coastal area management. ICLARM also published an international newsletter, "Tropical Coastal Area Management," which is circulated to 1500 scientists in 92 nations. ICLARM will be glad to cooperate with experts of the region and UNEP in future activities in the protection of the marine environment.

The representative of ICLARM elaborated on the forthcoming Management and Training Workshop on *Pyrodinium* Red Tides which is scheduled to take place in Bandar Seri Begawan during 23–30 May 1989. The objectives of the workshop are:

    (*a*) pooling of experience by biological, economic, and medical researchers;

    (*b*) recommendations for future research towards management and monitoring;

    (*c*) immediate and long-term monitoring and management considerations/recommendations;

    (*d*) practical training in all aspects of red tide management; and

    (*e*) preparations of reviews, media kits and extension material.

Report of the World Maritime University: 1989*

OPERATIONS OF THE WMU: ITS FUNCTIONS AND ITS MAIN CHARACTERISTICS

In a number of respects the World Maritime University is unique. It is an outstanding example of genuine international cooperation, manifested in a number of ways. Some of these are innovations, introduced perhaps with some boldness or hopeful confidence, but which have proven to be very successful.

The student body may be considered to be the most important source of strength of the World Maritime University. Since its establishment no less than 104 countries have enrolled students at the World Maritime University. Almost all have come from developing countries, but in recent years there have also been some students from developed countries as well. The University is unique in that it is the only place of education and training where students from so many countries can live and study together at one time.

Indeed it was a bold undertaking to assemble students from so many countries with diverse backgrounds, cultures, prior education and work experience, and this diversity has involved problems for the faculty, but the challenge has been met with great success. Because of this diversity among students, special attention has been paid to groups of students as well as to individuals, and the faculty's dedication to ensure a beneficial learning experience and satisfactory results has been remarkably successful. The diversity and universality of the student body has produced its own richness in creating an amalgam of a new kind of international personality and institutional loyalty stemming from shared experiences and friendships among professionals who will occupy important places in the future of world shipping. This has already added an entirely new dimension to international cooperation in the maritime sector.

One further word about the students at the World Maritime University. When establishing the University one could not be sure of how sponsoring countries would approach the problem, not only in terms of the number of those who would be interested in sending students, but also in the selection that they would make in nominating the candidates for study at WMU. The experience has shown that sponsoring governments and organizations take great care in sending a high quality of students and there has been evidence of greater selectivity each year as the new students have joined the University. This should not be surprising because it is the sponsors who wish to make the best use of these specialized personnel when they return to their home countries after graduation.

*Editor's Note.—The information for this report was supplied by Ambassador Bernard Zagorin, Senior Adviser to the Secretary-General of IMO (World Maritime University). For further information on previous activities of the WMU, see "Progress Report of the World Maritime University (WMU) 1987," *Ocean Yearbook 7,* ed. Elisabeth Mann Borgese, Norton Ginsburg, and Joseph R. Morgan (Chicago; University of Chicago Press, 1988), pp. 454–60.

Thus far about 400 students have completed their studies at the World Maritime University and another 100 will graduate in December 1989. It is a source of great satisfaction, as has been found in a careful follow-up review which is being kept up-to-date, to find that in a large majority of cases graduates have either been promoted or returned to their previous positions with increased responsibilities. Moreover, graduates of WMU have increasingly been attending the sessions of the IMO Assembly, Council, and Committees and Sub-Committees, as well as UNCTAD and regional and sub-regional international meetings. In this way alone a rich reward is being reaped through the existence of WMU.

The faculty of the University is also universal in nature, coming from all regions of the world. In addition to the Rector and Vice Rector, the resident faculty consists of professors of seven nationalities and an additional ten maritime and English language lecturers from different countries.

An unusual feature of WMU is its scheme of Visiting Professors. This innovation has proven to be very successful. The Secretary-General wrote to very eminent people in different lines of activity concerned with maritime affairs to invite them to become Visiting Professors at WMU and this roster consists of over 150 persons. More than 80 of these persons renowned in their specialized fields of interest have given lectures at the University, many on a regular basis more than once a year, and they add a significant extra dimension to the curriculum presented by the resident faculty. Noteworthy about the Visiting Professors is their willingness to offer their services to the University at no cost, since no fees are provided and the only expenses borne by the University are for their travel to and from Malmö and for a small per diem. Normally the Visiting Professors stay at the University hotel where the students reside and thus have close contact with students not only in the classroom and at campus but also in the residence halls. Here again one cannot think only of the benefits to the students which derive from their period of enrollment at WMU. In many cases strong friendships and professional contacts are forged between students and Visiting Professors, which prove to be of very considerable importance to the students in their later careers after graduation.

Another unusual part of the World Maritime University is its program of field training which offers valuable practical exposure outside the classroom and in diverse working conditions. As in the case of Visiting Professors, here again the cooperation of many governments, companies and individuals have made available to the University a very valuable resource in the form of practical field training. There is no question that effort, time and expense are involved for host countries and organizations in providing the opportunities for such field training and this is another manifestation of the international cooperative effort that has characterized the activities of the World Maritime University. In this respect also the students not only get a valuable learning experience and exposure to current working activities in the maritime field, but they also develop contacts and friendships which can and are drawn on from time-to-time after graduation.

In recognizing the universality of its purpose and the character of its student body, WMU has established a strong program for intensive training in the English language. While the language of instruction and most of the learning materials are in English, many students come from countries where that is not their mother tongue. Accordingly an Intensive English Language Program is offered for students prior to the conduct of the maritime classes, either for an 18-week period or for a 10-week period during which English is the sole focus of attention.

Another aspect which reflects broad international support for the World Maritime University are the donations in kind to WMU. The university has been the beneficiary of more than US$800,000 worth of equipment made available by donors for teaching purposes. These donations have come very largely from companies all over the world and provide access for students to up-to-date modern equipment. Also the library of the University has benefitted from grants from a number of sources which it has used effectively in addition to the University's budgetary provision to build up a good maritime library at WMU.

The WMU has been in operation six years. During that period the University has earned widespread recognition for the high quality and practical value of its education and training program. Each year there has been increasing interest on the part of sponsoring countries to nominate qualified candidates for enrollment at WMU. WMU training is being put to good use after graduates have returned to their home countries, and the WMU experience is widely considered, both by students and their sponsors alike, to be of great value and benefit.

Apart from such indications of success, it was felt that in addition there should be an objective appraisal of the University's experience from an academic point of view. The Board of Governors decided to have an external Academic Review of the University after about five years of its operation, and this was completed in the Spring of 1988 and its report submitted to the Board of Governors for its last session in June 1988. The Academic Review was conducted by a group of very eminent persons in different maritime fields and they spoke very favorably of the work of the University while at the same time proposing a large number of recommendations for further improvement. The Report of the Academic Review noted that not only does the University's training and education contribute to upgrading the skills and competence of its students but it has provided an important impetus to international cooperation in the maritime field. The Academic Review has given much satisfaction in passing favorable judgment on the work of the University in its first five years, but perhaps more important are the helpful and constructive suggestions for improvement which it has proposed and which have been endorsed by the Board of Governors.

It has been extremely timely and of great importance in the life of the World Maritime University to have this thorough and objective evaluation conducted by the external Academic Review, both for the many parties who have had confidence in and assisted generously in its establishment and early operations and for the future work of the University. It should add to the widely held recognition of the World Maritime University for its unique contribution to the international community and for the excellence of its advanced education and practical training for maritime specialists. It contributes further to the renown and global acceptance of the University which can already be seen by the fact that thus far some 600 students have been enrolled from 104 countries.

The success of the World Maritime University since its establishment can only bring a sense of satisfaction to those who have supported its establishment and to the many who have contributed to its activities in so many ways. It has been a remarkable example of partnership effort and truly international cooperation. It is now not only a question of hope and confidence in its potential value, as was felt before the University was established; for it has been made clear that the WMU has been making an important contribution to the development and implementation of global standards for the promotion of maritime safety and prevention of marine pollution. By its very existence, the excellence of its training program and the bringing together of highly qualified specialists from many countries, the World Maritime University is creating a body of internationally aware and concerned maritime specialists who will play a key role in international maritime affairs in future years, promoting peaceful cooperation among the nations of the world.

Continued adequate support for the World Maritime University by the international community would be money well and purposefully spent in enabling further enhancement of the good work already achieved by the WMU.

ANNEX

REPORT OF THE BOARD OF GOVERNORS OF THE WORLD MARITIME UNIVERSITY TO THE COUNCIL

The Board of Governors of the World Maritime University is pleased to present this fifth report to the Council of the International Maritime Organization on the University's further progress during the year April 1988 to March 1989. The Members of the Board are very satisfied with the confirmed development and consolidation of the University and with its program in providing advanced training to maritime students from many countries throughout the world. There is much gratification with the effective use generally being made of the graduates and their WMU training, as they return to their home countries to take up positions of responsibility. The Board of Governors is especially encouraged by the widespread and increasing recognition of the training and education offered at WMU, the obvious interest on the part of nominating countries and organizations to send students, and the support which is being given to the University in so many ways by so many parties.

Membership of the Board of Governors

The Board of Governors of the University, established by the Secretary-General in accordance with criteria and guidelines approved by the Council, at present consists of 59 Members.

Board of Governors' Meeting—Sixth Session—June 1988

The sixth session of the Board of Governors of the World Maritime University was held in Malmö, Sweden on June 28 and 29, 1988. Thirty-two Governors and representatives of heads of UN organizations attended the meeting.

A number of documents were prepared for the Board of Governors meeting including a progress report on the operations of the University during its fifth year, a report on the University's accounts and audit for 1987, the revised budget for 1988, budget estimates for 1989, a Five-Year Plan for the period 1989–1993, the Report of the Academic Review of WMU, and a report on the promulgation of additional Staff Rules and amendments of the Staff Regulations.

Among its main decisions, the Board approved the revised budget for 1988 and the budget estimates for 1989, which will be submitted to the Council separately as provided in the Charter. The Governors emphasized *inter alia* the need for the University to limit expenditure on student travel costs. The 1987 accounts of the University were submitted to and approved by the Board of Governors, taking into account reports of the external auditor whose representatives were present at the sixth session of the Board.

Academic Review

The Report of the Governors presented to the Council last year provided information about the terms of reference for the Academic Review of WMU and listed the names of the very distinguished and experienced persons who served as Members of the Academic Review Team. The Report of the Academic Review was presented to the Chancellor in May 1988 and forwarded to the Board of Governors for its consideration at its sixth session in June 1988. The Report was distributed at the 61st session of the council under Agenda item C 61/12 in November 1988. It is appropriate to quote the following extracts from the introduction to the Report:

> The Academic Review has clearly concluded that the academic program offered at the World Maritime University meets the aims as set forth in its Charter. The University, in its brief lifetime of five years, has shown itself not only to be an effective medium for imparting more advanced skills and greater knowledge to maritime specialists but an important contributor to international cooperation in the maritime field.
> At the same time the Academic Review has identified the need for improvements in general and in particular courses which warrant attention at an early date. The appraisal and the recommendations presented in this report in no way diminish the remarkable achievement of WMU in the first five years of its existence. In a wider perspective, developments in the maritime world require consideration and planning for the longer term.

The Academic Review was the major special topic discussed by the Board of Governors at its sixth session:

> The Board decided that points raised by Governors during debate should be recorded in order to assist the University during its study of the report and implementation of its recommendations. The principal points are listed below in the order in which they were made:
> 1. Financial implications of the recommendations should be identified.
> 2. Administration recommendations should be implemented without delay.
> 3. The Academic Council of WMU should proceed forthwith with a review of the Report's recommendations.

4. The University should introduce without delay a course concentrating on ports and shipping administration as recommended in the Report.

5. Endorsement was given to the Report's recommendation concerning the necessity of a comprehensive survey of the needs of developing countries in respect of courses of study at the University.

6. The role of visiting professors should be reviewed, and their lectures integrated with lectures offered by the resident faculty.

7. Field training must be adapted to the requirements and qualifications of students.

8. It was important to maintain close contact with all WMU graduates.

9. The Report could be a useful tool for the University in its public relations activities, and with its high level of readability could improve its image at a world level.

10. Consideration should be given to the organization of short courses for those candidates who may not be able to join the regular two-year courses.

11. Bridging courses, preferably to be offered at regional institutions rather than at WMU, would be useful in the upgrading of qualifications of students wishing to enroll at WMU.

12. A caution was expressed about introducing an age limit of 40 years for entering students, and a suggestion was made to consider a minimum of 10 years of service after graduation as a practical alternative.

In concluding the intensive deliberations on the Academic Review, the Board of Governors:

(i) recorded its profound gratitude to the Members of the Academic Review who performed such a valuable service for the University and for the international maritime community in presenting such a thorough, valuable and constructive report;

(ii) considered favorably the conclusions and approved the recommendations of the Report;

(iii) decided to implement the recommendations in an expeditious manner or progressively over time, as appropriate;

(iv) decided to request that a report on the program of implementation, on a time-targeted basis, be prepared for the seventh session of the Board of Governors and for subsequent sessions, and that the Executive Council be informed of developments in a timely manner;

(v) requested the University and the International Maritime Organization to disseminate copies of the Report, as deemed appropriate, to encourage and stimulate interest in and support for the University;

(vi) requested the University to give consideration to the points made by Governors noted in the paragraphs above.

The conclusions and recommendations of the Academic Review have provided valuable guidelines for the future work of the WMU and identified a number of ways in which WMU can achieve further growth and strengthening. Many of the recommendations have already been put into place and a number of others are being prepared for implementation. As requested by the Governors, an initial report on the implementation of the Academic Review's recommendations was given to the Executive Council at its 10th session in November 1988. A more thorough report will be presented to the next session of the Board of Governors in June 1989. In accordance with the decision of the Governors, further reports on the implementation of the Academic Review's recommendations will be presented in future years.

When the Board of Governors met at its last session near mid-year 1988 the University celebrated the fifth anniversary of its inauguration on July 4, 1983. The main theme prevailing was satisfaction with the success of WMU thus far and optimism for the future as the University focuses on the quality of its performance, having in mind particularly the recommendations of the Academic Review, and with the expectation of continued widespread support for the University by the donors who have made available financial contributions and fellowships, Visiting Professors, organizers of field training, donors of equipment and others.

Executive Council

The Executive Council of the World Maritime University met three times in 1988: at its eighth session in London on April 14, 1988, its ninth session in Malmö on June 27, 1988 and its tenth session in London on November 28, 1988. It is to hold its eleventh session in London on April 6–7, 1989 at which it will consider, in accordance with the Charter, the University's budget results for 1988, the revised budget for 1989, budget estimates for 1990, and other matters to be dealt with at the seventh session of the Board of Governors. The Executive Council will meet again in Malmö just prior to the seventh session of the Board which is scheduled to be held on June 13–14, 1989.

Graduation in December 1988

The World Maritime University held its fifth graduation on December 4, 1988. On that occasion 101 Master of Science degrees were awarded to students from 58 countries who satisfied the requirements for graduation on completion of their two-year courses of study. With this fifth graduation, a total of 384 students have received their Master of Science degrees on completion of two-year courses and 13 students have been given Diplomas for completion of their one-year courses (which were no longer given after 1985).

It has been particularly gratifying that the general experience has been for graduates of WMU to be assigned to positions of responsibility in the maritime field, often with promotions, when returning to their home countries. Comments by high officials of many countries have clearly shown the high respect accorded to the University and to the benefits gained by students from their training and education at WMU, and this has been further evidenced by the increased interest in sending students to WMU. Also, the participation of WMU graduates in international maritime affairs can be readily seen as a number of graduates have attended sessions of the Assembly, the Council and various committees of IMO as well as meetings of UNCTAD and regional and sub-regional meetings. Thus in the words of the Academic Review, the University is making "a special contribution in creating a new impetus to international cooperation in the maritime field".

Student Enrollment in 1988

The enrollment of 104 students in 1989 was again at the University's capacity limit. With the enrollment in this seventh class of students from seven new countries, a total of 104 countries have sent students to WMU. As in 1988, two students from developed countries were also enrolled.

The Governors are pleased to note that the quality of students entering the University is being maintained at a high level and in fact progressing year by year. Nominating Governments and organizations have been selecting students to attend WMU with great care.

The intake in 1989 again shows a wide distribution from all regions of the world.

A summary of the regional distribution of students for all seven classes thus far is [in table 1].

The course distribution of student enrollment at WMU for the years 1983–1989 is [in table 2].

Changes in Faculty

During 1988 and early 1989 the University has had further turnover of academic personnel. A changeover in the posts of two resident professors occurred during the past year. A new maritime lecturer was appointed in 1988 and a replacement appointment of another lecturer will be made in 1989. An additional lecturer in engineering was seconded to WMU by the US Coast Guard in September 1988.

The Governors wish to record their deep appreciation of the excellence of work and dedication of the Rector, Vice Rector, faculty and administrative staff of the University.

Visiting Professors

The scheme of Visiting Professors at WMU has continued with much success. More than 80 persons prominent in various fields of maritime affairs have given lectures at the University during the past year, in addition to 10 senior IMO officials, and a number returned during the year to give additional lectures. This program has very much broadened the scope of the subjects covered at WMU and enriched its curriculum. The Governors are deeply appreciative of the valuable contribution made by the Visiting Professors who make this effort for the University as a public service without remuneration.

Field Training

Field training at the World Maritime University is a unique and special feature of its curriculum. Concentrated during a period of several months in the third semester of a

TABLE 1.—REGIONAL DISTRIBUTION OF STUDENT ENROLLMENT AT WORLD MARITIME UNIVERSITY, 1983–1989

				No. of Countries				
	1983–1989	1983	1984	1985	1986	1987	1988	1989	Total
Africa	33	25	23	28	28	30	33	41	208
Arab states	17	11	6	18	17	20	16	19	107
Asia and Pacific	22	12	19	16	20	20	29	26	142
Latin America and the Caribbean	22	18	15	17	19	20	16	11	116
European countries with UNDP country programs	6	6	2	2	1	8	6	5	30
Developed countries	4	...	...	...	...	4	2	2	8
Total	104	72	65	81	85	102	102	104	611

TABLE 2.—NUMBER OF STUDENTS ENROLLED IN WORLD MARITIME UNIVERSITY, 1983–1989

Course	1983	1984	1985	1986	1987	1988	1989	Total
General maritime administration	22	27	31	39	40	47	21	227
(which included ports and shipping)	...	...	...	(14)	(16)	(25)	...	(55)
Ports and shipping administration	...	...	...	...	...	...	27	27
Maritime education and training (nautical)	9	4	4	11	11	11	11	61
Maritime education and training (engineering)	10	1	3	3	6	7	9	39
Maritime safety administration (nautical)	15	6	12	12	16	15	16	92
Maritime safety administration (engineering)	6	12	7	8	14	9	5	61
Technical management of shipping companies	10	12	14	12	15	13	15	91
Technical officers engaged in maritime safety administration*	...	...	7	...	...	...	...	7
Technical staff of shipping companies*	...	3	3	...	...	...	...	6
Total	72	65	81	85	102	102	104	611

*One-year courses no longer given at WMU.

four-semester program of study, on-the-job training and field trips organized by WMU add a particularly valuable, practical aspect to the advanced training provided at WMU. The success of this program is made possible by the generous help, hospitality and support given by many governments and organizations throughout the world. The Governors are most thankful to all those who have made available their facilities, offered their time and effort and in other ways provided the necessary support to enable WMU to have such a successful field training program.

Fellowship Support

During the period under review there have been further encouraging developments in the provision of fellowship financing to enable students from developing countries to attend WMU. The largest source of fellowship financing for the students who entered the new class in 1989 continued to be country and regional programs of the United Nations Development Programme, the Federal Republic of Germany, as well as self-financing by governments, companies and national organizations. In addition recurrent fellowship support in 1989 was made available by Norway, Denmark, the Commonwealth Secretariat and under bilateral programs of Sweden (SIDA) and Norway (NORAD). In 1989 the Government of Canada continued the program of fellowship financing which was initiated in 1987 by pledging support for an additional eight students as in 1988 and 1987, and the International Centre for Ocean Development in Canada also has continued its program of providing the two new WMU fellowships a year. In 1989 the European Economic Community decided to continue its WMU fellowship support program begun in 1988. The Sasakawa Fellowship Fund provided fellowships for nine new students who joined the new class in 1989 with the financing provided under the Fund established in September 1987 with a $1 million grant by the Sasakawa Foundation. The Endowment Fund established by the Patt Manfield Company of Hong Kong to finance fellowships with investment earnings of the Endowment Fund was enlarged in March 1989, by a second $100,000 grant in addition to the initial $100,000 grant in October 1986. The Henri Kummerman Foundation, the Scholarship in memory of Markos and Angeliki Lyras and the Gokal Foundation continued with their fellowship financing in 1989. As in the previous year, the Korea Maritime Foundation Scholarship was made available for a second student to enter in 1989. A new fellowship contributor in 1989 was the Liberian Shipowners' Council.

Capital Fund

It will be recalled that the total contribution to the Capital Fund of WMU amounted to $343,293 as of March 31, 1988, as given in last year's report to the Council. Since then, as of March 15, 1989, the total standing in the Capital Fund amounted to $353,086, comprising $275,357 in donations and $91,729 of interest earned less a disbursement of $14,000 (for a fellowship for one student in 1988). The Trustees of the Capital Fund have met several times and are scheduled to meet again in London on April 7, 1989.

Conclusion

The Governors wish to express again their deep appreciation to the Assembly of the Organization, the Council of the Organization, and all other international bodies whose decisions have brought the University into being and have provided it with continued and necessary support. The University is an institution established by IMO, and the unflagging interest in and support of the WMU by the IMO governing bodies and all IMO Member States is essential for its effective functioning in meeting its important mandate. The Board firmly believes that the confidence placed in the World Maritime University, as seen in the keen interest to send students to WMU and in the widespread support given to it in so many ways, is well deserved. The World Maritime University is not only a unique institution in the service of the international maritime community, it is a remarkable example of genuine international cooperation on a truly global scale. The Board of Governors trusts that necessary future support for the University will be made available in order for it to continue its record of steady growth and improvement in future years.

In conclusion, the Governors wish to place on record their profound thanks to all those governments, organizations, companies and individuals whose confidence in and support for the WMU, in so many ways, have made it possible for the World Maritime University to achieve such remarkable success in the first six years of its existence.

Appendix B

Selected Documents and Proceedings

The documents and proceedings included in this appendix represent a selection of international agreements, legislation, proceedings of international conferences, and other documents bearing on important ocean-related developments. Commentaries on certain of these documents appear in the *Yearbook* and are referred to in the footnotes and noted in the Index.

THE EDITORS

International Conference on the Suppression of Unlawful Acts against the Safety of Maritime Navigation—Agenda Item 8

ADOPTION OF THE FINAL ACT AND ANY INSTRUMENTS, RECOMMENDATIONS AND RESOLUTIONS RESULTING FROM THE WORK OF THE CONFERENCE

CONVENTION FOR THE SUPPRESSION OF UNLAWFUL ACTS AGAINST THE SAFETY OF MARITIME NAVIGATION

The States Parties to this Convention,

HAVING IN MIND the purposes and principles of the Charter of the United Nations concerning the maintenance of international peace and security and the promotion of friendly relations and co-operation among States,

RECOGNIZING in particular that everyone has the right to life, liberty and security of person, as set out in the Universal Declaration of Human Rights and the International Covenant on Civil and Political Rights,

DEEPLY CONCERNED about the world-wide escalation of acts of terrorism in all its forms, which endanger or take innocent human lives, jeopardize fundamental freedoms and seriously impair the dignity of human beings,

CONSIDERING that unlawful acts against the safety of maritime navigation jeopardize the safety of persons and property, seriously affect the operation of maritime services, and undermine the confidence of the peoples of the world in the safety of maritime navigation,

CONSIDERING that the occurrence of such acts is a matter of grave concern to the international community as a whole,

BEING CONVINCED of the urgent need to develop international co-operation between States in devising and adopting effective and practical measures for the prevention of all unlawful acts against the safety of maritime navigation, and the prosecution and punishment of their perpetrators,

RECALLING resolution 40/61 of the General Assembly of the United Nations of 9 December 1985 which, *inter alia*, "urges all States unilaterally and in co-operation with other States, as well as relevant United Nations organs, to contribute to the progressive elimination of causes underlying international terrorism and to pay special attention to all situations, including colonialism, racism and situations involving mass and flagrant violations of human rights and fundamental freedoms and those involving alien occupation, that may give rise to international terrorism and may endanger international peace and security,"

RECALLING FURTHER that resolution 40/61 "unequivocally condemns, as criminal, all acts, methods and practices of terrorism wherever and by whomever committed, including those which jeopardize friendly relations among States and their security,"

RECALLING ALSO that by resolution 40/61, the International Maritime Organization was invited to "study the problem of terrorism aboard or against ships with a view to making recommendations on appropriate measures,"

HAVING IN MIND resolution A.584(14) of 20 November 1985, of the Assembly of the International Maritime Organization, which called for development of measures to prevent unlawful acts which threaten the safety of ships and the security of their passengers and crews,

NOTING that acts of the crew which are subject to normal shipboard discipline are outside the purview of this Convention,

AFFIRMING the desirability of monitoring rules and standards relating to the prevention and control of unlawful acts against ships and persons on board ships, with a view to updating them as necessary, and, to this effect, taking note with satisfaction of the Measures to Prevent Unlawful Acts against Passengers and Crews on Board Ships, recommended by the Maritime Safety Committee of the International Maritime Organization,

AFFIRMING FURTHER that matters not regulated by this Convention continue to be governed by the rules and principles of general international law,

RECOGNIZING the need for all States, in combating unlawful acts against the safety of maritime navigation, strictly to comply with rules and principles of general international laws,

HAVE AGREED as follows:

ARTICLE 1

For the purposes of this Convention, "ship" means a vessel of any type whatsoever not permanently attached to the sea-bed, including dynamically supported craft, submersibles, or any other floating craft.

ARTICLE 2

1. This Convention does not apply to:
    (a) a warship; or
    (b) a ship owned or operated by a State when being used as a naval auxiliary or for customs or police purposes; or
    (c) a ship which has been withdrawn from navigation or laid up.
2. Nothing in this Convention affects the immunities of warships and other government ships operated for non-commercial purposes.

ARTICLE 3

1. Any person commits an offence if that person unlawfully and intentionally:
    (a) seizes or exercises control over a ship by force or threat thereof or any other form of intimidation; or
    (b) performs an act of violence against a person on board a ship if that act is likely to endanger the safe navigation of that ship; or

(*c*) destroys a ship or causes damage to a ship or to its cargo which is likely to endanger the safe navigation of that ship; or

(*d*) places or causes to be placed on a ship, by any means whatsoever, a device or substance which is likely to destroy that ship, or cause damage to that ship or its cargo which endangers or is likely to endanger the safe navigation of that ship; or

(*e*) destroys or seriously damages maritime navigational facilities or seriously interferes with their operation, if any such act is likely to endanger the safe navigation of a ship; or

(*f*) communicates information which he knows to be false, thereby endangering the safe navigation of a ship; or

(*g*) injures or kills any person, in connection with the commission or the attempted commission of any of the offences set forth in subparagraphs (*a*) to (*f*).

2. Any person also commits an offence if that person:

(*a*) attempts to commit any of the offences set forth in paragraph 1; or

(*b*) abets the commission of any of the offences set forth in paragraph 1 perpetrated by any person or is otherwise an accomplice of a person who commits such an offence; or

(*c*) threatens, with or without a condition, as is provided for under national law, aimed at compelling a physical or juridical person to do or refrain from doing any act, to commit any of the offences set forth in paragraph 1, subparagraphs (*b*), (*c*) and (*e*), if that threat is likely to endanger the safe navigation of the ship in question.

ARTICLE 4

1. This Convention applies if the ship is navigating or is scheduled to navigate into, through or from waters beyond the outer limits of the territorial sea of a single State, or the lateral limits of its territorial sea with adjacent States.

2. In cases where the Convention does not apply pursuant to paragraph 1, it nevertheless applies when the offender or the alleged offender is found in the territory of a State Party other than the State referred to in paragraph 1.

ARTICLE 5

Each State Party shall make the offences set forth in article 3 punishable by appropriate penalties which take into account the grave nature of those offences.

ARTICLE 6

1. Each State Party shall take such measures as may be necessary to establish its jurisdiction over the offences set forth in article 3 when the offence is committed:

(*a*) against or on board a ship flying the flag of the State at the time the offence is committed; or (*b*) in the territory of that State, including its territorial sea; or

(*c*) by a national of that State.

2. A State Party may also establish its jurisdiction over any such offence when;

(*a*) it is committed by a stateless person whose habitual residence is in that State; or

(*b*) during its commission a national of that State is seized, threatened, injured or killed; or

(*c*) it is committed in an attempt to compel that State to do or abstain from doing any act.

3. Any State Party which has established jurisdiction mentioned in paragraph 2 shall notify the Secretary-General of the International Maritime Organization (hereinafter referred to as "the Secretary-General"). If such State Party subsequently rescinds that jurisdiction, it shall notify the Secretary-General.

4. Each State Party shall take such measures as may be necessary to establish its jurisdiction over the offences set forth in article 3 in cases where the alleged offender is present in its territory and it does not extradite him to any of the State Parties which have established their jurisdiction in accordance with paragraphs 1 and 2 of this article.

5. This Convention does not exclude any criminal jurisdiction exercised in accordance with national law.

ARTICLE 7

1. Upon being satisfied that the circumstances so warrant, any State Party in the territory of which the offender or the alleged offender is present shall, in accordance with its law, take him into custody or take other measures to ensure his presence for such a time as it necessary to enable any criminal or extradition proceedings to be instituted.

2. Such State shall immediately make a preliminary inquiry into the facts, in accordance with its own legislation.

3. Any person regarding whom the measures referred to in paragraph 1 are being taken shall be entitled to:

(*a*) communicate without delay with the nearest appropriate representative of the State of which he is a national or which is otherwise entitled to establish such communication or, if he is a stateless person, the State in the territory of which he has his habitual residence;

(*b*) be visited by a representative of that State.

4. The rights referred to in paragraph 3 shall be exercised in conformity with the laws and regulations of the State in the territory of which the offender or the alleged offender is present, subject to the proviso that the said laws and regulations must enable full effect to be given to the purposes for which the rights accorded under paragraph 3 are intended.

5. When a State Party, pursuant to this article, has taken a person into custody, it shall immediately notify the States which have established jurisdiction in accordance with article 6, paragraph 1 and, if it considers it advisable, any other interested States, of the fact that such person is in custody and of the circumstances which warrant his detention. The State which makes the preliminary inquiry contemplated in para-

graph 2 of this article shall promptly report its findings to the said States and shall indicate whether it intends to exercise jurisdiction.

ARTICLE 8

1. The master of a ship of a State Party (the "flag State") may deliver to the authorities of any other State Party (the "receiving State") any person who he has reasonable grounds to believe has committed one of the offences set forth in article 3.
2. The flag State shall ensure that the master of its ship is obliged, whenever practicable, and if possible before entering the territorial sea of the receiving State carrying on board any person whom the master intends to deliver in accordance with paragraph 1, to give notification to the authorities of the receiving State of his intention to deliver such person and the reasons therefor.
3. The receiving State shall accept the delivery, except where it has grounds to consider that the Convention is not applicable to the acts giving rise to the delivery, and shall proceed in accordance with the provisions of article 7. Any refusal to accept a delivery shall be accompanied by a statement of the reasons for refusal.
4. The flag State shall ensure that the master of its ship is obliged to furnish the authorities of the receiving State with the evidence in the master's possession which pertains to the alleged offence.
5. A receiving State which has accepted the delivery of a person in accordance with paragraph 3 may, in turn, request the flag State to accept delivery of that person. The flag State shall consider any such request, and if it accedes to the request it shall proceed in accordance with article 7. If the flag State declines a request, it shall furnish the receiving State with a statement of the reasons therefor.

ARTICLE 9

Nothing in this Convention shall affect in any way the rules of international law pertaining to the competence of States to exercise investigative or enforcement jurisdiction on board ships not flying their flag.

ARTICLE 10

1. The State Party in the territory of which the offender or the alleged offender is found shall, in cases to which article 6 applies, if it does not extradite him, be obliged, without exception whatsoever and whether or not the offence was committed in its territory, to submit the case without delay to its competent authorities for the purpose of prosecution, through proceedings in accordance with the laws of that State. Those authorities shall take their decision in the same manner as in the case of any other offence of a grave nature under the law of that State.
2. Any person regarding whom proceedings are being carried out in connection with any of the offences set forth in article 3 shall be guaranteed fair treatment at all stages of the proceedings, including enjoyment of all the rights and guarantees

provided for such proceedings by the law of the State in the territory of which he is present.

ARTICLE 11

1. The offences set forth in article 3 shall be deemed to be included as extraditable offences in any extradition treaty existing between any of the States Parties. States Parties undertake to include such offences as extraditable offences in every extradition treaty to be concluded between them.
2. If a State Party which makes extradition conditional on the existence of a treaty receives a request for extradition from another State Party with which it has no extradition treaty, the requested State Party may, at its option, consider this Convention as a legal basis for extradition in respect of the offences set forth in article 3. Extradition shall be subject to the other conditions provided by the law of the requested State Party.
3. States Parties which do not make extradition conditional on the existence of a treaty shall recognize the offences set forth in article 3 as extraditable offences between themselves, subject to the conditions provided by the law of the requested State.
4. If necessary, the offences set forth in article 3 shall be treated, for the purposes of extradition between States Parties, as if they had been committed not only in the place in which they occurred but also in a place within the jurisdiction of the State Party requesting extradition.
5. A State Party which receives more than one request for extradition from States which have established jurisdiction in accordance with article 7 and which decides not to prosecute shall, in selecting the State to which the offender or alleged offender is to be extradited, pay due regard to the interests and responsibilities of the State Party whose flag the ship was flying at the time of the commission of the offence.
6. In considering a request for the extradition of an alleged offender pursuant to this Convention, the requested State shall pay due regard to whether his rights as set forth in article 7, paragraph 3, can be effected in the requesting State.
7. With respect to the offences as defined in this Convention, the provisions of all extradition treaties and arrangements applicable between States Parties are modified as between States Parties to the extent that they are incompatible with this Convention.

ARTICLE 12

1. States Parties shall afford one another the greatest measure of assistance in connection with criminal proceedings brought in respect of the offences set forth in article 3, including assistance in obtaining evidence at their disposal necessary for the proceedings.
2. States Parties shall carry out their obligations under paragraph 1 in conformity with any treaties on mutual assistance that may exist between them. In the absence of such treaties, States Parties shall afford each other assistance in accordance with their national law.

ARTICLE 13

1. States Parties shall co-operate in the prevention of the offences set forth in article 3, particularly by:
   (a) taking all practicable measures to prevent preparations in their respective territories for the commission of those offences within or outside their territories;
   (b) exchanging information in accordance with their national law, and co-ordinating administrative and other measures taken as appropriate to prevent the commission of offences set forth in article 3.
2. When, due to the commission of an offence set forth in article 3, the passage of a ship has been delayed or interrupted, any State Party in whose territory the ship or passengers or crew are present shall be bound to exercise all possible efforts to avoid a ship, its passengers, crew or cargo being unduly detained or delayed.

ARTICLE 14

Any State Party having reason to believe that an offence set forth in article 3 will be committed shall, in accordance with its national law, furnish as promptly as possible any relevant information in its possession to those States which it believes would be the States having established jurisdiction in accordance with article 6.

ARTICLE 15

1. Each State Party shall, in accordance with its national law, provide to the Secretary-General, as promptly as possible, any relevant information in its possession concerning:
   (a) the circumstances of the offence;
   (b) the action taken pursuant to article 13, paragraph 2;
   (c) the measures taken in relation to the offender or the alleged offender and, in particular, the results of any extradition proceedings or other legal proceedings.
2. The State Party where the alleged offender is prosecuted shall, in accordance with its national law, communicate the final outcome of the proceedings to the Secretary-General.
3. The information transmitted in accordance with paragraphs 1 and 2 shall be communicated by the Secretary-General to all States Parties, to Members of the International Maritime Organization (hereinafter referred to as "the Organization"), to the other States concerned, and to the appropriate international intergovernmental organizations.

ARTICLE. 16

1. Any dispute between two or more States Parties concerning the interpretation or application of this Convention which cannot be settled through negotiation within a reasonable time shall, at the request of one of them, be submitted to arbitration. If,

within six months from the date of the request for arbitration, the parties are unable to agree on the organization of the arbitration any one of those parties may refer the dispute to the International Court of Justice by request in conformity with the Statute of the Court.

2. Each State may at the time of signature or ratification, acceptance or approval of this Convention or accession thereto, declare that it does not consider itself bound by any or all of the provisions of paragraph 1. The other States Parties shall not be bound by those provisions with respect to any State Party which has made such a reservation.

3. Any State which has made a reservation in accordance with paragraph 2 may, at any time, withdraw that reservation by notification to the Secretary-General.

ARTICLE 17

1. This Convention shall be open for signature at Rome on 10 March 1988 by States participating in the International Conference on the Suppression of Unlawful Acts against the Safety of Maritime Navigation and at the Headquarters of the Organization by all States from 14 March 1988 to 9 March 1989. It shall thereafter remain open for accession.

2. States may express their consent to be bound by this Convention by:
   (a) signature without reservation as to ratification, acceptance or approval; or
   (b) signature subject to ratification, acceptance or approval, followed by ratification, acceptance or approval; or
   (c) accession.

3. Ratification, acceptance, approval or accession shall be effected by the deposit of an instrument to that effect with the Secretary-General.

ARTICLE 18

1. This Convention shall enter into force ninety days following the date on which fifteen States have either signed it without reservation as to ratification, acceptance or approval, or have deposited an instrument of ratification, acceptance, approval or accession in respect thereof.

2. For a State which deposits an instrument of ratification, acceptance, approval or accession in respect of this Convention after the conditions for entry into force thereof have been met, the ratification, acceptance, approval or accession shall take effect ninety days after the date of such deposit.

ARTICLE 19

1. This Convention may be denounced by any State Party at any time after the expiry of one year from the date on which this Convention enters into force for that State.

2. Denunciation shall be effected by the deposit of an instrument of denunciation with the Secretary-General.

3. A denunciation shall take effect one year, or such longer period as may be specified

in the instrument of denunciation, after the receipt of the instrument of denunciation by the Secretary-General.

ARTICLE 20

1. A conference for the purpose of revising or amending this Convention may be convened by the Organization.
2. The Secretary-General shall convene a conference of the States Parties to this Convention for revising or amending the Convention, at the request of one third of the States Parties, or ten States Parties, whichever is the higher figure.
3. Any instrument of ratification, acceptance, approval or accession deposited after the date of entry into force of an amendment to this Convention shall be deemed to apply to the Convention as amended.

ARTICLE 21

1. This Convention shall be deposited with the Secretary-General.
2. The Secretary-General shall:
   (a) inform all States which have signed this Convention or acceded thereto, and all Members of the Organization, of:
      (i) each new signature or deposit of an instrument of ratification, acceptance, approval or accession together with the date thereof;
      (ii) the date of the entry into force of this Convention;
      (iii) the deposit of any instrument of denunciation of this Convention together with the date on which it is received and the date on which the denunciation takes effect:
      (iv) the receipt of any declaration or notification made under this Convention;
   (b) transmit certified true copies of this Convention to all States which have signed this Convention or acceded thereto.
3. As soon as this Convention enters into force, a certified true copy thereof shall be transmitted by the Depositary to the Secretary-General of the United Nations for registration and publication in accordance with Article 102 of the Charter of the United Nations.

ARTICLE 22

This Convention is established in a single original in the Arabic, Chinese, English, French, Russian and Spanish languages, each text being equally authentic.

IN WITNESS WHEREOF the undersigned being duly authorized by their respective Governments for that purpose have signed this Convention.

DONE AT ROME this tenth day of March one thousand nine hundred and eighty-eight.

Selected Documents and Proceedings

International Labour Organization (ILO): Report of the Committee on Conditions of Work in the Fishing Industry, May 13, 1988*

The Committee on Conditions of Work in the Fishing Industry met at the International Labour Office in Geneva from 4–13 May, 1988 in accordance with decisions of the Governing Body of the International Labour Office taken at its 237th (June 1987) Session.

The Committee was attended by 20 members appointed by the Governing Body, accompanied by 4 advisers on the Government side, 1 adviser on the Employers' side and 4 advisers on the Workers' side. Representatives of two States Members and of several international governmental and non-governmental organisations were also present as observers.

The Committee elected its officers as follows:

Chairman: Mr. K. M. JOSEPH (Government member, India)
Vice-Chairmen: Mr. A. EIDSMO (Employers' member, Norway)
                     Mr. O. JACOBSEN (Workers' member, Denmark/Faroe Islands)
Reporter: Mr. T. HANSEN (Government member, Norway)

The Employers' and Workers' groups elected their officers as follows:

Employers' group
  Chairman and spokesman: Mr. D. HAMMOND (GHANA)
  Secretary: Mr. A. PEÑALOSA (International Organisation of Employers)

Workers' group
  Chairman and spokesman: Mr. O. JACOBSEN (Denmark/Faroe Islands)
  Secretary: Mr. A. SELANDER (International Transport Workers' Federation)

The representative of the Director-General was Mr. J. von Muralt, Director of the Sectoral Activities Department, and the deputy representative of the Director-General was Mr. B. Klerck Nilssen, Chief of the Maritime Industries Branch of the ILO. Messrs. T. O. Braida and D. Appave of the same branch acted as experts.

AGENDA

The agenda of the Committee, as fixed by the Governing Body, comprised the following three items:

(1) systems of remuneration and earnings;
(2) occupational adaptation to technical changes in the fishing industry; and
(3) social and economic needs of small-scale fishermen and of rural fishing communities.

*EDITORS' NOTE.—The documents from which this report is compiled are CFI/4/12 and GB.241/1A/5/1. For further information about the International Labour Organization, contact the International Labour Office, 4 Route des Morillons, CH-1211, Geneva, 22, Switzerland.

The International Labour Office had prepared a report dealing with the items on the agenda. . . .

. . . The Committee unanimously adopted the following conclusions and resolutions:[1]

(a) conclusions concerning systems of remuneration and earnings;

(b) conclusions concerning occupational adaptation to technical changes in the fishing industry;

(c) conclusions concerning social and economic needs of small-scale fishermen and of rural fishing communities;

(d) resolution on future action of the ILO for the fishing industry;

(e) resolution on protection of the livelihood of fishermen;

(f) resolution on working and living conditions in the fishing industry;

(g) resolution on hours of work and manning.

CONCLUSIONS CONCERNING OCCUPATIONAL ADAPTATION TO TECHNICAL CHANGES IN THE FISHING INDUSTRY[2]

The Committee on Conditions of Work in the Fishing Industry of the International Labour Organisation,

Having met in Geneva, in its Fourth Session, from 4 to 13 May 1988,

Adopts this thirteenth day of May 1988 the following conclusions:

1. In the interest of promoting the prosperity of the fishing industry and improving fishermen's working and living conditions, all possible measures should be taken, to the extent reasonable and practicable by the countries concerned, to modernise fishing operations through the application of new technology and techniques.

2. Government support should be integrated and take account of both the technical aspects of development and the socio-economic needs of fishermen, particularly for the development of small-scale fisheries, in consultation with representatives of employers and fishermen. For fisheries in general, the following are among the factors that all concerned should bear in mind when introducing technologies which are appropriate to local conditions and when evaluating the socio-economic factors which influence the success or failure of new technologies or innovations:

(a) proper recognition of the economic and social needs of the fishing industry;

(b) the need for development of fisheries' infrastructure;

(c) the level of organisation of fishing-vessel owners and of fishermen;

(d) the effect on the employment of fishermen and their families;

(e) the training needs of fishermen and those involved in traditional fish processing;

(f) the pricing and procurement of fuel, fishing gear and supplies, spare parts, boats, etc.;

(g) custom duties on imported equipment and materials;

(h) taxation of fishing enterprises and of fishermen;

(i) possibilities of over-fishing and the importance of fish stock conservation;

1. Because of page restrictions, it is only possible to include some excerpts of the Committee's report here. The Committee's resolution and conclusions republished here complement the article by John Fitzpatrick, "Fishing Technology," in this volume.

2. Adopted unanimously.

(*j*) technical co-operation between developed and developing countries and developing countries themselves.

3. Support from governments could include financing schemes for the renovation and structural adjustment of fishing fleets and the improvement of fishing techniques in both industrial and small-scale fisheries in consultation with the parties concerned. This support should take account of the technical and economic aspects of fisheries, as well as the living conditions, health, education and training of fishermen.

4. The expansion of aquaculture should be encouraged as a means of increasing the supply of fish, especially in rural areas, of securing alternative sources of income and employment for fishermen and of earning foreign exchange.

5. New technologies are indispensable for the survival of the fishing industry. However, when introducing new technologies which affect employment conditions, the fishermen concerned should be informed in advance and, after having been consulted, appropriate training should be provided in order that they will be able to make the best and safest use of the new technology. All parties concerned should seek to avert or minimise as far as possible termination of employment, without prejudice to the efficient operation of the industry, and to mitigate the adverse effects of any termination of employment for technological reasons. Where appropriate, and in accordance with national legislation and practice, the competent authority should assist the parties in seeking solutions to the problems raised by the terminations. Efforts should be made to raise funds in order to train and retrain fishermen for other occupations, such as aquaculture.

6. Fishermen who cannot find work should be provided, whenever possible and in accordance with national legislation, with subsistence payments during periods of unemployment, help in locating new jobs and assistance (through retraining in particular) in preparing for such jobs. Other action which could be taken by governments and employers' and workers' organisations in this regard includes: the creation of jobs through the development of aquaculture; the provision of financial and technical assistance for the funding of new businesses; and assistance to the unemployed in finding jobs in fishing in neighbouring countries.

7. Training schemes for fishermen should be based upon clearly defined needs and realistic assessments of current technology and the availability of existing trained or experienced manpower. In this connection, governments should determine which schemes might be undertaken using local resources and which ones need regional and extra-regional expertise. Training should be concurrent with the introduction of new technology and financially assisted by governments and/or through international technical co-operation. It should include training for new tasks and working methods on board ship, in management and in the reduction of post-harvest fish losses.

8. Improved methods of training fishermen for new technology should be developed and applied. In this regard, special training equipment and simple materials could be used to meet the needs of illiterate and semi-literate individuals. Training in schools and on-the-job for both industrial and artisanal fishermen should in particular make use of audio-visual equipment and extension services.

9. Attention should be given to the training of extension staff and training specialists, as well as to the design and monitoring of on-the-job training programmes for fishing communities.

10. Attention should also be given to the training of local fishermen in basic resource management, environmental protection, operation and management of

fishermen's organisations, fishing management and activities associated with social welfare and community development.

11. The transfer of technology should be promoted through pilot projects and assistance in building the necessary infrastructure, as well as through fishery product development and marketing.

12. Information and education should be given to fishermen and boat owners, which would help prepare them for modernisation in the fishing industry.

13. Fishermen should be encouraged to co-operate in ensuring successful adaptation to the application of new technology and work methods in fishing. Fishermen's organisations can serve as a useful channel through which this involvement of fishermen can be achieved.

14. Maritime fishing can be a dangerous occupation, and in this regard safety is an important factor. The proper training of fishermen in the use of safety equipment on board ships and of fishing gear, as well as in safe working methods and practices, should be encouraged as the primary means of reducing accidents to vessels and injury and loss of life of fishermen. The strengthening of national regulations regarding the certification of fishermen and their employment, safety inspections of vessels and the creation of fishing safety committees at the appropriate national level should be encouraged. Appropriate national level bodies should deal with living and working conditions on board ship.

CONCLUSIONS CONCERNING SOCIAL AND ECONOMIC NEEDS OF SMALL-SCALE FISHERMEN AND OF RURAL FISHING COMMUNITIES[3]

The Committee on Conditions of Work in the Fishing Industry of the International Labour Organisation,

Having met in Geneva, in its Fourth Session, from 4 to 13 May 1988,

Adopts this thirteenth day of May 1988 the following conclusions:

1. The existence of adequate communication between governments, fishermen and their communities will help to improve the working and living conditions of small-scale and artisanal fishermen and the standard of living in rural fishing communities which have been recognised as requiring the most urgent action. Such services can be improved if they are based on a sound knowledge of fishermen's labour, social and living conditions. The development of infrastructure facilities such as approach roads, power supply, radio and television services to rural fishing communities can also help to improve extension services. In this regard, governments must play a major role.

2. Living conditions in rural fishing communities of developing countries are usually very poor. Proper housing, sanitation, schools, water supply and health care often do not exist. Research at the regional level on the appropriate ways of improving the standard of living can contribute greatly to development efforts. Most governments are in need of financial and technical help to strengthen their research on fishermen's technical, social and economic conditions so as to make it more effective.

3. Increasing the pace of fisheries development and improving social and economic conditions in small-scale fishing communities require both the collection and

3. Adopted unanimously.

interpretation of statistics. Steps that can be taken to improve this collection and interpretation include the development of an internationally recommended procedure and format for obtaining the needed information.

4. The development of small-scale fisheries as well as aquaculture, wherever possible, should be part of a national development strategy for the fishing industry as a whole. Some governments have formulated such strategies which are reflected in their policies on fisheries development programmes. Such programmes should take account not only of the technical aspects of development but the expressed socio-economic needs of fishermen and their communities, as well as the requirements of the processing and marketing sectors of the fishing industry. National surveys to help identify problems and needs of the sector can be helpful in formulating appropriate strategies. The demarcation of exclusive fishing zones for artisanal fishermen and the surveillance and control of these zones should be given priority whenever needed and should be implemented in collaboration with the fishermen's and boatowners' organisations and other parties concerned. Special attention should be given to inland fishermen operating on the banks of mangroves, lagoons and lakes where aquaculture schemes are likely to deprive them of their livelihood if they are not given a chance to upgrade their skills and learn the new techniques.

5. There is a need to improve basic education and literacy in many developing countries. Among the most important ways of achieving these objectives in isolated fishing communities are the provision of: (*a*) adequate financing for education activities; (*b*) free primary and secondary education (for children, education should be free or at tuition fees which their parents can afford); and (*c*) vocational training of adults through fisheries extension services with properly trained staff.

6. In the field of training the long-term objective of governments should be to achieve national self-reliance through the development of the skills required for all aspects of the fishing industry. At the operational level (fishing, fish handling, processing and marketing), training efforts should focus on the training of teachers and extension workers to carry out programmes in general and technical education to support fishery development plans. The training of fishermen, processors and other fishery workers should be done at the local level by national fisheries staff. The strengthening of local training facilities and capabilities can be promoted through international bilateral technical assistance, including technical co-operation between developing countries. Regional workshops and financial help in meeting the cost of training activities are important forms of such assistance.

7. Developing countries can benefit from modernising their fishing industries with improved fishing vessels and gear with a view to increasing the efficiency of fishing operations and reducing the cost of catching fish. However, if not properly planned and carefully implemented, this could lead to overfishing and the consequent redundancy in fishing vessels, etc. Governments should therefore plan in advance for these effects through the rational exploitation and utilisation of fishery resources and control measures to ensure sustained production from small-scale fisheries. Consideration should be given to all aspects of the fisheries development viz. the resources and the continuous monitoring of the environment; the technology for harvesting and post-harvest operations, including handling, processing, distribution and marketing; and the economic and social aspects, including education, health and traditions.

8. The relatively high cost of vessels, engines, fishing gear and other equipment needed to modernise the fishing fleets of developing countries is often a barrier to

development. Among measures which can be taken to provide such equipment at reasonable cost to fishermen are:

(*a*) local production of possible equipment and materials used in fishing;

(*b*) local experimentation and development of suitable cost-effective fishing craft such as surf landing boats, ferro-cement and fibre glass vessels and craft built from local materials;

(*c*) application of simple/modern new technology such as outboard engines in traditional fishing craft;

(*d*) collaborating with foreign manufacturers to make available engines and equipment which are more appropriate to specific local needs;

(*e*) favourable financial credit arrangements for fishermen such as lower interest rates and credit guarantees provided by governments or institutional financing; and

(*f*) the reduction or elimination of taxation on imported equipment and fuel for the fishing industry.

9. The development of motorised surf boats or motorised beach landing craft is desirable since they dispense with the need for expensive fishing harbours and landing sites and also cater to the needs of boatless fishermen. Nationally sponsored schemes implemented regionally for experimentation with, and construction of, suitable standardised boats could help increase the pace of surf boat development.

10. Ice is the ideal means for short-term storage of fish but its relatively high cost in most developing countries limits its use in some situations. Some of the measures which could be taken to make ice cheaper to small-scale and artisanal fishermen and to assist them in improving the efficiency of its use in the preservation of fish are:

(*a*) the provision of small, compact ice boxes suitable for carriage in canoes and small boats;

(*b*) the introduction of small capacity ice-making plants at the community level financed by local industry or by governments;

(*c*) dissemination of information to fishermen on the basic principles of cold storage of fish and the efficient use of ice for this purpose;

(*d*) short-term price subsidisation of ice by producers until sufficient numbers of fishermen are convinced of the benefits which can be derived from its use; and

(*e*) non-conventional energy operated ice plants such as solar energy/wind power operated ice plants.

11. For some developing countries it has proved difficult to encourage fishermen to spend long periods at sea in order to fish the offshore grounds effectively. This difficulty can be overcome by such measures as:

(*a*) encouraging fishermen to land caught fish at suitable coastal locations distant from their home port;

(*b*) the carriage of suitably designed ice boxes in canoes and small boats;

(*c*) the fitting of canoes and small boats with diesel outboard engines which speed up passages to and from the fishing grounds;

(*d*) improving the design of small fishing vessels so as to provide more spacious and comfortable living and working arrangements; and

(*e*) training of fishermen.

12. Lack of credit and high interest rates are major problems in the fisheries of developing countries. Some of the means by which investment and credit schemes, including those of fishermen's co-operatives, can be made more effective include:

(*a*) increasing the flexibility of the terms of loans made to fishermen by banks and other financial institutions so as to make them more favourable than the loan conditions offered by middlemen and money lenders, who generally appreciate and understand the seasonal nature of fishermen's incomes and regulate the terms of their loan repayments accordingly;

(b) researching the issues and problems of fishermen's credit and interest rates on loans as a basis for action to overcome difficulties;

(c) establishment of a well-organised system of credit for small-scale fishermen from the local to the national level through institutional financing as government policy; and

(d) encouraging fishermen to run their own credit schemes by individual weekly or monthly contributions to be granted in turn to members of the schemes.

13. Efforts could be made to avoid exploitation by middlemen in the market chain between small-scale fishermen and the consumers of fish. In this regard fishermen's co-operatives and governments could assist in creating new fish marketing practices, for example by:

(a) the establishment of an official fish auction system in fish markets at which licensed agents carry out all wholesale fish trading; and

(b) the introduction by co-operatives of equally efficient but cheaper alternatives to the marketing of fish by middlemen, thereby strengthening the bargaining power of small-scale fishermen vis-à-vis middlemen.

14. Boatless fishermen in developing countries are not generally in a position to bargain for higher wages or a larger share of the catch since they are seldom organised in unions. Because of this they may also be readily hired or fired by boat owners. Feasible solutions to these problems are to facilitate boat ownership by boatless fishermen and to establish fishermen's organisations. Governments could provide assistance in establishing fishermen's co-operatives and associations which can afford to invest in fishing vessels crewed by boatless fishermen. Alternatively, government programmes for the local construction of fishing vessels as part of efforts to develop national fisheries can provide employment to boatless fishermen.

15. The ILO could provide technical assistance as a contribution to meeting the social and economic needs of artisanal and small-scale fishermen in the rural fishing communities of developing countries. Such assistance should include the establishment of their own organisations, the exchange of information and expertise among countries on issues and problems regarding the characteristics of fishermen's occupation, living standards and the various conditions prevailing in their communities. The assistance could also include the convening of an ILO tripartite meeting to draft specific guide-lines on practical steps which could be taken for small-scale fishermen at the national level to improve fishermen's labour and social conditions based on existing ILO international standards in this field. Governments, employers and fishermen should fully participate in such efforts, in particular by facilitating the exchange of information.

RESOLUTION ON FUTURE ACTION OF THE ILO FOR THE FISHING INDUSTRY[4]

The Committee on Conditions of Work in the Fishing Industry,
    Having met in Geneva from 4 to 13 May 1988,

---

4. Adopted unanimously.

Considering that the fishing industry is a key sector for world needs as a source of employment, food and wealth generation,

Noting that the International Labour Organisation is the international agency responsible for setting international standards for all workers, including fishermen, and that more instruments relating specifically to fishermen should be adopted,

Also considering that the ILO has not previously given this sector adequate attention;

Adopts this thirteenth day of May 1988 the following resolution:

The Committee on Conditions of Work in the Fishing Industry invites the Governing Body of the ILO to request the Director-General:

(1) to create urgently a standing committee for the fishing industry;
(2) to convene regular meetings of this committee at least every four years;
(3) to incorporate tripartite delegations when meetings or discussions are held with the FAO, relating to the fishing industry;
(4) to consider as possible agenda items for a future meeting of the fishing industry the following:
   —safety and health in the fishing industry;
   —vocational training in the fishing industry;
   —productivity in the fishing industry;
   —aquaculture in the fishing industry;
   —marine pollution and its impact on the fishing industry;
(5) to develop regional meetings for specific matters so as to complement the work of the standing committee;
(6) to pay increased attention to compiling and disseminating statistical information related to the fishing industry;
(7) to study the incidence and possible consequences of the use of flags other than those of beneficial ownership as regards employment and conditions of work in the fishing industry;
(8) to include on the agenda of a future session of the International Labour Conference items specific to the fishing industry, for example remuneration and earnings.

Selected Documents and Proceedings

Shipping and Maritime Technology*

International Maritime Organization

I. OBJECTIVES AND ACTIVITIES OF THE ORGANIZATION

1. The International Maritime Organization (IMO) is the Specialized Agency of the United Nations charged with the global mandate to formulate technical and related rules and standards in matters concerning maritime safety, efficiency of navigation and the prevention and control of marine pollution from ships.

2. The principal objective of the Organization, as specified in its Charter (the IMO Convention), is to provide machinery for co-operation among Governments in the field of governmental regulation and practices relating to technical matters of all kinds affecting shipping engaged in international trade. It has in this regard two main objectives:

    1. to encourage adoption of the highest practicable standards of maritime safety and efficiency of navigation; and

    2. to promote measures and procedures for the prevention and control of marine pollution from ships.

3. The Organization is also empowered and required to deal with administrative and legal matters related to its mandate.

4. In pursuing these twin objectives the Organization adopts or facilitates the adoption of international agreements and other suitable instruments establishing appropriate regulations, standards and procedures in all aspects of ship construction, operation and the prevention and control of marine pollution. The Organization also promotes the development of legal and administrative arrangements, at the international and national levels, for the effective implementation of the international regulations. It also provides a medium for channelling and directing assistance to enable all countries to participate in and contribute to this co-operative effort.

5. With a membership of 132 States and one associate Member, the Organization is fully representative of the global maritime community.

6.   The substantive work of IMO is carried out by the following bodies:

    —Maritime Safety Committee and its sub-committees;

    —Marine Environment Protection Committee;

    —Legal Committee; and

    —Facilitation Committee.

There is a Technical Co-operation Committee which is responsible for co-ordinating and reviewing the Technical Co-operation Programme of the Organization. The

*This document was presented at *Pacem in Maribus XVI*, Halifax, Nova Scotia, August 22–26, 1988. Further information on the work of International Maritime Organization may be obtained from the Secretariat at the Organization's Headquarters, 4 Albert Embankment, London SE1 7SR, England.

Technical Co-operation Programme is the main channel by which the results of the work of the substantive committees are made available to the developing countries.

7. The supreme governing body of the Organization, the Assembly, meets every two years. It receives, through the Council, reports of the technical committees and adopts the standards and other measures necessary to advance the work of IMO. In between sessions of the Assembly, the functions of the Assembly are performed by the Council which consists of thirty-two members elected by the Assembly for two-year terms. The Council normally meets twice a year.

8. While science and technology are relevant in general to all aspects of IMO's work, they are of primary importance in the work of the technical committees. These committees are intergovernmental in composition, and they function as highly specialized bodies of experts in which technicians and specialists of both developed and developing countries exchange views on matters of common interest and concern. This process can, and in many cases does, constitute an effective transfer of scientific and technological knowledge in the fields of shipping and prevention and control of pollution from ships, from the traditional developed maritime States to the States which are in the process of developing their maritime programmes and capabilities.

9. This process is reinforced and amplified in concrete situations through the Technical Co-operation Programme. The aim of the Programme is to help developing countries achieve self-reliance, on an individual or regional basis, by promoting or strengthening indigenous scientific, technological and managerial capacity in shipping and related matters.

10. Although technical in content and specialized in scope, the work of IMO's technical committees often leads to the adoption of international agreements or conventions: legal instruments by which governments agree to apply certain common standards and procedures on an international scale. There are now more than thirty such legal instruments, by reference to which countries organize or develop the various aspects of their national maritime programmes. In a large number of cases, however, the international standards are adopted and submitted to governments for implementation, without legal constraints, in accordance with the capabilities of individual States, from time to time. Such standards are contained in Codes, Recommendations, Guidelines, etc.

11. Ultimately, concrete implementation of these standards, whether of a legal nature or otherwise, depends on the political will and ability of countries to apply the high standards which they have established through IMO. Only the countries themselves can provide the political determination. But the technical ability, on the other hand, can be and is frequently acquired or enhanced by means of external help. It is in this area that IMO has consistently sought to play an ever-increasing role.

II. TECHNOLOGICAL AND RELATED CHANGES IN SHIPPING

12. Despite the tremendous changes that have taken place in transportation in recent years, shipping remains vital to international commerce and trade. Although aircraft now carry far more passengers than ships, freight movements between continents still go mainly by sea.

13. The movement of bulk commodities, such as oil, ores and grain, is carried out largely by ships. Developing countries are particularly dependent upon shipping to

export their products to international markets, since these cannot in many cases be reached by road, rail or other means. For geographical and historical reasons, sea communications between developing countries are also often more effective than other systems.

14. Shipping is, however, one of the most international of all industries. Not only do the fleets of developing nations have to compete with each other and with those from other maritime countries, but they also have to conform to the high standards laid down at an international level. Most of these regulations have been established by the International Maritime Organization, which is especially concerned with the improvement of safety at sea and the prevention of pollution from ships.

15. Since it commenced operation in January 1959, IMO has grown to become an Organization of 132 Member States—with virtually every maritime nation in the world as a Member. The Organization has adopted some thirty international conventions and protocols dealing with maritime safety, the prevention of pollution and related matters. The purpose of these instruments, and the many hundreds of codes and recommendations also developed by IMO, is to raise shipping standards in all countries to a level which is acceptable to and attainable by all. The Organization's success in achieving these objectives can be judged by the fact that several of the most important treaties now apply to more than 95 per cent of the world's fleet of merchant ships.

(a) Adequate Participation in World Shipping

16. As developing countries strive to increase their share in international shipping, their efforts must be assisted by strengthening their capacity to acquire or build up the essential infrastructure for safe, efficient and up-to-date shipping operations. IMO stands ready to play its proper role in this essential process; and pledges full support for the efforts of all countries to ensure the universal application of adequate standards of maritime safety and for the prevention of pollution from ships.

17. It is now generally accepted that the maritime transport of goods and passengers cannot be properly developed or safely operated without regard to the internationally established régime constituted by these agreements and standards.

(b) Development of Technical Maritime Infrastructure

18. During the preceding two decades, the shipping scene has undergone an unprecedented technological revolution. These years have seen the arrival of the mammoth tanker, the chemical carrier, the gas carrier, the container ship, the roll-on and roll-off vessel, etc. These vessels are provided with the most modern and sophisticated equipment and automation has made tremendous progress. Maritime development has therefore become far more difficult and far more complex than ever before.

19. The acquisition of ships is but one element and needs to be supplemented by the development of supportive on-board and on-shore services. Global technical standards for the construction and operation of vessels need to be implemented and enforced. Correspondingly, ports have to be built, modernized and managed properly in order to meet the requirements of today's shipping, ship repair yards need to be developed, radio and telecommunication networks must be operational to ensure

marine safety. All this, moreover, cannot be done without proper regard for the ecology of the ocean. Thus, pollution from vessels must be prevented and controlled, and, when it occurs, dealt with efficiently and expeditiously.

20. These are by no means insuperable problems but, to be successfully tackled, they should be approached in a pragmatic and co-ordinated manner so that progress may be achieved on a sound and viable basis. This fact, therefore, needs to be clearly recognized and adequately reflected in the national development plans of developing countries.

(c) Development of Maritime Expertise

21. The most serious handicap for the developing world is the inadequacy of trained national maritime personnel. For acquiring and operating ships in international trade it is essential for developing countries to have expert manpower in the form of master mariners, marine engineers, naval architects, communications specialists, and so on. And this personnel needs to be trained in conformity with current global standards. It is for this reason that IMO's inter-governmental bodies have accorded the highest priority to the establishment of adequate and modern training facilities in the developing countries.

22. Such training facilities require land and suitable buildings. But perhaps even more than the buildings, they require a vast amount of modern equipment on which the trainees have to be trained very intensively. Thus the establishment of training facilities of this kind is a fairly expensive undertaking.

23. The process of establishing modern maritime training facilities in the developing world has already commenced and has, indeed, made considerable headway. There are now many good national academies in the developing world as well as subregional or regional maritime academies which are being assisted by IMO. Examples are the Arab Maritime Transport Academy and the two regional academies in West Africa, one for French-speaking countries and another for English-speaking countries. But there is still so much more to be done. The developing countries still need assistance, in respect of equipment and foreign maritime experts, to name but a few. This is particularly so in the initial period while national personnel is being trained, whether as instructors or as operational personnel.

24. The manpower requirements of developing countries are very high even at their present low level of participation in shipping. With any significant increase in tonnage, these requirements will be critical. Many developing countries, therefore, realize that training has a much higher priority than the immediate formation or expansion of their national fleets. For it is now recognized that it is an essential starting-point to have at least a substantial nucleus of able and trained local people available before a country could have genuine independence and freedom of action in the shipping trade.

III. THE ROLE OF IMO

25. Although the task of developing technical conventions has been largely completed, existing instruments are updated regularly to keep pace with technological changes.

These have affected all branches of shipping, including the design, construction and equipment of ships; navigation; maritime communications; the stowage and handling of cargoes; and the equipment and operation of ports.

26. Because of the international and competitive nature of shipping it is important for all maritime countries to be aware of and involved in these developments, particularly as the pace of change seems likely to increase in the near future.

27. There are, for example, already indications that automation in shipping will increase during the next few years, possibly associated with reductions in crew manning levels. The introduction of electronic charts and satellite-based global positioning systems could have a profound effect upon maritime navigation, while the revolution in communications at sea is already well under way. This involves the phasing-out of radiotelegraphy, the introduction of telex-type systems and the use of satellite communications.

28. The introduction of these and other technologies will need careful control at the international level and IMO provides a forum for its Member States to develop acceptable measures.

29. IMO, through its technical co-operation programme, which places primary emphasis on training, seeks to assist all States, and particularly the developing States, in the acquisition, development and transfer of technology in accordance with its own Convention, the Charter of the United Nations and the Convention on the Law of the Sea. This function is performed in a variety of ways:

1. The exchange of information through the work of its inter-governmental bodies. These bodies utilize the expertise of industry and professional bodies and relevant international bodies, as well as the organizations and agencies of the UN system.
2. The provision of advice and assistance to developing countries on the establishment of appropriate institutions at the national, regional and global level.
3. The development of human resources through training and education of national personnel in the various disciplines of marine science and technology.
4. The promotion of co-operation between States at the regional, sub-regional and global levels, through conferences, seminars, etc.

30. The main subjects with which IMO will be concerned in the next few years are listed in the annex to this document.

IV. THE CHALLENGES FACING DEVELOPING STATES

31. One of the most important concerns of the developing countries in the maritime field has been, and continues to be, their low participation in world shipping tonnage. This has been a problem of constant concern and attention within the United Nations system. The General Assembly, at various ordinary and special sessions, as well as UNCTAD and the regional economic commissions, have all recognized as legitimate the aspiration of developing countries to increase without delay their share in the world merchant fleet. These aspirations, moreover, have been identified as indispensable elements in the attainment of the new international economic order.

32. Unfortunately, past efforts by developing countries to develop their merchant

marines have not always been fully successful. For example, in spite of the targets set by the General Assembly for the Second Development Decade, recent studies indicate little improvement in the position of developing countries in terms of ownership of the world's merchant fleets.

33. As far as shipping is concerned, the international position of developing countries reveals three major areas of urgent concern. In the first place, the share of developing countries in world shipping tonnage is less than is desirable. Secondly, the technical infrastructure for modern shipping needs considerable improvement. And third, most developing countries lack the requisite indigenous expert maritime personnel, without which no meaningful and long-term national programme can expect to be effectively implemented.

V. POSSIBLE ASSISTANCE FROM OR THROUGH IMO

34. It is clear that in practice the challenges confronting the developing countries require a comprehensive and well-thought-out approach. Not only training and manpower requirements, but all and every factor in shipping deserves attention and needs to be carefully evaluated. In the end, an overall assessment of the situation as a whole should precede the formulation of plans and the execution of specific programmes. In this respect, through its various activities and work programmes, IMO can help developing countries gain a head-start or move further along towards the harmonious development of their national shipping sector.

35. Assistance from IMO falls mainly within two categories. One results from the work of meetings and conferences sponsored by IMO. The other is by means of the technical co-operation programme, as such, of the Organization.

36. As stipulated in the IMO Convention, the Organization deals with "technical matters of all kinds affecting shipping engaged in international trade." These and closely related subjects are reflected in IMO current and long term programmes of work, whose execution is entrusted to its Organs and subsidiary bodies, i.e., the Assembly, the Council, the Maritime Safety Committee, the Legal Committee, the Marine Environment Protection Committee, the Facilitation Committee and the Technical Co-operation Committee. As comprehensive as possible, but flexible in approach, the work programme is subject to adjustments in the light of the exigencies of new shipping technology and other realities affecting maritime trade. When, as in the case of an accident, or as the result of an international conference, new problems or novel aspects of old ones come to light, IMO has shown the ability to respond with deliberate speed and efficiency. On a number of occasions in the recent past, the work programme has been adjusted and priority given to emergency problems and situations. Following the *Amoco Cadiz* disaster, for instance, the IMO Council established an *Ad Hoc* Working Group to conduct a study into the relevance to maritime safety and pollution prevention and control of the relationship between the shipmaster, the shipowner and the maritime administration. The Maritime Safety Committee gave urgent attention to the revision of steering gear standards as contained in a prior resolution of the IMO Assembly and the Marine Environment Protection Committee undertook an examination and made some recommendations on reporting systems for incidents while reaffirming its commitment to the promotion of regional co-operation in combating

marine pollution. For its part the Legal Committee has undertaken a comprehensive study of the legal problems brought to light by the disaster.

37. On the institutional side, the Maritime Safety Committee, IMO's principal organ in the field of maritime safety, has been transformed from a restricted Committee of sixteen Members to a Committee composed of all Members of the Organization. With the amendments adopted in the 1970s to the IMO Convention, the other Committees of IMO, the Marine Environment Protection Committee, the Legal Committee and the Technical Co-operation Committee—have been institutionalized in the Constitution of IMO as principal organs, and are open to participation by every Member of IMO. The Secretariat has also been restructured in line with the institutional changes in order to make it even more responsive to the changed requirements of the IMO organs and Member Governments of IMO.

38. It is the membership of IMO, now standing at 132 full Members and one Associate Member, which contributes the wealth of specialized expertise and technical knowledge that is brought to bear on the questions under consideration. Member Governments provide a broad range of specialized information, knowledge and relevant experience by sending to meetings, as required, naval architects and engineers, master mariners, technicians in radiocommunications and in the packaging and transport of dangerous goods, maritime safety administrators, experts in marine pollution, maritime insurance specialists and maritime lawyers, etc. Together they exchange information, discuss, adopt and amend standards and procedures, establish contacts and effectively participate in a process of transfer of technology. Furthermore, this work is assisted, as appropriate, by the contributions of observers from technical and specialized organizations associated with IMO, to whom individual items or questions may be of special interest and concern. The joint efforts of Member States, in what constitutes a true partnership between developed and developing countries, have resulted in positive developments in almost all areas of IMO's work. Those mentioned below will help illustrate the type of questions which affect shipping development in today's world.

39. In the field of maritime safety, work under the responsibility of the Maritime Safety Committee has achieved considerable success. Two important conventions adopted in 1972 for preventing collisions at sea and for promoting safety of container transport have entered into force. The revision of the Safety of Life at Sea Convention has progressed successfully with the adoption of a new Convention in 1974 and a major Protocol thereto in 1978. Further improvements were adopted in 1981 and 1983. In the related field of maritime search and rescue, a convention adopted in April 1979, entered into force in June 1985.

40. A further achievement in the field of maritime safety was the adoption in 1978 of the Convention on Standards of Training, Certification and Watchkeeping for Seafarers, the first such international instrument. The Maritime Safety Committee and its subsidiary bodies have also elaborated many important recommendations and codes of practice in the fields of navigation and marine technology. One of the most important of these codes is the International Maritime Dangerous Goods Code, which continues to grow in popularity and acceptability throughout the international maritime community. Substantial progress has also been made in the preparatory work for the Conference on Maritime Safety to be held in November 1988 for the harmonization of survey and certification requirements and the introduction of the Global Maritime

Distress and Safety System (GMDSS). The Maritime Safety Committee prepared "Measures to prevent unlawful acts against passengers and crews on board ships" pursuant to the request made by the Assembly by resolution A.584(14).

41. In the environmental field, developments have been no less significant. The adoption of the 1973 Marine Pollution Convention underlined IMO's concern with the prevention and control of marine pollution resulting from all ship-borne substances and not just pollution from oil; and the establishment of the Marine Environment Protection Committee (MEPC) provided the Organization with an inter-governmental organ with the appropriate status and expert membership to co-ordinate the Organization's activities in this important and sensitive area. In addition to considering matters relating to the implementation of the 1954 Convention for the Prevention of Pollution of the Sea by Oil, as amended, and to preparing for the early entry into force and effective application of the 1973 Marine Pollution Convention, the MEPC played a key role, with the Maritime Safety Committee, in developing and organizing the 1978 Conference on Tanker Safety and Pollution Prevention and the adoption of the 1978 Protocols relating respectively to the 1974 Safety of Life at Sea Convention and the 1973 Marine Pollution Convention.

42. Attention continues to be given to the implementation of MARPOL 73/78, in particular with respect to the regulations for the control of pollution by oil and noxious liquid substances carried in bulk. Work is currently under way to include maritime pollutants in the International Maritime Dangerous Goods Code (IMDG Code), so as to provide for the acceptance and implementation of Annex III of MARPOL 73/78.

43. In evidence of the universal acceptability of IMO as a positive force in the global environmental effort, the Contracting Parties to the 1972 London Dumping Convention decided unanimously to entrust to IMO the Secretariat functions under that Convention, a function which IMO has been discharging since 1976. The London Dumping Convention is generally recognized as the global regulatory framework for the prevention of such a form of pollution. A matter of far-reaching significance currently under consideration is the question of disposal of low-level radioactive waste. Further studies are being undertaken and a resolution calling for the suspension of this form of disposal has been extended. In addition, IMO provides the Administrative Secretary and supporting facilities for the Group of Experts on Scientific Aspects of Marine Pollution (GESAMP), the co-operative venture involving the United Nations and all the Specialized Agencies concerned with marine pollution.

44. In the legal field IMO has been responsible for the adoption of several major treaties on liability and compensation for damage caused by the maritime carriage of substances. Work has recently been completed in the preparation of a draft convention on salvage for submission to a diplomatic conference. Consideration of the subject of maritime liens and mortgages is proceeding in co-operation with the United Nations Conference on Trade and Development (UNCTAD), through an Intergovernmental Group of Experts established jointly by IMO and UNCTAD.

45. The Facilitation Committee has been made fully operational again after a lull in its work. The Committee has produced two sets of amendments to the Facilitation Convention, the first set was adopted by a Conference in March 1986 and the second set was adopted, for the first time, by the Committee itself, in September 1987 in accordance with the revised Article VII of the Convention.

46. Another major development in respect of an extremely important and urgent matter of interest to all Member States of the Organization is the preparation of

international treaty instruments for the suppression of unlawful acts against the safety of maritime navigation. On the initiative of the Governments of Austria, Egypt and Italy, the Council unanimously agreed, in November 1986, that the matter was appropriate for consideration within IMO and that it required urgent attention. Accordingly, the Council decided to establish an *Ad Hoc* Preparatory Committee to prepare a draft convention on the matter on a priority basis. This Preparatory Committee concluded its work most expeditiously and has prepared a draft Convention as well as a draft Protocol for submission to a diplomatic conference. At the request of the Council, these draft treaty instruments were examined by the Legal Committee, and the Committee's comments and observations were presented to a diplomatic conference held in March 1988. The Conference adopted, in March 1988, a Convention on the Suppression of Unlawful Acts Against the Safety of Navigation and a Protocol for the Suppression of Unlawful Acts Against Fixed Platforms Located on the Continental Shelf.

47. These developments—the increase in the membership of the Organization, the widening of its scope of activities and the acceptance of IMO as a universal body of competence and achievement in its specialized field—have, in their turn, necessitated a change in orientation and in emphasis in IMO and its work. After the successful conclusion of so many important conventions and standard-setting instruments, there is general agreement that much more emphasis should now be placed on action to obtain more effective and wider implementation of the standards and procedures provided for in the instruments. Thus, whilst the standard-setting function remains an essential part of the Organization's overall programme, much greater attention is now being given to ensuring that all Member States and other bodies concerned are able and willing to implement the established standards and procedures.

48. In 1981, the IMO Assembly adopted a resolution which proved to be a landmark instrument in the history of the Organization. For the first time, by resolution A.500(XII), the highest priority was given, not to the development and adoption of new conventions and technical measures, but to the promotion and implementation of existing international standards for maritime safety and for the prevention and control of marine pollution from ships.

49. The resolution therefore constituted a mandate for the future which was applicable to every aspect of the Organization's work, including its Technical Co-operation Programme.

50. Through this Programme, IMO fielded many advisory missions to practically every developing country and it was concluded that the common desire to build up maritime activities and to observe international standards was seriously handicapped by a shortage of skilled national personnel. Accordingly, it became apparent that if the resolution was to be followed, the Programme's future emphasis must lie in the transfer of specialized knowledge.

51. Since the early 1980s, IMO's Technical Co-operation Programme has been managed with one principal objective: to assist the developing countries to acquire the expertise and technology on technical and legal matters in shipping, with particular emphasis on implementation of internationally agreed standards for maritime safety and the prevention of marine pollution.

52. The Programme is designed to bridge the gap between developed and developing countries and thereby ensure benefits to both. The former gain from a steady improvement in safety and pollution prevention standards, while developing countries

have access to a unique springboard into the complex, high technology world of international shipping.

53. Accordingly, the aim of IMO's Technical Co-operation Programme is to promote change through the transfer of maritime expertise and technology.

54. The Programme is financed entirely through donor contributions, including the generous support of some developing countries, and it includes national, regional and interregional projects providing expert missions, fellowship training, educational material and equipment.

55. These are complemented by additional programmes for the award of fellowships, the organization of seminars, workshops and symposia as well as a range of associated advisory services to developing countries.

56. It is generally recognized that in the developing countries maritime transport is a fundamental tool for socio-economic well-being. Simply put, without maritime transport the developing countries would be unable to trade with their partners around the world. Indeed, it has been established that in Africa, for example, over 90% of the continent's international trade is carried by sea and passes through the ports.

57. Such wide use of shipping requires adequate infrastructure and expertise for the efficient management and operation of national maritime administrations, merchant and shipping fleets, maritime training centres, ports and associated shore-based industries.

58. With this in mind IMO has given the highest priority to the human element. For the developing countries to build up their maritime infrastructure, they must also have trained, experienced people.

59. Accordingly, IMO's policy is to ensure self-reliance through the training of national experts in the various fields required by the developing countries. In turn, highly-trained nationals will ensure the proper running of maritime transport and this will give rise to:
  –greater efficiency in international seaborne trade; and
  –the generation of economic growth.

60. At the heart of IMO's policy lies the need to provide the highest quality of training to the developing countries so that their nationals may become:
  –deck, engineering and radio officers for service on board merchant and fishing vessels;
  –maritime safety administrators;
  –surveyors and inspectors;
  –casualty investigators;
  –general maritime administrators;
  –ship designers and ship repairers;
  –maritime lawyers;
  –technical managers of shipping companies;
  –maritime teachers;
  –technical port managers;
  –harbour masters and pilots, etc.;
  –personnel for pollution control.

61. IMO has therefore put into effect a comprehensive strategy for the development of human resources by promoting the highest quality of training at all levels, the objective being the implementation of global standards by nationals who are as highly

trained as any of their counterparts world-wide. Through its different components, the strategy now covers all the elements required for a modern maritime infrastructure thus providing scope for developing countries to have experts in all the necessary fields.

62. IMO's maritime training strategy aims at ensuring the global and effective implementation of the standards adopted by its intergovernmental bodies.

63. To achieve this, IMO's Technical Co-operation Programme provides the highest quality of assistance in the field of training to the developing countries through an integrated network of maritime centres.

64. The strategy that has grown from IMO's experience has four phases:

(a) the provision of basic training;

(b) the provision of training for the award of certificates of competency;

(c) the provision of specialized training; and

(d) the provision of post-graduate training.

VI. THE ROLE OF IMO UNDER UNCLOS 1982

65. The 1982 United Nations Convention on the Law of the Sea affirms and recognizes that the rights and responsibilities of States to regulate navigation and to prevent and control pollution from vessels or by dumping must be exercised and discharged by reference to international standards. The Convention thus makes the relevant regulations and standards of IMO an integral part of the guidelines by reference to which the Convention's provisions are to be implemented. Furthermore, the Convention on the Law of the Sea recognizes IMO as the appropriate forum ("the competent international organization") for the development of international regulations and rules which are necessary for the effective implementation of the provisions of the Convention relating to navigation and the prevention, control and reduction of marine pollution from vessels or by dumping. Indeed many important provisions of the Convention envisage and expect continuing or new activities by IMO in these fields, and some of the provisions in the Convention will depend on the results of IMO's work for their interpretation and application.

66. In a number of the provisions dealing with safety of navigation and the prevention, reduction and control of marine pollution, the Convention on the Law of the Sea enjoins or empowers States to "take account of" or "to conform to" or "give effect to" or "implement" certain international standards developed by or through IMO. These are variously referred to as "applicable international rules and standards" or "internationally agreed rules, standards and recommended practices and procedures" or "generally accepted international rules and standards" or "generally accepted international regulations," "applicable international standards," or "applicable international instruments" or "generally accepted international regulations, procedures and practices."

67. The Convention on the Law of the Sea does not give formal definitions for these expressions, and no clear guidelines are provided as to how the "international regulations and rules, etc." referred to in the Articles may be identified. However, it appears to be generally accepted that the international regulations and standards adopted by IMO constitute a major component of the "generally accepted" international regulations and standards in matters relating to safety of navigation and the

prevention and control of marine pollution from vessels and by dumping. Therefore, these IMO regulations and standards will be of relevance to States and other entities involved in the interpretation or application of provisions of the Convention on the Law of the Sea dealing with such matters.

68. However, since there are no express provisions in the Convention identifying the regulations and rules which may be considered as "generally accepted" or "applicable" in particular contexts, States and other interested entities will expect some guidance with regard to the status of IMO regulations and standards in relation to the provisions of the Convention on the Law of the Sea. The need for guidance will apply not only in respect of the conventions and treaty instruments of IMO, but also in relation to the large body of important international rules, regulations, standards and recommended practices which have been adopted by IMO and embodied in Codes, Guidelines and Manuals, etc.

69. It is, of course, to be noted that formal and authoritative interpretations of the 1982 Convention's provisions can only be undertaken by the States Parties to that Convention or, in appropriate cases, by judicial or arbitral tribunals provided for that purpose in the Convention itself. Nevertheless the views and suggestions of IMO on these matters may be useful to States, particularly developing countries, in ascertaining the nature and extent of their obligations under the Convention on the Law of the Sea relating to maritime safety and the preservation of the marine environment.

70. It may also be necessary and useful for IMO to examine the interpretation and application of some of its own rules and standards in the light of the relevant provisions of the Convention on the Law of the Sea. Such an examination may assist Governments in taking measures to implement regulations of IMO while also discharging their responsibilities under the Convention on the Law of the Sea.

(a) The Development and Transfer of Marine Technology and International
    Co-operation (Articles 202 to 203 and 266 to 269)

71. The basic objectives of international co-operation, as spelt out in Articles 202 and 268, and especially "the development of human resources through training and education of nationals of developing States and countries" are already part of the fundamental aims of IMO and its Technical Co-operation Programme, as provided for in the IMO Convention and in the relevant decisions of its inter-governmental bodies. In implementing these aims, IMO may find it useful to refer, in appropriate cases, to some of the specific arrangements and measures suggested or envisaged in the relevant articles of the Convention on the Law of the Sea, particularly those relating to the transfer of technology and the provision of assistance to developing countries in the maritime field.

(b) Promotion of Global and Regional Co-operation for the Protection and
    Preservation of the Marine Environment (Articles 197 to 201)

72. IMO, together with other organizations, co-operates in the Regional Seas Programme of the United Nations Environment Programme (UNEP). In particular IMO has played a key role in the establishment of Regional Arrangements for Combating

Marine Pollution. These arrangements are directly pertinent to the provisions of the Convention on the Law of the Sea dealing with global co-operation. Also worth mentioning is the significance of IMO's participation in and contribution to the Group of Experts on Scientific Aspects of Marine Pollution (GESAMP) which brings together several agencies within the United Nations for the expert consideration and the undertaking of appropriate studies on scientific aspects of marine pollution. As the Organization which provides administrative secretariat services to GESAMP, IMO can make a significant contribution to the work of the participating organizations in furthering the objectives and purposes outlined in Articles 204 to 206.

(c) Development of National and Regional Marine Scientific and Technological Centres

73. The provisions relating to the development of national and regional centres as set out in Articles 275 to 277 of the Convention reflect in many respects the programmes which IMO has been promoting for some time in many areas of the world. In this connection, the World Maritime University constitutes a prime example at the global level of the kind of institution envisaged in the Articles, and the experience of the World Maritime University will be of direct relevance in this context. Also of relevance are the new Institutes being established by IMO in various maritime fields including the two new Institutes for Maritime Training and Maritime Law to be located respectively in Trieste and in Malta.

(d) Co-operation among International Organizations

74. The Law of the Sea Convention enjoins, in its Article 278, the competent international organizations to take all appropriate measures to ensure, either directly or in close co-operation among themselves, the effective discharge of their functions and responsibilities. In accordance with its Constitution and pursuant to decisions of its governing organs, IMO has established very co-operative and fruitful arrangements for collaboration with the United Nations and the other agencies and organizations within the United Nations system. However, IMO may need to explore appropriate avenues to promote and facilitate further co-operation with all international organizations whose activities may affect, or be affected by, the measures taken by the Organization with regard to matters dealt with by the Convention. In particular, it may be necessary to review the existing liaison with the Secretary-General of the United Nations and the existing organizations of the United Nations system in respect to matters which pertain to the field of responsibility or interests of the respective organizations. This is particularly so in the light of the views expressed at the IMO Assembly, at its thirteenth regular session, regarding possible "assistance by IMO to Member States and other agencies in respect of the provisions of the Convention on the Law of the Sea dealing with matters within the competence of IMO," and the development of "suitable and necessary collaboration with the Secretary-General of the United Nations on the provision of information, advice and assistance to developing countries on the law of the sea matters within the competence of IMO." Effective and co-ordinated liaison will also be needed with the International Sea-Bed Authority and the Interna-

tional Tribunal for the Law of the Sea when these bodies are established. Any such liaison and co-operation will be subject to the relevant provisions of the Convention on the Law of the Sea, and in accordance with the view of the IMO Assembly that IMO might provide "advice and assistance" to the Preparatory Commission for the International Sea-Bed Authority "on matters falling within the competence of IMO."

* * *

ANNEX

Maritime Safety

1. Implementation, technical interpretation and improvement of conventions, codes, recommendations and guidelines;
2. procedures for the control of ships including deficiency reports;
3. casualty statistics and investigations into serious casualties;
4. harmonization of surveys and certification requirements and authorization granted to non-governmental organizations to conduct surveys;
5. training, watchkeeping and operational procedures for maritime personnel, including seafarers, fishermen, maritime pilots and those responsible for maritime safety in mobile offshore units;
6. shipboard and shore-based management for the safe operation of ships;
7. measures to improve navigational safety, including ships' routing, requirements and standards for navigational aids and ship reporting systems;
8. the global maritime distress and safety system and other maritime radiocommunication matters including navigational warning services, shipborne radio equipment and operational procedures;
9. survival in case of maritime casualty or distress, and the provision of maritime search and rescue services;
10. safe carriage of solid bulk cargoes, timber, grain and other cargoes by sea, including containers and vehicles;
11. carriage of dangerous goods in packaged form, portable tanks, unit loads, other transport units, shipborne barges and intermediate bulk containers (IBCs);
12. carriage of bulk chemicals in offshore support vessels;
13. carriage of irradiated nuclear fuel in purpose-built and non-purpose-built ships;
14. emergency procedures and safety measures for ships carrying dangerous goods, medical first aid in case of accidents involving dangerous goods and the safe use of pesticides in ships;
15. safe handling and storage of dangerous goods in port areas;
16. intact stability, subdivision, damage stability and load lines of ships;
17. tonnage measurement of ships;
18. safety considerations for machinery and electrical installations in ships;
19. manoeuvrability of intact and disabled ships;
20. control of noise and related vibration levels on board ships;
21. matters pertaining to fire safety on board ships;

22. safety aspects of the design, construction, equipment and operation of specific types of ships, such as fishing vessels, oil tankers, chemical tankers, gas carriers, dynamically supported craft, mobile offshore drilling units, special purpose ships, offshore supply vessels, nuclear merchant ships, roll-on/roll-off ships, barge carriers, barges carrying dangerous chemicals in bulk and diving systems;
23. prevention of piracy and unlawful acts against ships;
24. IMO ship identification number scheme;
25. revision of the 1977 International Convention on the Safety of Fishing Vessels.

Legal

1. Offshore mobile craft;
2. novel types of craft, such as air-cushion vehicles, operating in the marine environment;
3. wreck removal and related issues;
4. ocean data acquisition systems (ODAS).

Marine Environment Protection

1. Promotion of technical co-operation, including the development of regional arrangements on co-operation to combat pollution in cases of emergency;
2. reception facilities for residues;
3. oily-water separators and oil discharge monitoring and control systems including those for light refined oils and oil-like substances;
4. surveys and certification of ships under MARPOL 73/78;
5. casualty investigations in relation to marine pollution;
6. arrangements for combating major incidents or threats of marine pollution;
7. promotion of regional arrangements for combating marine pollution;
8. development and updating of anti-pollution manual;
9. identification of particularly sensitive sea areas;
10. categorization of noxious liquid substances and harmful substances;
11. prevention of pollution by noxious solid substances in bulk;
12. measures on board ships to minimize the escape of pollutants in the case of accidents;
13. shipboard and shore-based management for the prevention of marine pollution.

Technical Co-operation

1. Advice and assistance to Governments of developing countries for the development of well co-ordinated and more efficient maritime transport systems;
2. the promotion of adequate infrastructure in the shipping and ports sector;
3. the promotion of appropriate self-reliance at the national level, the encouragement of subregional and regional co-operation, particularly technical co-operation among developing countries (TCDC);

4. advice and assistance to Governments of developing countries in taking appropriate measures to ratify important conventions and instruments of IMO and to implement their provisions;
5. the planning and organization of courses, workshops and seminars as and when the occasion arises to emphasize the importance of maritime training, maritime safety and prevention of marine pollution from ships.

Facilitation

1. Facilitation activities including:
   (a) promotional activities carried out in co-operation with Member Governments, Contracting Governments and organizations concerned; and
   (b) facilitation aspects of forms and certificates emanating from other activities of the Organization;
2. automatic data processing of shipping documents and documents used for clearance of ships;
3. formalities connected with the arrival, stay and departure of ships, persons and cargo.

Selected Documents and Proceedings

Memorandum of Understanding concerning an Environmental Management Plan for the North Fraser Harbour between the Department of Fisheries and Oceans (D.F.O.) and the North Fraser Harbour Commission (N.F.H.C.)

September 1988

MEMORANDUM OF UNDERSTANDING CONCERNING NORTH FRASER
HARBOUR ENVIRONMENTAL MANAGEMENT PLAN
BETWEEN: THE NORTH FRASER HARBOUR COMMISSION*

a body corporate established pursuant to the Harbour Commission Act, Chapter H-1, Revised Statutes of Canada, 1970, represented by the Chairman, North Fraser Harbour Commission (herein referred to as "N.F.H.C.")

of the FIRST PART

AND: THE DEPARTMENT OF FISHERIES AND OCEANS†

represented by the Director General, Department of Fisheries and Oceans, Pacific Region (herein referred to as "D.F.O.")

of the SECOND PART

WHEREAS N.F.H.C. is mandated, under section 9.0 of the Harbour Commissions Act, to regulate and control the use and development of all land, buildings and property within the limits of the Harbour, and under section 2.1 to manage and operate the Harbour in a manner that ensures the integrity and efficiency of the port system and the optimum deployment of resources, which are of critical importance to the well being of the B.C. and Canadian economies;

AND WHEREAS D.F.O. is responsible, pursuant to sections 20, 28, 31, and 33 of the Fisheries Act, for the protection of fish habitats, which are essential for the maintenance of viable B.C. and Canadian commercial, domestic and sport fisheries;

AND WHEREAS both N.F.H.C. and D.F.O. are signatories to an Agreement respecting a Fraser River Estuary Management Program; wherein Section 2(f) defines the program as including those activities designed to guide the management of the estuary's resources and any activities which may occur in place of, in addition to, or as a result of continual efforts over time to implement the aforementioned program;

AND WHEREAS both D.F.O. and N.F.H.C. are implementing members of the Fraser

*Contact address: Mr. George Colquhoun, Port Manager, North Fraser Harbour Commission, 2020 Airport Road, Richmond, B.C. Canada, V7B 1C6; telephone (604) 273-1866; FAX 273-3772.

†Contact Address: Mr. Otto Langer, Head, Habitat Management Unit, Fraser River, Northern B.C., Yukon Division, Department of Fisheries and Oceans, 80 6th Street, Room 330, New Westminster, B.C., Canada V3L 5B3; telephone (604) 666-0315; FAX 666-6627.

River Estuary Management Program and as such share a responsibility for coordinating management committee activities and implementing the Program;

AND WHEREAS N.F.H.C. and D.F.O. under the mandates of the Ministers of Transport and Fisheries and Oceans, respectively, desire to maintain and improve the quality of aquatic environments of the North Fraser Harbour in order to protect and conserve the fishery resources of the Fraser River while at the same time, accommodate socio-economic developments required to support a growing population and economy;

AND WHEREAS, in support of their mutual commitment to greater cooperation, a joint N.F.H.C./D.F.O. inventory was completed in July/August 1986, and described in the report entitled "North Fraser Harbour shoreline habitat inventory" (September 1986), which presents a classification system for fish habitats in the North Fraser Harbour which would serve as a basis for development of a joint North Fraser Harbour Environmental Management Plan;

AND WHEREAS D.F.O., in consultation with other agencies, wishes to proceed with implementation of a North Fraser Harbour Environmental Management Plan with the N.F.H.C.;

AND WHEREAS under the D.F.O. "Policy for the Management of Fish Habitat," the strategy of Integrated Resource Planning encourages D.F.O. to enter into agreements to achieve resource planning and management objectives and to carry out joint programs;

AND WHEREAS N.F.H.C. and D.F.O. have agreed to establish a joint North Fraser Harbour Environmental Management Plan which will constitute an integral part of the Fraser River Estuary Management Program implementation strategy;

NOW THEREFORE:

N.F.H.C. and D.F.O. agree to proceed with the implementation of a joint North Fraser Harbour Environmental Management Plan as described in this memorandum and associated attachments.

DEFINITIONS

For the purposes of this memorandum, it is agreed by and between the Parties as follows:

(a) "Parties" means the North Fraser Harbour Commission (N.F.H.C.) and Department of Fisheries and Oceans (D.F.O.);

(b) "North Fraser Harbour" includes all lands and water areas under the legal jurisdiction of the North Fraser Harbour Commission;

(c) "Productive capacity" is a habitat's maximum ability to support healthy fish or other biological material upon which fish depend. This maximum productive capacity is set by physical, chemical and biological characteristics of the site. Productive capacity is often confused with productivity which is a measure of a habitat's current yield of fish and other biological material such as fish food organisms or organic matter which are necessary to directly or indirectly support that yield of fish. For example, a site can have very low productivity due to extensive smothering by wood debris and stranded logs, but because elevations and soil quality are suitable for marsh growth, the site's productive capacity is high;

(d) "Productivity" is an index of a habitat's present ability to produce fish, fish food organisms, and detrital matter which are directly or indirectly necessary to support that fish production;

(e) "Mitigation" means those actions taken during the planning, design, construction and operation of works and undertakings to alleviate potential adverse effects on the productive capacity of fish habitats;

(f) "Compensation" means the replacement of natural habitat alienated by development by creating 'new' habitat and/or the increase in productive capacity of existing habitat to offset unavoidable habitat losses where mitigation and other strategies are not adequate. Examples include lowering upland elevations to create an intertidal marsh area, or raising mudflat elevations to permit intertidal marsh establishment, both of which increase the site's productive capacity;

(g) "Habitat restoration" means the cleanup or other action taken for the purpose of increasing the productivity of fish habitats which have been altered, disrupted or degraded;

(h) "Habitat development" refers to the creation of new habitat by converting upland into shallow subtidal or intertidal areas capable of sustaining marsh and/or other aquatic communities, or the improvement of the productive capacity of existing habitats;

(i) "Habitat compensation bank" is an agreed upon mechanism or process where habitat is developed for the purpose of providing compensation for future commercial or industrial development projects requiring compensation;

(j) "Habitat zones" are the major habitat types within the North Fraser Harbour situated between, or immediately adjacent to low and high water. They include riparian vegetation, intertidal high and low marsh, intertidal mud and sandflats and shallow subtidal areas;

(k) "Harbour" refers to the North Fraser Harbour;

(l) "Plan" refers to the North Fraser Harbour Environmental Management Plan which is described in this Memorandum of Understanding and associated attachments;

(m) "Reach" is a discrete geographic unit within the North Fraser Harbour, with the specific boundaries determined from biological and administrative considerations.

PURPOSE

The purpose of this memorandum is to describe an environmental management plan for the North Fraser Harbour intended to facilitate a pro-active approach toward port development and environmental management by:

(a) establishing a shoreline rating or classification system to assist port administrators and habitat managers in maintaining an economically and environmentally viable harbour;

(b) developing a more detailed project review and assessment process, which will be integrated with existing Fraser River Estuary Management Program (FREMP) coordinated project review procedures;

(c) creating a habitat compensation banking system to facilitate compensation for industrial, port, commercial, or recreational developments where on-site mitigation or compensation would be impractical or inadequate and where off-site compensation would be considered acceptable;

(d) establishing a cooperative management program designed to make fish management and fish habitat concerns an integral component of Harbour operations and development planning processes.

IMPLEMENTATION

1.0 Administration

1.1 *General:* It is intended by the Parties hereto that:
(a) The Parties will cooperate fully to implement the Plan by exchanging information, and meeting to discuss management plans and issues as required.
(b) The Parties will work together in a constructive manner with other government agencies, municipalities, industries and the public, to ensure that the most appropriate actions are taken under the Plan.
(c) The Parties will make use of existing mechanisms (e.g., FREMP coordinated project review) to implement projects under the plan.
(d) The N.F.H.C. will be responsible for liaising with all levels of government and non-government development interests in the harbour to obtain and provide initial advice on suitable siting and appropriate mitigation requirements.
(e) The N.F.H.C. will be responsible for general administration and record keeping for programs initiated under the Plan, preparing annual reports on the status of the Plan and acting as Lead Agency for coordinating the review of development proposals in the Harbour.
(f) D.F.O., in concert with other environmental agencies, will be responsible for providing technical guidance and support and policy advice for habitat development initiatives in the Harbour.
(g) The Port Manager will be the representative of N.F.H.C. for negotiations and management of the Plan.
(h) The Head, Habitat Management Unit, Fraser River, Northern B.C., and Yukon Division will be the representative of D.F.O. for negotiations and management of the Plan.
(i) The representatives of the N.F.H.C. and D.F.O. identified in sections 1.1 (g) (h) will constitute a permanent management steering committee which will be responsible for implementation of this plan, and will advise the Chairman N.F.H.C. and Director D.F.O., Fisheries Branch.
(j) The Parties may utilize consultants to assist in implementation of the Plan.
(k) Program initiatives will be developed on an annual basis during meetings between N.F.H.C. Port Manager and D.F.O. Head, Habitat Management Unit, their appointed assistants and (as appropriate) other environmental agencies.
(l) Program plans and shoreline classifications developed pursuant to this Agreement will be consistent with or complementary to FREMP activity programs and area designations.

1.2 *Financial*
(a) The N.F.H.C. will contribute $75,000.00 annually towards the Plan. This fund should be dedicated to habitat development, restoration, program coordination and the Cooperative Management Program.
(b) Each Party shall bear those costs of salaries and travel, and any administrative costs for implementing the Plan which are incurred by their representatives.

(c) The Parties may initiate cost sharing sub-agreements to further the purposes of this Agreement.

2.0 Environmental Habitat Management Plan Components

2.1 *Habitat Inventory and Shoreline Classification System*
(a) An inventory and shoreline classification map, which is referenced in the "North Fraser Harbour shoreline habitat inventory" report (September 1986) will serve as the basis for determining the acceptability of shoreline areas for future industrial or commercial development and the extent of mitigation and/or compensation which may be required (see fig. 1).
(b) The Parties will continue to work cooperatively to refine and update the inventory and develop additional biological criteria in support of habitat classifications and ratings.

2.2 *Project Review and Assessment Procedures*
(a) Project assessment procedures for future development within the Harbour will:
   i) follow the FREMP coordinated project review process, and
   ii) require proponents to prepare and submit a habitat map which is supplemental to the FREMP application and which provides additional project specific information.
(b) Consistent with "A Policy for the Management of Fish Habitat," D.F.O. will apply an established hierarchy of preferences to proposed industrial or commercial developments to ensure "no net loss" of fisheries habitat productive capacity.
(c) D.F.O. guidelines for industrial activities such as dredging, blasting, instream construction, and foreshore development, will be applied to all proposed developments in the Harbour.

2.3 *Habitat Compensation Bank*
(a) A Habitat Compensation Bank administered by the N.F.H.C. will be established as part of the Plan.
(b) The Habitat Compensation Bank may be used (where approved by D.F.O.) as compensation for unavoidable habitat losses resulting from commercial or industrial developments in yellow classification areas.
(c) The Habitat Compensation Bank will be composed of habitat credits accumulated by the development of aquatic habitats in the North Arm. These will be drawn upon and used as habitat compensation only when all feasible mitigation, relocation or on-site compensation alternatives have been thoroughly reviewed and considered insufficient.
(d) The N.F.H.C., as administrator of the Habitat Bank, will issue habitat credits to qualifying developers, subject to recommendations of the FREMP Environmental Review Committee (ERC) and DFO approval.
(e) The N.F.H.C. shall maintain complete records concerning the operation of the Bank and produce an annual report outlining the activities and habitat balance of the Bank.
(f) The N.F.H.C. will be responsible for general monitoring of habitats which have been developed for banking purposes.

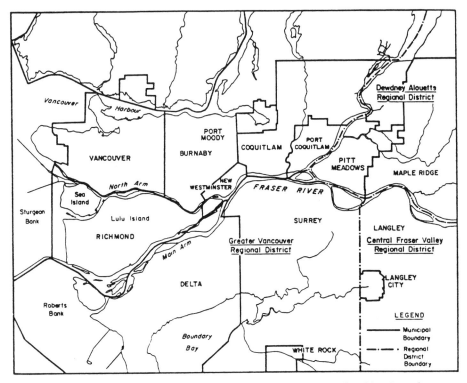

FIG. 1.—Greater Vancouver Regional District, British Columbia, Canada

(g) D.F.O. may undertake comprehensive site-specific monitoring studies of selected habitats to audit habitat development.

2.4 *Cooperative Management Programs*

(a) The Parties will establish a cooperative management program as outlined in Appendix I[1] to:

i) improve fish habitat and contribute to improved water quality of the Harbour;

ii) improve harbour keeping and industrial operating practices and develop mechanisms to actively improve degraded shoreline habitats;

iii) encourage dialogue and increase habitat awareness among harbour user groups;

iv) increase joint surveillance of activities in the Harbour which have the potential to adversely affect fish habitat and the related environment;

v) initiate applied research studies to address topics of mutual concern.

1. Editors' note.—Appendix I is not included here.

AMENDMENTS AND TERMINATION

(*a*) Amendments to the Plan may be made by mutual consent of the Parties.

(*b*) This memorandum will continue in force for 5 years whereupon it may be renewed for a further term as mutually agreed upon by the parties, or until one party formally provides 1 year notification to the other of its intent to terminate the memorandum.

MISCELLANEOUS

(*a*) Other government agencies may become parties to the Plan if this is agreeable to both the N.F.H.C. and D.F.O.

(*b*) This memorandum and all programs initiated pursuant to this memorandum will be reviewed annually.

(*c*) This memorandum will not operate to vest in either of the Parties any proprietary right, interest or obligation that it otherwise would not have.

(*d*) This memorandum and none of its conditions shall in any way abrogate, replace, substitute, or derogate from any of the rights, powers or jurisdictions of Canada including those pursuant to the *Fisheries Act*, R.S.C., any other Act of Parliament or which Canada may have otherwise have had at law or in equity prior to the execution of this memorandum.

(*e*) No member of the Parliament of Canada or Commissioner of the North Fraser Harbour Commission will hold, enjoy or be admitted to any share, part or benefit from this memorandum or any agreement, contract or benefit arising therefrom.

(*f*) This memorandum will be governed by and construed in accordance with the laws of Canada.

(*g*) In this memorandum, wherever the singular or masculine is used, it will be construed as if the plural or feminine or body corporate, as the case may be, had been used where the context or the parties hereto so require.

(*h*) The Parties hereto agree to cooperate on the release of any announcements concerning the undertaking of this memorandum or any related agreements, and to provide due credit and recognition.

IN WITNESS WHEREOF the Parties hereto execute this Memorandum of Understanding the 7th day of September 1988.

The CORPORATE SEAL of the North Fraser Harbour Commission was hereunto affixed in the presence of:

(signed) I. M. Frith
(Chairman)

(signed) T. Hurschman
(Secretary)

Signed on behalf of the Department of Fisheries and Oceans and in the presence of:

(signed) D. L. Deans
(Witness)

(signed) Thomas E. Sidden
Minister of Fisheries and Oceans

Selected Documents and Proceedings

Pacem in Maribus XVII

MOSCOW, U.S.S.R., JUNE 26–30, 1989

Introductory Remarks by Elisabeth Mann Borgese

Let me begin with a quote from *Perestroika:*

> And if the Russian word "Perestroika" has easily entered the international lexicon, this is due to more than just interest in what is going on in the Soviet Union. Now the whole world needs restructuring, i.e., progressive development, a fundamental change. [MIKHAIL GORBACHEV][1]

The message we hope to elaborate during these next few days is that the new regime for the oceans emerging from the United Nations Convention on the Law of the Sea is indeed a first piece of international *perestroika*, a restructuring, a progressive development, a fundamental change in an area that covers over 70 percent of the surface of our globe, on whose resources national economies will increasingly depend, and which is a crucial component of common and comprehensive security. Common in the sense that the security of each can only be founded on the security of all; comprehensive in the sense that it must comprise economic security as well as environmental security together with military security.

Nowhere are these three aspects of comprehensive security more inseparably linked than in the oceans. It is in the oceans that denuclearization is a *must* for the protection of the marine environment, which is a *must* for the development of marine resources.

Perestroika suggests the building of regional security systems: in the Pacific, the Mediterranean, the Indian Ocean, the Arctic, and the Antarctic. Regional cooperation and new forms of regional organization constitute one of the major developments triggered by the new Law of the Sea. Clarifying and utilizing the linkages between environmental security and denuclearization and peace, the UNEP-initiated Regional Seas Programme has indeed a great potential for making a major contribution to such regional security systems.

There are yet other ways in which the new Law of the Sea can be considered a pilot experiment for *perestroika*. *Perestroika* stresses the importance of new forms of economic/technological, scientific/industrial international cooperation, such as joint ventures, new forms of mutually advantageous cooperation, and the implementation of joint projects in high technology and information technology. Such undertakings, between East, West, North, and South, could serve not only the cause of economically,

---

1. Mikhail Gorbachev, *Perestroika: New Thinking for Our Country and the World* (San Francisco: Harper & Row, 1987).

socially, and environmentally sustainable development. They would also constitute measures of confidence building of considerable importance.

The Law of the Sea provides the most advanced institutional framework for the implementation of such proposals.

A joint venture for research and development in sea-bed mining technology that would cover practically the whole range of high technologies could be established under the aegis of the International Sea-Bed Authority and, until the coming into force of the Convention, under the aegis of the Preparatory Commission.[2] Proposals to this effect have been put forward, over the years, by the Delegations of Austria and Colombia.[3] If these were to be followed up, there would be a solid first piece of international *perestroika*.

The Regional Centers, which should be established under Articles 276 and 277 of the Convention on the Law of the Sea, could be organized on the basis of the same new principles of Joint R&D Ventures. A proposal by the Government of Malta is before the United Nations and has been acted upon by UNIDO and UNEP and supported by Unesco and FAO.[4] We hope that the first Center will be established for the Mediterranean shortly. This could serve as a pilot experiment for the establishment of similar centers in other oceanic regions. These would advance the broad internationalization of marine scientific research and the normalization of international relations in the economic, information, and ecological areas that *perestroika* seeks to achieve.

The linkage between environment and development, unbreakable in ocean economics, may in fact become the real point of convergence between free-market and centrally planned economies and economic theories. Clearly, both systems are now transcending traditional limits in their search for new parameters, new indicators for sustainable growth and real wealth, and human welfare. The result, conceivably, could be what we call the Economics of the Common Heritage of Mankind, common and acceptable to East, West, North, and South.

Perestroika calls for a binding dispute settlement process within the United Nations system. The Law of the Sea Convention provides the most comprehensive, flexible, and binding system ever devised by the international community. It might provide a pattern for the United Nations as a whole.

Last, not least, *Perestroika* states:

2. Editors' note.—Each volume of *Ocean Yearbook* contains reports on the developments of the Preparatory Commission. For example, in this volume, refer to the statements in Appendix B by the chairmen of the Special Commissions 1 and 2 to the Plenary of the Preparatory Commission, and see the article by Elisabeth Mann Borgese, Mahinda Perera, and Aldo Chircop, "The UN Convention on the Law of the Sea: The Cost of Ratification"; and, in *Ocean Yearbook 7*, ed. Elisabeth Mann Borgese, Norton Ginsburg, and Joseph R. Morgan (Chicago: University of Chicago Press, 1988), Elisabeth Mann Borgese, "Implementing the Convention: Developments in the Preparatory Commission," pp. 1–7, and "Guidelines for a Training Programme: Plans for Training Staff for the Enterprise," pp. 462–68, and "The International Venture: Study Submitted by the Republic of Colombia to the Preparatory Commission for the International Sea-Bed Authority and for the International Tribunal for the Law of the Sea, Special Commission 2," pp. 469–79, both in App. B of *Ocean Yearbook 7*.

3. Editors' note.—For the Colombian proposal, see "The International Venture."

4. The proposal is outlined in the recommendations of *Pacem in Maribus XVII*; see recommendation 34.

The necessity of effective, fair, international procedures and mechanisms which would ensure rational utilization of our planet's resources as the property of all mankind becomes ever more pressing.

The Law of the Sea has set the first precedent by declaring the deep sea-bed and its resources to be the Common Heritage of Mankind and has provided precise mechanisms to ensure rational utilization of these resources for the benefit of all mankind. The technical implementation of these provisions constitutes a great common challenge and a unique opportunity.

We are here to explore the potential of the new Law of the Sea for the building of an international order based on these new principles and for enhancing new forms of international cooperation and organization in the economic, environmental, and technological fields that are apt to strengthen comprehensive and common security in the oceans, *Pacem in Maribus,* and in the world.

We thank our Soviet hosts, not only for their generous hospitality to this Conference, but for providing with *perestroika* an inspiration for our discussions and a hope for their success.

PACEM IN MARIBUS XVII

"Peace in the Oceans: The New Era"

DECLARATION

Pacem in Maribus XVII convened in Moscow from June 26 to June 30, 1989. Nearly 600 persons, including representatives of international organizations and of the media, from nearly 50 countries, took part in the discussions. The conference took note of the considerable interest accorded to its deliberations in Moscow. Ideas, to become reality, require their own time. *Pacem in Maribus XVII* took place at the right time at the right place, given its theme, "Peace in the Oceans: The New Era."

Pacem in Maribus noted that Mikhail Gorbachev's book, *Perestroika,* calls for the restructuring of the international order and the pursuit of common and comprehensive security, including economic and environmental security, together with military security. The new order for the seas and oceans emerging from the 1982 United Nations Convention on the Law of the Sea should be considered as a first step in international *perestroika*—a restructuring, a fundamental change in an area covering over two thirds of the surface of the Earth.

Pacem in Maribus XVII calls for the early ratification and implementation of the United Convention on the Law of the Sea, the progressive development of the law of the sea, and its optimal utilization for the enhancement of common and comprehensive security and the peaceful settlement of disputes.

Pacem in Maribus XVII urges the inclusion of the naval arms race in the ongoing negotiations on arms control and disarmament. The greatest danger to world security is the growing naval arms race, especially its nuclear aspect. Urgent measures are needed for the disengagement of naval forces, including submarines, the prevention of naval accidents on the high seas, and the withdrawal of naval vessels from positions from which they could strike at the territories of foreign states.

The Conference urged the United Nations General Assembly to convene, as soon as possible, a World Conference to deal with naval disarmament in a comprehensive manner.

The Conference appealed to member States of the United Nations to accelerate the establishment of nuclear weapon free zones in the Baltic, the Sea of Japan, and the East China Sea, and of zones of peace in the Indian Ocean and in the South Atlantic in implementation of the Declarations by the United Nations General Assembly, as well as in the South Pacific, the Mediterranean, and the Arctic. *Pacem in Maribus XVII* urged all States to fully implement the Treaty provisions declaring Antarctica and the surrounding waters a nuclear weapon free zone.

Pacem in Maribus XVII considered confidence-building measures and measures of openness in the naval sphere as an important way to improve the situation in the seas and oceans. Confidence-building measures must become not only a starting point for future steps of limitation and reduction of naval armaments, but also an important independent factor contributing to the strengthening of stability and predictability in the naval sphere.

Technology is one of the dynamic forces in the world today. Rapid advances in this field and unequal structures for absorbing high technology are leading to a widening technological gap which could become a potent destabilizing force threatening the economic security of States in all parts of the world. There is an urgent need, therefore, for all States, East, West, North, and South, to cooperate in the sphere of technology development through international joint ventures and new forms of scientific/technological/industrial cooperation.[5] This would not only be economically beneficial and enhance the stability of the international system; it would be an important confidence-building measure as well.

Development in the past has led to growing stress on the biosphere, threatening the environmental security of States and people everywhere. The solution to this problem obviously cannot be sought in stopping economic development but in finding a synthesis between development and environmental concerns and in establishing strategies that lead to sustainable development. This requires the development of environmentally sound technologies, the establishment of international environmental impact assessment systems, and the adoption of appropriate value systems that regard the ecosystem as the Common Heritage of Mankind.

Pacem in Maribus XVII examined in depth the economic, environmental, military, cultural, and institutional implications of the concept of the Common Heritage of Mankind and urged its expansion and the full utilization of its potential for the strengthening of common and comprehensive security in the oceans, joint technology development and, in the words of *Perestroika*, "the rational utilization of our planet's resources as the property of all mankind."

Pacem in Maribus XVII expressed its support for the ideals and goals of "Mir na Moriach," the Soviet organization "Peace to the Oceans."

Pacem in Maribus XVII addressed this Declaration and the Recommendations annexed to it, to the Governments of all States, to the United Nations and its competent

5. See "Guidelines for a Training Programme"; "Report of the International Ocean Institute: 1989," App. A, in this volume; and M. C. W. Pinto, "Transfer of Technology under the UN Convention on the Law of the Sea," *Ocean Yearbook 6,* ed. Elisabeth Mann Borgese and Norton Ginsburg (Chicago: University of Chicago Press, 1986), pp. 241–70.

Agencies and Institutions, to the Non-Aligned Movement, to the OECD and COM-ECON and other relevant intergovernmental and nongovernmental organizations.

RECOMMENDATIONS OF *PACEM IN MARIBUS XVII*

Panel I

1. States should progressively reduce their naval armaments and eliminate all nuclear weapons from the oceans.

2. The most urgent naval disarmament measure is the elimination of tactical nuclear weapons.

3. Maritime powers should be encouraged to abolish naval bases on the territories of foreign states. Naval ports should be converted to civilian use.

4. The establishment of nuclear weapon free zones, zones of peace, and zones of naval disengagement in various areas, particularly in the Mediterranean, the Baltic, the North Atlantic, the Arctic, the Sea of Japan and other waters adjacent to the Korean Peninsula, the Indian Ocean, and the Southern Ocean, should be encouraged.

5. Naval arms control measures should be integrated with other arms control measures in a comprehensive approach to disarmament.

6. Support should be given to the proposal to convene in 1990–1991 a world conference on the demilitarization of ocean space and the stopping of the naval arms race.

7. The proposal put forward by the Government of Iceland, and made in the Disarmament Conference by Bulgaria, Finland, GDR, Indonesia, Sweden, and USSR, to hold an international conference on the limitation of naval activities in the Northern Atlantic should be supported.

8. States should introduce legislation ensuring the peaceful uses of the oceans in the interest of all mankind and to promote security, mutual confidence, and economic development in the world. States should endeavor to contribute constructively to the progressive development of international maritime law.

9. The establishment of national and regional Marine Research and Development Centers should be encouraged.

10. Nongovernmental organizations and public opinion should be mobilized for citizens' diplomacy action to press on governments to adopt confidence-building measures and move toward the denuclearization of the oceans.

Panel II

11. The concept of the Common Heritage of Mankind should be accepted as a fundamental principle of international law.

12. Declarations on the Common Heritage of Mankind in relation to new subject areas, such as the genetic heritage, Antarctica, and the global climate need to be supplemented by the establishment of authoritative structures and legal instruments.

13. The concept of the Common Heritage of Mankind should not be construed as contradicting universally recognized principles of international law.

Panel III

14. The phrases "reserved for peaceful purposes" and "exclusively for peaceful purposes," contained in the 1982 Convention on the Law of the Sea, should receive agreed international definition for the promotion of peace.[6]

15. The Conference on Disarmament, in consultation with States Parties to the Sea-Bed Treaty of 1971, should, in implementation of Articles V and VII of the Treaty, proceed without further delay with consideration of measures in the field of disarmament for the prevention of an arms race on the sea-bed, the ocean floor, and the subsoil thereof.

16. A global agreement should be made, and bilateral arrangements entered into, for the prevention of collisions and other incidents at sea between military vessels. This would provide predictability and be an important confidence-building measure.

17. Port states should extend the regime covering safety of life and property at sea, and continue to devise appropriate measures to implement it.

Panel IV

18. Measures should be taken, within the framework of the 1982 Convention of the Law of the Sea, to facilitate access of scientific research vessels to foreign ports.

19. Measures should be taken in accordance with the recommendations of the Brundtland Commission to balance the responsible use of marine resources with the preservation of the marine environment.

20. Measures should be taken to promote maritime tourism while eliminating its deleterious effects.

21. There should be cooperation between all nations, whether North, South, East, or West, in the development of marine technology, especially through measures aimed at codevelopment of technology.

22. New forms of cooperation between Western and Eastern Europe, such as a multinational consortium for the transportation of goods between Europe and other parts of the world should be developed. This could constitute an important contribution to a more efficient, equitable, and sustainable global transport system.

Panel V

23. States lacking legislation for the control of land-based sources of marine pollution should enact legislation which is no less stringent than is provided for in the 1982 Convention on the Law of the Sea.

24. The U.S.S.R. and the U.S.A. should cooperate more fully than hitherto in matters concerning the protection and preservation of the marine environment.[7]

6. Editors' note.—See Boleslaw Boczek, "The Peaceful Purposes Reservation of the UN Convention on the Law of the Sea," in this volume.

7. Editors' note.—As, e.g., in the case of the U.S.-Soviet working group, Conservation and Management of Natural and Cultural Heritage, which is exploring the possibility of jointly

Panel VI

25. States Parties should be encouraged to peacefully resolve their disputes. To this end, disputes should not be left unsettled, and parties should avail themselves of the dispute settlement procedures of the 1982 Convention on the Law of the Sea.

26. The use of arbitration should be encouraged as a preferred procedure of dispute settlement within the scheme of the 1982 Convention on the Law of the Sea.

27. The use of the UNCITRAL Model Arbitration Act and rules of procedure should be encouraged for both public and private international commercial arbitration.[8]

28. Non-signatory States to the 1982 Convention on the Law of the Sea should be encouraged to make use of the UNCLOS dispute settlement system.

Regional Security Systems

The Mediterranean[9]

29. The Mediterranean should be proclaimed a zone of peace and cooperation in the same spirit and with the same aims as those contained in the relevant Resolutions of the UN General Assembly.

30. Foreign fleets, including submarines, should be withdrawn from the Mediterranean.

31. Mediterranean coastal states should endeavor to establish a collective system of multi-purpose monitoring and surveillance, and common and comprehensive security.

32. Mediterranean states should cooperate in the establishment of a Mediterranean Centre for Research and Development in Marine Industrial Technology to enhance North-South cooperation in the joint development of socially relevant and environmentally safe technology. Such a center could become a model for centers to be established in other regions in pursuance of Articles 276 and 277 of the 1982 Convention on the Law of the Sea.

33. Mediterranean states should agree, through an additional protocol to the Barcelona Convention or otherwise, to levy a small tax on tourists. The tax should not constitute an unreasonable burden, and a cost-effective collection system should be established. The proceeds should go to the Trust Fund of the Mediterranean Action Plan and thereafter be utilized to strengthen environmental security in the Mediterranean.

---

managed parks in the eastern Siberian/Alaskan Bering Strait region. The areas are so huge and remote that several would probably be designated international biosphere reserves under the Unesco Man and the Biosphere Program. These biospheres complement the Beringian Heritage Project. For further information, see, e.g., William Brown, "Beringia: A Common Border— Soviets and Americans work to create a joint park in the Bering Strait," *National Parks* (November/December 1988), pp. 18–23.

8. Editors' note.—See "UNCITRAL Arbitration Rules," *Ocean Yearbook 7*, pp. 537–49.

9. Editors' note.—See Aldo Chircop, "Participation in Marine Regionalism: An Appraisal in a Mediterranean Context," in this volume, which has references to further studies on the Mediterranean Sea region.

34. An international conference under the auspices of the UN, and with the participation of the members of the Security Council, should be held to find a solution to the question of Palestine. The aim should be the establishment of a Palestinian State which allows all Palestinians and all populations in the region to enjoy their human rights. *Pacem in Maribus XVII* sees this process as an important contribution to the restoration of peace and to the strengthening of cooperation among the states of the region.

The Arctic

35. States bordering the Arctic Ocean should elaborate a Plan of Action for the preservation of the environment, taking into account the fragility of the Arctic ecosystem and its impact on the world's climate.

36. To underpin the Arctic Action Plan, a multilateral international mechanism, including additional protocols, should be elaborated to regulate specific types of Arctic environmental pollution from various sources, and a special international body with coordinating functions should be established.

37. An international Arctic system of ecological monitoring and exchange of data should be established.

38. To tackle ecological incidents, an international mechanism providing for urgent reciprocal help needs to be established for the Arctic region.

39. International cooperation in marine scientific research should be strengthened.

40. Joint ventures should be established for research and development in technologies which are socially relevant, appropriate to Arctic conditions, and suitable for environmentally sustainable development.

41. In the context of the process of nuclear disarmament and the limitation and reduction of naval forces, agreements should be concluded for the establishment of a collective multifunctional security system.[10]

The Baltic

42. The Baltic Sea should be declared a nuclear weapons free zone.

The Antarctic

43. The Antarctic should be declared a zone of peace and a Common Heritage of Mankind.

The South Pacific

44. State members of the South Pacific Forum, and nuclear powers who have not yet ratified the South Pacific Nuclear-Free Zone Treaty of 1 December 1986 and its protocols, are urged to do so as speedily as possible.[11]

10. Editors' note.—See Gary Luton, "Strategic Issues in the Arctic Region," *Ocean Yearbook 6*, pp. 399–416.

11. Editors' note.—For the text of the Treaty and Protocols, see "South Pacific Nuclear Free Zone Treaty," *Ocean Yearbook 6*, pp. 594–605; and "Entry into Force of the South Pacific Nuclear-free Zone Treaty on 11 December 1986," *Ocean Yearbook 7*, p. 529.

The Indian Ocean

45. An international conference should be convened in 1990 on the Indian Ocean as a Zone of Peace to implement the Resolutions of the General Assembly requesting the holding of such a conference to contribute to strengthening the security of States in the region and to international peace and security as a whole.[12]

Southeast Asia and the Korean Peninsula

46. Nuclear weapon free zones should be established in Southeast Asia and around the Korean peninsula.

PACEM IN MARIBUS XVII

"Peace in the Oceans: The New Era"

RESOLUTION OF THANKS

Pacem in Maribus XVII expresses its deep appreciation to the Chairman of the USSR Council of Ministers, Mr. N. I. Ryzhkov, for his inspiring message, presented by the Minister of the Merchant Marine of the USSR, at the inauguration of the international conference entitled "Peace in the Oceans: The New Era."

The Conference received with gratitude and particular interest the message of support sent, through his special envoy, Dr. Jean-Pierre Levy, by the Secretary-General of the United Nations, Mr. Javier Perez de Cuellar.

Pacem in Maribus XVII had the privilege of being addressed by Mr. Federico Mayor, the Director-General of Unesco, who referred to the various matters of importance before the conference. The prospect of an enlarged cooperation between Unesco and the International Ocean Institute (IOI) and *Pacem in Maribus* conferences, on the basis of proposals put forward by the Director-General, are expected to have a positive impact on the diverse activities of the marine science and technology sector of Unesco.

The conference expresses its profound thanks for their cosponsorship of *Pacem in Maribus XVII* to:

—the Soviet Maritime Law Association;
—the International Ocean Institute; and
—The Soviet Peace Fund.

The cosponsors were given indispensable support by:

—The Peace to the Oceans Commission; and
—The USSR State Research Project Development Institute of the Merchant Marine ("Soyuzmorniiprojekt").

The conference records its great appreciation to the sponsors who gave generous financial assistance, making its success possible. These were:

—UNEP;
—the Canadian Centre for International Peace and Security;

---

12. Editors' note.—The use of the Indian Ocean as a Zone of Peace is discussed, e.g., by Stanley D. Brunn and Gerald L. Ingalls, "Voting Patterns in the UN General Assembly on Uses of the Seas," *Ocean Yearbook 7*, pp. 42–64.

—the International Ocean Institute;
—the Murmansk Shipping Company;
—the Estonian Shipping Company;
—the Estonian Production Association of the Fishing Industry "Estribprom"; and
—The Fishing Kolkhoz "Mayak."

Gratitude is also acknowledged to other intergovernmental and nongovernmental organizations whose participation contributed materially to the success of the conference.

Selected Documents and Proceedings

Statement to the Plenary by the Chairman of Special Commission 1 on the Progress of Work in That Commission[1]

On behalf of the Special Commission and its Bureau, I am pleased to report that during this session the Special Commission has made extremely significant progress in its work. The ultimate mandate of the Special Commission is to submit recommendations to the International Sea-Bed Authority on ways and means for minimizing the difficulties of developing land-based producer States likely to be most seriously affected by future sea-bed production and for helping such States to make the necessary economic adjustment. Since the beginning of its deliberations in 1984, the Special Commission has discussed a number of issues related to that mandate. During this session, at the request of the Special Commission, a list of sixty-six provisional conclusions was prepared, based on these discussions, which can form the basis of its recommendations to the Authority.

This list (LOS/PCN/SCN.1/1989/CRP.16) covers, with very few exceptions, all the subjects falling under the programme of work of the Special Commission. It reflects areas of agreement, areas where agreement is emerging, and areas that will be important for the Authority in discharging its responsibilities towards developing land-based producer States. The list also includes certain areas where negotiations are still continuing in the Special Commission's *Ad Hoc* Working Group. Thus, while the list is neither exhaustive nor definitive, it constitutes the first and the most important concrete step towards the formulation of the Special Commission's recommendations to the Authority. I would like to express my appreciation to the delegations and the members of the Bureau for their co-operation and guidance which made it possible for us to take this step. It is my hope that delegations will take this document to their capitals, consult with their Governments during the intersessional period, and come back at the next session ready and willing to finalize the list and formulate the recommendations to the Authority.

In tandem with the Special Commission, the *Ad Hoc* Working Group carried out intensive negotiations on certain unresolved matters in a business-like manner. As is evident from the report of the Chairman of the *Ad Hoc* Working Group, included below, a compromise solution is imminent with regard to at least one matter—criteria for the identification, after sea-bed production occurs, of the developing land-based producer States actually *affected,* and before sea-bed production occurs, of the developing land-based producer States *likely to be* affected.

---

1. United Nations, "Statement to the Plenary by the Chairman of Special Commission 1 on the Progress of work in that Commission," Preparatory Commission for the International Sea-Bed Authority for the International Tribunal for the Law of the Sea, Seventh Session, Kingston, Jamaica, 27 February–23 March 1989, Document LOS/PCN/L.68, 17 March 1989.

PROGRAMME OF WORK FOR THE SESSION

The Special Commission, its *Ad Hoc* Working Group, and its Bureau held 9, 5, and 3 meetings, respectively. The Special Commission devoted its meetings to the consideration of (a) the report on the work of the Third Regional Conference on the Development and Utilization of Mineral Resources in Africa, held in Kampala, Uganda, in June 1988, under the auspices of the Government of Uganda and the Economic Commission for Africa (ECA); (b) updated data on copper, nickel, cobalt and manganese; (c) recycling of these four minerals; (d) substitution for and by these four minerals; (e) application to the Authority for remedial measures by a developing land-based producer State actually affected or likely to be affected by sea-bed productions; (f) other matters, especially effects of subsidized sea-bed mining, and finally, (g) the issue of recommendations to the Authority. The *Ad Hoc* Working Group concentrated on the matter of a compensation system; its meetings, however, were utilized in considering a related matter—criteria for the identification of the developing land-based producer State actually affected or likely to be affected by sea-bed production. The Bureau made recommendations on (a) the programme of work for the session, (b) the note verbale and the enclosed questionnaire to be transmitted by the Secretariat to Member States requesting statistics on certain items relevant to the work of the Special Commission, and later, of the Authority, and (c) the progress report for the session.

Kampala Conference

Document LOS/PCN/SCN.1/WP.5/Add.3 contains the statements made at the sixth session by Uganda, the Chairman of the Conference, and by the Secretariat, and a report by ECA. At the conclusion of the Kampala Conference, the Ministers of African States responsible for the mineral sector made a number of recommendations, which primarily addressed ways and means of alleviating the problems of the mineral sector in Africa, especially in the context of falling prices and demand during the last decade.

The Special Commission felt that all the recommendations were very pertinent to its own work, and discussed three specific recommendations which it saw as having direct relevance thereto. The first of these was the African Ministers' recommendation that it would be premature to form African associations concerned with specific minerals or groups of minerals. It was explained by some delegations that this did not imply opposition by African Ministers to commodity agreements; rather they wished to evaluate the benefits to African States from such agreements. The Ministers felt that intensified interaction among Member States and African producers of mineral raw materials and mineral-based products under the auspices of the economic groupings would be a better alternative at this time.

Another recommendation addressed the issue of financing new mineral resources development projects and rehabilitation of existing production facilities from the resources of various funding agencies. Some delegations welcomed this approach; others, however, pointed out that the demand for capital for mineral resources projects remained very high.

A third recommendation which was extremely relevant for the Special Commission addressed the issue of adverse effects of future sea-bed production on the African mining sector. The African Ministers urged African States to participate in the on-going negotiations of the Preparatory Commission to ensure (i) the preservation of the fundamental concept of equitable sharing of the financial benefits accruing from the commercial exploitation of the common heritage resources; and (ii) the establishment of modalities for the payment of financial compensation or the implementation of other means of compensation including economic adjustment schemes so as to miti-gate the adverse effects of mining the resources of the common heritage of mankind on the economies of developing land-based producer States. Co-operation between ECA and the United Nations Office for Ocean Affairs and the Law of the Sea in assisting Member States to achieve the above objectives was stressed, and the need to draw on expertise within the United Nations Office for Ocean Affairs and the Law of the Sea in dealing with the question of sea-bed mining was emphasized.

Updated Data

The discussion centered around the statistics presented by the Secretariat on produc-tion, consumption, export and import of copper, nickel, cobalt and manganese, by country and world total, on an annual basis for the period 1981–1986, and on the importance of these minerals in the total export earnings and GDP of developing land-based producer States (document LOS/PCN/SCN.1/Wp.2/Add.5). This document is an update of document LOS/PCN/SCN.1/WP.2/Add.1, which presented statistics for the period 1973–1982.

Both the Secretariat and delegations drew attention to a few errors and omissions in the statistics; these will be corrected by the Secretariat in due course.

There was extensive discussion with regard to the statistics. Some delegations were of the view that since Special Commission 1 deals with developing land-based producer States, the statistics of other countries were not required. Other delegations felt that world-wide statistics were important to provide an overview of the world mineral situation which was relevant for developing land-based producer States.

Some delegations were of the opinion that statistics on production and exports are sufficient for the Special Commission. Other delegations emphasized the need for statistics on consumption and imports in order to be able to view the world supply-demand situation in its totality. In this context, some delegations felt that statistics on stocks, by country and world total, would be useful. Others thought that statistics on reserves could also be of use.

Some delegations felt that export and import statistics at a highly disaggregated level, by stages of processing, can be confusing and may in some cases incorporate double or even triple counting; statistics at an aggregated level would therefore be preferable. Attention should also be paid to distinguishing between exports and re-exports, they added.

Some delegations questioned the usefulness of collecting and disseminating statis-tics at this time. Others pointed out the need for statistics to facilitate the work of the Special Commission itself, as well as the need for historical data for the purposes of the Authority.

The lack of availability of some types of statistics, especially on cobalt and manga-nese consumption and trade, was also discussed. Some delegations suggested that certain commodity-related institutes, e.g., the Cobalt Development Institute and the

Iron and Steel Institute (for manganese statistics) might be good sources of data in this respect.

Firm statistics for certain countries, some of which are developing land-based producer States, were not available in the United Nations sources. It was suggested that in such cases, the respective national Governments should be requested to provide statistics, subject to any restrictions for the sake of confidentiality.

The Secretariat was once again instructed to develop and maintain a database by compiling data from sources in the public domain—databases and publications of international organizations, national organizations, commodity associations/study groups/institutes—and from national Governments.

It may be recalled that a note verbale requesting such information was transmitted to national Governments in 1984, and adequate responses were received by the Secretariat. The Bureau of the Special Commission held a meeting for the specific purpose of giving guidance to the Secretariat with regard to the design of the questionnaire enclosed with the note verbale. The Secretariat was instructed to develop a simple questionnaire with clearer explanatory notes and indications of priority items.

Finally, some delegations emphasized that there was a danger of getting too involved in the intricacies of data-collection exercises, and that what was important was to obtain sufficient statistics for the purpose of assisting developing land-based producer States who would be affected by future sea-bed production.

Recycling

The discussion was based on document LOS/PCN/SCN.1/WP.2/Add.6 presenting information on the current pattern of recycling of the four metals concerned, and on the trends and developments which may have a bearing on the pattern of recycling in the future.

The supply of recycled metal has an impact on the demand for primary metal, and can thus affect developing land-based producer States.

Changes in demand for metals and alloys, in fabrication techniques, in the technology for scrap identification and refining as well as environmental considerations are important factors influencing the pattern of recycling. The use of scrap for recycled metal is determined by a range of factors, including the availability and quality of scrap, price, cost of processing, purity requirements, technology, user standards, and government policies with regard to environment and conservation.

The current pattern and future potential of recycling vary from metal to metal. In the case of copper, between 20 and 40 percent of demand in developed countries is met by recycled copper. In the case of nickel, the proportion is about 25 percent. Cobalt is recycled, but to a lesser extent than copper and nickel. The recycling of manganese through the recycling of steel is quite high.

Some delegations provided information on recycling activities in their own countries as well as on national policies and regulations for promoting recycling and conservation of natural resources.

A need was felt for statistics on secondary production. Some delegations suggested that commodity study groups, e.g., the proposed International Nickel Study Group and the proposed International Copper Study Group, would be able to provide such data. In this connection, it was pointed out that developing land-based producer

States should be encouraged to participate in such study groups in order to keep abreast of developments in the mineral situation.

It was generally agreed that the Authority should monitor the trends and developments with regard to the recycling of the four metals concerned, in order to evaluate the impact of recycling on the mineral situation and on developing land-based producer States and to isolate its effects from those of sea-bed mining.

Substitution

Since substitution can have a significant impact upon the copper, nickel, cobalt, and manganese industries and thus, on developing land-based producer States, discussions were carried out on substitution on the basis of a paper prepared by the Secretariat (document LOS/PCN/SCN.1/WP.2/Add.7). The paper provided information on the current situation with respect to substitution for and by these metals, and also on future substitution potentials in certain key end-use sectors.

Substitution can take three forms: direct substitution for one metal by another; substitution for a metal by "engineered" materials; and substitution for a conventional technology that requires a metal by a new technology with different metal/material requirements. All three forms of substitutions may affect the four metals concerned.

Copper faces competition from aluminum in electrical and automotive industries. There has also been growing competition from non-metallic materials in building and construction industries. Fibre glass is competing with copper in the telecommunications industry. Nickel use is strongly tied with steel use. The changing composition of various classes of steel in end-uses may, in fact, increase the use of nickel in the steel industry. Cobalt faces competition from "engineered" materials in the production of superalloys used in jet engines. Cobalt may also have to compete with alternative magnetic materials. Manganese is essential in the production of iron and steel; however, steel is subject to competition from other materials. Manganese requirements in the production of steel may also decline in the newer steel-making technology.

It was generally agreed that in view of the significant effect of substitution on future metal demand and therefore on developing land-based producer States, the Authority should monitor the trends and developments with regard to substitution, study the effects of substitution on developing land-based producer States and attempt to isolate the effects of substitution from those of sea-bed mining.

Application by Developing Land-based Producer States for Remedial Measures

A developing land-based producer State affected by the impact of sea-bed mining will be able to apply to the Authority for remedial measures. At the sixth session there was some discussion on the content of such applications and the procedural matters related to their processing. Discussions were held at the present session to elaborate on this matter.

The first action would be taken by the developing land-based producer State which considers itself affected by sea-bed production. In its application to the Authority, it would:

(a) Identify itself as a developing land-based producer State by providing statistics on production, volume of exports, export earnings from one or more of the four metals concerned;

(b) Identify the changes that it feels occurred because of sea-bed production: decrease in price, decrease in volume of exports, decrease in export earnings, other effects on its economy (i.e., decrease in GDP or GNP, decrease in rate of growth of GDP or GNP, decrease in level of employment, decrease in foreign exchange reserves, etc.);

(c) Indicate why it considers that the above effects have resulted from sea-bed production and not as a result of other factors; and

(d) Indicate what kind of remedial measure it requires, and to what extent.

Having received the application, the Authority would:

(a) Determine whether the applicant State's production, volume of exports and export earnings from one or more of the four metals concerned exceed a certain specified level called the dependency threshold;

(b) If the dependency threshold is exceeded, determine whether the effects that are considered to have resulted from sea-bed production exceed a certain specified level that is called the trigger threshold;

(c) If the effects exceed the trigger threshold, undertake studies on a case-by-case basis in order:

    (i) to determine whether and to what extent the effects resulted from sea-bed production in the Area; and

    (ii) to decide on the necessary remedial measures.

At the sixth session, it was also pointed out that the Convention provides for certain actions by a developing land-based producer State which considers itself *likely* to be affected. It was generally agreed that in such cases the first action would again be taken by the developing land-based producer State itself. In its application to the Authority, it would:

(a) Identify itself as a developing land-based producer State by providing statistics on production, volume of exports, export earnings from one or more of the four metals concerned;

(b) Present projections of its production, volume of exports and export earnings from one or more of the four metals concerned in a scenario not involving sea-bed production; and

(c) It must present projections of the variables above, in a scenario including sea-bed production.

Having received the application, the Authority would:

(a) Undertake studies on the *potential effects* on the applicant State under various projections for the levels of sea-bed production; and

(b) If the above studies warrant it, recommend measures to minimize the potential adverse effects.

Some delegations pointed out that according to the Convention, the applications should be submitted to the Economic Planning Commission which will then examine the applications, undertake the necessary studies and if the studies warrant it, make recommendations to the Council with regard to remedial measures.

Issue of Subsidized Sea-Bed Mining

The delegation of Australia introduced a proposal (LOS/PCN/SCN.1/1989/CRP.15) addressing the issue of subsidized sea-bed mining. Recognizing that subsidized sea-bed mining is likely to exacerbate the adverse effects of sea-bed mining on developing land-based producer States, the proposal includes a draft recommendation to the Authority to the effect that the Authority should take all steps necessary to ensure an orderly and economically rational development of the resources of the Area, pursuant to the relevant provisions of the Convention, that there should be adequate surveillance of States Parties' compliance with their obligations in this connection, and that the Authority should also develop rules, regulations and procedures to ensure that the exploitation of the resources is carried out on an economically viable basis and in accordance with sound commercial principles.

Some delegations, supporting the proposal, said that in addition to minimizing adverse effects on developing land-based producer States, sea-bed mining in accordance with sound commercial principles would be in the best long-term economic interest of producers and consumers.

Some delegations felt that subsidization is only one aspect of unfair economic practices, a subject which is within the Convention, as well as by other international fora, e.g., GATT. The formulation of rules, regulations and procedures, if at all necessary, should be the domain of Special Commission 3, they maintained.

Some delegations felt that the issue of subsidized sea-bed mining should be linked with the issue of compensation to developing land-based producer States.

There seemed to be general agreement that the matter should be taken up in the Working Group of Special Commission 1, whose mandate already includes the issue of subsidized sea-bed mining.

Proposal by Pakistan

The delegation of Pakistan made a proposal (LOS/PCN/SCN.1/1989/CRP.17) addressing the issues of (*a*) criteria for identification of developing land-based producer States, (*b*) criteria to determine adverse effects of sea-bed mining on developing land-based producer States, and (*c*) a system of compensation.

The proposal included concrete criteria for identifying developing land-based producer States likely to be affected by sea-bed production, and for characterizing what would be considered adverse effects of sea-bed mining on such States. It called for the establishment of a compensation fund to be financed by contributions from those States/entities, including the Enterprise, which would derive benefits from sea-bed mining.

There was some discussion on the proposal, and delegations agreed that the proposal would be deliberated upon in depth in the Working Group.

Issue of Recommendation

The Special Commission conveys its appreciation to the former Chairman of the Ad Hoc Working Group, Ambassador Karl Wolf (Austria), who has retired. The Working Group continued its deliberations under the Chairmanship of Mr. Luis Giotto Preval Paez (Cuba). The report of the Chairman of the *Ad Hoc* Working Group, reads as follows:

> Pursuant to the priority established by the Working Group at the sixth session, it concentrated on the issue of a system of compensation. It proceeded in a

step-by-step fashion, dealing as a first step with the criteria for identifying developing land-based producer States likely to be affected by sea-bed production.

The Group seemed to agree that the characterization of adverse effects of sea-bed production on the export earnings or economics of developing land-based producer States involved a combination of two aspects. The first aspect is that the individual developing land-based producer States have to be dependent to a certain degree on one or more of the four minerals concerned. The second aspect is that there has to be a decrease of somewhat substantial extent in the export earnings of individual developing land-based producer States from the exports of these minerals, or an unfavourable change of somewhat substantial extent in the economies of such States.

The Group then attempted to characterize both the dependency aspect and the aspect of somewhat substantial unfavourable change after sea-bed production occurs, in a quantifiable manner with the use of some figures, that are called dependency levels and trigger levels, respectively.

Certain suggestions about dependency levels and trigger levels were put forward by myself (document LOS/PCN/SCN.1/1989/CRP.18). The proposal by Pakistan, mentioned earlier, also contained certain suggestions.

The suggestions about the dependency levels involved various factors, the two primary ones being expressed in terms of percentages of total export earnings of particular developing land-based producer States contributed by one or more of the four minerals concerned, and in terms of value of exports of these minerals.

There were also suggestions that developing land-based producer States should be categorized in three groups: (i) affected, (ii) seriously affected, and (iii) most seriously affected.

The Group had intensive discussions based on these suggestions. Some delegations expressed reservations that the use of dependency levels and trigger levels, and categorization in three groups may run the risk of excluding some developing land-based producer States which may require assistance. Other delegations felt that for the purpose of the Authority's preparedness to assist and of minimizing the costs of administering assistance measures, the above ideas were extremely useful.

There was also intensive discussion about the figures to be used for the purpose of dependency levels and trigger levels.

It is my assessment that the differences among delegations are narrowing down, and with further efforts, we should be able to reach compromise solutions.

PROGRAMME OF WORK FOR THE FUTURE

Special Commission 1 has completed discussions on all the topics under its mandate except three: (i) bilateral trade, including barter trade in the four minerals concerned, (ii) commodity agreements or commodity associations/study groups related to the four minerals, and (iii) projection of future supply-demand-price of the four minerals. These three topics will be deliberated upon at the next session, and the Secretariat is requested to prepare, if possible and as time and resources permit, background papers on them to facilitate our deliberations. Over the period 1985–1987, the Special Commission received extensive information about a number of existing international and multilateral economic measures which can be of relevance in alleviating the problems of developing land-based producer States. That information needs to be supplemented by information on new developments since 1987, especially in the programmes of Common Fund, the United Nations Industrial Development Organization (UNIDO) and United Nations Department of Technical Co-operation for Development (UN/DTCD). The Secretariat is requested to disseminate such information to facilitate our discussion on the relevant existing measures.

Finally, since the basis of our recommendations to the Authority has been provided in document LOS/PCN/SCN.1/1989/CRP.16, we should make utmost efforts to build upon that basis in order to finalize the recommendations. The outcomes of the deliberations of the Working Group which will continue during the next session will be incorporated in the recommendations following the due process.

Selected Documents and Proceedings

Statement to the Plenary by the Chairman of Special Commission 2 on the Progress of Work in That Commission[1]

Special Commission 2 held 11 of its scheduled 13 meetings, most of this time being devoted to its consideration of the structure and organization of the Enterprise and to the drafting of a Training Programme under paragraph 12(a)(ii) of resolution II, in accordance with the programme of work for the seventh session set forth in paragraph 29 of LOS/PCN/L.65. The *Ad Hoc* Working Group on Training, established by Special Commission 2 in 1987 to prepare draft principles, policies, guidelines and procedures for the Training Programme, completed its work at this session. Its recommendations are contained in two Conference Room Papers: LOS/PCN/SCN.2/1988/CRP.3 of 15 June 1988 and LOS/PCN/SCN.2/1989/CRP.4 of 20 March 1989.

ASSUMPTIONS

The Chairman's Advisory Group on Assumptions, set up in 1986 to advise the Special Commission on when it might be possible to revise the current model of sea-bed mining contained in LOS/PCN/SCN.2/WP.10 and Add.1, again met during the course of the session to review this possibility.

As at prior sessions, the Group reviewed the current movements of and long-term projections for metal prices. The information and statistical data received from the experts of Australia, the EEC and the Secretariat showed that current prices for nickel and copper had again escalated dramatically during November and December 1988 and January 1989. The Group was, however, of the view that despite the cyclical upswing in metal prices over the last two years, long-term trend projections remain essentially the same. It is unlikely therefore that major increases in real prices of any of the sea-bed metals can be sustained over the next decade.

At this session, the Group did not limit its attention to metal prices and forecasts and their implications for sea-bed mining viability, but turned to the question of preparing a new list of assumptions that could be agreed to by the Special Commission and would in that event justify revision of the current model contained in LOS/PCN/SCN.2/WP.10. Its methodology is considered to be satisfactory so that the model does not need to be changed in this respect. The Group estimates that the work involved will take at least two or three sessions to complete.

In keeping therefore with its general assumption that sea-bed mining will prove to be economically viable when metal prices improve and/or *when capital and operating costs can be reduced with the introduction of highly efficient technologies,* the Group has now

1. United Nations, "Statement to the Plenary by the Chairman of Special Commission 2 on the Progress of Work in that Commission," Preparatory Commission for the International Sea-Bed Authority and for the International Tribunal for the Law of the Sea, Seventh Session, Kingston, Jamaica, February 27–March 23, 1989, Document LOS/PCN/L.70, March 22, 1989.

decided to examine all the economic and technical factors and parameters which are relevant to sea-bed mining viability.

To facilitate the work involved in preparing a new set of assumptions, members of the Group will examine the list of assumptions used in the current model and come to the next session with their proposed lists of assumptions, including suggested priorities, and their proposals on the figures to be used. In this connection, the Secretariat was asked to collect as much information as possible on new technological developments in sea-bed mining and processing, as well as on any other developments which might justify changing the capital and operating costs used in the current model.

The expert from Australia reiterated that his country would be prepared to revise LOS/PCN/SCN.2/WP.10 for the Preparatory Commission as and when Special Commission 2 agrees on a new set of assumptions formulated by the Group.

The Special Commission noted that the Group has been expanded with the addition of an expert from the delegation of Canada.

TRAINING

The Special Commission wishes to place on record its special thanks to Mr. Baidy Diene of Senegal, Co-ordinator of the *Ad Hoc* Working Group on Training, for his efforts in bringing the work of that Group to a successful conclusion. The members of that Group are also to be commended for the practical approach they adopted to the resolution of the major substantive issues that were involved in drawing up the Training Programme.

The Group's recommendations on the principles and policies to govern the Programme (LOS/PCN/SCN.2/1988/CRP.3) were submitted to the Special Commission at the last meeting in New York. Its recommendations on the guidelines and procedures for implementation of the Programme, and on additional policies, have now been submitted in LOS/PCN/SCN.2/1989/CRP.4. There is general agreement on the substance of the Programme, so that all that remains to be done by the Special Commission is to finalize its recommendation to the Plenary. It is to be noted that the Group did not make any recommendation on the question of financial responsibility, nor did it do so on the number of persons to be trained in the interim period. These elements of the Programme have necessarily been left to the Preparatory Commission for decision.

As stated in my last report (LOS/PCN/L.65, paragraph 28), we have to ensure that the drafting of our recommendation is coherent and that its different elements are harmonized. The Secretariat will therefore circulate intersessionally a Working Paper which integrates the two Conference Room Papers and takes care of any obvious drafting problems. It is understood that this final process will require no more than two meetings at the next session.

STRUCTURE AND ORGANIZATION OF THE ENTERPRISE

The Working Paper (LOS/PCN/SCN.2/WP.16) prepared by the Secretariat on this item proved to be a useful means of focusing attention on the constituent elements of the Enterprise. While it is understood to be merely a convenient guide to the provisions of the Convention dealing with the Enterprise, certain problems in rendering these provisions in an organized fashion were noted, so that the Secretariat was asked to keep a record of those instances where drafting might have to be corrected in the future.

Two uses have been found for the Working Paper. Firstly, it will serve to identify where provisions of the Convention might require annotations by the Special Commission with a view to assisting the Authority and subsequently the Enterprise itself in making their initial decisions. In general, the Special Commission would address the need to preserve maximum autonomy for the Enterprise, in keeping with its basic commercial character and the need to ensure efficient management and a streamlined structure. It was agreed that while the Special Commission could not amend the Convention, it did not rule out liberal interpretations of its provisions as long as this did not disturb their basic purpose and intent. Secondly, the Working Paper will serve to identify those elements that would form the substance of a special recommendation to the Authority on transitional arrangements to deal with the pre-operational stage of the Enterprise.

Although the Special Commission has only reached article 15 in its reading of the Working Paper, it has already identified several places where annotations might be in order. Some focus on general concerns such as the relationship between the Governing Board and the Council/Assembly, where it will be important to avoid unnecessary constraints on the Enterprise. Here, it was also emphasized that the directives given to the Enterprise should be limited to those of a political or general economic character, transcending commercial considerations which are the exclusive concern of the Governing Board. It was also noted that it would be important for the Mining Code to avoid introducing constraints on the Enterprise, e.g. with respect to its procedures for the acquisition of technology. Other possible annotations would deal directly with cases where some additional guidance is indicated, e.g. for the nomination process referred to in article 12 of the Working Paper.

Considerable progress has been made at the same time in understanding the issues involved in dealing more concretely with the question of a nucleus Enterprise. It is already understood that this question is best dealt with by means of a transitional arrangement which does not disturb the constitution of the Enterprise laid down in the Convention, and which is limited to a specified period that may be lengthened or shortened by the Authority depending on the circumstances at the time. Much of the discussion in this context focused on what functions a nucleus Enterprise would perform, in particular whether or not, even in the pre-operational phase, they should encompass research and development activities. It was concluded that any such function would have to be directly linked to actual operations and it would be limited at the transitional stage to the monitoring of technological developments and to the development of a strategy for technology acquisition.

EXPLORATION

The Special Commission did not have time to begin its discussion of exploration under paragraph 12(a)(i) of resolution II, for which the Secretariat had prepared a new Working Paper (LOS/PCN/SCN.2/WP.17). The main purpose of that Paper, as pointed out by the Secretariat, was to determine, in relation to the standard sequence of events in any mining industry, what the current status of pre-exploitation activities is and what would remain to be done in respect of "exploration." The Secretariat has concluded that the next step would be a "detailed regional appraisal of claimed sites," i.e. the stage which follows upon prospecting. It has emphasized that there is a basic sequence involved and that, in order for the Special Commission to deal more effectively with the issues relating to the implementation of paragraph 12(a)(i) of resolution II, it would be necessary to make clear separations between the remaining phases required to establish the exploitability of a reserved area.

PROGRAMME OF WORK FOR THE SUMMER SESSION

In the light of the provisional allocation of 10 meetings to Special Commission 2, the programme is as follows:

(i) Two meetings, at the outset of the session, for finalization of the recommendation to Plenary on the Training Programme;
(ii) Five meetings for the item "Structure and organization of the Enterprise," entailing firstly, completion of the article-by-article examination of LOS/PCN/SCN.2/WP.16, and then identification of the elements which relate to the nucleus Enterprise;
(iii) Two meetings for examination of the implementation of paragraph 12 of resolution II in respect of "exploration."

The final meeting for the session will be devoted to such matters as the progress made by the Chairman's Advisory Group on Assumptions and the programme of work for the eighth session.

The Special Commission had made significant progress at this session on the Training Programme that will be established by the Preparatory Commission. It has also made major strides in coming to grips with the kind of constructive recommendations that can be forwarded to the Authority on the structure and organization of the Enterprise, including those that would address an Enterprise in a nucleus or transitional stage. Also of great significance is the work now planned to carry forward the study of assumptions. In time, this will allow the Special Commission to proceed to a new examination of the most viable options for the first operation of the Enterprise.

Selected Documents and Proceedings

Proclamation 5928 of December 27, 1988[1]

Territorial Sea of the United States of America

By the President of the United States of America

A Proclamation

International law recognizes that coastal nations may exercise sovereignty and jurisdiction over their territorial seas.

The territorial sea of the United States is a maritime zone extending beyond the land territory and internal waters of the United States over which the United States exercises sovereignty and jurisdiction, a sovereignty and jurisdiction that extend to the airspace over the territorial sea, as well as to its bed and subsoil.

Extension of the territorial sea by the United States to the limits permitted by international law will advance the national security and other significant interests of the United States.

NOW, THEREFORE, I, RONALD REAGAN, by the authority vested in me as President by the Constitution of the United States of America, and in accordance with international law, do hereby proclaim the extension of the territorial sea of the United States of America, the Commonwealth of Puerto Rico, Guam, American Samoa, the United States Virgin Islands, the Commonwealth of the Northern Mariana Islands, and any other territory or possession over which the United States exercises sovereignty.

The territorial sea of the United States henceforth extends to 12 nautical miles from the baselines of the United States determined in accordance with international law.

In accordance with international law, as reflected in the applicable provisions of the 1982 United Nations Convention on the Law of the Sea, within the territorial sea of the United States, the ships of all countries enjoy the right of innocent passage and the ships and aircraft of all countries enjoy the right of transit passage through international straits.

Nothing in this Proclamation:

(a) extends or otherwise alters existing Federal or State law or any jurisdiction, rights, legal interests, or obligations derived therefrom; or

[1]*Federal Register*, vol. 54, no. 5 (Monday, January 9, 1989), Presidential Documents.

(b) impairs the determination, in accordance with international law, of any maritime boundary of the United states with a foreign jurisdiction.

IN WITNESS WHEREOF, I have hereunto set my hand this 27th day of December, in the year of our Lord nineteen hundred and eighty-eight, and of the Independence of the United States of America the two hundred and thirteenth.

(signed) Ronald Reagan

Appendix C

Tables, Living Resources

The six tables of this appendix follow the typology of those in *Ocean Yearbook 6*. Data from previous years have been updated wherever possible. Tables compiled before Burma officially adopted the name Myanmar may list the country as Burma. For additional information on living resources see the articles by John Fitzpatrick and Kenneth Sherman in this volume.

THE EDITORS

TABLE 1C. -- WORLD NOMINAL MARINE CATCH BY CONTINENT* (1,000 Metric Tons)

	1970	1980	1984	1985	1986	1985-86 (% change)
Africa	3,131	2,725.0	2,630.2	2,692.7	2,887.9	+7.2
America, N.	4,750	6,698.1	7,499.0	7,932.7	8,141.1	+2.6
America, S.	14,629	7,438.2	10,009.3	11,344.7	13,698.6	+20.7
Asia	19,453	26,258.0	30,651.2	30,609.7	32,550.0	+6.3
Europe	11,815	12,169.4	12,598.6	12,351.0	12,098.1	-2.0
Oceania	194	448.3	589.0	589.9	636.3	+7.9
USSR	6,399	8,722.6	9,711.4	9,711.4	10,333.0	+6.4
World Totals§	61,432	64,459.6	73,688.7	75,138.0	80,345.0	+6.9

SOURCE. -- FAO, <u>Yearbook of Fishery Statistics</u>.
*Continental Classification follows FAO usage.
§Exceeds the sum of the figures by continent due to the inclusion of catches not elsewhere included.

TABLE 2C. -- WORLD NOMINAL MARINE CATCH, BY MAJOR FISHING AREA

	Million Metric Tons				1985-86
	1970	1980	1985	1986	(% change)
Atlantic, N.W.	4.23	2.87	2.85	2.91	+2.11
Atlantic, N.E.	10.70	11.81	11.00	10.49	-4.64
Atlantic, N.	14.93	14.68	13.85	13.40	-3.25
Atlantic, W.C.	1.42	1.80	2.26	2.11	-6.64
Atlantic, E.C.	2.77	3.43	2.85	3.02	+5.96
Mediterranean & Black	1.15	1.64	1.97	1.99	+1.02
Atlantic, C.	5.34	6.87	7.08	7.12	+0.56
Atlantic, S.W.	1.10	1.27	1.57	1.71	+8.92
Atlantic, S.E.	2.52	2.17	2.10	2.11	+0.48
Atlantic, S.	3.62	3.90	3.90	4.28	+9.74
Atlantic	23.53	25.45	24.83	24.80	-0.12
Indian, W.	1.72	2.09	2.63	2.64	+0.38
Indian, E.	.81	1.46	1.76	1.82	+3.41
Indian*	2.53	3.55	4.42	4.50	+1.81
Pacific, N.W.	13.01	18.76	23.78	25.84	+8.66
Pacific, N.E.	2.65	1.97	2.88	3.21	+11.46
Pacific, N.	15.60	20.73	26.66	29.05	+8.96
Pacific, W.C.	4.22	5.49	6.38	6.68	+4.70
Pacific, E.C.	.91	2.42	2.65	2.62	-1.13
Pacific C.	5.12	7.91	9.03	9.30	+2.99
Pacific, S.W.	.17	.38	.57	.73	+28.07
Pacific, S.E.	13.76	6.23	9.63	11.95	+24.09
Pacific, S.	13.93	6.62	10.20	12.68	+24.31
Pacific	34.71	35.26	45.89	51.03	+11.20
World Total	61.43	64.39	75.14	80.45	+7.07
Northern Regions	30.59	35.41	40.51	42.45	+4.79
Central Regions	13.00	18.33	20.50	20.88	+1.85
Southern Regions	17.54	10.65	14.13	17.00	+20.31

SOURCE. -- FAO.
NOTE. -- Totals and subtotals may differ from the sum of included figures due to rounding and the inclusion of catches not elsewhere included.
*Temperate and tropical.

TABLE 3C. -- WORLD NOMINAL FISH CATCH, DISPOSITION*   (Million Metric Tons)

	1970	1980	1984	1985	1986	1985-86 (% change)
Human Consumption	44.6 (100.0)	53.2 (100.0)	59.2 (100.0)	60.5 (100.0)	64.5 (100.0)	+6.6
Marketing Fresh	18.6 (41.7)	14.5 (27.3)	15.2 (25.6)	15.7 (25.9)	18.3 (28.3)	+16.6
Freezing	9.8 (22.0)	16.3 (30.6)	20.1 (34.0)	20.3 (33.6)	21.4 (33.2)	+5.4
Curing	8.1 (18.1)	11.6 (21.8)	12.3 (20.8)	13.1 (21.5)	13.5 (20.9)	+3.1
Canning	8.1 (18.2)	10.8 (20.3)	11.6 (19.6)	11.4 (19.0)	11.4 (17.6)	0.0
Other Purposes	26.0 (100.0)	18.9 (100.0)	24.3 (100.0)	25.1 (100.0)	26.9 (100.0)	+7.2
Reduction	25.0 (96.2)	18.1 (96.0)	23.4 (96.3)	24.1 (96.0)	25.9 (96.3)	+7.5
Miscellaneous	1.0 (3.8)	0.8 (4.0)	0.9 (3.7)	1.0 (4.0)	1.0 (3.7)	0.0
World Total	70.6	72.1	83.4	85.6	91.5	+6.9

SOURCE. -- FAO.
NOTE. -- Percentages of subtotals shown in parentheses.
*The figures for disposition are based on "live weight" and include freshwater catches.

TABLE 4C. -- WORLD NOMINAL MARINE CATCH, BY COUNTRY* (1,000 Metric Tons)

	1970	1980	1985	1986	1985-86 (% change)
Anglo-America:					
Canada	1,290.1	1,292.7	1,374.6	1,421.6	+3.4
Greenland	39.8	103.6	149.4	155.4	+4.0
U.S.A.	2,729.3	3,565.0	4,692.5	4,871.2	+3.8
Other	0.9	20.5	9.6	8.1	-15.2
Total	4,060.1	4,981.8	6,226.1	6,456.3	+3.7
Latin America:					
Argentina	186.1	376.9	396.8	411.8	+3.8
Brazil	432.7	547.4	628.1	632.9#	+0.8
Chile	1,200.3	2,815.1	4,803.8	5,570.6	+16.0
Colombia	21.3	29.3	22.4	25.6	+14.3
Costa Rica	7.0	19.7	18.7	20.6	+10.2
Cuba	105.3	180.1	203.0	228.5	+12.6
Dominican Rep.	5.0	8.2	15.8	16.3	+3.2
Ecuador	91.4	643.5	946.1	1,018.4	+7.6
Guyana	17.4	36.6	45.0	43.8	-2.7
Mexico	344.1	1,212.6	1,113.2	1,210.8	+8.8
Panama	52.2	216.4	280.1	129.0#	-53.9
Peru	12,532.9	2,696.1	4,108.1	5,581.3	+35.9
Uruguay	13.2	120.1	138.4	140.5	+1.5
Venezuela	122.6	169.4	249.5	267.6	+7.3
Other	64.2	83.4	82.3	85.5	+3.9
Total	15,215.5	9,154.8	13,051.3	15,383.2	+17.9
Western Europe:					
Belgium	53.0	45.6	44.6	39.0	-12.6
Denmark	1,217.1	2,010.5	1,728.6	1,847.1	+6.9
Faeroe Is.	207.8	274.7	372.9	353.7	-5.1
Finland	62.6	111.0	126.2	126.2	0.0
France	764.4	761.3	814.5	820.4#	+0.7
Germany, F.R.	597.9	288.7	201.3	178.2	-11.5
Greece	91.5#	95.1	105.4	106.3#	+0.9
Iceland	733.3	1,514.4	1,679.9	1,656.4	-1.4
Ireland	78.9	149.4	229.3	228.9	-0.2
Italy	379.3	467.3	533.7	503.7	-5.6
The Netherlands	298.8	338.4	500.3	450.5	-10.0
Norway	2,906.2	2,408.6	2,118.6	1,898.0	-10.4
Portugal	464.4	270.4	298.6	389.6	+30.5
Spain	1,517.0	1,231.8	1,311.5	1,276.3	-2.7
Sweden	284.2	230.7	236.8	211.5	-10.7
U.K.	1,091.3	844.1	885.1	842.7	-4.8
Other	1.4	2.6	4.0	2.6	-34.9
Total	10,749.1	11,044.6	11,191.3	10,931.1	-2.3
Socialist Eastern Europe:					
Bulgaria	86.8	114.0	88.3	95.2	+7.8
German D.R.	308.2	223.1	178.2	189.3	+6.2
Poland	451.3	621.9	654.6	615.8	-5.9
Romania	24.8	120.9	179.2	205.3	+14.6
U.S.S.R.	6,386.5	8,722.7	9,617.2	10,333.0	+7.4
Yugoslavia	26.7	34.9	49.3	51.4	+4.3
Other	4.0	10.0	10.0	10.0	0.0
Total	7,288.3	9,847.4	10,776.8	11,500.0	+6.7
Near East:					
Algeria	25.7	48.0#	66.0#	70.0#	+6.1
Egypt	27.2	32.2	26.4	26.4#	0.0
Iran	18.0#	40.0	96.4	121.8	+26.3
Morocco	246.8	329.6	471.8	594.5	+26.0
Oman	180.0#	79.0#	101.2	96.3	-4.8
Saudi Arabia	21.1	26.4	43.7#	45.5#	+4.1
Tunisia	24.0	60.1	88.9	92.6	+4.2
Turkey	165.3	394.6	532.6	539.6	+1.3
U.A. Emirates	40.0	64.4	72.4	72.4#	0.0
Yemen, A.R.	7.6#	17.0	20.6	22.3	+8.3
Yemen, Dem.	120.0	80.2	85.1	91.2	+7.2
Other	35.7	68.3	41.5	50.1	+20.9
Total	911.4	1,239.8	1,646.6	1,822.7	+10.7

TABLE 4C. -- (continued)

	1970	1980	1985	1986	1985-86 (% change)
Sub-Saharan Africa:					
Angola	368.2	77.6	66.5	50.4	-24.2
Cameroon	19.2	73.0	66.0	64.0	-3.0
Cape Verde	5.1	8.8	10.2	10.2	0.0
Congo	15.2	21.0	14.3	18.0	+25.9
Gabon	4.5#	18.0#	19.2	18.6#	-3.1
Ghana	141.5	191.9	234.2	269.2	+14.9
Guinea	5.6#	18.9#	28.0#	28.0#	0.0
Ivory Coast	66.5#	67.9	84.2	76.2	-9.5
Madagascar	13.1	12.2	17.6	17.6#	0.0
Mauritania	47.6	15.6#	103.2#	98.1#	-4.9
Mozambique	7.6	30.4	33.3	31.2	-6.3
Namibia	711.2	252.6	185.4	201.2	+8.5
Nigeria	105.9	292.4	154.3	161.5	+4.7
Senegal	169.2	217.8	240.4	240.4#	0.0
Sierra Leone	29.6	34.2	36.5#	37.0#	+1.4
Somalia	30.0#	14.3	16.5#	16.5#	0.0
South Africa	511.1	601.9	600.1	627.9	+4.6
Tanzania	18.3	38.4	42.7	44.1	+3.3
Togo	6.4	8.4	14.8	14.1	-4.7
Other#	53.5	54.5	63.9	71.9	+12.5
Total	2,341.5	2,049.3	2,031.3	2,096.1	+3.2
South Asia:					
Bangladesh	90.0#	122.0	187.6	207.5	+10.6
India	1,085.6	1,554.7	1,734.2	1,720.5	-0.8
Maldives	34.5	34.6	51.3	45.8	-10.7
Pakistan	149.3	232.9	333.3	331.7	-0.5
Sri Lanka	89.8	165.2	146.4	141.5	-3.3
Total	1,449.2	2,109.4	2,452.8	2,447.0	-0.2
East Asia:					
China	2,192.5	2,995.4	3,835.1	4,636.6	+20.9
Hong Kong	133.3	187.5	192.4	207.8	+8.0
Japan	8,658.4	10,213.3	11,203.7	11,768.1	+5.0
Korea, D.P.R.	447.0	1,330.0#	1,590.0#	1,600.0#	+0.6
Korea, Rep.	725.5	2,052.0	2,598.1	3,045.6	+17.2
Taiwan£	540.4	761.3	787.0	828.5	+5.3
Other	9.6	6.9	12.4	8.0	-35.5
Total	12,706.7	17,546.4	20,218.7	22,094.6	+9.3
Southeast Asia:					
Burma	311.4	428.8	497.0	497.0	0.0
Indonesia	804.0	1,387.0	1,758.8	1,914.2	+8.8
Malaysia	338.5	733.2	622.4	607.0	-2.5
Philippines	844.1	1,134.3	1,330.9	1,377.8	+3.5
Singapore	17.3	15.5	22.8	20.3	-11.0
Thailand	1,343.4	1,653.0	2,057.8	1,951.1	-5.2
Vietnam	668.3	423.0#	566.0#	570.0#	+0.7
Other	21.7	46.6	14.9	15.4	+3.0
Total	4,348.7	5,821.4	6,870.6	6,952.8	+1.2
Australasia:					
Australia	100.7	130.3	157.7	154.4	-2.1
Fiji	3.9	18.0	25.2	23.9	-5.2
Kiribati	9.7	18.9	29.6	33.6	+13.5
New Zealand	59.2	190.9	304.6	339.6	+11.5
Solomon Is.	1.0	34.8	44.0	54.9	+24.8
Other#	20.7	17.7	22.2	23.4	+5.5
Total	195.2	410.5	583.3	629.8	+8.0
Other NEI§	240.6	253.4	98.3	38.3	-61.1
World total	59,495.0	64,459.6	75,138.0	80,345.0	+6.9

SOURCE. -- FAO, unless otherwise noted.

*Nominal marine catch = nominal catch in marine areas. Countries which reported marine catches of less than 10,000 metric tons in 1986 are included under "Other" for all years.

#FAO estimate.

£Based on Republic of China, Council for Economic Planning and Development, Taiwan Statistical Data Book: 1988.

§Not Elsewhere Included. These values differ from those found in the source due to the exclusion of Taiwan's catch from the FAO sum, the inclusion of the catches for the French Southern and Antarctic Territories in the FAO sum, and small discrepancies due to rounding.

TABLE 5C. -- TRADE OF FISHERIES COMMODITIES,* BY MAJOR IMPORTING AND
            EXPORTING COUNTRIES (million US $)#

	1970	1980	1985	1986	1985-86 (% change)
Imports:					
Japan	291.9	3,158.7	4,744.3	6,593.5	+39.0
U.S.A.	835.8	2,633.2	4,051.8	4,748.7	+17.2
France	203.9	1,131.2	1,039.8	1,510.4	+45.3
Italy	159.6	831.7	985.0	1,264.5	+28.4
UK	294.0	1,033.7	940.6	1,216.0	+29.3
Germany, F.R.	264.8	1,023.9	819.6	1,113.2	+35.8
Spain	46.6	544.0	412.1	722.0	+75.2
Hong Kong	56.0	361.4	471.6	624.7	+32.5
Demark	47.1	330.7	370.4	596.0	+60.9
Canada	51.0	301.6	355.9	433.1	+21.7
Belgium	86.3	408.3	303.6	427.7	+40.9
The Netherlands	93.2	389.4	308.4	387.9	+25.8
Sweden	98.8	325.2	245.2	333.1	+35.8
Thailand	3.9	23.4	138.3	283.7	+105.1
Switzerland	47.5	211.8	193.3	264.9	+37.0
Singapore	28.9	142.1	204.4	257.7	+26.1
Portugal	32.6	100.0	202.0	256.4	+26.9
Australia	41.6	178.5	217.5	225.7	+3.8
USSR	16.6	91.2	157.1	156.0	-0.7
Brazil	34.7	89.6	47.8	130.5	+173.0
Subtotal	2,734.8	13,309.6	16,208.7	21,545.7	+32.9
Other	540.2	2,647.8	2,296.2	2,574.6	+12.1
World Total	3,275.0	15,957.4	18,504.9	24,120.3	+30.3

TABLE 5C. -- (continued)

	1970	1980	1985	1986	1985-86 (% change)
Exports:					
Canada	257.3	1,094.5	1,359.2	1,744.2	+28.3
USA	111.9	1,001.7	1,162.4	1,481.0	+27.4
Denmark	165.6	1,000.0	952.7	1,381.5	+45.0
Korea, Rep.	42.0	681.8	796.9	1,188.4	+49.1
Norway	300.2	974.7	922.5	1,171.2	+27.0
Thailand	17.7	358.3	675.1	1,011.9	+49.9
Japan	335.5	905.2	819.8	897.9	+9.5
Iceland	112.9	708.6	617.4	858.0	+39.0
The Netherlands	111.8	524.6	543.7	766.4	+41.0
China	NA£	348.4	366.9	645.8	+76.0
USSR	90.4	307.9	383.9	587.1	+52.9
Chile	27.4	323.0	438.6	516.0	+17.6
France	37.0	320.3	359.0	501.2	+39.6
U.K.	55.1	365.2	342.0	481.7	+40.8
Mexico	71.5	580.0	371.0	423.9	+14.3
Spain	95.5	344.4	353.8	398.7	+12.7
Hong Kong	18.4	165.3	277.3	396.9	+43.1
Ecuador	6.1	200.0	260.9	383.6	+47.0
India	41.1	268.5	298.8	362.5	+21.3
Germany, F.R.	62.9	317.1	285.8	358.4	+25.4
Subtotal	1,919.2	10,789.5	11,587.7	15,556.3	+34.2
Other	975.8	4,511.0	5,515.7	6,929.8	+25.6
World Total	2,895.0	15,300.5	17,103.4	22,486.1	+31.5

SOURCE. -- FAO.
*The term "Fishery Commodities" follows FAO usage and includes the seven principle fishery commodity groups. Countries are ranked according to 1986 figures.
‡Data in metric tons is no longer available.
£1970 date for China not available.

TABLE 6C. -- FISHING FLEETS, BY COUNTRY, 1987*

	Trawlers and Fishing Vessels			Factory Ships and Carriers	
	grt	no.	over 500 grt (%)§	grt	no.
Albania	300	2	0.0		
Algeria	2,731	23	0.0		
Angola	15,712	73	6.7		
Antigua & Barbuda	263	1	0.0		
Argentina	87,169	182	64.4	4,264	2
Australia	47,618	265	3.3	291	1
Bahamas	1,550	10	0.0		
Bahrain	1,004	7	0.0		
Bangladesh	10,496	47	31.7	263	2
Barbados	3,368	27	0.0		
Belgium	21,405	115	2.6		
Benin	1,078	8	0.0		
Bermuda	1,481	3	77.4	706	1
Brazil	13,747	82	6.3		
Bulgaria	73,541	32	99.2	29,056	6
Burma	5,578	27	0.0		
Cameroon	5,993	36	8.7		
Canada	153,673	477	49.6	698	3
Cape Verde	1,567	7	0.0		
Cayman Islands	33,814	67	77.7		
Chile	73,191	159	54.1	1,257	1
China, P.R.	39,940	103	47.1	24,354	21
Colombia	2,258	17	0.0		
Congo	7,682	18	71.0		
Costa Rica	5,485	12	59.4		
Cuba	141,006	264	84.2		
Cyprus	5,012	7	87.8		
Denmark	178,154	543	43.1	312	1
Dominica	103	1	0.0		
Ecuador	25,249	85	24.0		
Egypt	4,184	6	86.8		
El Salvador	3,514	12	65.2		
Ethiopia	218	2	0.0		
Fiji	620	5	0.0		
Finland	2,543	18	0.0		
France	132,250	372	59.3	3,393	1
Gabon	1,108	8	0.0		
Gambia	605	3	0.0		
German D.R.	87,554	109	90.5	62,016	9
Germany, F.R.	41,147	94	69.6		
Ghana	55,942	102	65.1	5,602	3
Greece	29,975	103	35.5		
Guatemala	377	3	0.0		
Guinea	2,333	6	46.4		
Guinea-Bissau	1,941	6	66.8		
Guyana	5,117	48	0.0	100	1
Haiti	280	1	0.0		
Honduras	22,323	127	4.1	1,720	2
Hong Kong	990	4	50.5		
Iceland	103,656	339	25.1		
India	21,779	127	12.2		
Indonesia	49,433	230	9.0	1,510	5
Iran	8,799	27	60.4		
Iraq	19,439	16	91.7	10,413	2
Ireland	17,355	63	37.1		
Israel	2,908	3	100.0		
Italy	67,674	240	46.3		
Ivory Coast	9,994	33	52.1	499	1
Jamaica	1,028	6	0.0		
Japan	886,705	2,632	29.8	165,436	144
Kenya	1,352	6	0.0		
Kiribati	121	1	0.0	401	1
Korea, D.R.	5,982	13	43.5	36,190	6
Korea, Rep.	380,523	999	37.3	67,012	41
Kuwait	10,964	71	6.2	788	1
Lebanon	560	4	0.0	738	1
Libya	5,322	32	0.0		

TABLE 6C. -- (continued)

	Trawlers and Fishing Vessels			Factory Ships and Carriers	
	grt	no.	over 500 grt (%)§	grt	no.
Madagascar	6,252	38	0.0		
Malaysia	2,588	11	0.0		
Maldive Islands	1,602	3	62.3	3,751	4
Malta	9,180	30	48.1	1,311	1
Mauritania	26,753	90	11.6	1,100	1
Mauritius	6,854	20	31.9		
Mexico	123,743	407	57.5		
Morocco	73,350	221	30.0		
Mozambique	15,682	74	21.0		
Nauru	948	1	100.0		
The Netherlands	127,844	417	30.2		
New Zealand	18,970	43	57.2		
Nicaragua	1,971	17	0.0		
Nigeria	22,929	101	35.1	4,923	3
Norway	238,323	586	50.0	5,682	13
Oman	945	4	0.0		
Pakistan	1,419	5	0.0		
Panama	141,114	369	40.9	24,914	20
Papua New Guinea	2,282	16	0.0		
Peru	156,727	550	34.8		
Philippines	66,725	266	8.7	1,374	6
Poland	217,456	305	89.6	86,377	13
Portugal	118,739	181	77.2		
Qatar	696	5	0.0		
Romania	131,555	50	99.3	94,350	12
St. Lucia	105	1	0.0		
St. Vincent	1,074	4	0.0		
Sao Tome & Principe	993	2	53.4		
Saudi Arabia	3,005	15	0.0	354	1
Senegal	31,278	132	27.9		
Seychelles				1,827	1
Sierre Leone	3,920	21	0.0		
Singapore	3,824	15	0.0		
Solomon Islands	979	5	0.0		
Somalia	5,188	14	79.0		
South Africa	68,491	161	51.9	303	1
Spain	500,231	1,533	35.8	3,440	3
Sri Lanka	3,156	12	22.1		
Surinam	1,260	7	0.0		
Sweden	20,335	100	14.5		
Taiwan	85,779	277	13.5	1,656	1
Tanzania	911	4	0.0		
Thailand	6,147	12	57.2		
Togo	446	3	0.0		
Tonga	764	4	0.0		
Trinidad & Tobago	2,758	20	0.0		
Tunisia	2,685	16	0.0		
Turkey	3,935	10	44.2	668	2
Turks & Caicos Is.	124	1	0.0		
Tuvalu	173	1	0.0		
U.S.S.R.	3,673,752	2,802	95.5	3,155,604	520
U.A.E.	2,388	3	93.7		
U.K.	86,723	348	24.1		
U.S.A.	615,348	3,127	29.0	24,507	13
Uruguay	16,293	56	21.1		
Vanuatu	6,324	6	100.0		
Venezuela	35,044	97	54.1	199	1
Vietnam	7,626	30	36.8	1,996	2
Virgin Is. (Br.)	705	4	0.0		
Western Samoa	213	1	0.0		
Yemen, P.D.R.	4,611	10	60.4		
Yugoslavia	2,204	18	0.0	113	1
Zaire	4,793	14	53.5		
World Total	9,666,065	21,267	62.9	3,831,468	875

SOURCE. -- Lloyd's Register of Shipping Statistical Tables.
*Data exclude vessels of less than 100 grt. The smaller vessels used in artisanal fishing are therefore excluded.
§Percent of grt.

Appendix D

Tables, Nonliving Resources

The four tables of this appendix are modeled on those in *Ocean Yearbook 6* and have been updated wherever possible. Tables compiled before Burma officially adopted the name Myanmar may list the country as Burma. The reader is also referred to the article in this volume by Irene Baird for additional data on the development of offshore petroleum resources.

THE EDITORS

TABLE 1D. --   WORLD PRODUCTION OF CRUDE OIL, TOTAL AND OFFSHORE
                (Barrels per Day, in Thousands)

Year	World Production	Offshore Production	Offshore as % of World
1973	55,212.70	10,067.28	18.2
1974	56,772.00	9,268.62	16.3
1975	53,850.00	8,278.36	15.4
1976	57,210.00	9,431.91	16.5
1977	56,567.00	11,436.75	20.2
1978	60,337.00	11,480.75	19.0
1979	62,768.00	12,491.93	19.9
1980	59,812.00	13,587.49	22.7
1981	55,886.20	13,664.61	24.5
1982	53,191.00	13,541.25	25.5
1983	53,259.00	13,791.04	26.7
1984	54,090.00	15,311.50	28.3
1985	53,391.00	15,128.33	28.0
1986	55,864.00	13,498.78	24.2
1987	55,954.00	14,749.98	26.4

SOURCES. -- <u>Offshore</u> (June 20, 1977, 1978, 1979, 1980, 1981, 1982, 1983, July 20, 1984, and May 1986, 1987, 1988, and 1989).

NOTE. -- 6.998 barrels of crude petroleum approximately equal 1.0 metric ton (ASTM-1P Petroleum Measurement Tables).

TABLE 2D. -- OFFSHORE CRUDE OIL PRODUCTION, BY REGION AND COUNTRY (Barrels per Day, in Thousands)

Area and Country	1970	1975	1980	1986	1987	1988	1987-88 (% change)
Total	7,532.0	8,278.4	13,587.5	13,498.8#	14,750.0	14,271.3	-3.2
Anglo-America:							
Canada						0.1	
U.S.A.	1,557	909.6	1,038.1	1,257.0*	1,212.3	1,129.0	-6.9
Subtotal				1,257.0	1,212.3	1,129.1	-6.9
Latin America:							
Brazil	8.0	19.0	73.0	376.0	396.0	376.0	-5.1
Chile				11.3	10.0	9.0	-9.6
Mexico	35.0	45.0	500.2	1,700.0	1,671.4	1,666.3	-0.3
Peru		28.9	29.9	116.6	115.6	103.5	-10.5
Trinidad & Tobago	76	174.0	166.5	127.8	128.3	117.2	-8.6
Venezuela	2,460	1,737.1	1,095.6	900.0	1,061.6	989.0	-6.8
Subtotal	2,579	2,004.0	1,865.2	3,231.6	3,382.9	3,261.0	-3.6
Western Europe:							
Denmark		3.3	6.6	55.1	56.1	58.4	4.2
Greece				27.2	26.2	24.2	-7.5
Italy	12	10.4	6.3	28.9	57.4	62.3	8.6
The Netherlands				20.7	19.2	18.6	-3.2
Norway		189.6	628.8	780.6	864.9	907.0	4.9
Spain		32.9	31.2	36.1	35.1	30.2	-13.9
U.K.		83.0	1,650.0	2,236.6	2,314.7	2,094.7	-9.5
Subtotal	12	319.2	2,322.9	3,185.1	3,373.6	3,195.5	-5.3
Socialist Eastern Europe:							
U.S.S.R.	258	228.0	200.0	165.0*	207.0	201.6	-2.6
Near East:							
Divided Zone	257	315.1	403.0	266.0	252.0	255.1	1.3
Egypt	322	165.0	390.3	589.6	602.1	591.3	-1.8
Iran		481.2	150.0	505.0	515.5	323.0	-37.3
Libya						20.0	
Qatar	172		247.6	158.0	191.3	190.0	-0.7
Saudi Arabia	1,251	1,385.8	2,958.0	1,107.0	1,510.3	1,481.3	-1.9
Tunisia		43.0	43.6	26.1	28.4	28.3	-0.1
UAE	339.0	750.4	1,676.9	615.3	591.7	590.8	-0.1
Subtotal	2,341	3,140.5	5,769.4	3,267.0	3,691.2	3,479.9	-5.7

TABLE 2D. -- (continued)

Area and Country	1970	1975	1980	1986	1987	1988	1987-88 (% change)
Sub-Saharan Africa:							
Angola/Cabinda	96	143.2	99.0	185.6	185.0	450.9	143.7
Cameroon				125.0	201.3	187.1	-7.0
Congo		37.3	27.0	115.0	116.8	133.7	14.5
Gabon	29	179.9	177.9	105.0	92.6	108.3	16.9
Ghana			2.0	0.3	0.3	0.2	-23.3
Ivory Coast			5.8	19.5	17.4	15.4	-11.1
Nigeria	275	431.3	579.1	289.5	456.2	471.6	3.4
Zaire			21.6	16.6	17.5	13.5	-22.5
Subtotal	400	791.7	910.4	856.5	1,087.0	1,380.7	27.0
South Asia:							
India			142.1	621.0	584.0	618.2	5.9
East Asia:							
China			2.0	1.0	1.3	2.9	123.8
Japan	3	0.9	1.6	1.4	1.3	1.2	-4.6
Subtotal	3	0.9	3.6	2.4	2.6	4.1	57.7
Southeast Asia:							
Brunei	146§	141.2	192.2	98.7	95.7	89.8	-6.2
Indonesia		246.4	533.0	391.9	451.5	443.6	-1.8
Malaysia		84.5	280.3	320.9	204.3	262.7	28.6
Philippines			4.0	7.8	4.7	6.0	25.9
Thailand				15.9	14.3	19.3	34.9
Subtotal	146	472.1	1,009.5	835.2	770.5	821.3	6.6
Australasia:							
Australia	216	412.5	323.2	384.1	434.6	404.0	-7.0
New Zealand			3.2	14.0	20.8	21.1	1.6
Subtotal	216	412.5	326.4	398.1	455.3	425.1	-6.6

SOURCE. -- OFFSHORE.
*Estimate.
§Brunei/Malaysia combined production.
#Subtotals do not equal total because of rounding.

TABLE 3D. -- WORLD PRODUCTION OF NATURAL GAS, TOTAL AND OFFSHORE
(Cubic Feet, in Billions)

Year	World Production	Offshore Production	Offshore as % of World
1973	56,992.3	7,697.0	13.5
1974	47,253.3	8,088.7	17.1
1975	47,029.9	9,932.1	20.3
1976	50,407.5	10,847.0§	21.5
1977	53,883.7	6,663.3§	12.4
1978	53,859.5	9,509.0§	17.8
1979	57,194.6	9,369.0§	16.4
1980	58,636.4	10,160.9§	17.3
1981	57,816.0	10,085.1§	17.4
1982	55,893.9	10,326.7§	18.5
1983	55,066.5	10,360.1§	18.8
1984	59,932.3	12,196.8§	20.4
1985	62,721.4	12,451.4§	19.9
1986	63,683.2	12,348.6	19.4
1987	68,168.0	11,688.1	17.1

SOURCES. -- <u>Basic Petroleum Data Book</u> for 1973-75, <u>Oil and Gas Journal</u> and <u>Offshore</u> for 1976-88.
§Based on extrapolation from average daily rate.

TABLE 4D. -- OFFSHORE NATURAL GAS PRODUCTION BY REGION AND COUNTRY
(Cubic Feet per Day, in Millions)

Area and Country	1970	1975	1980	1986	1987	1988	1987-88 (% change)
Total	10,315.1	20,967.0	27,838.0	33,831.9	32,022.2	30,843.2	-3.7
Anglo-America:							
U.S.A.	8,591.8	11,664.3	14,703.7	11,917.0	11,363.7	10,444.4	-8.1
Latin America:							
Brazil		25.4	100.0	215.0	389.0	454.0	16.7
Chile				92.0	96.7	90.3	-6.6
Colombia			64.7	127.0	212.8	245.0	15.2
Mexico			25.0*	950.0	1,063.8	1,099.0	3.3
Peru	64.7	77.0					
Trinidad & Tobago	11.0	123.0	420.0	367.5	386.8	374.6	-3.2
Venezuela				628.6	609.3	601.1	-1.3
Subtotal	75.7	225.0	609.7	2,380.1	2,758.4	2,864.1	3.8
Western Europe:							
Denmark				145.0	164.4	171.0	4.0
Greece				9.1	9.7	9.0	-6.8
Ireland			125.0	230.0	232.1	297.0	28.0
Italy			22.0	415.1	420.7	103.2	-75.5
The Netherlands		186.0	1,170.0	1,480.0	1,645.0	1,591.0	-3.3
Norway		16.5	2,426.0	3,011.0	2,862.7	2,800.2	-2.2
Spain				43.5	91.9	82.9	-9.8
U.K.		3,600.0	3,610.0	5,600.0	4,468.1	4,423.0	-1.0
Subtotal		3,802.5	7,353.0	10,933.7	9,894.5	9,477.3	-4.2
Socialist Eastern Europe:							
U.S.S.R.		774.0	1,225.0	1,300.0	1,400.0	1,330.0	-5.0
Near East:							
Egypt				110.0	120.9	115.0	-4.9
Qatar				94.0	96.3	98.4	2.2
Saudi Arabia	494.8	3,825.4	0.0$	501.0	531.9	526.7	-1.0
UAE			1,434.0	696.3	735.0	701.2	-4.6
Subtotal	494.8	3,825.4	1,491.9	1,401.3	1,484.1	1,441.4	-2.9

Sub-Saharan Africa:							
Angola/Cabinda				30.2	33.8	39.2	15.8
Congo				3.0			
Ghana			8.0				
Ivory Coast			500.0	1.9	1.9	1.3	-31.1
Nigeria		314.0		98.0	96.9	114.1	17.7
Subtotal		314.0	508.0	133.1	132.7	154.6	16.5
East Asia:							
Japan			47.7	55.0	51.0	46.0	-9.8
Taiwan					19.3	20.2	4.5
South Asia:							
India			2.7	372.3	430.4	433.0	0.6
Southeast Asia:							
Brunei		158.8	984.8	815.0	822.0	814.0	-1.0
Indonesia			440.0	663.7	793.0	830.0	4.7
Malaysia				1,440.3	1,412.0	1,320.0	-6.5
Thailand				341.9	353.0	550.0	55.8
Subtotal		158.8	1,424.8	3,260.9	3,380.0	3,514.0	4.0
Australasia:							
Australia	60.5	203.0	388.0	725.3	821.0	840.2	2.3
New Zealand			84.0	369.0	287.0	271.0	-5.6
Subtotal	60.5	203.0	472.0	1,094.3	1,108.0	1,111.2	0.3

SOURCES. -- <u>Offshore</u> and <u>Oil and Gas Journal</u>.
*Estimate.
§All gas flared.

Appendix E

Tables, Transportation and Communication

The four tables of this appendix are continuations of the shipping tables that appeared in previous volumes of the *Yearbook*. The two tables showing the member countries of INMARSAT and INTELSAT, which first appeared in *Ocean Yearbook 5,* have not been repeated in this volume. Tables compiled before Burma officially adopted the name Myanmar may list the country as Burma. For additional information on shipping see the articles by Bernhard Abrahamsson, Satyesh Chakraborty, Toufiq Siddiqi, and Jan van Ettinger.

THE EDITORS

627

TABLE 1E. -- WORLD SHIPPING TONNAGE, BY TYPE OF VESSEL (Million grt as of July 1)

	1970	1980	1985	1986	1987	1986-87 (% change)
Oil tankers*	86.1	175.0	138.4	128.4	124.7	-2.9
Liquified gas carriers#	1.4	7.4	10.0	9.8	9.8	0.0
Chemical carriers	0.5	2.2	3.4	3.6	3.5	-3.8
Miscellaneous tankers		0.2	0.3	0.3	0.3	0.0
Bulk/oil carriers§	8.3	26.2	23.7	21.3	20.5	-3.9
Ore and bulk carriers	38.3	83.4	110.3	111.6	110.6	-0.9
General cargo£	72.4	82.6	75.8	73.2	72.2	-1.5
Miscellaneous cargo ships						0.0
Container ships (fully cellular)	1.9	11.3	18.4	19.6	21.1	7.6
Barge-carrying vessels						0.0
Vehicle carriers		1.8	3.9	4.4	4.5	1.5
Fishing factories, carriers, and trawlers	7.8	12.8	13.2	13.4	13.5	0.9
Passenger liners	3.0					0.0
Ferries and other passenger vessels		7.6	8.3	8.8	9.2	5.0
All other vessels•	7.8	8.6	10.6	10.5	10.8	3.3

SOURCE. -- Lloyd's Register of Shipping Statistical Tables.
NOTE. -- grt = gross registered tons.
*Including oil/chemical tankers.
#Including oil/chemical tankers.
#I.e., ships capable of liquid natural gas or liquid petroleum gas or other similar hydrocarbon and chemical products which are all carried at pressures greater than atmosphere or at subambient temperature or a combination of both.
§Including ore/oil carriers.
£Including passenger/cargo.
•Including livestock carriers, supply ships and tenders, tugs, cable ships, dredgers, icebreakers, research ships, and others.

TABLE 2E. -- ESTIMATED AVERAGE SIZE OF SELECTED TYPES OF VESSELS: EXISTING WORLD FLEETS, MIDYEAR (grt)

	1970	1980	1985	1986	1987	1986-87 (% change)
Oil tankers (100 grt and above)	14,110	24,606	21,009	20,742	20,635	-0.5
Ore/bulk carriers (6,000 grt and above)*	18,450	23,408	24,853	25,201	25,697	2.0
Container ships (100 grt and above)	11,420	17,030	18,164	18,430	19,295	4.7
Liquified gas carriers (grt)	4,690	11,624	12,841	12,769	12,691	-0.6

SOURCE. -- Lloyd's Register of Shipping Statistical Tables.
NOTE. -- grt = gross registered tons.
*Including bulk/oil carriers.

TABLE 3E. -- WORLD MERCHANT FLEETS, BY REGION AND COUNTRY (grt as of July 1)

	1980	1985	1986	1987	1986-87 (% change)
Anglo-America:					
Bermuda	3,180,126	980,707	1,208,276	1,925,297	59.3
Canada		3,343,823	3,160,043	2,971,155	-6.0
U.S.A.	18,464,271	19,517,571	19,900,843	20,178,236	1.4
Subtotal	21,644,297	23,842,101	24,269,162	25,074,688	3.3
Latin America:					
Anguilla	399	3,966	4,106	3,705	-9.8
Antigua & Barbuda	410	559	1,048	51,875	4849.9
Argentina	2,546,305	2,457,337	2,117,017	1,901,026	-10.2
Bahamas	87,320	3,907,267	5,985,011	9,105,182	52.1
Barbados	5,257	8,408	7,572	8,348	10.2
Belize	620	620	620	620	0.0
Bolivia	15,130	14,913	14,913	13,824	-7.3
Brazil	4,533,633	6,057,364	6,212,287	6,324,059	1.8
Caymen Is.	256,715	413,752	1,389,903	706,160	-49.2
Chile	614,425	454,484	566,881	546,745	-3.6
Colombia	283,457	365,638	380,074	423,631	11.5
Costa Rica	20,333	19,936	13,325	14,781	10.9
Cuba	881,260	965,077	958,613	966,288	0.8
Dominica		1,390	2,013	1,724	-14.4
Dominican Rep.	37,659	46,676	42,241	43,560	3.1
Ecuador	275,142	443,893	437,682	421,361	-3.7
El Salvador	501	3,501	3,819	3,819	0.0
Falkland Is.	7,907	6,907	6,907	6,907	0.0
Grenada	226	425	425	550	29.4
Guatemala	13,626	16,046	9,432	4,694	-50.2
Guyana	18,261	23,368	22,731	22,310	-1.9
Haiti	1,120	2,688	2,688	512	-81.0
Honduras	213,421	356,610	555,202	506,374	-8.8
Jamaica	13,307	9,419	9,419	13,118	39.3
Mexico	1,006,417	1,467,191	1,520,246	1,532,485	0.8
Monserrat	1,010	711	711	711	0.0
Nicaragua	15,726	17,956	22,930	12,739	-44.4

Panama	24,190,680	40,674,201	41,305,009	43,254,716	4.7
Paraguay	23,019	42,851	43,298	41,670	-3.8
Peru	740,510	818,103	754,179	788,171	4.5
St. Kitts-Nevis	256	556	256	556	117.2
St. Lucia	2,378	1,786	2,766	2,092	-24.4
St. Vincent	19,679	235,183	509,878	699,947	37.3
Suriname	14,921	15,222	12,655	11,457	-9.5
Trinidad & Tobago	17,456	18,969	19,381	18,527	-4.4
Turks & Caicos Is.	2,408	3,095	3,583	3,469	-3.2
Uruguay	198,478	173,423	149,811	144,394	-3.6
Venezuela	848,540	984,881	998,296	999,195	0.1
Virgin Is. (U.K.)	5,826	8,592	8,077	8,179	1.3
Subtotal	36,910,961	60,042,964	64,095,005	68,609,481	7.0*
Western Europe:					
Austria	88,784	134,225	124,794	193,513	55.1
Belgium	1,809,829	2,400,292	2,419,661	2,268,383	-6.3
Denmark	5,390,365	4,942,175	4,651,224	4,754,837	2.2
Faeroe Is.	66,085	103,077	115,394	118,628	2.8
Finland	2,530,091	1,974,008	1,469,927	1,122,249	-23.7
France	11,924,557	8,237,418	5,936,268	5,371,273	-9.5
Germany, F.R.	8,355,638	6,177,032	5,565,214	4,317,616	-22.4
Gibraltar	2,291	583,270	1,612,948	2,827,098	75.3
Greece	29,471,744	31,031,544	28,390,800	23,559,852	-17.0
Iceland	188,215	180,323	176,409	173,618	-1.6
Ireland	208,926	194,022	149,308	153,637	2.9
Italy	11,095,694	8,843,181	7,896,569	7,817,353	-1.0
Malta	132,861	1,855,807	2,014,947	1,725,984	-14.3
Monaco	31,422	3,268			
The Netherlands	5,723,845	4,301,324	4,324,135	3,908,231	-9.6
Norway	22,007,490	15,338,557	9,294,630	6,359,349	-31.6
Portugal	1,355,989	1,436,892	1,114,444	1,048,197	-5.9
Spain	8,112,245	6,256,188	5,422,002	4,949,387	-8.7
Sweden	4,233,977	3,161,939	2,516,614	2,269,541	-9.8
Switzerland	310,775	341,972	346,220	354,614	2.4
U.K.	27,135,155	14,343,512	11,567,117	8,504,605	-26.5
Subtotal	150,175,978	111,840,026	95,108,625	81,797,965	-14.0

TABLE 3E. -- (continued)

	1980	1985	1986	1987	1986-87 (% change)
Socialist Eastern Europe:					
Albania	56,127	56,133	56,133	56,133	0.0
Bulgaria	1,233,307	1,322,231	1,385,009	1,551,176	12.0
Czechoslovakia	155,319	184,299	197,868	156,791	-20.8
German D.R.	1,532,197	1,434,428	1,518,944	1,494,039	-1.6
Hungary	74,997	77,182	86,395	77,377	-10.4
Poland	3,639,078	3,315,285	3,457,242	3,469,670	0.4
Romania	1,856,292	3,023,770	3,233,906	3,263,823	0.9
U.S.S.R.	23,433,534	24,745,435	24,960,888	25,232,091	1.1
Yugoslavia	2,466,574	2,699,302	2,872,613	3,164,893	10.2
Subtotal	34,475,421	36,858,065	37,768,998	38,465,993	1.8
Sub-Saharan Africa:					
Angola	65,310	90,978	92,285	91,712	-0.6
Benin	4,557	4,887	4,887	4,665	-4.5
Cameroon	62,080	76,433	76,660	57,871	-24.5
Cape Verde	11,426	14,095	14,095	14,579	3.4
Comoros	1,116	1,302	1,261	1,795	42.3
Congo	6,784	8,458	8,458	8,458	0.0
Djibouti	3,135	2,868	3,051	3,051	0.0
Equatorial Guinea	6,412	6,412	6,412	6,412	0.0
Ethiopia	23,811	56,744	66,926	73,456	9.8
Gabon	77,095	97,528	97,967	23,843	-75.7
Gambia	3,907	2,588	2,588	3,878	49.8
Ghana	250,428	162,593	165,644	142,421	-14.0
Guinea	5,648	7,179	7,179	7,179	0.0
Guinea-Bissau	577	3,677	4,070	4,070	0.0
Ivory Coast	186,127	141,674	120,679	118,952	-1.4
Kenya	17,371	8,052	9,040	7,872	-12.9
Liberia	80,285,176	58,179,717	52,649,444	51,412,029	-2.4
Madagascar	91,211	74,235	73,715	64,162	-13.0
Malawi		424	424	424	0.0
Mali	200				
Mauritania	874	17,103	22,752	29,644	30.3
Mauritius	37,675	37,716	151,978	162,749	7.1

Mozambique	37,887	40,850	42,801	35,957	-16.0
Nigeria	498,202	443,384	563,912	593,582	5.3
St. Helena	3,150	3,640	3,640	3,640	0.0
Sao Tome & Principe		1,488	1,488	1,488	0.0
Senegal	34,499	50,991	50,429	46,448	-7.9
Seychelles	4,602	1,740	3,813	3,233	-15.2
Sierra Leone	3,738	5,913	6,979	8,756	25.5
Somalia	45,553	29,340	15,719	17,896	13.8
South Africa	728,926	632,455	599,509	533,092	-11.1
Sudan	104,803	95,742	95,742	96,699	1.0
Tanzania	55,916	50,576	50,726	31,551	-37.8
Togo	14,886	53,988	54,882	59,690	8.8
Uganda	5,510	3,394	5,091	5,091	0.0
Zaire	91,894	84,720	65,833	56,393	0.0
Zambia					-14.3
Subtotal	82,772,023	60,492,884	55,140,079	53,732,738	-2.6#
Near East:					
Algeria	1,218,621	1,347,398	881,670	892,553	1.2
Bahrain	10,248	47,552	51,713	43,833	-15.2
Cyprus	2,091,089	8,196,056	10,616,809	15,650,207	47.4
Egypt	555,786	952,644	1,063,020	1,074,192	1.1
Iran	1,283,629	2,379,957	2,911,359	3,976,873	36.6
Iraq	1,465,949	1,011,864	1,016,343	1,002,236	-1.4
Israel	450,216	549,732	556,628	514,815	-7.5
Jordan	496	48,300	42,365	32,884	-22.4
Kuwait	2,529,491	2,349,904	2,580,924	2,087,856	-19.1
Lebanon	267,787	504,956	484,624	460,876	-4.9
Libya	889,908	853,782	825,231	816,570	-1.0
Morocco	359,552	460,927	416,482	418,451	0.5
Oman	6,954	17,495	14,793	25,321	71.2
Qatar	91,934	353,221	306,673	306,443	-0.1
Saudi Arabia	1,589,668	3,137,178	2,978,016	2,692,044	-9.6
Syria	39,255	58,000	63,142	63,077	-0.1
Tunisia	131,079	284,314	285,535	285,483	-0.0

TABLE 3E. -- (continued)

	1980	1985	1986	1987	1986-87 (% change)
Turkey	1,454,838	3,684,357	3,423,745	3,336,093	-2.6
United Arab Emirates	158,210	868,564	653,525	732,013	12.0
Yemen, A.R.	2,979	3,203	7,115	199,909	2709.7
Yemen, P.D.R.	12,230	11,931	12,543	12,278	-2.1
Subtotal	14,609,918	27,121,335	29,192,255	34,624,007	18.6
South Asia:					
Bangladesh	353,586	358,071	378,563	410,721	8.5
India	5,911,367	6,604,548	6,540,121	6,725,776	2.8
Maldives	136,037	133,254	84,808	100,200	18.1
Pakistan	478,019	450,996	434,079	394,407	-9.1
Sri Lanka	93,471	634,658	622,226	594,491	-4.5
Subtotal	6,972,480	8,181,527	8,059,797	8,225,595	2.1
East Asia:					
China	6,873,608	10,568,236	11,566,974	12,341,477	6.7
Hong Kong	1,717,230	6,858,099	8,179,670	8,034,668	-1.8
Japan	40,959,683	39,940,135	38,487,773	35,932,177	-6.6
Korea, D.P.R.	230,695	512,568	407,253	406,647	-0.1
Korea, Rep.	4,344,114	7,168,940	7,183,617	7,214,070	0.4
Taiwan	2,039,123	4,327,487	4,272,795	4,512,749	5.6
Subtotal	56,164,453	69,375,465	70,098,082	68,441,788	-2.4
Southeast Asia:					
Brunei	899	1,235	1,973	352,276	17754.8
Burma	78,519	116,556	125,524	239,261	90.6
Indonesia	1,411,688	1,936,420	2,085,635	2,120,531	1.7
Kampuchea	3,558	3,558	3,558	3,558	0.0
Malaysia	702,145	1,773,115	1,743,629	1,688,523	-3.2
Philippines	1,927,869	4,593,979	6,922,499	8,681,227	25.4
Singapore	7,664,229	6,504,582	6,267,627	7,098,116	13.3
Thailand	391,456	586,288	533,138	510,991	-4.2
Vietnam	240,895	298,584	338,668	360,470	6.4
Subtotal	12,430,258	15,814,317	18,022,251	21,054,953	16.8

Australasia:					
Australia	1,642,594	2,088,349	2,368,462	2,404,559	1.5
Fiji	14,733	30,608	29,954	35,324	17.9
Kiribati	980	2,051	3,197	3,332	4.2
Nauru	54,004	66,725	66,725	65,777	-1.4
New Zealand	263,543	295,899	314,206	334,193	6.4
Papua New Guinea	24,904	28,517	30,922	36,346	17.5
Solomon Islands		5,811	6,002	6,387	6.1
Tongo	25,395	17,252	16,349	18,295	11.9
Tuvalu		526	526	526	0.0
Vanuatu	12,541	138,025	164,953	540,088	227.4
Western Samoa	4,765	26,087	26,087	26,087	0.0
Subtotal	2,046,167	2,699,850	3,027,403	3,470,914	14.6

SOURCE. -- Lloyd's Register of Shipping Statistical Tables.

NOTE. -- grt = gross registered tons.

*Excluding Panama, the 1986-87 change for Latin America is 11.3%.

#Excluding Liberia, the 1986-87 change for Sub-Saharan Africa is -6.8%.

TABLE 4E. -- VESSELS LOST, BY COUNTRY (grt)

	1980	1982	1983	1984	1985	1986
Argentina	14,405 (5)	3,122 (3)		442 (1)	2,122 (1)	132 (1)
Australia	186 (1)	1,265 (2)	141 (1)	138 (1)	124 (1)	
Belgium		3,729 (4)				
Brazil	12,044 (1)	11,513 (9)		6,466 (2)	8,395 (2)	11,372 (1)
Canada	6,926 (6)	77,903 (18)	2,437 (3)	1,541 (4)	726 (2)	807 (2)
Cyprus	41,054 (17)		39,614 (9)	243,285 (10)	177,804 (13)	361,760 (21)
Denmark	2,979 (7)	279 (2)	2,002 (7)	3,355 (6)	857 (2)	896 (2)
Finland				1,323 (1)	499 (1)	2,326 (1)
France	729 (2)		447 (3)	5,644 (4)	458 (3)	641 (1)
German D.R.				27,389 (2)	1,743 (1)	8,227 (1)
Germany, F.R.	2,620 (6)	119 (1)	6,324 (5)	2,179 (5)	5,020 (3)	1,286 (3)
Greece	327,161 (39)	352,822 (49)	418,173 (38)	480,115 (21)	342,828 (20)	217,055 (13)
Hong Kong	139 (1)			41,238 (2)		92,630 (2)
India	18,680 (6)	10,050 (2)	2,939 (1)	39,741 (2)	33,969 (4)	299 (1)
Italy	64,195 (10)	43,401 (11)	3,607 (6)	21,196 (8)	6,994 (7)	4,705 (6)
Japan	55,518 (46)	16,990 (42)	9,872 (26)	13,890 (47)	19,181 (46)	10,006 (27)
Korea, Rep.	127,840 (21)	27,644 (20)	43,377 (14)	20,965 (11)	89,699 (17)	127,851 (13)
Liberia	516,534 (12)	212,493 (6)	174,458 (7)	424,300 (10)	107,251 (6)	849,663 (7)
The Netherlands	8,766 (2)	4,297 (5)		5,146 (1)	125 (1)	5,945 (4)
Norway	5,329 (11)	12,883 (7)	16,259 (10)	115,549 (6)	829 (4)	
Panama	142,190 (46)	292,855 (70)	341,746 (63)	231,123 (49)	295,895 (46)	178,309 (38)
Philippines	49,473 (10)	4,737 (6)	22,041 (8)	8,318 (8)	13,307 (7)	4,189 (5)
Poland		927 (1)	1,991 (1)	8,750 (2)	1,974 (1)	3,008 (1)
Portugal	7,804 (5)			803 (1)		320 (1)
Singapore	19,330 (3)	2,881 (3)	11,836 (3)	10,867 (3)	3,575 (2)	19,593 (7)
Spain	138,186 (17)	17,371 (5)	151,010 (14)	11,687 (18)	13,032 (7)	113,845 (15)
Sweden		6,142 (9)	500 (1)	349 (1)	500 (1)	1,949 (3)
Turkey	3,933 (6)	14,796 (4)	775 (2)		118,211 (5)	80,580 (3)
U.S.S.R.	109,587 (9)	8,069 (3)	6,912 (3)	43,623 (3)	10,944 (1)	47,317 (3)
U.K.	25,066 (27)	23,655 (10)	10,292 (9)	28,907 (7)	3,818 (8)	20,220 (7)
U.S.A.		72,245 (32)	34,753 (24)	64,510 (18)	10,979 (14)	16,578 (8)
Yugoslavia	15,605 (2)	1,445 (2)		36,034 (5)	12,162 (1)	2,175 (1)
Others*	87,887 (70)	408,058 (75)	171,105 (82)	455,068 (68)	368,144 (79)	425,051 (67)
Total	1,804,027(387)	1,631,930(402)	1,472,611(340)	2,353,941(327)	1,651,210(307)	2,608,735(265)

SOURCE. -- Lloyd's Register of Shipping Statistical Tables.
NOTE. -- grt = gross registered tons; no. of vessels shown in parentheses.
*Others includes, inter alia, China, P.R.; Indonesia; Kuwait; Mexico; Saudi Arabia; and Taiwan.

Tables, Military Activities

The seven tables of this appendix were begun by Andrzej Karboszka in *Ocean Yearbook 2* and have been updated in each of the subsequent volumes of the *Yearbook*. Tables compiled before Burma officially adopted the name Myanmar may list the country as Burma. For additional information on military activities the reader is referred to the articles in this volume by Joseph Morgan and Boleslaw Boczek.

THE EDITORS

TABLE 1F. -- WORLD STOCK OF AIRCRAFT CARRIERS, BY COUNTRY

	1950	1960	1970	1980	1986	1987	1988
World total:							
Attack	44	30	25	22	24	24	25
Other	75	42	30	21	25	26	27
U.S.A.:							
Attack	27	14	15	16	18	18	18
Other	75	40	24	14	12	12	13
U.S.S.R.:							
Attack					1	1	2
Other				3	4	6	6
U.K.:							
Attack	12	8	4	1			
Other		1	2	2	3	3	3
France:							
Attack	2	3	2	2	2	2	2
Other			1		1	1	1
Australia:							
Attack	1	2	1	1			
Other			1				
Canada:							
Attack	1	1					
Italy:							
Other					2	2	2
Netherlands:							
Attack	1	1					
Spain:							
Other			1	1	1	2	2
Argentina:							
Attack		1	2	1	1	1	1
Brazil:							
Other		1	1	1	1	1	1
India:							
Attack			1	1	2	2	2

SOURCE. -- J.E. Moore, ed. Jane's Fighting Ships 1988/89 (London: MacDonald & Jane's, 1988).

NOTE. -- Other = antisubmarine, amphibious helicopter assault, and utility.

TABLE 2F. -- WORLD STOCK OF STRATEGIC SUBMARINES,* BY GROUPS OF COUNTRIES

	1960	1970	1980	1985	1986	1987	1988
World total:							
Nucl.	7	70	124	114	115	116	118
Conv.	10	26	30	17	17	16	16
U.S.A.:							
Nucl.	3	41	41	37	38	39	39
Conv.							
Other NATO:							
Nucl.		5	9	10	11	11	11
Conv.		1	1	1	1		
Total NATO:							
Nucl.	3	46	50	47	49	50	50
Conv.		1	1	1	1		
U.S.S.R.:							
Nucl.	4	24	74	65	63	62	63
Conv.	10	25	19	15	15	15	14
Other WTO:							
Nucl.							
Conv.							
Total WTO:							
Nucl.	4	24	74	65	63	62	63
Conv.	10	25	19	15	15	15	14
China:							
Nucl.				2	3	4	5
Conv.				1	1	1	2

SOURCE. -- See table 1F.
NOTE. -- Nucl. = nuclear powered; Conv. = conventionally powered;
WTO = Warsaw Treaty Organization.
*Equipped with medium- or long-range ballistic missiles.

TABLE 3F. -- WORLD STOCK OF NUCLEAR-POWERED ATTACK SUBMARINES,* BY COUNTRY

	1955	1960	1965	1970	1976	1980	1986	1987	1988
World total	1	15	48	108	157	189	242	250	256
U.S.A.	1	11	22	46	68	81	97	97	99
France							3	4	5
U.K.				4	9	12	14	15	16
U.S.S.R.		4	26	58	80	96	124	128	132
China							4	4	4

SOURCE. -- See table 1F.
*Includes nuclear-powered cruise missile submarines.

TABLE 4F. -- WORLD STOCK OF PATROL SUBMARINES,* BY GROUPS OF COUNTRIES

	1950	1960	1970	1980	1986	1987	1988
World total:							
Nucl.		15	108	189	243	250	257
Conv.	355	535	535	495	501	497	472
Developed:							
Nucl.		15	108	180	239	244	252
Conv.	351	502	459	326	289	294	283
U.S.A.:							
Nucl.		11	46	81	97	97	99
Conv.	194	158	52	6	4	4	4
Other NATO:							
Nucl.			4	12	18	19	21
Conv.	105	89	75	82	86	82	80
Total NATO:							
Nucl.		11	50	93	115	116	120
Conv.	299	247	127	88	90	86	84
U.S.S.R.:							
Nucl.		4	58	96	124	128	132
Conv.	46	238	283	155	145	151	140
Other WTO:							
Nucl.							
Conv.			7	6	6	8	9
Total WTO:							
Nucl.		4	58	96	124	128	132
Conv.	46	238	290	161	151	159	149
Other Europe:§	3	12	27	32	24	21	22
Other developed:§	3	5	15	24	24	28	28
Total developing countries:							
Nucl.					4	4	5
Conv.	4	33	76	169	212	203	189
Middle East§		10	16	12	16	15	15
South Asia							
Nucl.							1
Conv.			8	13	15	17	19
China#							
Nucl.					4	4	4
Conv.		12	27	83	114	102	86
Other East and Southeast Asia§		2	14	20	23	22	22

TABLE 4F. -- (continued)

	1950	1960	1970	1980	1986	1987	1988
Sub-Saharan Africa§							
North Africa§				6	6	8	10
Central America§				2	4	4	3
South America§	4	9	11	33	34	35	34

SOURCE. -- See table 1F.

NOTE. -- Nucl. = nuclear powered; Conv. = conventionally powered; WTO = Warsaw Treaty Organization; Other Europe = Albania, Finland, Ireland, Malta, Sweden, and Yugoslavia (Spain included in NATO beginning 1982); Other developed = Australia, Japan, New Zealand, South Africa, and Taiwan; Middle East = Bahrain, Cyprus, Egypt, Iran, Iraq, Israel, Jordon, Kuwait, Lebanon, Oman, Qatar, Saudi Arabia, Syria, and the Yemens; South Asia = Bangladesh, India, Maldives, Pakistan, and Sri Lanka; Other East & S.E. Asia = Burma, Indonesia, Kampuchea, Malaysia, North Korea, Singapore, South Korea, Thailand, the Philippines, Vietnam, and the Pacific island nations; Sub-Saharan Africa = Angola, Cameroon, Cape Verde, Comoros, Congo, Ethiopia, Gabon, Ghana, Guinea, Ivory Coast, Kenya, Madagascar, Mauritania, Mozambique, Nigeria, Senegambia, Seychelles, Sierra Leone, Somalia, Sudan, Tanzania, Togo, and Zaire; North Africa = Algeria, Libya, Morocco, and Tunisia; Central America includes the Caribbean island nations; South America = Argentina, Brazil, Chile, Colombia, Ecuador, Guyana, Peru, Suriname, Uruguay, and Venezuela.

*Post-World War II submarines displacing 700 tons or more.

§All conventionally powered.

#Figures for China may vary considerably between years because of changes in the quality of available data.

TABLE 5F. -- WORLD STOCK OF COASTAL SUBMARINES,* BY GROUPS OF COUNTRIES

	1950	1960	1970	1980	1985	1986	1987	1988
World total	313	179	72	76	58	59	61	51
Developed	299	162	64	60	46	47	47	38
Developing	14	17	8	16	12	12	14	13
U.S.A.								
Other NATO		20	34	37	45	46	43	34
Total NATO		20	34	37	45	46	43	34
U.S.S.R.	273	127	22	20				
Other WTO								
Total WTO	273	127	22	20				
Other Europe	26	15	8	3	1	1	4	4
Other developed								

SOURCE. -- See table 1F.

NOTE. -- WTO = Warsaw Treaty Organization.

*Submarines displacing less than 700 tons; all conventionally powered.

TABLE 6F. -- WORLD STOCK OF MAJOR SURFACE WARSHIPS,* BY GROUPS OF COUNTRIES

	1950	1960	1970	1980	1986	1987	1988
World total:							
Miss.		18	191	560	770	812	880
Conv.	1,783	1,789	1,414	501	332	326	259
Developed:							
Miss.		18	189	480	615	641	697
Conv.	1,600	1,520	1,128	335	218	214	160
U.S.A.							
Miss.		15	77	148	206	210	233
Conv.	817	704	478	67	21	24	5
Other NATO:							
Miss.			57	160	204	214	220
Conv.	520	402	294	105	72	73	51
Total NATO:							
Miss.		15	134	308	410	428	453
Conv.	1,337	1,106	772	172	93	97	56
U.S.S.R.:							
Miss.		1	39	107	131	134	140
Conv.	150	260	206	86	76	74	74
Other WTO:							
Miss.			1	3	4	4	6
Conv.	5	14	9	2	6	6	7
Total WTO:							
Miss.		1	40	110	135	138	146
Conv.	155	274	215	88	82	80	81
Other Europe:							
Miss.		2	6	16	2	3	4
Conv.	67	75	71	23	1		
Other developed:							
Miss.			9	46	68	72	94
Conv.	41	63	64	52	42	42	23
Total developing:							
Miss.			2	80	155	171	183
Conv.	183	269	286	166	114	112	99
Middle East:							
Miss.			1	8	19	22	21
Conv.	18	15	21	9	7	3	5

TABLE 6F. -- (continued)

	1950	1960	1970	1980	1986	1987	1988
South Asia:							
Miss.				8	18	22	23
Conv.	16	26	32	27	23	16	14
China:							
Miss.				22	44	46	46
Conv.	4	23	21	6	6	10	10
Other East and S.E. Asia:§							
Miss.			1	6	25	29	31
Conv.	50	71	109	48	33	37	31
Sub-Saharan Africa:							
Miss.				1	1	1	1
Conv.			1	2	2	3	3
North Africa:							
Miss.				1	4	4	7
Conv.			2	1	1	1	1
Central America:							
Miss.					2	2	4
Conv.	30	29	22	31	10	10	12
South America:							
Miss.				34	42	45	50
Conv.	65	105	78	42	32	32	23

SOURCE. -- See table 1F.

NOTE. -- Miss. = missile armed (both SSMs and SAMs); Conv. = conventionally armed.   See table 4F for definitions of groups of countries.

*Cruisers, destroyers, frigates, and escorts (over 1,000 tons displacement).

§Taiwan included under "other developed" beginning 1980.

TABLE 7F. -- WORLD STOCK OF LIGHT NAVAL FORCES,* BY GROUPS OF COUNTRIES

	1950	1960	1970	1980	1986	1987	1988
World total:							
Miss.		5	281	822	1,212	1,134	1,130
Conv.	987	1,849	2,457	3,696	3,497	3,653	3,529
Developed:							
Miss.		5	188	368	548	481	483
Conv.	822	1,422	1,290	1,145	1,106	1,073	1,001
U.S.A.:							
Miss.			1	1	6	6	6
Conv.	147	35	35	21	21	25	24
Other NATO:							
Miss.			1	106	152	142	136
Conv.	190	230	241	249	292	278	262
Total NATO:							
Miss.			2	107	158	148	142
Conv.	337	265	276	270	313	303	286
U.S.S.R.:							
Miss.		5	150	183	198	148	156
Conv.	395	769	600	422	345	327	309
Other WTO:							
Miss.			28	37	57	59	59
Conv.	16	141	164	183	200	195	163
Total WTO:							
Miss.		5	178	220	255	207	215
Conv.	411	910	764	605	545	522	472
Other Europe:							
Miss.			8	34	70	55	55
Conv.	60	180	198	212	162	154	154
Other developed:							
Miss.				7	65	71	71
Conv.	14	67	54	58	86	94	89
Total developing:							
Miss.			93	454	664	653	647
Conv.	156	427	1,141	2,556	2,381	2,580	2,528
Middle East:							
Miss.			32	105	156	163	160
Conv.	11	77	140	325	360	346	338

TABLE 7F. -- (continued)

	1950	1960	1970	1980	1986	1987	1988
South Asia:							
Miss.				20	28	30	31
Conv.		1	7	52	70	94	106
China:							
Miss.			15	181	247	237	228
Conv.		150	408	762	594	754	690
Other East & S.E. Asia:							
Miss.			12	61	91	90	98
Conv.	55	149	403	717	706	729	752
Sub-Saharan Africa:							
Miss.				15	27	29	37
Conv.		5	54	228	258	259	262
North Africa:							
Miss.			16	31	52	52	49
Conv.		2	45	67	67	70	68
Central America:							
Miss.			18	26	34	23	18
Conv.	16	18	72	191	246	251	243
South America:							
Miss.				15	29	29	26
Conv.	42	26	38	109	80	77	69

SOURCE. -- See table 1F.

NOTE. -- Miss. = missile armed; Conv. = conventionally armed. WTO = Warsaw Treaty Organization. See table 4F for definitions of groups of countries.

*This category includes corvettes, fast patrol boats, torpedo boats, and large and coastal patrol crafts (gun armed). Riverine craft are excluded. Corvettes are not included for the years 1950-1976.

Appendix G

Tables, General Information

Since the signing and general acceptance of the 1982 United Nations Law of the Sea Convention many states have brought their maritime claims into line with the Convention. These tables provide general information about coastal states and their maritime claims and the status of the Convention. The table on maritime claims, coastline length, and shipping ports is updated from a table that appeared in *Ocean Yearbook 3*. The table on the status of the International Maritime Organization's conventions has been included for the first time. Tables compiled before Burma officially adopted the name Myanmar may list the country as Burma.

THE EDITORS

TABLE 1G. -- STATUS OF THE UN CONVENTION ON THE LAW OF THE SEA

A. Table of signatures and ratifications as of October 31, 1989

STATES	FINAL ACT SIGNATURE	CONVENTION SIGNATURE*	CONVENTION RATIFICATION
Afghanistan		3/18/83	
Albania			
Algeria* ᵇ	x	x	
Angola*	x	x	
Antigua and Barbuda		2/7/83	2/2/89
Argentina*		10/5/84	
Australia	x	x	
Austria	x	x	
Bahamas	x	x	7/29/83
Bahrain	x	x	5/30/85
Bangladesh	x	x	
Barbados	x	x	
Belgium*	x	12/5/84	
Belize	x	x	8/13/83
Benin	x	8/30/83	
Bhutan	x	x	
Bolivia*		11/27/84	
Botswana	x	12/5/84	
Brazil*	x	x	12/22/88
Brunei Darussalam		12/5/84	
Bulgaria	x	x	
Burkina Faso	x	x	
Burundi	x	x	
Byelorussian SSR*	x	x	
Cameroon	x	x	11/19/85
Canada	x	x	
Cape Verde* ᵓ ᶜ	x	x	8/10/87
Central African Republic		12/4/84	
Chad	x	x	
Chile*	x	x	
China	x	x	
Colombia	x	x	
Comoros		12/6/84	
Congo	x	x	
Costa Rica*	x	x	
Côte d'Ivoire	x	x	3/26/84
Cuba* ᵓ	x	x	8/15/84
Cyprus	x	x	12/12/88
Czechoslovakia	x	x	
Democratic Kampuchea		7/1/83	
Democratic People's Rep. of Korea	x	x	
Democratic Yemen**	x	x	7/21/87
Denmark	x	x	
Djibouti	x	x	
Dominica		3/28/83	
Dominican Republic	x	x	
Ecuador	x		
Egypt**	x	x	8/26/83
El Salvador		12/5/84	

TABLE 1G. -- (continued)

STATES	FINAL ACT SIGNATURE	CONVENTION SIGNATURE*	CONVENTION RATIFICATION
Equatorial Guinea	x	1/30/84	
Ethiopia	x	x	
Fiji	x	x	12/10/82
Finland*	x	x	
France*	x	x	
Gabon	x	x	
Gambia	x	x	5/22/84
German Democratic Republic*	x	x	
Germany, Federal Republic of	x		
Ghana	x	x	6/7/83
Greece*	x	x	
Grenada	x	x	
Guatemala		7/8/83	
Guinea*		10/4/84	9/6/85
Guinea-Bissau**	x	x	8/25/86
Guyana	x	x	
Haiti	x	x	
Holy See	x		
Honduras	x	x	
Hungary	x	x	
Iceland**	x	x	6/21/85
India	x	x	
Indonesia	x	x	2/3/86
Iran (Islamic Republic of)*	x	x	
Iraq*	x	x	7/30/85
Ireland	x	x	
Israel	x		
Italy*	x	12/7/84	
Jamaica	x	x	3/21/83
Japan	x	2/7/83	
Jordan	x		
Kenya	x	x	3/2/89
Kiribati			
Kuwait**	x	x	5/2/86
Lao People's Democratic Republic	x	x	
Lebanon		12/7/84	
Lesotho	x	x	
Liberia	x	x	
Libyan Arab Jamahiriya	x	12/3/84	
Liechtenstein		11/30/84	
Luxembourg*	x	12/5/84	
Madagascar		2/25/83	
Malawi		12/7/84	
Malaysia	x	x	
Maldives	x	x	
Mali*		10/19/83	7/16/85
Malta	x	x	
Mauritania	x	x	
Mauritius	x	x	
Mexico	x	x	3/18/83

TABLE 1G. -- (continued)

STATES	FINAL ACT SIGNATURE	CONVENTION SIGNATURE*	CONVENTION RATIFICATION
Monaco	x	x	
Mongolia	x	x	
Morocco	x	x	
Mozambique	x	x	
Myanmar[d]	x	x	
Nauru	x	x	
Nepal	x	x	
The Netherlands	x	x	
New Zealand	x	x	
Nicaragua*		12/9/84	
Niger	x	x	
Nigeria	x	x	8/14/86
Norway	x	x	
Oman*	x	7/1/83	8/17/89
Pakistan	x	x	
Panama	x	x	
Papua New Guinea	x	x	
Paraguay	x	x	9/26/86
Peru	x		
Philippines* **	x	x	5/8/84
Poland	x	x	
Portugal	x	x	
Qatar*		11/27/84	
Republic of Korea	x	3/14/83	
Romania*	x	x	
Rwanda	x	x	
Saint Kitts and Nevis		12/7/84	
Saint Lucia	x	x	3/27/85
St. Vincent and the Grenadines	x	x	
Samoa	x	9/28/84	
San Marino			
Sao Tome and Principe*		7/13/83	11/3/87
Saudi Arabia		12/7/84	
Senegal	x	x	10/25/84
Seychelles	x	x	
Sierra Leone	x	x	
Singapore	x	x	
Solomon Islands	x	x	
Somalia	x	x	7/24/89
South Africa*		12/5/84	
Spain*	x	12/4/84	
Sri Lanka	x	x	
Sudan*	x	x	1/23/85
Suriname	x	x	
Swaziland		1/18/84	
Sweden*	x	x	
Switzerland	x	10/17/84	
Syrian Arab Republic			
Thailand	x	x	
Togo	x	x	4/16/85
Tonga			

TABLE 1G. -- (continued)

STATES	FINAL ACT SIGNATURE	CONVENTION SIGNATURE[a]	CONVENTION RATIFICATION
Trinidad and Tobago	x	x	4/25/86
Tunisia[**]	x	x	4/24/85
Turkey			
Tuvalu	x	x	
Uganda	x	x	
Ukrainian SSR[*]	x	x	
Union of Soviet Socialist Republics[*]	x	x	
United Arab Emirates	x	x	
United Kingdom of Great Britain and Northern Ireland	x		
United Republic of Tanzania[**]	x	x	9/30/85
United States of America	x		
Uruguay[*]	x	x	
Vanuatu	x	x	
Venezuela	x		
Viet Nam	x	x	
Yemen[*]	x	x	
Yugoslavia[**]	x	x	5/5/86
Zaire	x	8/22/83	2/17/89
Zambia	x	x	3/7/83
Zimbabwe	x	x	
TOTAL FOR STATES	140	155	41

OTHERS
(Art. 305(1)(b),(c),(d),(e) and (f))

Cook Islands	x	x	
European Economic Community[*]	x	12/7/84	
Namibia (United Nations Council for)	x	x	4/18/83
Niue		12/5/84	
Trust Territory of the Pacific Is.	x		
West Indies Associated States			
TOTAL FOR STATES AND OTHERS	144	159	42

OTHER ENTITIES WHICH SIGNED THE FINAL ACT OF THE CONFERENCE

African National Congress of South Africa
Netherlands Antilles
Palestine Liberation Organization
Pan Africanist Congress of Azania
South West Africa People's Organization

Notes

[a]Those States which signed the Final Act and/or the Convention on December 10, 1982 are indicated by an "x". Those which signed at a later date are indicated by that date.

[b]Those States which made declarations at the time of signature of the Convention are indicated by an asterisk (*).

[c]Those States which made declarations at the time of ratification of the Convention are indicated by a double asterisk (**).

[d]The name of Burma was officially changed to "Myanmar" as of June 18 1989.

TABLE 1G. -- (continued)

B. <u>List of ratifications in chronological order and by regional groups</u>

	Date	State/Entity	Regional Group
1.	December 10, 1982	Fiji	Asian
2.	March 7, 1983	Zambia	African
3.	March 18, 1983	Mexico	Latin American
4.	March 21, 1983	Jamaica	Latin American
5.	April 18, 1983	Namibia (United Nations Council for Namibia)	African
6.	June 7, 1983	Ghana	African
7.	July 29, 1983	Bahamas	Latin American
8.	August 13, 1983	Belize	Latin American
9.	August 26, 1983	Egypt	African
10.	March 26, 1984	Côte d'Ivoire	African
11.	May 8, 1984	Philippines	Asian
12.	May 22, 1984	Gambia	African
13.	August 15, 1984	Cuba	Latin American
14.	October 25, 1984	Senegal	African
15.	January 23, 1985	Sudan	African
16.	March 27, 1985	Saint Lucia	Latin American
17.	April 16, 1985	Togo	African
18.	April 24, 1985	Tunisia	African
19.	May 30, 1985	Bahrain	Asian
20.	June 21, 1985	Iceland	West European and other States
21.	July 16, 1985	Mali	African
22.	July 30, 1985	Iraq	Asian
23.	September 6, 1985	Guinea	African
24.	September 30, 1985	United Republic of Tanzania	African
25.	November 19, 1985	Cameroon	African
26.	February 3, 1986	Indonesia	Asian
27.	April 25, 1986	Trinidad and Tobago	Latin American
28.	May 2, 1986	Kuwait	Asian
29.	May 5, 1986	Yugoslavia	Eastern European
30.	August 14, 1986	Nigeria	African
31.	August 25, 1986	Guinea-Bissau	African
32.	September 26, 1986	Paraguay	Latin American
33.	July 21, 1987	Democratic Yemen	Asian
34.	August 10, 1987	Cape Verde	African
35.	November 3, 1987	Sao Tome and Principe	African
36.	December 12, 1988	Cyprus	Western European and other States
37.	December 22, 1988	Brazil	Latin American/Caribbean
38.	February 2, 1989	Antigua and Barbuda	Latin American/Caribbean
39.	February 17, 1989	Zaire	African
40.	March 2, 1989	Kenya	African
41.	July 24, 1989	Somalia	African
42.	August 17, 1989	Oman	Asian

= 41 States and 1 entity (42)

TABLE 2G. -- TABLE OF MEMBERS, OBSERVERS AND PARTICIPANTS OF THE
           PREPARATORY COMMISSION[a]

Seventh Session (Kingston and New York)

STATES	Kingston[b] Member/ Observer	Participant	New York[c] Member/ Observer	Participant
Afghanistan	M		M	
Albania				
Algeria	M	x	M	x
Angola	M	x	M	x
Antigua and Barbuda	M		M	
Argentina	M	x	M	x
Australia	M	x	M	x
Austria	M	x	M	x
Bahamas	M		M	
Bahrain	M		M	
Bangladesh	M	x	M	x
Barbados	M		M	
Belgium	M	x	M	x
Belize	M		M	
Benin	M		M	x
Bhutan	M		M	
Bolivia	M	x	M	x
Botswana	M		M	
Brazil	M	x	M	x
Brunei Darussalam	M		M	
Bulgaria	M	x	M	x
Burkina Faso	M		M	x
Burundi	M		M	
Byelorussian SSR	M	x	M	x
Cameroon	M	x	M	x
Canada	M	x	M	x
Cape Verde	M	x	M	x
Central African Republic	M		M	
Chad	M		M	
Chile	M	x	M	x
China	M	x	M	x
Colombia	M	x	M	x
Comoros	M		M	
Congo	M	x	M	
Costa Rica	M	x	M	
Côte d'Ivoire	M	x	M	x
Cuba	M	x	M	x
Cyprus	M	x	M	
Czechoslovakia	M	x	M	x
Democratic Kampuchea	M		M	
Democratic People's Rep. of Korea	M	x	M	x
Democratic Yemen	M	x	M	
Denmark	M	x	M	x
Djibouti	M		M	
Dominica	M		M	
Dominican Republic	M		M	
Ecuador	O	x	O	x
Egypt	M	x	M	x
El Salvador	M		M	
Equatorial Guinea	M		M	
Ethiopia	M		M	
Fiji	M		M	x
Finland	M	x	M	x
France	M	x	M	x

TABLE 2G. -- (continued)

STATES	Kingston[b] Member/ Observer	Participant	New York[c] Member/ Observer	Participant
Gabon	M	x	M	x
Gambia	M		M	
German Democratic Republic	M	x	M	x
Germany, Federal Republic of	O	x	O	x
Ghana	M	x	M	x
Greece	M	x	M	x
Grenada	M		M	
Guatemala	M		M	
Guinea	M		M	x
Guinea-Bissau	M	x	M	x
Guyana	M		M	
Haiti	M	x	M	x
Holy See	O		O	
Honduras	M		M	
Hungary	M	x	M	x
Iceland	M		M	
India	M	x	M	x
Indonesia	M	x	M	x
Iran (Islamic Republic of)	M	x	M	x
Iraq	M		M	x
Ireland	M	x	M	x
Israel	O		O	
Italy	M	x	M	x
Jamaica	M	x	M	x
Japan	M	x	M	x
Jordan	O		O	
Kenya	M	x	M	x
Kiribati				
Kuwait	M	x	M	x
Lao People's Democratic Rep.	M		M	
Lebanon	M		M	
Lesotho	M		M	
Liberia	M	x	M	x
Libyan Arab Jamahiriya	M	x	M	x
Liechtenstein	M		M	
Luxembourg	M		M	
Madagascar	M	x	M	x
Malawi	M		M	
Malaysia	M	x	M	x
Maldives	M		M	
Mali	M		M	x
Malta	M	x	M	x
Mauritania	M		M	
Mauritius	M		M	x
Mexico	M	x	M	x
Monaco	M		M	
Mongolia	M		M	
Morocco	M	x	M	x
Mozambique	M	x	M	x
Myanmar	M	x	M	x
Nauru	M		M	
Nepal	M		M	
The Netherlands	M	x	M	x
New Zealand	M	x	M	x
Nicaragua	M		M	
Niger	M		M	

TABLE 2G. -- (continued)

STATES	Kingston[b] Member/ Observer	Kingston[b] Participant	New York[c] Member/ Observer	New York[c] Participant
Nigeria	M	x	M	x
Norway	M	x	M	x
Oman	M	.	M	x
Pakistan	M	x	M	x
Panama	M	x	M	
Papua New Guinea	M	x	M	
Paraguay	M		M	
Peru	O	x	O	x
Philippines	M	x	M	x
Poland	M	x	M	x
Portugal	M	x	M	x
Qatar	M		M	x
Republic of Korea	M	x	M	x
Romania	M		M	
Rwanda	M		M	
St. Kitts and Nevia	M		M	
Saint Lucia	M		M	
St. Vincent and the Grenadines	M		M	
Samoa	M		M	
San Marino				
Sao Tome and Principe	M		M	
Saudi Arabia	M	x	M	x
Senegal	M	x	M	x
Seychelles	M		M	
Sierra Leone	M		M	
Singapore	M		M	
Solomon Islands	M		M	
Somalia	M	x	M	x
South Africa	M		M	
Spain	M	x	M	x
Sri Lanka	M	x	M	x
Sudan	M	x	M	x
Suriname	M	x	M	x
Swaziland	M	x	M	x
Sweden	M	x	M	x
Switzerland	M	x	M	x
Syrian Arab Republic				
Thailand	M	x	M	x
Togo	M	x	M	x
Tonga				
Trinidad and Tobago	M	x	M	x
Tunisia	M	x	M	x
Turkey				
Tuvalu	M		M	
Uganda	M	x	M	x
Ukrainian SSR	M	x	M	x
Union of Soviet Socialist Reps.	M	x	M	x
United Arab Emirates	M	x	M	x
United Kingdom of Great Britain and Northern Ireland	O	x	O	x
United Republic of Tanzania	M	x	M	x

TABLE 2G. -- (continued)

STATES	Kingston[b] Member/ Observer	Kingston[b] Participant	New York[c] Member/ Observer	New York[c] Participant
United States of America	O		O	
Uruguay	M		M	x
Vanuatu	M	x	M	x
Venezuela	O	x	O	x
Viet Nam	M		M	x
Yemen	M		M	
Yugoslavia	M	x	M	x
Zaire	M	x	M	x
Zambia	M	x	M	x
Zimbabwe	M	x	M	x

ENTITIES
(Under Art. 305 (1)(b),(c),(d),(e) and (f))

Cook Islands	M		M	
European Economic Community	M	x	M	x
Namibia (United Nations Council for Namibia)	M	x	M	x
Netherlands Antilles	O		O	
Niue	M		M	
Trust Territory of the Pacific Islands	O		O	
West Indies Associated States				

NATIONAL LIBERATION MOVEMENTS

African National Congress of South Africa	O	x	O	x
Palestine Liberation Organization	O		O	
Pan Africanist Congress of Azania	O	x	O	x
South West Africa People's Organization	O		O	

OTHER OBSERVERS

Asian-African Legal Consultative Committee			O	x
Council on Ocean Law	O	x		
Organization of African Unity	O	x	O	x
International Bauxite Association	O	x		
International Labour Organization	O	x	O	x
International Maritime Organization	O		O	x
International Ocean Institue	O	x		
League of Arab States			O	x
United Nations Educational, Scientific & Cultural Organization	O	x	O	x
United Nations Development Programme	O	x		
United Nations Secretariat	O	x	O	x

TOTAL OF MEMBERS	159	92	159	96
TOTAL OF OBSERVERS	23	15	23	14
GRAND TOTAL	182	107	182	110

[a]States and other entities which are members or observers of the Preparatory Commission as defined in resolution I, paragraph 2, of the Third United Nations Conference on the Law of the Sea are indicated by an "M" for members or an "O" for observers. States or entities which did not sign the Convention and the Final Act of the United Nations Conference on the Law of the Sea are left blank. Those States or entities indicated by an "x" participated in the session or the meeting.
    [b]Held from February 27-March 23, 1989, at Kingston, Jamaica.
    [c]Held from August 14, 1989, in New York.

Table 3G -- MARINE JURISDICTIONAL CLAIMS, BY COUNTRY

Country or Territory	Area (km2)	Coastline (km)	Ports (no.) Major	Ports (no.) Minor	Territorial Sea	Contiguous Zone	Exclusive Economic Zone	Exclusive Fishing Zone	Continental Shelf	Security Zone	Pollution Zone
Albania	28,750	362	1	3	15	-	-	-	-	15	-
Algeria	2,381,740	998	12	11	12	-	-	-	-	-	-
Angola	1,246,700	1,600	3	5	20	-	-	200	-	-	-
Antigua & Barbuda	440	153	1	1	12	24	200	-	-	-	-
Argentina	2,766,890	4,989	7	30	200/a	-	-	-	b	-	c
Australia	7,686,850	25,760	12	-	3	-	-	200	b	-	c
Bahamas, The	13,940	3,542	2	9	3	-	-	200	-	-	-
Bahrain	620	161	1	1	3	-	-	-	d	-	-
Bangladesh	144,000	580	2	7	12	18	200	-	e	f	200c
Barbados	430	97	1	2	12	-	200	-	-	-	-
Belgium	30,510	64	6	1	12	-	-	200	d	-	c
Belize	22,960	386	2	6	3	-	-	-	-	-	-
Benin	112,620	121	1	-	200	-	-	-	-	-	-
Bermuda	50	103	3	-	3	-	-	200	-	-	-
Brazil	8,511,970	7,491	8	23	200	-	-	-	b	-	c
Brunei Darussalam	5,770	161	1	4	12	-	-	200	-	-	-
Bulgaria	110,910	354	3	6	12	-	200	-	b	-	-
Burma	676,550	3,060	4	6	12	24	200	-	g	f	200c
Cameroon	475,440	402	1	3	50	-	-	-	g	-	-
Canada	9,976,140	243,791	25	225	12	-	-	200	b	-	c(1)

TABLE 3G -- (continued)

Country or Territory	Area (km2)	Coastline (km)	Ports (no.) Major	Minor	Territorial Sea	Contiguous Zone	Jurisdictional Claims (nm) Exclusive Economic Zone	Exclusive Fishing Zone	Continental Shelf	Security Zone	Pollution Zone
Cape Verde	4,030	965	2	2	12/h	-	200	-	-	-	-
Chile	756,950	6,435	10	13	12	24	200	-	i/j	-	-
China	9,596,960	14,500	15	180	12	-	-	-	d	-	c
Colombia	1,138,910	3,208	5	5	12	-	200	-	b	-	-
Comoros	2,170	340	1	2	12/h	-	200	-	-	-	-
Congo	342,000	169	2	-	200	-	-	-	-	-	-
Cook Islands	240	120	-	2	12	-	200	-	g	-	-
Costa Rica	50,900	1,290	1	4	12	-	200	-	i	-	-
Cuba	110,860	3,735	20	50	12	-	200	-	i	-	-
Cyprus	9,250	648	3	11	12	-	-	-	k	-	-
Dem. Kampuchea	181,040	443	2	5	12	24	200	-	i	-	-
Denmark	43,070	3,379	19	41	3	4	-	200	1	-	-
Djibouti	22,000	314	1	-	12	24	200	-	-	-	-
Dominica	750	148	1	1	12	24	200	-	-	-	-
Dominican Rep.	48,730	1,288	4	17	6	24	200	-	g	-	-
Ecuador	283,560	2,237	4	6	200	-	-	-	i/m(2)	-	-
Egypt	1,001,450	2,450	5	15	12	18	200	-	b	f	-
El Salvador	21,040	307	2	1	200/a	-	-	-	-	-	-
Eq. Guinea	28,050	296	1	3	12	-	200	-	-	-	-
Ethiopia	1,221,900	1,094	2	-	12	-	-	-	-	-	-

Falkland Is.	12,170	1,288	1	4	3	–	–	200/n	b	–	–
Faero Is.	1,400	764	2	8	3	4	–	200	b	–	–
Fiji	18,270	1,129	1	6	12	–	200	–	b	–	–
Finland	337,030	1,126	11	34	4	6	–	12	b	–	–
France	547,030	3,427	26	6	12	24	200	–	b	–	–
French Guiana	91,000	378	1	7	12	–	200	–	b	–	–
Fr. Polynesia	3,941	2,525	1	6	12	–	200	–	b	–	–
Gabon	267,670	885	2	3	12	24	200	150	–	–	–
Gambia, The	11,300	80	1	–	12	18	–	200	i	–	–
German Dem. Rep.	108,330	901	4	13	12	–	–	200	b	–	–
Germany, F. Rep.	248,580	1,488	10	11	3/p	–	–	200	b	–	–
Ghana	238,540	539	2	–	12	24	200	–	q	–	–
Gibraltar	7	12	1	–	3	–	–	–	b	–	–
Greece	131,940	13,676	15	42	6	–	–	–	b	f	–
Greenland	2,175,600	44,087	8	14	3	4	–	200	b	–	–
Grenada	340	121	1	1	12	–	200	–	–	–	–
Guadeloupe	1,780	306	1	3	12	–	–	–	b	–	–
Guatemala	108,890	400	2	3	12	–	200	–	b	–	–
Guinea	245,860	320	1	2	12	–	200	–	–	–	–
Guinea-Bissau	36,120	350	1	–	12	–	200	–	–	–	–
Guyana	214,970	459	1	6	12	–	–	200	g	–	–
Haiti	27,750	1,771	2	12	12	24	200	–	b	–	–
Honduras	112,090	820	1	4	12	24	200	–	b	–	–
Hong Kong	1,040	733	1	–	3	–	–	–	b	–	–

TABLE 3G -- (continued)

Country or Territory	Area (km2)	Coastline (km)	Ports (no.) Major	Ports (no.) Minor	Territorial Sea	Contiguous Zone	Jurisdictional Claims (nm) Exclusive Economic Zone	Exclusive Fishing Zone	Continental Shelf	Security Zone	Pollution Zone
Iceland	103,000	4,988	4	50	12	-	200	-	g/a	-	-
India	3,287,590	7,000	9	79	12	24	200	-	g/a	f	-
Indonesia	1,919,440	54,716	15	70	12/h	-	200	-	b	-	-
Iran	1,648,000	3,180	6	12	12	-	-	50/r	f	-	-
Iraq	434,920	58	3	-	12	-	-	-	d	-	-
Ireland	70,280	1,448	8	38	12	-	-	200	d	-	c
Israel	20,770	273	3	5	6	-	-	-	b/s	-	-
Italy	301,230	4,996	20	40	12	-	-	-	b	-	-
Ivory Coast	322,460	515	2	1	12	-	200	-	i	-	c
Jamaica	10,990	1,022	2	10	12	-	-	-	b	-	-
Japan	372,310	13,685	132	2,000	12/t	-	-	200	-	-	-
Jordan	91,880	26	1	-	3	-	-	-	-	-	-
Kenya	582,650	536	1	-	12	-	200	-	b	-	-
Kiribati	717	1,143	2	-	12	-	200	-	-	-	-
Korea D.P.R.	120,540	2,495	6	26	12	-	200	-	-	50	-
Korea, Rep.	98,480	2,413	11	32	12/u	-	200	200	d	-	c
Kuwait	17,820	499	3	6	12	-	-	-	d	-	-
Lebanon	10,400	225	2	6	12	-	-	-	-	f	-
Liberia	111,370	579	1	6	200	-	-	-	b	-	-
Libyan Arab	1,759,540	1,770	6	21	12	-	-	-	-	(4)	-

Jamahiriya											
Macau	16	40	1	–	6	–	12	–	–	–	–
Madagascar	587,040	4,828	4	–	12	–	200	200	i/m	–	–
Malaysia	329,750	4,675	6	26	12	–	–	–	b	–	–
Maldives	300	644	–	2	12	–	–	–	v	–	–
Malta	320	140	3	1	12	24	–	25	b	–	–
Martinique	1,100	290	1	5	12	–	–	–	b	–	–
Mauritania	1,030,700	754	2	–	12	–	200	–	g	–	–
Mauritius	1,860	177	1	–	12	–	200	–	g	–	–
Mexico	1,972,550	9,330	11	20	12	–	200	–	b	–	–
Monaco	2	4	–	1	12	–	–	–	–	–	c
Morocco	446,550	1,835	9	15	12	24	200	–	b	–	c
Mozambique	801,590	2,470	3	2	12	–	200	–	–	–	–
Namibia	824,290	1,489	1	–	6	–	–	12	–	–	–
Nauru	20	24	–	1	12	–	–	200	–	–	–
Netherlands	37,310	451	10	2	12	12	–	200	b	–	–
Neth. Antilles	960	364	3	6	12	–	–	–	–	–	–
New Caledonia	19,060	2,254	1	21	12	–	200	–	b	–	c
New Zealand	268,680	15,134	3	3	12	–	200	–	g	–	–
Niue	260	64	–	–	12	–	200	–	–	–	–
Nicaragua	129,494	910	6	11	200	–	–	–	i	–	–
Nigeria	923,770	853	6	–	30	–	200	–	b	–	–
Norway	324,220	21,925	20	58	4	10	200	–	b	–	–
Oman	212,460	2,092	2	5	12	–	200	–	b	–	–

TABLE 3G -- (continued)

Country or Territory	Area (km2)	Coastline (km)	Ports (no.) Major	Ports (no.) Minor	Territorial Sea	Contiguous Zone	Jurisdictional Claims (nm) Exclusive Economic Zone	Exclusive Fishing Zone	Continental Shelf	Security Zone	Pollution Zone
Pakistan	803,940	1,046	2	4	12	–	200	–	g/q	–	–
Panama	78,200	2,490	2	8	200	–	–	–	–	–	–
Papua New Guinea	461,690	5,152	5	9	12/h	–	200	200	b	–	c
Peru	1,285,220	2,414	7	25	200	–	–	–	g	–	–
Philippines	300,000	36,289	10	–	w/h	–	200	–	d/s	–	c
Poland	312,680	491	4	12	12	–	–	200	b	–	–
Portugal	92,080	1,793	7	34	12	–	200	–	b	f	c
Qatar	11,000	563	2	1	3	–	x	200	d	–	–
Reunion	2,510	201	1	–	12	–	200	–	b	–	–
Romania	237,500	225	4	7	12	–	200	–	b	–	–
St. Lucia	620	158	1	1	3	–	200	12	a/g	–	–
St. Vincent and the Grenadines	340	84	1	1	3	24	200	–	–	–	c
São Tomé & Príncipe	960	209	1	–	12/h	–	200	–	–	–	–
Saudi Arabia	2,149,690	2,510	7	17	12	18	–	–	d	–	–
Senegal	196,190	531	1	2	12	24	200	–	g	–	–
Seychelles	455	491	1	–	12	–	200	–	g	–	c
Sierra Leone	71,740	402	1	2	200	–	–	–	b	–	–
Singapore	580	193	3	–	3	–	–	12	–	–	c
Solomon Islands	28,450	5,313	–	5	12/h	–	200	–	–	–	–

Somalia	637,660	3,025	3	-	200	-	-	-	-	-	-
South Africa	1,221,040	2,881	8	-	12	-	-	200	b	-	-
Spain	504,750	4,964	23	175	12	200	200	-	b	-	-
Sri Lanka	65,610	1,340	3	9	12	24	200	-	g	f	-
Sudan	2,505,810	853	1	-	12	18	-	-	b	f	-
Suriname	163,270	386	1	6	12	-	200	-	-	-	-
Sweden	449,960	3,218	17	30	12	-	-	200	b	-	-
Syrian Arab Rep.	185,180	193	3	2	35	-	-	-	b	-	-
Taiwan	35,980	1,448	5	4	12	-	200	-	-	-	-
Tanzania, United Rep.	945,090	1,424	3	-	12	-	-	-	-	-	-
Thailand	514,000	3,219	2	16	12	-	200	-	b	-	-
Togo	56,790	56	1	1	30	-	200	-	-	-	-
Tonga	700	419	-	2	12	-	200	-	b	-	-
Trinidad & Tobago	5,130	362	1	8	12/h	-	200	-	b	-	-
Tunisia	163,610	1,148	7	14	12	-	200/x	-	-	-	-
Turkey	780,580	7,200	14	18	6/x	-	200	-	-	-	-
Tuvalu	26	24	-	2	12	-	200	-	b	-	c
USSR	22,402,200	42,777	53	180	12	-	200	-	-	-	-
United Arab Emirates	83,600	1,448	7	25	3(4)	-	200	/	b	-	-
United Kingdom	244,820	12,429	25	180	12/y	-	-	200	b	-	-
USA	9,372,610	19,924	44	-	12	12	200	-	b	-	c
Uruguay	176,220	660	1	9	200/a	-	-	-	b	-	c
Vanuatu	14,760	2,528	-	3	12	24	200	-	g	-	-
Venezuela	912,050	2,800	6	17	12	15	200	-	b	-	-

TABLE 3G -- (continued)

Country or Territory	Area (km2)	Coastline (km)	Ports (no.) Major	Ports (no.) Minor	Territorial Sea	Contiguous Zone	Jurisdictional Claims (nm) Exclusive Economic Zone	Exclusive Fishing Zone	Continental Shelf	Security Zone	Pollution Zone
Vietnam	329,560	3,444	9	23	12	24	200	-	g	f	-
Wallis & Futuna	274	129	-	2	12	-	200	-	b	-	-
Western Sahara	266,000	1,110	-	2	z	-	-	-	-	-	-
Western Samoa	2,860	403	1	1	12	-	200	-	-	-	-
Yemen, Arab Rep. (North Yemen/Sanaa)	195,000	523	1	3	12	18	-	-	i	-	-
Yemen, Dem. Rep. (South Yemen/Aden)	332,970	1,383	1	5	12	24	200	-	g	f	c
Yugoslavia	255,800	3,935	9	24	12	-	-	-	b	-	c
Zaire	2,345,410	37	2	1	12	-	-	200	-	-	-

Sources:

Central Intelligence Agency, The World Factbook 1988 (Washington, D.C.: US Government Printing Office, 1988), CPAS WF 88-001, May 1988.

U.S. Department of State, Bureau of Intelligence and Research, Office of the Geographer, National Claims to Maritime Jurisdiction, 5th Revision, Limits of the Seas, no. 36. edited by Robert W. Smith (Washington, D.C.: Department of State, May 1985).

Law of the Sea Bulletin (New York: United Nations, Office for Ocean Affairs and the Law of the Sea, 1988), Nos. 11-13.

U.S. Department of State, Bureau of Oceans and International Environmental and Scientific Affairs, "Notice to Research Vessel Operators No. 61 (Rev. 6): Claimed Maritime Jurisdictions" (Washington, D.C.: Department of State, June 14, 1989).

NOTES:

a   Overflight and navigation permitted beyond 12 nm

b   Claims continental shelf to 200 meters (m) or depth of exploitability

c   Claims pollution zone

d   Claims continental shelf without specific limits

e   Claims continental shelf to the outer limits of the continental margin

f   Claims security zone (breadth given if known)

g   Claims Continental Shelf to the edge of the continental margin or 200 nm

h   Claims archipelagic waters

i   Claims continental shelf to 200 nm

j   Continental shelf of 350 nm applies to Sala y Gomez and Easter Island

k   Continental shelf beyond 200 m if part of the natural prolongation of the land territory

l   1958 Continental Shelf Convention definition

m   Continental shelf to 200 nm or 100 nm from the 2,500 m isobath

n   Exclusive Fishing Zone of 200 nm enforced to only 150 nm (February 1, 1987)

p   Territorial Sea of 3 nm but extends to 16 nm at one point in the Helgolander Bucht (by decree of November 12, 1984 for the prevention of tanker casualties)

q   Claims continental shelf to 100 fathoms or exploitability

r   Exclusive Fishing Zone of 50 nm in the Sea of Oman and to median-line boundaries in the Persian Gulf

s   Continental Shelf claimed to depth of exploitability

t   Territorial Sea of 12 nm, but 3 nm in the international straits – La Perouse or Soya, Tsugaru, Osumi, and Eastern and Western Channels of Tsushima or Korea Strait

u   Territorial Sea of 12 nm, and 3 nm in the Korea Strait

v   Exclusive Economic Zone of from 37 nm to 310 m

w   Territorial Sea of an irregular polygon extending up to 100 nm from the coastline as defined by 1898 treaty; since late 1970s has also claimed polygonal-shaped area in South China Sea up to 285 nm in breadth

x   Territorial Sea of 6 nm (12 nm in the Black Sea and Mediterranean Sea); Exclusive Economic Zone of 200 nm in the Black Sea only

y   Territorial Sea delimited by loxodromes between England and France (November 2, 1988)

z   Maritime claims contingent upon resolution of sovereignty issue

(1)   Claims a pollution zone in the Arctic region

(2)   Continental shelf of 100 nm from the 2,500 meter isobath applies to the Galapagos Islands

(3)   Claims non-specific security zone

(4)   Claims Gulf of Sidra as Libyan sovereign territory; Gulf of Sidra closing line at 32° 30' N

(5)   Territorial Sea of 12 nm claimed by Sharjah

Table 4G--International Maritime Organization's Conventions: Status as of June 1, 1989

Title and year of adoption (Conventions which are in force are in bold type)	Year of initial entry into force	Status: Number of Contracting States	Number of ratifications, etc. required for entry into force
Safety of Life at Sea (SOLAS) 1974	1980	105	-
(SOLAS) Protocol 1978	1981	69	-
(SOLAS) Protocol 1988	-	-	15 with 50% + 1 year
Regulations for Preventing Collisions at Sea 1972	1977	105	-
Prevention of Pollution from Ships 1973 as modified by the Protocol of 1978	1983	56	-
Facilitation of International Maritime Traffic 1965	1967	58	-
Load Lines 1966	1968	116	-
(Load Lines) Protocol 1988	-	-	15 with 50% + 1 year
Tonnage Measurement of Ships 1969	1982	85	-
Intervention on the High Seas in Cases of Oil Pollution Casualties 1969	1985	55	-
(Intervention) Protocol 1973	1983	24	-
Civil Liability for Oil Pollution Damage 1969	1975	65	-
(CLC) Protocol 1976	1981	32	-
(CLC) Protocol 1984	-	-	10 (with certain conditions)
Civil Liability in the Field of Maritime Carriage of Nuclear Material 1971	1975	11	-
Establishment of an International Fund for Compensation for Oil Pollution Damage 1971	1978	42	-
(Fund) Protocol 1976	-	16	8 (with certain conditions)
(Fund) Protocol 1984	-	2	8 (with certain conditions)
Special Trade Passenger Ships Agreement 1971	1974	14	-
(Space Requirement) Protocol 1973	1977	12	-

Safe Containers 1972	1977	50	–
Carriage of Passengers and Their Luggage by Sea 1974 (PAL) Protocol 1976	1987 1989	12 10	– –
International Maritime Satellite Organization (INMARSAT) 1976 + Operating Agreement	1979	56	–
Limitation of Liability for Maritime Claims 1976	1986	17	–
Safety of Fishing Vessels 1987	–	16	(with certain conditions)
Standards of Training, Certification and Watchkeeping for Seafarers 1978	1984	73	–
Maritime Search and Rescue 1979	1985	36	–
Suppression of Unlawful Acts (SUA) 1988 SUA Protocol	– –	2 2	15 (+ 90 days) 3 (+ 90 days)
Salvage 1989	–	–	15 (+ 1 year)

SOURCE:

International Maritime Organization, "IMO's Conventions: Status on 1 June 1989," *IMO News* No. 2 (1989):2

Bernhard J. Abrahamsson, a native of Sweden, received his Ph.D. in economics from the University of Wisconsin—Madison. In addition, he holds a Swedish Master Mariner's (unlimited) license. Prior to assuming his current position as chairman of the Division of Business and Economics at the University of Wisconsin—Superior, he served as head of the Department of Marine Transportation at the U.S. Merchant Marine Academy. In October 1988 Dr. Abrahamsson was detailed for one year to the Federal Maritime Commission to participate in the 5-year review of the 1984 Shipping Act. Earlier academic affiliations include the University of Denver's Graduate School of International Studies (1969–85) where he also served a period as dean of the school. Dr. Abrahamsson has taught and researched extensively in maritime economics both in the U.S. and abroad. He served as scientific director of the Israel Shipping Research Institute (1971–73), was a fellow at the Canadian Marine Transportation Research Centre (1978–79), and was senior fellow in the Marine Policy Program at the Woods Hole Oceanographic Institution (1984–85). He has served as consultant on shipping economics and policy to various government and international agencies.

Steinar Andresen, program director and senior research fellow at the Fridtjof Nansen Institute (FNI) in Norway, is a political scientist from the University of Oslo (1979). He has worked with various aspects of ocean policy, law of the sea issues, and international resource management. Among his publications are *Power and Law of the Oceans* (Oslo: Norwegian University Press, 1987) and *Norwegian Petroleum Politics and the Interests of the Fishing Industry* (Oslo: Aschehoug, 1983). Together with Director Willy Østreng (FNI) he is editor of *International Resource Management* (London: Pinter Publishers, 1989). He has also published a number of articles in journals and books. He has been affiliated with the FNI since 1979. In 1987/88 he was a visiting research scholar at the Institute for Marine Studies (IMS), University of Washington, Seattle. He is currently working on a joint research program between FNI, IMS, and the University of Oslo entitled "The Greenhouse Effect: Potential for Effective Global and Regional Response."

Bjorn Aune is a doctoral research candidate at the London School of Economics and Political Science. His fields of specialty are drug trafficking by sea, piracy, and juridical topics in navigation as related to the Law of the Sea. In addition, he is a licensed officer in the United States Merchant Marine and has served on a variety of vessels engaged in the bulk trades. He has published articles on the above topics in six languages.

Irene M. Baird was born and grew up in Newfoundland. She holds a B.A. in sociology from Memorial University of Newfoundland and a master of public health from the University of North Carolina, Chapel Hill. Following a lengthy career in the health field, she was appointed director of social policy in the Newfoundland cabinet secretariat. Later, as assistant secretary to cabinet for offshore petroleum impacts, she participated in the policy development and planning for offshore petroleum exploitation. When the government of Newfoundland commenced negotiations regarding ownership and management of offshore petroleum resources with the government of Canada, she coordinated the socioeconomic research conducted by the various provincial government departments to assist in the establishment of government policy positions for the negotiations. In 1983, she was appointed assistant deputy minister in the Newfoundland Petroleum Directorate, where she was responsible for policy and planning. She continued in the oil field until May 1989, when she retired from the position of assistant deputy minister, Newfoundland Department of Energy, where, in addition to petroleum matters, she was responsible for the province's conservation and alternative energy programs. Ms. Baird has been active in environmental conservation and has been a member of a number of environmental assessment panels. She has also served in several volunteer capacities, principally in the education and hospital fields. In August 1989 Ms. Baird commenced consulting work and operates IMB Associates, St. John's, Newfoundland.

Boleslaw Adam Boczek is professor of international law and political science at Kent State University, Kent, Ohio. A lawyer and political scientist by training, he has in the past been a research associate at Harvard Law School and adviser to the United Nations Economic Commission for Latin America. As professor at Kent, he has been invited as visiting professor of international law to a number of universities abroad, including those in Frankfurt am Main, Mexico, D.F., Leuven, Utrecht, and Thessaloniki. His main area of interest is the Law of the Sea and marine policy. The author of the now classic study *The Flags of Convenience,* he has written several other books, among them *The Transfer of Marine Technology to Developing Nations in International Law* and (as coauthor) *The International Law Dictionary.* He has also been a frequent contributor to professional journals, including *American Journal of International Law, German Yearbook of International Law, Ocean Yearbook,* and *Ocean Development and International Law.*

Elisabeth Mann Borgese is professor of political science at Dalhousie University, chairman of the Planning Council of the International Ocean Institute, Malta, and chairman of the Board of Directors of the International Centre for

Ocean Development (ICOD), Canada, and was an adviser to the delegation of Austria at UNCLOS III. For some years she was a senior fellow at the Center for the Study of Democratic Institutions, Santa Barbara. She has written numerous books, monographs, and essays on international ocean affairs and marine resource management, including *The Ocean Regime* (1968), *The Drama of the Oceans* (1976), *The New International Economic Order and the Law of the Sea* (with Arvid Pardo, 1976), *Seafarm* (1980), *The Mines of Neptune* (1984), and *The Future of the Oceans* (1986).

Roger Charlier, professor emeritus, studied at universities in Belgium, France, Germany, and Canada (M.S., Sc.D., and Litt.D.) and has been the recipient of numerous national and international academic, military, and professional awards and scholarships. He has been an exchange scientist and visiting lecturer at many universities worldwide. For the past 7 years he has been a scientific adviser to firms involved in ocean-related activities— currently HAECON N.V., Harbour and Engineering Consultants, Ghent, Belgium. He is retained by Unesco and UNITAR as an expert consultant, and chairs a subcommittee of the European Economic Community Scientific Directorate. The author of over 600 articles and books (over half are scientific), his most significant recent books are *Tidal Energy* (Van Nostrand Reinhold, 1982), and *Ocean Energies* (Elsevier, 1989).

Aldo E. Chircop is of Maltese nationality. He received his undergraduate education in law at the University of Malta and he pursued graduate studies at Dalhousie University, Halifax, Nova Scotia, Canada, where he received his LL.M. (1984) and J.S.D. (1988). Previously director of the Mediterranean Institute at the Foundation for International Studies, Malta, he is now academic director at the International Ocean Institute, and coordinator of the Marine Sciences Network in the Malta Council for Science and Technology. He is an IOI training programme alumnus and an associate of the Oceans Institute of Canada.

Daniel Decroo is project head and field engineer of the Knokke-Heist beach restoration project for HAECON N.V., Harbour and Engineering Consultants, Ghent, Belgium.

Charles De Meyer is a lecturer at the faculty of engineering (IFAQ Program), University of Brussels, and is the President and C.E.O. of HAECON N.V., Harbour and Engineering Consultants, Ghent, Belgium.

Satyesh C. Chakraborty was born in 1931 in a small town which is now in Bangladesh. He received a B.A. and an M.A. in geography from Calcutta University and a Ph.D. in economic geography from the London School of Economics and Political Science. Presently a professor at the Indian Institute of Management, Calcutta, he has taught in various universities in India, the U.S.A., Canada, and Japan and traveled extensively in the Asia-Pacific region, Western Europe, and the USSR on academic visits. He is interested in economic development and cultural change and the management of environmental risks arising from developmental actions. He has held a political appointment in the government as a member of the West Bengal State Planning Board, and was a director of the Indian Institute of Port Management, Calcutta. He is actively involved with voluntary social organizations for action research for developing workers' cooperatives in the subsectors of agriculture, tea plantation, and handicraft industry, and maintains close relationships with the nongovernmental organizations of similar interest in Western Europe, New Zealand, and North America. He has published many papers in the subject areas of his interest, and acts as a consultant in the area of public systems management. He is fond of music, both Western and Indian classical, is married, and has two sons.

John A. Dixon is a research associate at the Environment and Policy Institute of the East-West Center. His interest in coastal resources is partly due to growing up in Puerto Rico and living in a series of island settings—Taiwan, Penang, Java, and now Hawaii. Educated in economics at the University of California, Berkeley (B.A.), and Harvard (M.A. and Ph.D.), he spent 8 years working in Malaysia and Indonesia before coming to the East-West Center in 1981. His research and writing in Hawaii have focused on methodology and applied economic analysis of natural resources including ground and surface water, coastal resources, and mangroves. He has coauthored or edited a number of books on benefit-cost analysis, project appraisal methodology, watershed management, and dryland management. A frequent consultant to the World Bank and regional development banks, he is presently involved in a number of international research projects focusing on the economics of resources management.

John Fitzpatrick was born in Dundee, Scotland, where he graduated with degrees in Marine Engineering (C.Eng., M.I.Mar.E., and F.I.Diag.E.) and received diplomas in naval architecture and electrotechnology. His early career was directly related to the design, construction, and operation of cargo vessels, and in the early 1960s he moved on to chemical engineering. He became a director of the Caley Fisheries Group, Ltd., based at Peterhead, Scotland, and managed the group's boatbuilding and engineering activities.

In 1971 he joined the Food and Agriculture Organization of the United Nations as a consultant. Between 1975 and 1984 he managed the UNDP/FAO fleet of research, fishing, and training vessels and was also active in the execution of special projects such as building self-propelled barges for the food security network in Bangladesh and the rehabilitation of the inland water fishery in Kampuchea. He has developed a worldwide program entitled "Cooperative Use of Vessels for Fisheries Research, Development, and Training," which is now fully operational. In 1984 he was transferred to his present post, chief of FAO's Fishing Technology Service, Fishing Industry Division, where he has responsibility for fishing vessel design and construction, fishing gear technology, training, fishermen's organizations, and the technical and social aspects of small-scale fisheries development, and retains control over the vessel operation program. He represents FAO on matters concerning fishing vessel safety and conditions of work and service in the fishing industry at the relevant committee sessions of the IMO and ILO.

Lawrence S. Hamilton has been a research associate at the Environment and Policy Institute, East-West Center, Honolulu, Hawaii, since 1980. He holds B.Sc.F. and M.S. degrees from Toronto and Syracuse Universities and a Ph.D. from the University of Michigan. He is professor emeritus of Cornell University, a visiting professor at various Asian and Pacific universities, and a zone forester for the Ontario Department of Lands and Forests, Canada. He has done Unesco and IUCN forest consultancies in Australia, rainforest conservation studies in Guatemala and Venezuela, and a watershed consultancy in Costa Rica. He was cited as one of 100 Global Environmental Achievers by Friends of UNEP in 1987. He is currently engaged in collaborative research on biological diversity and protected area management, tropical forest land use, and watershed management in the Asia-Pacific region.

Sally McDonald graduated B.A. (honors) from the Department of Geography in the University of Melbourne in 1988. Her honors thesis was entitled "Difficulties in Interpreting the Rules for the Use of Straight Baselines on Unstable Coasts." She is currently working as a planner while studying for the degree of master of urban planning.

Glenys Owen Miller is a research fellow with the Environment and Policy Institute, East-West Center, and assistant editor of *Ocean Yearbooks 7* and *8*. She is an ardent environmentalist with a lifelong passion for the ocean and seashore, marine mammals and seabirds. Educated in zoology and law at the University of Otago and the University of Auckland, New Zealand (B.Sc. and LL.B.), she is currently writing her Ph.D. in environmental law.

Joseph R. Morgan spent 25 years in the U.S. Navy, retiring with the rank of captain. He then obtained his M.A. and Ph.D. degrees in geography from the University of Hawaii. He is currently associate professor of geography at the University of Hawaii, a research associate in the Environment and Policy Institute at the East-West Center, and coeditor of *Ocean Yearbook*. His primary academic interest is marine political geography.

Mahinda Perera is a Ph.D. candidate at Dalhousie University, Halifax, Nova Scotia, Canada.

Victor Prescott holds a personal chair in geography at the University of Melbourne. He has studied maritime issues in the Pacific and Indian oceans and has benefited from regular visits to the Environment and Policy Institute of the East-West Center in Honolulu. His latest book is entitled *Political Frontiers and Boundaries* (Unwin Hyman, 1987).

Kenneth Sherman is in charge of the Narragansett Laboratory and is chief of the Ecosystems Dynamics Branch, National Marine Fisheries Service Northeast Fisheries Center, National Oceanic and Atmospheric Administration (NOAA), and an adjunct professor of oceanography, University of Rhode Island. His specialty is the recruitment process within fish populations, and he is the author of numerous papers on fisheries ecology. He conducts studies on the production of fish biomass of the Northeast Continental Shelf. His studies include comparisons among large marine ecosystems' productivity and biomass yields. He has served as chairman of the Biological Oceanography Committee of the International Council for the Exploration of the Sea, chief scientist of the Antarctic Program of the National Marine Fisheries Service, the U.S. representative to the Scientific Committee of the Commission for the Conservation of Antarctic Marine Living Resources, a scientific consultant to FAO on assignments in West Africa and South America, and U.S. project officer in joint U.S.-Polish studies on marine productivity. He and Dr. Lewis M. Alexander coedited the first large marine ecosystems volume, *Management and Variability of Large Marine Ecosystems*, AAAS Symposium Series 99 (Westview Press, 1986).

Toufiq A. Siddiqi has been on the research staff of the East-West Center since 1977, is an affiliate faculty member of the geography department of the University of Hawaii at Manoa, and has served as assistant director of the Environment and Policy Institute and special assistant to the president of the East-West Center. He initiated and coordinated a multicountry project entitled "The Environmental Dimensions of Energy Policies." He is a member of the executive council of the Hawaiian Academy of Science. He also serves as an associate editor of *Energy—The International Journal,* and of *The Energy-Environment Monitor.* He received his early education in India and Pakistan. He then joined Trinity College, Cambridge, where he read mathematics and physics and received a B.A. (honors) in 1959. Subsequently, he received a doctorate in nuclear physics from the Johann Wolfgang Goethe University at Frankfurt am Main (Germany). He is the author, coauthor, or editor of a number of books and other publications dealing primarily with energy and environmental issues in Asia and the Pacific. His recent research has been on China-U.S. cooperation in science and technology and on a possible "Law of the Atmosphere" for dealing with such concerns as climate change and ozone depletion.

Jon M. Van Dyke is professor of law at the University of Hawaii Law School, director of the University of Hawaii Institute for Peace, and adjunct research associate of the East-West Center's Environment and Policy Institute. He has been on the University of Hawaii Law School faculty since 1976, and served as associate dean between 1980 and 1982. He previously taught at the University of California's Hastings College of the Law in San Francisco and at Catholic University Law School in Washington, D.C. He earned his J.D. at Harvard University and his B.A. from Yale University, both cum laude. He served as law clerk for Justice Roger J. Trainer, Chief Justice of the California Supreme Court, in 1969–70. He has written or edited four books on constitutional law and international law topics and has authored and coauthored many articles in these fields, focusing particularly on international human rights and ocean law and international and environmental law. He is a member of the executive council of the American Society of International Law and is a member of the international editorial board of *Marine Policy.*

Jan van Ettinger graduated from the Technical University, Delft, The Netherlands (M.Sc. [M.E.]). He has worked both within The Netherlands and abroad as a governmental adviser and trainer of participants from developing countries in the fields of housing, building, and physical planning. From 1974 to 1976 he directed the "Reshaping of International Order (RIO)" project initiated by the Club of Rome, which resulted in *The RIO Report,* published in 11 languages. Following this he was director of the RIO Foundation for five years. Since 1982 he has free-lanced as an adviser at both national and international levels and, in 1988, founded van Ettinger and Associates—a consultant company with a focus on such transnational problems as the interlinking between energy, environment, and development. He is a member of the

International Ocean Institute Planning Council and is the coordinator of *Pacem in Maribus XVIII,* which will take place in Rotterdam August 27–31, 1990. Its theme is "Ports as 'nodal points' in an emerging global transport system."

Philomène A. Verlaan is a member of the Florida bar and practiced EEC and international trade law for many years in Brussels, Belgium, before succumbing to the siren song of Law of the Sea. Convinced that the best ocean law can only be made upon a rigorous foundation of ocean science, she is currently on sabbatical to obtain an M.Sc. (anticipated December 1989) in oceanography at the University of Hawaii. Her thesis requires field work on coral reefs and seamounts using a variety of ocean equipment ranging from the simple snorkel to the submersible *Pisces V.* She is also editor of the professional correspondence series *LOS Lieder* of the Law of the Sea Institute.